/900

Powell's Amphibians & Reptiles of Pakistan.

OP / 9.98 PC

Science & Natural History   117503

Amphibians and Reptiles
of Pakistan

Muhammad Sharif Khan
Herpetological Laboratory, Rabwah, Pakistan

Amphibians and Reptiles of Pakistan

Muhammad Sharif Khan

KRIEGER PUBLISHING COMPANY
Malabar, Florida
2006

Original Edition 2006

Printed and Published by
**KRIEGER PUBLISHING COMPANY
KRIEGER DRIVE
MALABAR, FLORIDA 32950**

Copyright © 2006 by Krieger Publishing Company

All rights reserved. No part of this book may be reproduced in any form or by any means, electronic or mechanical, including information storage and retrieval systems, without permission in writing from the publisher.
No liability is assumed with respect to the use of the information contained herein.
Printed in China.

FROM A DECLARATION OF PRINCIPLES JOINTLY ADOPTED BY A COMMITTEE OF THE AMERICAN BAR ASSOCIATION AND A COMMITTEE OF PUBLISHERS:
This publication is designed to provide accurate and authoritative information in regard to the subject matter covered. It is sold with the understanding that the publisher is not engaged in rendering legal, accounting, or other professional service. If legal advice or other expert assistance is required, the services of a competent professional person should be sought.

Library of Congress Cataloging-in-Publication Data

Khan, Muhammad Sharif, 1939–
    Amphibians and reptiles of Pakistan / Muhammad Sharif Khan.
      p. cm.
    Includes bibliographical references and index.
    ISBN 1-89464-952-3 (hardcover : alk. paper)
    1. Amphibians—Pakistan.   2. Reptiles—Pakistan.   I. Title.

QL661.P18K49 2005
597.9'095491—dc22
2004048656

10 9 8 7 6 5 4 3 2

Dedication

I cherish the memories of the following persons who helped to nurture my interest in the fascinating world of herpetology: my parents, Dr. and Mrs. Habib Ullah Khan, who allowed me to rear amphibian tadpoles, dissect and prepare skeletons of frogs in the backyard of our house in the remote Pakistani village of Chaksan, District Gujranwala, and helped me to understand the morphology, way of life and locomotion of tadpoles; Professor Dr. Ahsanul Islam (1927–1974), Government College, Lahore, my university professor and guide, suggested amphibian morphology and embryology for my lifelong study; Professor Dr. Robert Mertens (1894–1975), curator, Natur-Museum and Forschungs-Institute Senckenberg, Frankfurt am Main, Germany, who helped me with my first identifications and understanding of various morphological terms used in herpetological studies; Professor Sherman A. Minton, Jr. M.D. (1919–1999), Department of Microbiology, Indiana University, Medical School, Indianapolis, who helped me to identify the Pakistani reptiles and provided me with literature from his library. Moreover, he edited this book before his death. In a way, it is his last contribution to the herpetology of Pakistan. His suggestions and comments helped to make this book up to date and useful.

Contents

Foreword ix

Preface xi

Acknowledgments xiii

Introduction xiv

**Chapter 1. Checklist of Amphibians
and Reptiles of Pakistan** 1
    Amphibians: Toads and Frogs 1
    Chelonians: Turtles and Tortoises 1
    Crocodiles: Muggers and Gharials 2
    Lizards 2
    Ophidia: Snakes 4

**Chapter 2. Keys to the Identification of
Families, Species, and Subspecies** 6
    Keys to Families 6
    Keys to Species 7
    Figures (2 through 32) 19

Chapter 3. Amphibians: Toads and Frogs 42
    Family Bufonidae 42
    Family Megophryidae 52
    Family Microhylidae 54
    Family Ranidae 56

**Chapter 4. Chelonians: Turtles
and Tortoises** 68
    Family Cheloniidae 68
    Family Dermochelyidae 72
    Family Emydidae 73
    Family Testudinidae 77
    Family Trionychidae 80

Chapter 5. Crocodiles: Muggers and Gharials 85
    Family Crocodylidae 85
    Family Gavialidae 87

Chapter 6. Lizards 89
    Family Agamidae 89
    Family Chamaeleonidae 117
    Family Eublepharidae 118
    Family Gekkonidae 119
    Family Lacertidae 153
    Family Scincidae 162
    Family Uromastycidae 174
    Family Varanidae 176

Chapter 7. Ophidia: Snakes 180
    Family Leptotyphlopidae 180
    Family Typhlopidae 181
    Family Boidae 186
    Family Colubridae 190
    Family Elapidae 222
    Family Hydrophiidae 227
    Family Viperidae 234
    Family Crotalidae 241

**Chapter 8. Distribution and Affinities
of Herpetofauna** 243
    Herpetological Survey 243
    Affinities of Herpetofauna 243
    Migration Routes and Distribution 246
    Zoogeographical Analysis 248

Chapter 9. Herpetology of Habitat Types 250
    Plants and Their Herpetological
        Associations 250
    Mountain Region 252
    Foothill Region 254
    Indus Plains 258

**Chapter 10. Altitudinal Distribution
of Amphibians and Reptiles** 269

Chapter 11. Snakebite Problem in Pakistan ... 278
    Key to the Identification of Terrestrial
        Venomous Snakes 279

Reference List of Plates and Figures
   Pertaining to Terrestrial Venomous
   Snakes 279

**Chapter 12. Threats to the Herpetofauna
   of Pakistan** 282

Bibliography 288

Suggested Readings 302

Index 304

Foreword

Pakistan is a land of stunning transitions. It is the connection between the Middle East and the Indian subcontinent, or, in biogeographic terms, the Palearctic and the Oriental. Pakistan covers territory from the snowy ranges of the world's highest peaks through the arid plains of the Punjab and Sind to the flood plains of the Indus Valley. As such, Pakistan has an unusually wide range of habitats, most of which are inhabited by amphibians and reptiles. It is therefore not surprising that Pakistan's herpetofauna is also highly diverse, comprising 24 species of amphibians and about 200 of reptiles, representing 26 families.

The herpetofauna of Pakistan has long been the subject of scientific study, originally by resident and nonresident Europeans in the 19th and early 20th centuries—Albert Günther, Ferdinand Stoliczka, William Blanford, James A. Murray, George A. Boulenger, Frank Wall, and Malcolm A. Smith—and in the late 20th century by Sherman A. Minton, an American, and Robert Mertens, a German. Only in 1965 did a native Pakistani—Muhammad Sharif Khan—begin herpetological work in the country. He was trained at Punjab University in Lahore, receiving his bachelor's and master's degrees in the early 1960s and achieving a doctorate from the same institution in 1996. He is, unquestionably, the best trained herpetologist of his country, with years of personal experience in the field and at his own "Herp Laboratory" in Rabwah, a little-known city in northern Punjab. He has published extensively on Pakistan's amphibians and reptiles, including about 250 papers beginning in 1965, and four books, two of them in his native Urdu language and one that was also translated into German. These contributions have dealt with nearly every aspect of these animals' biology: systematics, distribution, morphology, development, and natural history. He has also described some two dozen new species and subspecies of Pakistani amphibians and reptiles. Dr. Khan retired as professor of zoology at Talimul Islam College in Rabwah in 1999 when he emigrated to the United States.

This book represents the distillation of nearly 40 years of research on the amphibians and reptiles of Pakistan by its foremost expert. It includes an annotated checklist, keys to species, descriptions of all of the species with color photographs of nearly all of them, discussions of zoogeography, habitat types and distribution, snakebite, and conservation, together with an extensive bibliography. It is, in short, the most authoritative illustrated compendium on Pakistan's amphibians and reptiles, a work with no peer in its geographic realm. This book, therefore, is a most welcome and useful contribution to herpetological science. It will be useful not only in Pakistan but in all the surrounding countries, the general region, and beyond.

KRAIG ADLER
CORNELL UNIVERSITY

Preface

In writing this guide, the impetus has been to promote interest in this neglected branch of zoology among both the general public and Pakistani zoologists. It is hoped that this book will provide a better understanding of the beautiful world of Pakistani amphibians and reptiles to the novice as well as to the experienced herpetologist. Moreover, the summarized herpetological knowledge will serve a broader audience, including visiting field workers, biologists working on various aspects of biodiversity, and particularly, researchers in universities and colleges in Pakistan and neighboring countries, who have little access to the basic herpetological literature about Pakistan. To fulfill these purposes, I have tried to make this book as understandable and as accessible as possible.

Presentation
A checklist is presented with all species of amphibians and reptiles that so far have been recorded and described from Pakistan. The species accounts are arranged alphabetically for easy access at all taxonomic levels, disregarding phylogenetic relationships among them. Genera are defined before the description of the taxa contained in them.

Species name
The scientific name of the species is followed by the English common name. I have picked up appropriate common English names of some species from Smith (1935, 1943), Minton (1966) and Szczerbak and Golubev (1996). However, wherever an appropriate name is not available for a species, I have coined one myself, basing it on some distinctive feature of the taxon, generally avoiding naming taxa after their authors. Despite considerable effort (Das, 1998), no acceptable names for amphibians and reptiles are available in the vernacular languages of the subcontinent. I coined Urdu names for some species in my Urdu books (Khan, 1993d, 2000b); in this book I have named in Urdu every species which occurs in Pakistan. I hope it will help to popularize the common English and Urdu names throughout Pakistan, since Urdu is the national language of the country.

Synonymy
I have restricted myself to the primary synonymy for each taxon.

Diagnosis
The salient external morphological features of each taxon are enumerated in order of importance, with range of snout-vent and tail length.

Natural History Notes
Notes on habitat preferences, behavior, diet, prey, predators, and reproductive activity are given, where available.

Taxonomic Notes
Recent taxonomic changes are discussed in this section, wherever necessary.

Distribution
My emphasis has been on the ranges of species in Pakistan; however, total worldwide range is also given for widely distributed species.

Distribution Maps
There are not many collection reports for amphibians and reptiles from the various parts of Pakistan, so the sources for construction of distribution maps are, firstly, from my collecting tours to different parts of the country and secondly from the literature. The dots represent localities from which the species were collected or have been reported to have been sighted. However, the range of a species can be drawn by encircling freely around dots lying at the perimeter. In such cases dots included indicate the localities within the drawn range from where the species is reported to have been sighted or collected.

Figures
I have provided clear line drawings of salient morphological features of most of the species to assist in identification.

Color Plates
Color photographs of living animals record natural coloration and stance which an animal assumes in the presence of man; they are helpful in proper identification of taxa. Photographing a live animal to get an appropriate pose is a laborious job; just when you are ready to push the button, the animal suddenly changes its pose and all of your efforts are ruined. The photographs are mostly from my collection and that of Dr Minton, to whom I am grateful.

Photographs of different biotopes are presented, usually from areas where the collections were made.

Layout of the Book
Over the past decade, the list of the herpetofauna has been the most frequently revised part of the fauna of Pakistan (Khan, 1991b, f). Several new species have been discovered and new species have been described and added, and there has been a steady increase in the number of families, genera, and species (Chapter 8) that have been recorded from Pakistan. Minton (1966) recorded 144 species and subspecies, Mertens (1969a) 178, and Khan (1980b) 195 species, while present reports count more than 223 species and subspecies. An up-to-date checklist of the Pakistani herpetofauna is presented in Chapter 1.

The taxonomy of Pakistani amphibians and reptiles forms the bulk of the book. In Chapter 2, keys to the identification of families and species are presented. The keys are based on the most apparent and easily observable external morphological characteristics of species which are illustrated by several line drawings. Chapter 3 deals with amphibians; Chapter 4, chelonians; Chapter 5, crocodiles; Chapter 6, lizards; and Chapter 7, snakes. In the species accounts, the most distinctive taxonomic characteristics of species are enumerated and explained with the help of clearly labelled figures. Detailed notes on color, natural history, and distribution of every species are provided. Moreover, almost every species is illustrated by a clear color photograph. Wherever necessary, remarks on taxonomic changes are provided. Latest taxonomic changes have been incorporated, for family Ranidae (Dubois, 1984, 1986a, b, 1992), genus *Rana* (Dubois, 1975, 1976, 1981, 1983a, b; 1984, 1986a, b, 1992); family Testudinidae (Khozatsky and Mlynarsky, 1966); family Agamidae (Moody, 1980, 1987; Queiroz and Gauthier, 1994); family Lacertidae (Salvador, 1982; Arnold, 1983), genus *Acanthodactylus* (Arnold, 1983, 1986); geckos (Szczerbak and Golubev, 1986; Szczerbak, 1988; Khan, 2003a); scincid lizards (Greer, 1970; Pasteur 1981; Eremchenko and Szczerbak 1986; Griffith et al., 2000; Mausfeld and Schmitz, 2003); family Colubridae, genus *Coluber* (Schatti and Utiger, 2001); and for Crotalidae (Hoge and Romano-Hoge, 1981).

Zoogeographically, Pakistan occupies a peculiar position in Southeast Asia. It is sandwiched between Sclater's (1858) Oriental and Palearctic zoogeographical regions. Naturally, the Pakistan herpetofauna is a mixture of species from Palearctic and Oriental Regions. The zoogeographical affinities of the Pakistani herpetofauna are discussed in Chapter 8.

Despite its relatively small size, Pakistan is rich in its habitat types (see Figure 33). Its varied topography and bioclimates are reflected in the diversity of its soil types, climates, habitat types, flora, and fauna (Khan 1980b, 1999c; Roberts, 1977, 1991, 1992; F.K. Khan 1996; Mufti et al., 1997). It is surprising to note that most of the world's major bioclimates are represented within Pakistan and it is known as "the land of many lands" (Ahmad, 1951; Khan, 1980b; F.K. Khan, 1996). The amalgamation of physical geography with climates has played a key role in the creation of diverse habitats of the country (Mufti et al., 1997). Chapter 9 correlates the Pakistani herpetofauna to the 15 habitat types (Roberts, 1991; F.K. Khan, 1996; Khan, 1999c).

The study of altitudinal distribution of Pakistani amphibians and reptiles reveals interesting facts about the adaptation of various taxa to varying temperature and altitudes. This aspect is discussed in Chapter 10.

Snakebite has always been a problem in the warmer parts of the world. There are nine species of venomous land snakes in Pakistan, as well as several species of sea snakes in the coastal waters. To facilitate identification of venomous snakes, a special key is provided in Chapter 11. Since correct administration of antivenin necessitates correct identification of the snake involved in snakebite, a special key is provided to identify the venomous snake by its way of biting. A table of nonvenomous species of Pakistani snakes that are often confused with venomous species is also provided. Elapid and viperid venoms differ in their effects; a chart is provided comparing the symptomology of snakebite, which may help in identification of the species of snake involved in the bite.

Due to the increasing human interference in ecosystems the world over, several animal and plant species are threatened with extinction. This problem is becoming more acute day by day, resulting in the destruction of fast-receding natural habitat and disrupting animal and plant assemblages. Moreover, amphibians and reptiles are increasingly hunted for food, medicine, and hides. Hazards faced by Pakistani amphibians and reptiles are discussed in Chapter 12. The bibliography contains all references cited in the text pertaining to Pakistani taxa, while publications not dealing directly with Pakistan are placed in the Suggested Reading section of the bibliography.

Acknowledgments

I am extremely thankful to Professors Kraig Adler, Cornell University, Ithaca, New York, and Steven C. Anderson, University of the Pacific, Stockton, California. They have been my constant source of literature and encouragement. Moreover, Professor Anderson's technical remarks have helped to make this book more scientific.

My indebtedness to my wife, Rashida Tasnim, is understandable for several reasons: Despite her neat and clean habits, she tolerated the smell of formalin, the scattered jars and papers at my table, and the dust everywhere in my study, over several years of our relationship. Moreover, she allowed me to devote much of my time in the field and laboratory. Her protests have always been short-lived. I am thankful to her for her immense understanding and meticulous care.

MUHAMMAD SHARIF KHAN
SECANE, PENNSYLVANIA
4 APRIL, 2005

Introduction

The Indo-Pakistan subcontinent occupies a peculiar geographical position in Asia. It was originally a part of eastern Gondwana, the southern supercontinent, which fragmented into several plates in the remote past: India, Africa, South America, etc. According to the theory of Plate Tectonics, the plates moved across ancient oceans to their present position (King, 1967; Courtillot, 1983; Sengör, 1985). The Indian Plate moved across the Indian Ocean, to abut against the Asian land mass.

In August 1947, the subcontinent was partitioned politically into two sovereign states, the northwestern part as Pakistan and rest of the subcontinent as India.

Pakistan covers an area of 796,095 sq. km, sandwiched between two main zoogeographical regions, the Palearctic in the west and the Oriental in the east (Figure 1). It extends between longitude 60° 52′ to 75° 22′ E, and latitude 24° to 37° N. Its southwestern border touches the arid coast of the Arabian Sea, while its northernmost tip lies in the permanent snowfields of Pamir in the greater Himalayas.

Pakistan is comprised of three major geographical units. The first, mountains and plateaus in the north, northwest, and west, represent a confluence of three mighty mountain ranges, the Himalayas, the Karakorams, and the Hindu Kush. Some of the world's highest peaks lie in these ranges. This mountainous rampart has profoundly affected the geology, physiography, hydrology, and climate, and has played an important role in the present-day composition and distribution of the fauna and flora of the subcontinent. The second unit is the Salt Range at the foot of the Himalayas. It is one of the most ancient parts of the subcontinent and cradles a record of 600 million years of history of deposits of debris washed down from the Himalayas. It represents the ancient Tethys Sea (Powell, 1979). It is here that the great heights of the Himalayas intergrade with the plains of Punjab and Sind, thus affecting distribution of highland and plains flora and fauna.

The third unit is the vast plain that is really the watershed of the mighty River Indus. The Indus and its drainage basin form the dominant physiographic feature of the fertile plains of Punjab and Sind. This wide stretch of plain is formed mainly by the alluvium which has been washed down by the Indus. It is the most fertile part of Pakistan. Originally, it was covered with vast fields of grassland and tropical thorn forests. Large populations of diverse mammals and reptiles inhabited the area. Steatite seals of the Harappan civilization (circa 2000 BC) show rhinoceros, elephant, tiger, and gharial. This rich ecosystem has been drastically changed due to extensive human intervention over the centuries, with complete elimination of several species. The Indus Valley is now the most extensively cultivated and densely populated part of Pakistan.

The arid subtropical climate of Pakistan is ideal for reptiles but unsuitable for most amphibians. Variations in altitude, latitude, and proximity to main mountain ranges all affect Pakistan's climate. Higher mountains in the north and west are climatically moist alpine to subalpine with low soothing temperatures in the summer, while temperatures in the plains often exceed 54°C. The northern and southeastern parts of the country receive 40–140 cm of rain annually, resulting in yearly flooding of the plains. However, the western and southern parts of the country rarely receive more than 10 cm of rain annually and remain desolate and dry.

Politically the Indian subcontinent came under British dominion in the eighteenth century. In view of the history of the subcontinent, it is not surprising that European workers pioneered studies of amphibians and reptiles in southern Asia. The first general work was *The Reptiles of British India* by Albert Günther (1864), followed by George Boulenger's volume in the *Fauna of British India* series (1890). It was later revised by Malcolm Smith in three volumes (1931, 1935, 1943). The revision updated the herpetological knowledge

Introduction

Figure 1. Pakistan, geographical position and political divisions.

which quickly accumulated between 1889 and 1943, but it fell short of the coverage of the area in actual collections of the animals these publications claimed to have covered. Large sections of land in the subcontinent are even now not well-known herpetologically.

William Blanford (1874a, 1876b), James A. Murray (1886), and Ferdinand Stoliczka (1872b) are among those who published herpetological papers prior to 1900 on the region that became Pakistan. After the 1947 partition, Minton (1962, 1966) and Mertens

(1969a, 1970, 1974) were the only contributors to the herpetology of Pakistan, covering large areas not covered by previous works. Minton's collections were made mainly in lower Sind and Baluchistan, while Mertens's were of wider scope. However, large parts of Punjab and Northwest Frontier Province (NWFP) are not yet known herpetologically.

My interest in amphibians and reptiles dates back to 1965. Since then my field studies have resulted in the publication of several reports from different parts of Punjab, NWFP, and Baluchistan (Khan, 1968a, 1972a, 1979, 1985a, 1986; Khan and Ahmed, 1987: Khan and Baig, 1988). Several species have been recorded for the first time from Pakistan (Khan 1974, 1977, 1984a, b, 1985b, 1986, 1989, 1992) and new species of frogs, geckos, and snakes have been described (Dubois and Khan, 1979; Khan, 1980a, 1985b, 1988, 1991a, b, 1993a, 1997c, d, 1998c, 1999d, e; Khan and Tasnim, 1989, 1990b; Khan and Baig, 1992; M.S. Khan and A.R.Z. Khan, 1997; M.S. Khan and A.Q. Khan, 2000).

Herpetology has never been the favorite field with Pakistani zoologists. In India, however, there have always been several active renowned herpetologists. Herpetology is taught as a subject in several Indian universities. The world-renowned Bombay Natural History Society, with its well-established museum and journal, actively promotes interest in herpetological studies. Lack of interest by Pakistani biologists is due largely to lack of incentive, unavailability of pertinent literature, and absence of an institution where herpetological material can be deposited and kept available for comparison. To partially fill this gap, the author, in collaboration with Rashida Tasnim, has published easy-to-use field guides for amphibians (Khan and Tasnim, 1987a) and chelonians (Khan and Tasnim, 1990a), while work on lizards and snake guides is in preparation. Moreover, checklists of amphibians (Khan, 1976), chelonians, and crocodilians (Khan and Mirza, 1976), lizards (Khan and Mirza, 1977), and snakes (Khan, 1982a) of Pakistan have been published. To popularize herpetology with the local peoples, the author wrote two chapters in the recently published (1991c) Urdu book *Pakistan ki jangli hayat* (*Wildlife of Pakistan*). Moreover, a book in Urdu, on *Sar Zameen-a-Pakistan kay Saamp* (*Snakes of Pakistan*), was published in 1993; another book, also in Urdu, *Sar Zameen-a-Pakistan kay mandak aur khazanday* (*Frogs and lizards of Pakistan*), was published in 2000. Recently (2002), my *A guide to the snakes of Pakistan* was published both in English and German.

Chapter 1
Checklist of Amphibians and Reptiles of Pakistan

The following checklist includes almost all species that have been recorded from the areas now included in Pakistan in major works on the herpetology of the subcontinent (Boulenger, 1890; Smith, 1931, 1935, 1943; Minton, 1966; Mertens, 1969a, 1970, 1971, 1972, 1974; Khan, 2002b, 2003a, b). Remarks on the validity of these taxa are presented in respective species sections.

Amphibians: Toads and Frogs

Family BUFONIDAE
Bufo Laurenti, 1768
Bufo himalayanus Günther, 1864
Bufo latastii Boulenger, 1882
Bufo melanostictus Schneider, 1799
Bufo melanostictus hazarensis Khan, 2000
Bufo olivaceus Blanford, 1874
Bufo pseudoraddei pseudoraddei Mertens, 1971
Bufo pseudoraddei baturae Stock, Schmid, Steinlein, and Grosse, 1999
Bufo siacheninsis Khan, 1997
Bufo stomaticus Lütkin, 1862
Bufo surdus Boulenger, 1891
Bufo viridis Laurenti, 1768
Bufo viridis zugmayeri Eiselt and Schmidtler, 1973

Family MEGOPHRYIDAE
Stutiger Theobald, 1868
Scutiger nyingchiensis Fei, 1977

Family MICROHYLIDAE
Microhyla Tschudi, 1828
Microhyla ornata (Duméril and Bibron, 1841)
Uperodon Dúmeril and Bibron, 1841
Uperodon systoma (Schneider, 1799)

Family RANIDAE
Euphlyctis Fitzinger, 1843
Euphlyctis cyanophlyctis cyanophlyctis (Schneider, 1799)
Euphlyctis cyanophlyctis microspinulata Khan, 1997
Euphlyctis cyanophlyctis seistanica (Nikolsky, 1900)
Fejervarya (Bolkay, 1915)
Fejervarya limnocharis (Boie, 1834)
Fejervarya syhadrensis (Annandale, 1919)
Hoplobatrachus Peters, 1863
Hoplobatrachus tigerinus (Daudin, 1802)
Paa Dubois, 1975
Paa barmoachensis (Khan and Tasnim, 1989)
Paa hazarensis (Dubois and Khan, 1979)
Paa sternosignata (Murray, 1885)
Paa vicina (Stoliczka, 1872)
Sphaeroteca (Duméril and Bibron, 1841)
Sphaeroteca breviceps (Schneider, 1799)

Chelonians: Turtles and Tortoises

Family CHELONIIDAE
Caretta Rafiuesque, 1814
Caretta caretta (Linnaeus, 1758)
Chelonia Brongniart, 1800
Chelonia mydas (Linnaeus, 1758)
Eretmochelys Fitzinger, 1843
Eretmochelys imbricata (Linnaeus, 1766)
Lepidochelys Fitzinger, 1843
Lepidochelys olivacea (Eschscholtz, 1824)

Family DERMOCHELYIDAE
Dermochelys Blainville, 1816
Dermochelys coriacea (Vandelli, 1761)

Family EMYDIDAE
Geoclemys (Gray, 1821)
Geoclemys hamiltonii (Gray, 1821)
Hardella Gray, 1870
Hardella thurjii Gray, 1870
Kachuga Gray, 1856
Kachuga smithii (Gray, 1863)
Kachuga tecta (Gray, 1831)

Family TESTUDINIDAE
Agrionemys **Khozatsky and Mlynarsky, 1966**
Agrionemys horsfieldii (Gray, 1844)
Geochelone **Fitzinger, 1835**
Geochelone elegans (Schopff, 1792)

Family TRIONYCHIDAE
Aspideretes **Hay, 18353**
Aspideretes gangeticus (Cuvier, 1825)
Aspideretes hurum (Gray, 1831)
Chitra **Gray, 1844**
Chitra indica (Gray, 1831)
Lissemys **Smith, 1931**
Lissemys punctata andersoni Webb, 1980

Crocodiles: Muggers and Gharials

Family CROCODYLIDAE
Crocodylus **Laurenti, 1768**
Crocodylus palustris Lesson, 1831

Family GAVIALIDAE
Gavialis **Oppel, 1811**
Gavialis gangeticus (Gmelin, 1789)

Lizards

Family AGAMIDAE
Brachysaura
Brachysaura minor (Hardwicke and Gray, 1827)
Calotes **Cuvier, 1817**
Calotes versicolor versicolor (Daudin, 1802)
Calotes versicolor farooqi Auffenberg and Rehman, 1995
Japalura **Gray, 1853**
Japalura kumaonensis (Annandale, 1907)
Laudakia **Gray, 1845**
Laudakia agrorensis (Stoliczka, 1872)
Laudakia badakhshana (Anderson and Leviton, 1969)
Laudakia caucasia (Eichwald, 1831)
Laudakia fusca (Blanford, 1876)
Laudakia himalayana (Steindachner, 1869)
Laudakia lirata (Blanford, 1874)
Laudakia melanura Blyth, 1854
Laudakia melanura melanura Blyth, 1854
Laudakia melanura nasiri Baig, 1999
Laudakia microlepis (Blanford, 1874)
Laudakia nupta (de Filippi, 1843)
Laudakia nuristanica (Anderson and Leviton, 1969)
Laudakia pakistanica (Baig, 1989)
Laudakia pakistanica pakistanica Baig, 1989
Laudakia pakistanica auffenbergi Baig and Böhme, 1996
Laudakia pakistanica khani Baig and Böhme, 1996
Laudakia tuberculata (Hardwicke and Gray, 1827)
Phrynocephalus **Kaup, 1825**
Phrynocephalus clarkorum (Anderson and Leviton, 1967)
Phrynocephalus euptilopus Alcock and Finn, 1896
Phrynocephalus luteoguttatus Boulenger, 1887
Phrynocephalus maculatus Anderson, 1872
Phrynocephalus ornatus Boulenger, 1887
Phrynocephalus scutellatus Olivier, 1807
Trapelus **Cuvier, 1816**
Trapelus agilis Olivier, 1804
Trapelus agilis agilis (Olivier, 1804)
Trapelus agilis pakistanensis Rastegar-Pouyani, 1999
Trapelus megalonyx Günther, 1864
Trapelus rubrigularis Blanford, 1876
Trapelus ruderatus baluchianus (Smith, 1935)

Family CHAMAELEONIDAE
Chamaeleo **Laurenti, 1768**
Chamaeleo zeylanicus Laurenti, 1768

Family EUBLEPHARIDAE
Eublepharis **Gray, 1827**
Eublepharis macularius (Blyth, 1854)

Family GEKKONIDAE
Agamura **Blanford, 1874**
Agamura persica (Duméril, 1856)
Altigekko **Khan, 2003**
Altigekko baturensis (Khan and Baig, 1992)
Altigekko boehmei (Szczerbak, 1991)
Altigekko stoliczkai (Steidachner, 1869)
Bunopus **Blanford, 1874**
Bunopus tuberculatus Blanford, 1874
Crossobamon **Boettger, 1888**
Crossobamon lumsdeni (Boulenger, 1887)
Crossobamon maynardi (Smith, 1933)
Crossobamon orientalis (Blanford, 1876)
Cyrtopodion **Fitzinger, 1843**
Cyrtopodion agamuroides (Nikolsky, 1900)

Cyrtopodion kachhense kachhense (Stoliczka, 1872)
Cyrtopodion kachhense ingoldbyi Khan, 1997
Cyrtopodion kohsulaimanai (Khan, 1991)
Cyrtopodion montiumsalsorum (Annandale, 1913)
Cyrtopodion potoharensis Khan, 2001
Cyrtopodion scabrum (Heyden, 1827)
Cyrtopodion watsoni (Murray, 1892)
Hemidactylus Oken, 1817
Hemidactylus brookii Gray, 1845
Hemidactylus flaviviridis Rüppell, 1835
Hemidactylus frenatus Schlegel, 1836
Hemidactylus leschenaultii Duméril and Bibron, 1836
Hemidactylus persicus Anderson, 1872
Hemidactylus triedrus (Daudin, 1802)
Hemidactylus turcicus (Linnaeus, 1758)
Indogekko Khan, 2003
Indogekko fortmunroi (Khan, 1993)
Indogekko indusoani (Khan, 1980)
Indogekko rhodocaudus (Baig, 1998)
Indogekko rohtasfortai (Khan and Tasnim, 1990)
Mediodactylus Szczerbak and Golubev, 1977
Mediodactylus walli (Ingoldby, 1922)
Ptyodactylus Goldfuss, 1820
Ptyodactylus homolepis Blanford, 1876
Rhinogekko de Witte, 1973
Rhinogekko femoralis (Smith, 1933)
Rhinogekko misonnei de Witte, 1973
Siwaligekko Khan, 2003
Siwaligekko battalensis (Khan, 1993)
Siwaligekko dattanensis (Khan, 1980)
Siwaligekko mintoni (Golubev and Szczerbak, 1981)
Teratolepis Günther, 1870
Teratolepis fasciata (Blyth, 1853)
Teratoscincus Strauch, 1863
Teratoscincus microlepis Nikolsky, 1899
Teratoscincus scincus keyserlingii Strauch, 1863
Tropiocolotes Peters, 1880
Tropiocolotes depressus Minton and Anderson, 1965
Tropiocolotes persicus persicus (Nikolsky, 1903)
Tropiocolotes persicus euphorbiacola Minton, Anderson, and Anderson, 1970

Family LACERTIDAE
Acanthodactylus Wiegmann, 1834
Acanthodactylus blanfordii Boulenger, 1918
Acanthodactylus cantoris Günther, 1864
Acanthodactylus micropholis Blanford, 1874
Eremias Wiegmann, 1834
Eremias acutirostris (Boulenger, 1887)
Eremias aporosceles (Alcock and Finn, 1896)
Eremias fasciata Blanford, 1874
Eremias persica Blanford, 1874
Eremias scripta (Strauch, 1867)
Mesalina Gray, 1838
Mesalina brevirostris Blanford, 1874
Mesalina watsonana (Stoliczka, 1872)
Ophisops Ménétriés, 1832
Ophisops elegans Ménétriés, 1832
Ophisops jerdonii Blyth, 1853

Family SCINCIDAE
Ablepharus Fitzinger, 1823
Ablepharus grayanus (Stoliczka, 1872)
Ablepharus pannonicus (Fitzinger, 1823)
Chalcides Laurenti, 1768
Chalcides ocellatus (Forskål, 1775)
Eutropis Fitzinger, 1826
Eutropis dissimilis (Hallowell, 1860)
Eutropis macularia (Blyth, 1853)
Eurylepis Blyth, 1854
Eurylepis taeniolatus taeniolatus (Blyth, 1854)
Lygosoma Hardwick and Gray, 1827
Lygosoma punctata (Linnaeus, 1766)
Novoeumeces Griffith, Ngo, and Murphy, 2000
Novoeumeces blythianus (Anderson, 1871)
Novoeumeces indothalensis (Khan and Khan, 1997)
Novoeumeces schneiderii zarudnyi (Nikolsky, 1900)
Ophiomorus Duméril and Bibron, 1839
Ophiomorus blanfordi Boulenger, 1887
Ophiomorus brevipes (Blanford, 1874)
Ophiomorus raithmai Anderson and Leviton, 1966
Ophiomorus tridactylus (Blyth, 1853)
Scincella Mittleman, 1950
Scincella himalayana (Günther, 1864)
Scincella ladacensis (Günther, 1864)

Family UROMASTYCIDAE
Uromastyx Merrem, 1820
Uromastyx asmussi (Strauch, 1863)
Uromastyx hardwickii Gray, 1827

Family VARANIDAE
Varanus Merrem, 1820
Varanus bengalensis (Daudin, 1802)

Varanus flavescens (Hardwicke and Gray, 1827)
Varanus griseus caspius (Eichwald, 1831)
Varanus griseus koniecznyi Mertens, 1954

Ophidia: Snakes

Family LEPTOTYPHLOPIDAE
***Leptotyphlops* Fitzinger, 1843**
Leptotyphlops blanfordii (Boulenger, 1890)
Leptotyphlops macrorhynchus (Jan, 1862)

Family TYPHLOPIDAE
***Ramphotyphlops* Fitzinger, 1843**
Ramphotyphlops braminus (Daudin, 1803)
***Typhlops* Oppel, 1811**
Typhlops ahsanuli Khan, 1999
Typhlops diardii platyventris Khan, 1998
Typhlops ductuliformes Khan, 1999
Typhlops madgemintonae madgemintonae Khan, 1999
Typhlops madgemintonae shermanai Khan, 1999

Family BOIDAE
***Eryx* Daudin, 1803**
Eryx conicus (Schneider, 1801)
Eryx johnii (Russell, 1801)
Eryx tataricus speciosus Zarevsky, 1915
***Python* Daudin, 1803**
Python molurus (Linnaeus, 1758)

Family COLUBRIDAE
***Amphiesma* Duméril, Bibron, and Duméril, 1854**
Amphiesma platyceps (Blyth, 1854)
Amphiesma sieboldii (Günther, 1860)
Amphiesma stolatum (Linnaeus, 1758)
***Argyrogena* Werner, 1924**
Argyrogena fasciolata (Shaw, 1802)
***Boiga* Fitzinger, 1826**
Boiga melanocephala (Annandale, 1904)
Boiga trigonata (Schneider, 1802)
***Coluber* Linnaeus, 1758**
Coluber karelini karelini Brandt, 1838
Coluber karelini mintonorum Mertens, 1969
***Enhydris* Sonnini and Latreille, 1802**
Enhydris pakistanica Mertens, 1959
***Hemorrhois* Boie, 1826**
Hemorrhois ravergieri (Ménétriés, 1832)
***Lycodon* Boie, 1826**
Lycodon aulicus aulicus (Linnaeus, 1758)
Lycodon striatus Shaw, 1802
Lycodon striatus striatus (Shaw, 1802)
Lycodon striatus bicolor (Nikolsky, 1903)
Lycodon travancoricus (Beddome, 1870)
***Lytorhynchus* Peters, 1862**
Lytorhynchus maynardi Alcock and Finn, 1896
Lytorhynchus paradoxus (Günther, 1875)
Lytorhynchus ridgewayi Boulenger, 1887
***Oligodon* Boie, 1827**
Oligodon arnensis arnensis (Shaw, 1802)
Oligodon taeniolatus taeniolatus (Jerdon, 1853)
***Platyceps* Blyth, 1860**
Platyceps rhodorachis rhodorachis (Jan, 1865)
Platyceps rhodorachis kashmirensis (Khan and Khan, 2000)
Platyceps rhodorachis ladacensis (Anderson, 1871)
Platyceps ventromaculatus ventromaculatus (Gray and Hardwicke, 1834)
Platyceps ventromaculatus bengalensis (Khan and Khan, 2000)
Platyceps ventromaculatus indusai (Khan and Khan, 2000)
***Psammophis* Fitzinger, 1826**
Psammophis condanarus condanarus (Merrem, 1820)
Psammophis leithii leithii Günther, 1869
Psammophis lineolatus lineolatus (Brandt, 1838)
Psammophis schokari schokari (Forskål, 1775)
***Pseudocyclophis* Boettger, 1888**
Pseudocyclophis persicus (Anderson, 1872)
***Ptyas* Fitzinger, 1843**
Ptyas mucosus mucosus (Linnaeus, 1758)
***Sibynophis* Fitzinger, 1843**
Sibynophis sagittarius (Cantor, 1839)
***Spalerosophis* Jan, 1865**
Spalerosophis arenarius (Boulenger, 1890)
Spalerosophis diadema diadema (Schlegel, 1837)
Spalerosophis diadema var. *atriceps* (Fisher, 1885)
Spalerosophis schirazianus (Jan, 1865)
***Telescopus* Wagner, 1830**
Telescopus rhinopoma (Blanford, 1874)
***Xenochrophis* Günther, 1864**
Xenochrophis cerasogaster cerasogaster (Cantor, 1839)
Xenochrophis piscator piscator (Schneider, 1799)
Xenochrophis sanctijohannis (Boulenger, 1890)
Xenochrophis tessellata (Laurenti, 1768)

Family ELAPIDAE
***Bungarus* Daudin, 1803**
Bungarus caeruleus caeruleus (Schneider, 1801)
Bungarus sindanus sindanus Boulenger, 1847
Bungarus sindanus razai Khan, 1985
***Naja* Laurenti, 1768**
Naja naja naja (Linnaeus, 1758)
Naja oxiana (Eichwald, 1831)

Family HYDROPHIIDAE
***Astrotia* Fisher, 1856**
Astrotia stokesii (Gray, 1846)
***Enhydrina* Gray, 1849**
Enhydrina schistosa (Daudin, 1803)
***Hydrophis* Latreille, 1802**
Hydrophis caerulescens (Shaw, 1802)
Hydrophis cyanocinctus Daudin, 1803
Hydrophis fasciatus (Schneider, 1799)
Hydrophis lapemoides (Gray, 1849)
Hydrophis mamillaris (Daudin, 1803)
Hydrophis ornatus (Gray, 1842)
Hydrophis spiralis (Shaw, 1802)
***Lapemis* Gray, 1835**
Lapemis curtus (Shaw, 1802)
***Microcephalophis* Lesson, 1834**
Microcephalophis cantoris (Günther, 1864)
Microcephalophis gracilis (Shaw, 1802)
***Pelamis* Daudin, 1803**
Pelamis platurus (Linnaeus, 1766)
***Praescutata* Wall, 1921**
Praescutata viperina (Ph. Schmidt, 1852)

Family VIPERIDAE
***Daboia* Gray, 1842**
Daboia russelii russelii (Shaw and Nodder, 1797)
***Echis* Merrem, 1820**
Echis carinatus (Schneider, 1820)
Echis carinatus astolae Mertens, 1969
Echis carinatus multisquamatus Cherlin, 1981
Echis carinatus sochureki Stemmler, 1964
***Eristicophis* Alcock and Finn, 1896**
Eristicophis macmahonii Alcock and Finn, 1897
***Macrovipera* Reuss, 1927**
Macrovipera lebetina obtusa (Dwigubsky, 1832
***Pseudocerastes* Boulenger, 1896**
Pseudocerastes bicornis Wall, 1913
Pseudocerastes persicus (Duméril, Bibron, and Duméril, 1854)

Family CROTALIDAE
***Gloydius* Hoge and Romano-Hoge, 1981**
Gloydius himalayanus (Günther, 1864)

Chapter 2
Keys to the Identification of Families, Species, and Subspecies

The following keys will provide an aid to accurate identifications with a minimum of effort. To shorten the number of steps, the keys are written in a dichotomous way, the original and subsequent sets of taxa are divided into equal units. Where distribution of useful characters among taxa does not allow realization of this method, a choice has been made between two alternatives, each defined by one or more characters, taking care not to mask relationships among the taxa (i.e., as would be the case if similar appearing forms came out together, but closely related forms were set apart). The keys show patterns of common characters distributed among closely related taxa.

The keys are based on the most easily observable morphological characters on the bodies of animals, needing at the most a hand lens. (A watchmaker's monocular lens or loupe is of great help in counting and observing scales, etc.)

Keys to Families

Amphibia
1. Parotid gland present (Fig. 2A5).. Bufonidae
   Parotid gland not present (Fig. 4A) 2
2. Pupil vertical...3
   Pupil horizontal ....Ranidae
3. Head and mouth narrow; body smooth with few smooth small tubercles ....Microhylidae
   Head and mouth broad; body heavily worty; a distinct elevated post orbital tuberculated ridge ......Megophryidae

Amphibian Tadpoles
1. Body transparent, tail with a long vibratile flagellum; oral disc without keratinized structures
   (Fig. 4B, C) ....Microhylidae
   Body opaque at least dorsally, tail not produced in a flagellum; oral disc with keratinized structures
   (Fig. 3A) 2
2. Tadpole small, not exceeding 20 mm total length; labial tooth row formula always 2(2)/3 (Fig. 3B) ....Bufonidae
   Tadpole large, exceeding 25 mm; labial tooth row formula very variable ....Ranidae

Reptilia
1. Body enclosed in a bony shell (Fig. 10) 2
   Body not enclosed in a bony shell 6
2. Shell covered with epidermal plates 3
   No epidermal plates on shell..5
3. Limbs oar-shaped, with 1–3 claws (Fig. 8)...Cheloniidae
   Limbs not oar-shaped, forelimb with 4–5 claws (Fig. 8D)..4
4. Digits with vestiges of web; posterior limbs more elongated, plantar surface longer than broad ......Emydidae
   Digits without web; feet stump-shaped, planter surface as broad as long ....Testudinidae
5. Limbs with distinct digits, 3 claws present (Fig. 8E) ....Trionychidae
   Limbs without external evidence of digits, oar-shaped, clawless (Fig. 8C) ....Dermochelyidae
6. Limbs present 7
   Limbs absent..16

Chapter 2. Keys to the Identification of Families, Species, and Subspecies

7. Body large, heavy, often exceeds 2 m in length; hand with 5, foot with 4 digits, only inner 3 digits clawed; cloacal opening longitudinal; body covered with rows of sculptured scutes 8
   Body does not exceed 2 m in length; hand and foot with 5 digits, all clawed; cloacal opening transverse; body covered with thin scales...................9
8. Snout long, slender, at least 3 times as long as broad at base; 27–29 teeth on each side of the upper jaw ....Gavialidae
   Snout broad, not more than twice as long as broad at the base; 19–19 teeth in upper jaw............Crocodylidae
9. Head with small, irregularly arranged scales 10
   Head with large, regularly arranged scales...15
10. Head compressed, with an elevated median casque; digits fused in 2 bundles (Fig. 19A, B)......Chamaeleonidae
    Head depressed, no casque; digits free 11
11. Head with granular scales; eye pupil vertical, with pinholes when contracted (Fig. 12A) 12
    Head with flat scales; round pupil 13
12. Eyelid moveable ....Eublepharidae
    Eyelid immovable ....Gekkonidae
13. Head rectangular; nostrils very close to the tip of snout 14
    Head elongated; nostrils close to eye or midway between eye and snout (Fig. 21)...........Varanidae
14. Tail round, elongated, with flat keeled scales ....Agamidae
    Tail flat with dorsal transverse rows of long spinose scales.............Uromastycidae
15. Scales rough with a keel; tail much longer than body; digits long ....Lacertidae
    Scales keeled or keelless, smooth; if keeled, with multiple low keels; tail as long as or a little longer than body; digits short ....Scincidae
16. Tail round, tapering behind to a point 20
    Tail compressed, flat (Fig. 29A).................Hydrophiidae
17. Ventrals and other body scales alike 18
    Ventrals distinct from rest of body scales; body serpentine 19
18. Nasal and ocular scales both border mouth; an enlarged precloacal scale present; 14 scales around the body (Fig. 22C)..................Leptotyphlopidae
    Nasal and ocular scales do not border mouth; no enlarged precloacal scale; 11–36 scales around the body (Fig. 22A) ....Typhlopidae
19. Ventrals extending to half the width of body (Fig. 23D) ....Boidae
    Ventrals extending to the whole width of body 20
    Maxillary teeth not uniform, at least a pair of anterior maxillary teeth enlarged, hollowed, modified into fangs 21
    Maxillary teeth uniform in size, except sometimes posterior enlarged ....Colubridae
20. Pupil round; fangs small, not folded back, not enclosed in a membrane; no loreal scale (Fig. 30Ai,Bi)...............Elapidae
    Pupil vertical; fangs long, folded, enclosed in a membrane; a loreal scale present 22
21. Top of head with smooth large scales; a loreal pit, between eye and nostril (Fig. 32A,B) ....Crotalidae
    Top of head with keeled small scales; no loreal pit (Fig. 31A) ....Viperidae

Keys to Species

Amphibians
Family Bufonidae
1. Head with cranial crests (Fig. 2)..............................2
   Head without cranial crests............................3

2. Supraorbital crest only is present; tympanum indistinct .... *Bufo himalayanus*
   Supraorbital, canthal, postorbital, and orbitotympanic crests present,
   tympanum distinct .... *Bufo melanostictus*
3. Tympanum distinct 4
   Tympanum indistinct .... *Bufo surdus*
4. Tibial gland absent 5
   Tibial gland present 6
5. Dorsum with green pattern 7
   Dorsum uniformly olive .... *Bufo olivaceus*
6. Tarsal fold indicated by a weak spinulated line .... *Bufo stomaticus*
   Tarsal fold present .... *Bufo latastii*
7. Dorsal pattern of scattered green spots .... *Bufo viridis zugmayeri*
   Dorsal pattern of coalesced green blotches 8
8. Dorsal pattern of longitudinal stripes, three on each side .... *Bufo siachinensis*
   Dorsum heavily green with occasional light spots .... *Bufo pseudoraddei*

Family Microhylidae
1. Tongue elliptical; a dermal ridge between internal naris; adult does not exceed 30 mm
   in body length; body dorsum with elongated light brown large branched blotch .... *Microhyla ornata*
   Tongue oval; a pair of tubercles between internal naris; adult size 50–60 mm;
   dorsum with brown reticulation .... *Uperodon systoma*

Family Ranidae
1. Tympanum indistinct 2
   Tympanum distinct 3
2. Body dorsum with thick broken heavily spinulated longitudinal folds .... *Scutiger nyingchiensis*
   Body dorsum smooth with a few tubercles on flanks .... *Paa vicina*
3. Toes half webbed (Fig. 5C) 4
   Toes extensively webbed (Fig. 5B) 5
4. Habitus toadlike, inner metatarsal tubercle shovel-shaped (Fig. 5C) .... *Sphaeroteca breviceps*
   Habitus froglike, inner metatarsal tubercle elongate 6
5. First finger hardly extending beyond second; tibiotarsal joint reaching to
   anterior border of eye or a point between eye and tip of snout .... *Fejervarya syhadrensis*
   First finger longer than second; tibiotarsal joint reaching tympanum or naris .... *Fejervarya limnocharis*
6. Body dorsum pustulate 7
   Body dorsum with longitudinal folds .... *Hoplobatrachus tigerinus*
7. Nuptial spines on at least first 2 fingers 8
   Nuptial spines absent .... *Euphlyctis cyanophlyctis*
8. Pustules large, multispinulate; belly spiny .... *Paa sternosignata*
   Pustules small, unispinulate; belly spineless 9
9. Spinules on longitudinal ridges .... *Paa hazarensis*
   Spinules on pustules .... *Paa barmoachensis*

Amphibian Tadpoles
1. Body transparent, tail with a long vibratile flagellum; oral disc without keratinized
   structures (Fig. 4B, C) .... Microhylidae
   Body opaque at least dorsally, tail not produced in a flagellum; oral disc with keratinized
   structures (Fig. 3A) 2

Chapter 2. Keys to the Identification of Families, Species, and Subspecies

2. Tadpole small, not exceeding 20 mm in total length; labial tooth row formula always 2(2)/3 (Figure 3B) ....Bufonidae
   Tadpole large, exceeding 25 mm; labial tooth row formula very variable..3
3. No spiracle tube, dorsal fin brown, with spots; dental formula 1(4)/1(4) ....Megophryidae
   Distinct spiracle tube, dental formula variable ....Ranidae
4. Anterior labium with one dental row (Fig. 6B) ....Euphlyctis cyanophlyctis
   Anterior labium with more than 2 dental rows..5
5. Tooth row on anterior labium less than 3 6
   Tooth rows on anterior labium more than 3 7
6. Labial papillae continuous around the posterior labium ....Sphaeroteca breviceps
   Labial papillae confined to lateral sides of the oral disc..................................genus Fejervarya
7. Labial tooth row formula 5(4)/(3)5 ....Hoplobatrachus tigerinus
   Labial tooth row formula 8(6)/(3)2 ....genus Paa

Chelonians: Turtles and Tortoises

Family Cheloniidae
1. Costals 4 pairs..3
   Costals 5 or more pairs 2
2. Inframarginals 3, without pores ....Caretta caretta
   Inframarginals 4, some pierced with pores ....Lepidochelys olivacea
3. Prefrontals 2 pairs; dorsal plates imbricate; jaws hooked ....Eretmochelys imbricata
   Prefrontals 1 pair; dorsal plates juxtaposed; jaws not hooked ....Chelonia mydas

Family Dermochelyidae
1. Limbs paddle-shaped, indistinct digits; shell with smooth skin; 7 longitudinal ridges on carapace ....Dermochelys coriacea

Family Emydidae
1. Alveolar surface of jaws broad with a median ridge ....Geoclemys hamiltonii
   Alveolar surface with 1 to 2 ridges..2
2. Fourth vertebral not longer than broad, not longer than third..................................Hardella thurjii
   Fourth vertebral much longer than broad, longer than third..3
3. Vertebral much longer than broad; third elongate, quadrangular, with straight posterior border, its keel ends in a knob..................................Kachuga smithii
   Vertebrals not longer than broad, third pentagonal, its keel ends in a backward-directed spine........Kachuga tecta

Family Testudinidae
1. Forelimb with 4 claws; head with symmetrical shields; carapace with flat plates (Fig. 8D) ....Argrionemys horsfieldii
   Forelimb with 5 claws; head with asymmetrical shields; carapace with umbovate plates ....Geochelone elegans

Family Trionychidae
1. Plastron with cutaneous femoral valves; marginal bones present; 7 plastral callosities (Fig. 11) ....Lissemys punctata
   Plastron without femoral valves; no marginal bones; 4 plastral callosities 2
2. Head broad, massive, dorsally convex; nasal septum with lateral ridges..3
   Head long, narrow, flat above; nasal septum without lateral ridges ....Chitra indica

3. Alveolar surface of jaw raised at its inner marginal edge, forms a projection
    at the joint; head with black streaks; no ocelli on young disc ....*Aspideretes gangeticus*
    Alveolar surface not raised, grooved at the symphysis; head marked with black
    and yellow; in young, disc with 4 or more ocelli ....*Aspideretes hurum*

Crocodiles: Muggers and Gharials
Family Crocodylidae
Snout broad, not more than twice as long as broad at the base; 19–19 teeth
in upper jaw...*Crocodylus palustris*

Family Family Gavialidae
Snout long, slender, at least 3 times as long as broad at base; 27–29 teeth on each
side of the upper jaw..*Gavialis gangeticus*

Lizards
Family Agamidae
1. Body laterally compressed 2
    Body dorsoventrally depressed..3
2. Median dorsal row of distinct, pointed, elevated scales, extending to tail ....*Calotes versicolor*
    Median dorsal row of indistinct elevated scales to midbody ....*Japalura kumaonensis*
3. Tympanum distinct..4
    Tympanum concealed (Fig. 20) 20
4. Tympanum large, superficial; fifth toe extends beyond second; caudals in distinct annuli 5
    Tympanum small, deeply sunk, caudal scale irregular..16
5. Middorsum of body with several rows of homogeneous enlarged scales 6
    Middorsum of body with several rows of heterogeneous enlarged scales.......................*Laudakia nuristanica*
6. Scales of dorsal rows smooth 7
    Scales of dorsal rows keeled...8
7. Patch of strongly enlarged scales on flanks; male with a patch of callous
    abdominal scales (Fig. 18,1) ....*Laudakia. badakhshana*
    Patch of enlarged scales on flanks absent; male without callous abdominal
    scales ....*Laudakia himalayana*
8. Caudal scales large, less than 30 at the base of tail 9
    Caudal scales small, more than 30 at the base of tail 10
9. Head with smooth scales; 8 rows of median dorsal large scales ....*Laudakia pakistanica*
    Head with keeled scales; 10 or more rows of median dorsal scales...11
10. Large dorsal scale, larger than ventrals; flanks with numerous enlarged scales ....*Laudakia agrorensis*
    Largest dorsal scales, smaller than ventrals, flanks with few enlarged scales
    Laudakia tuberculata
11. Caudal segments distinct, with 2 whorls of scales; tail short, thick, not exceeding
    1.5 times the body length 12
    Caudal whorls 1 or 3 in a segment, tail long, slender, longer than 1.5 times the body 13
12. Scales around the midbody 115–188 ....*Laudakia caucasia*
    Scales around the midbody 177–259...*Laudakia microlepis*
13. Enlarged dorsals in 12 or more rows; tail segment with 3 annuli; adult with spiny
    excrescences around ear opening 14
    Enlarged dorsals in 10 or fewer rows; tail segments with single whorl; spiny
    excrescences around ear opening small or absent 15

Chapter 2. Keys to the Identification of Families, Species, and Subspecies

14. Supralabials 13–16, enlarged dorsals in 13–16 rows, dorsum yellow-brown to sooty black with few yellow specks on back ....*Laudakia fusca*
    Supralabials 14–18, enlarged dorsals in 16–20 rows, dorsum olive, thickly mottled, limbs with yellow spots ....*Laudakia nupta*
15. Bright, sandy head, with yellow spotted anterior part of the body ....*Laudakia lirata*
    Brown olive head, with dark reticulation; dorsum with dark spots interspersed with bright yellow spots ....*Laudakia melanura*
16. Dorsal scales subequal in size, disposed in irregular rows..17
    Dorsal scales larger, about twice the size of smaller and are irregular in arrangement 18
17. Tail exceeds body in length; males with callous precloacal scales ....*Trapelus agilis*
    Tail equals or is slightly less than body length; males without callous precloacal scales..........*Brachysaura minor*
18. Enlarged dorsal scales rounded; about 100 scales around the midbody ....*Trapelus rubrigularis*
    Enlarged dorsal scales pointed; fewer than 100 scales around the body..19
19. Largest dorsals about twice the size of smallest; dorsum with reddish or orange ocelli, with dark borders...*Trapelus megalonyx*
    Largest dorsal scales more than twice the size of smallest; dorsal ocelli absent or without dark borders ....*Trapelus ruderatus baluchianus*
20. Dorsal scales markedly unequal in size 21
    Dorsal scales subequal 22
21. Enlarged dorsal scales broad, nail-like, with free posterior borders; sides of head and neck without long spinose scales ....*Phrynocephalus scutellatus*
    Enlarged dorsal scales without posterior free border; sides of head and neck with long spinose scales ....*Phrynocephalus luteoguttatus*
22. Spinose scales on head and neck absent..23
    Spinose scales on head and neck present.. *Phrynocephalus euptilopus*
23. Nasal scales in contact with each other 24
    Nasal scales not in contact with each other..*Phrynocephalus maculatus*
24. Suborbital scale single and elongated ....*Phrynocephalus clarkorum*
    Suborbital scales 2–3 .... *Phrynocephalus ornatus*

Family Chamaeleonidae
Head compressed, with an elevated median casque; digits fused in 2 bundles (Fig. 19A,B) ....*Chamaeleo zeylanicus*

Family Eublepharidae
Eyelids movable; digits not dilated, straight ....*Eublepharis macularius*

Family Gekkonidae
1. Digits dilated..2
   Digits not dilated..10
2. Dilated part of the digit confined to the terminal phalanx ....*Ptyodactylus homolepis*
   Dilated part of the digit extends along the whole digit..3
3. Tail swollen, covered with large flat imbricate scales; subdigital lamellae undivided ....*Teratolepis fasciata*
   Subdigital lamellae divided 4
4. Dorsum with keeled enlarged tubercles, arranged in regular rows 5
   Dorsum with hemispherical keelless tubercles irregularly arranged or absent 8
5. Dorsal pattern of clearly defined broad dark saddles ....*Hemidactylus triedrus*
   Dorsal pattern of small spots, or uniformly colored 6

6. Six to ten lamellae under 4th toe, males with precloacal and femoral pores .... *Hemidactylus brookii*
   Nine to fifteen lamellae under 4th toe; males with precloacal pores only 7
7. Eleven to fifteen lamellae under 4th toe; 6 to 9 precloacal pores................................. *Hemidactylus persicus*
   Nine to twelve lamellae under 4th toe; 4 to 6 precloacal pores .... *Hemidactylus turcicus*
8. Inner toe less than half the length of second toe; a continuous series of 23 to
   33 precloaco-femoral pores .... *Hemidactylus frenatus*
   Inner toe more than half the length of second toe; precloacal and femoral pores
   separated by at least six scales 9
9. Dorsum with tubercles; 20 or more femoral pores; 12 or fewer lamellae under
   4th toe .... *Hemidactylus leschenaultii*
   No dorsal tubercles; femoral pores 15 or less; 12 to 15 lamellae under 4th toe .... Hemidactylus flaviviridis
10. Digits straight 11
    Digits angularly bent between last and penultimate phalanx 16
11. Toes fringed on sides with pointed, flexible, long scales 12
    Toes not fringed so .... *Bunopus tuberculatus*
12. Several series of large, thin scales on tail dorsum; habitus robust 13
    Tail dorsum with small scales; habitus slender 14
13. Body with large cycloid scales, 30–35 around midbody .... *Teratoscincus scincus keyserlingii*
    Body scales small, 100 or more around midbody .... *Teratoscincus microlepis*
14. Unregenerated tail shorter than body; in male fewer than 5 precloacal pores .... *Crossobamon orientalis*
    Tail longer than body; precloacal pores 6 or more 15
15. Dorsum with numerous tubercles; dorsal pattern of transverse bands .... *Crossobamon lumsdeni*
    Few or no dorsal tubercles; dorsal pattern of longitudinal stripes .... *Crossobamon maynardi*
16. Body and tail depressed; tail longer than body...19
    Body and tail cylindrical, equal or subequal in length 17
17. Three nasal scales; dorsal pattern of transverse bands which are much
    narrower than interspaces, tending to break in spots on sides .... *Siwaligekko mintoni*
    Two nasal scales; dorsal pattern of transverse bands, band as broad or
    broader than interspaces 18
18. Dorsal bands broader than interspaces; mid-ventrals 85 to 162 .... *Siwaligekko dattanensis*
    Dorsal bands breaking into a reticulum; mid-ventrals 194 to 205 .... *Siwaligekko battalensis*
19. Tail with even taper; limbs small, heel not reaching axilla 22
    Tail tapering abruptly; limbs long and slender; heels reaching axilla or beyond...................................20
20. Nasal scales strongly projecting vertically carrying naris at higher level .... *Rhinogekko misonnei*
    Nasal scales not as above 21
21. A row of enlarged scales under the thigh; tail longer than snout-vent length .... *Rhinogekko femoralis*
    No enlarged scales under thighs; tail shorter than snout-vent length .... *Agamura persica*
22. Body non tuberculated 23
    Body tuberculated..24
23. Internasals not differentiated from surrounding scales; 4 scales border naris.............. *Tropiocolotes depressus*
    Internasals well differentiated, followed by a second pair of large scales;
    5 scales border naris .... *Tropiocolotes persicus*
24. Trihedral tubercles on body and tail; body moderately depressed 25
    Trihedral tubercles on tail only; body much depressed 31
25. Interspaces between tubercles much smaller than size of the tubercles 26
    Interspaces as large as or larger than size of the tubercles 27
26. Interorbital scales more than 14; dorsal tubercles usually in contact with
    each other; midventrals more than 120; snout-vent length less than 48 mm .... *Cyrtopodion montiumsalsorum*

Interorbital scales less than 14; dorsal tubercles always separated by 1 to 3 granular imbricate scales; midventral scales less than 120; snout-vent length more than 50 mm ....*Cyrtopodion kohsulaimanai*
27. Two whorls of subcaudals to a caudal segment 28
    Three whorls of subcaudals to a caudal segment ....*Cyrtopodion agamuroides*
28. Subcaudals small, as broad as long, in 2 rows ....*Cyrtopodion kachhense*
    Subcaudals broader than long, in a single row 29
29. Scales across mid-abdomen less than 25 ....*Cyrtopodion scabrum*
    Scales across mid-abdomen more than 25 30
30. 25–33 scales across mid-abdomen ....*Cyrtopodion potoharensis*
    30–40 scales around mid-abdomen ....*Cyrtopodion watsoni*
31. Caudal tubercles trihedral, arising from last annulus of caudal segment 32
    Caudal tubercles non-trihedral, arising from center of caudal segment 35
32. Only precloacal pores present in males 33
    Precloacal and femoral pores present in males..34
33. Flat dorsal tubercles keeled ....*Indogekko indusoani*
    Dorsal tubercles feebly keeled or keelless...*Indogekko fortmunroi*
34. 16–18 scales across mid-abdomen 92–106 midventral scales ....*Indogekko rhodocaudus*
    21–25 scales across mid-abdomen; 102–132 midventral scales ....*Indogekko rohtasfortai*
35. Dorsal tubercles round with raised center 36
    Dorsal tubercles flat, with or without a keel 37
36. Unregenerated tail flat, laterally deeply sected, subcaudals indistinct ....*Altigekko stoliczkai*
    Tail quadrangular, a distinct row of subcaudals..*Mediodactylus walli*
37. Number of mid-ventrals 158–171 ....*Altigekko baturensis*
    Number of mid-ventrals 109 ....*Altigekko boehmei*

Family Lacertidae
1. Naris in contact with first supralabial..2
   Naris not in contact with first supralabial..4
2. Dorsals scarcely larger than laterals; usually 7 light stripes on body;
   tail tip yellow ....*Acanthodactylus micropholis*
   Dorsals much larger than laterals; usually 6 stripes on dorsum; tail tip blue,
   gray, or pink 3
3. Median dorsals and laterals are of almost equal size; middorsals 40–46 across
   midbody; gulars 18–20 ....*Acanthodactylus blanfordii*
   Median dorsals distinctly larger than laterals, 26–36; gulars 26 to 36 ......*Acanthodactylus cantoris*
4. Eyelids immovable, forming spectacle; dorsals pointed, imbricate, and keeled;
   no collar 5
   Eyelids movable, lower with transparent disc; dorsals granular, subimbricate
   or juxtaposed; collar present (Fig. 16A)..6
5. Head scales rugose; 25 to 35 scales around midbody ....*Ophisops jerdonii*
   Head scales smooth; 31–38 scales around the midbody ....*Ophisops elegans*
6. Ventrals in straight longitudinal rows (Fig. 17B); an occipital scale present
   (Fig. 16B) 7
   Ventrals in oblique longitudinal rows; occipital scale absent (Fig. 17A).............................8
7. Occipital scale in contact with interparietals; transparent scale in lower eyelid,
   edged with black ....*Mesalina watsonana*

Occipital scale not in contact with interparietals; transparent scale in lower
eyelid not edged with black ....*Mesalina brevirostris*
8. Fringe of pointed scales distinct on the 4th toe 9
   Fringe on the 4th toe absent 11
9. Fringe is only on the outer side of 4th toe...*Eremias scripta*
   Fringe is on both sides of the 4th toe 10
10. Femoral pores present ....*Eremias acutirostris*
    Femoral pores absent...*Eremias aporosceles*
11. Dorsals 55 or more at midbody ....*Eremias persica*
    Dorsals 50 or fewer at midbody..*Eremias fasciata*

Family Scincidae
1. Body serpentine; limbs short and vestigial..2
   Body not markedly serpentine; limbs well developed 5
2. Fingers 4; toes 3..3
   Fingers and toes 3...4
3. Scale rows at midbody 20..*Ophiomorus blanfordi*
   Scale rows at midbody 22 ....*Ophiomorus brevipes*
4. Parietals in contact with anterior temporal..*Ophiomorus tridactylus*
   Parietal and anterior temporal separated by posterior temporal ....*Ophiomorus raithmai*
5. Eyelids fused to form spectacles..6
   Eyelids movable 7
6. Ear hidden beneath scales...*Ablepharus grayanus*
   Ear opening small but distinct ....*Ablepharus pannonicus*
7. Supranasal scale present 8
   Supranasal scale absent..9
8. Scales of middorsal rows much wider than laterals 10
   Scales of dorsal side of body equal in size throughout..13
9. Scales at midbody 24–30 ....*Scincella himalayana*
   Scales at midbody 32–36 ....*Scincella ladacensis*
10. Broad dorsals in a single row ....*Eurylepis taeniolatus*
    Broad dorsals in a double row 11
11. Middorsals 26–30; dorsum uniform brown or with 3 more or less
    distinct dark stripes..12
    Middorsals 52–56; 5–7 dark brown dorsal stripes ....*Novoeumeces indothalensis*
12. Dorsum uniform brown or with 3 more or less distinct dark stripes ....*Novoeumeces schneiderii zarudnyi*
    Dorsum pale gray, vermillion stripes from temporal to groin,
    scattered orange scales on body ....*Novoeumeces blythianus*
13. Naris piercing nasals..14
    Naris between nasals and rostral..*Chalcides ocellatus*
14. Limbs well developed, pentadactyl 15
    Limbs short and vestigial ....*Lygosoma punctata*
15. Supranasals in contact with each other; dorsals with 3 keels ....*Eutropis dissimilis*
    Supranasals separated from each other, dorsals with 3–7 keels ....*Eutropis macularia*

Family Uromastycidae
1. Body dorsum with uniform granular scales, caudal spines smaller,
   20–24 in a row at the base of the tail ....*Uromastyx hardwickii*

Body dorsum with granular scales interspersed with transverse rows of spiny scales; caudal spines larger, 8–10 in a row at the base of the tail ....*Uromastyx asmussi*

Family Varanidae

1. Tail compressed with a median dorsal ridge; scales on side of neck keeled 2
   Tail round in cross-section, ridge slightly indicated along the middle of tail; scales on the sides of neck conical ....*Varanus griseus*
2. Naris nearer to the tip of snout than eye (Fig. 21A) ....*Varanus flavescens*
   Naris nearer to eye than the tip of snout (Fig. 21B) ....*Varanus bengalensis*

Ophidia: Snakes

Family Leptotyphlopidae

1. Rostral large, hooked; total length of body 80–110 times its diameter (Fig. 22C) ....*Leptotyphlops macrorhynchus*
   Rostral normal, round; total length 55–70 times its diameter............................*Leptotyphlops blanfordii*

Family Typhlopidae

1. Midbody scales 25 or more ....*Typhlops diardii*
   Midbody scales less than 25 2
2. Scales around midbody 20 (Fig. 22A)..*Ramphotyphlops braminus*
   Scales around the midbody 18 3
3. Midbody diameter does not exceed 2 mm ....*Typhlops ductuliformes*
   Midbody diameter exceeds 2 mm...4
4. Nasal scale completely divided ....*Typhlops madgemintonai*
   Nasal scale incompletely divided 5
5. Preocular in contact with 3rd supralabial only...*Typhlops ahsanai*
   Preocular in contact with 3rd and 4th supralabials ....*T. madgemintonai shermanai*

Family Boidae

1. Premaxillary toothed; head with large symmetrically arranged scales; subcaudals in double row (Fig. 23E)..*Python molurus*
   Premaxillary not toothed; head with small scales; subcaudals in a single row (Fig. 23D) 2
1. Mental groove present 3
   Mental groove absent..*Eryx conicus*
3. All body scales on anterior half smooth; cloacal and caudal scales keeled; 49 scale rows at midbody ....*Eryx tataricus*
   All body scales keeled; 51–61 midbody scale rows ....*Eryx johnii*

Family Colubridae

1. Ventrals extend across abdomen; rostral deeply grooved (Fig. 24A,B)................................2
   Ventrals do not extend across abdomen; rostral not grooved, projected downward ....*Enhydris pakistanica*
2. Subocular scale present; prefrontal fragmented (Fig. 25D) 3
   Subocular scale absent; prefrontal not fragmented...5
3. Rostral long, wedged between, but not completely separating internasals; midbody scales 25 ....*Spalerosophis arenarius*

	Rostral of normal size; midbody scale rows more than 25	4
4.	Belly white with gray smudges; subcaudals less than 100	*Spalerosophis schirazianus*
	Belly with reddish tint; subcaudals more than 100	*Spalerosophis diadema*
5.	Dorsals smooth throughout	6
	At least some rows of dorsals keeled	29
6.	Lower jaw countersunk (Fig. 25C)	*Argyrogena fasciolata*
	Lower jaw normal	7
7.	Pupil of eye dark, hardly visible in life	8
	Pupil of eye large, clearly visible in life	10
8.	Loreal not or slightly in contact with internasals; subcaudals in a single row	*Lycodon travancoricus*
	Loreal extensively in contact with internasals; subcaudals in pairs (Fig. 26E)	9
9.	Posterior nasal distinctly smaller than anterior; 8 supralabials	*Lycodon striatus*
	Posterior and anterior nasal scales equal in size; 9 supralabials	*Lycodon aulicus*
10.	Rostral large, projecting, concave below; naris narrow slits	11
	Rostral and naris not as above	13
11.	Rostral truncate anteriorly	*Lytorhynchus ridgewayi*
	Rostral pointed	12
12.	5th labial touching eye	*Lytorhynchus paradoxus*
	No labial touching eye (Fig. 26A, I)	*Lytorhynchus maynardi*
13.	Pupil elliptical; head much wider than neck (Fig. 26 D)	14
	Pupil round, head slightly wider than neck	16
14.	Head black, labials and chin scales dark; ventrals white, without dark speckling	*Boiga melanocephala*
	Head and rest of the characters not as above	15
15.	Supralabials 8; nostrils large in a pair of nasal scales	*Boiga trigonata*
	Supralabials 9–10; nostril in a partially divided nasal scale	*Telescopus rhinopoma*
16.	Head and nape with a pair of oblique or dark crossbars; fewer than 60 subcaudals	17
	Head and nape not as above; more than 60 subcaudals	19
17.	Scales rows at midbody 15; a thick dark crossbar at nape (Fig. 26B)	*Oligodon taeniolatus*
	Scale rows at midbody 17	18
18.	Head with a thick, inverted V-shaped, dark stripe, another on nape (Fig. 26C)	*Oligodon arnensis*
	Head and nape dark brown or black, a large yellow mark on each side of head	*Sibynophis sagittarius*
19.	Ventrals fewer than 190	20
	Ventrals 190 or more	23
20.	Cloacal scale not divided; head elongated (Fig. 27B)	*Psammophis leithii*
	Cloacal scale divided	21
21.	Nasal scale completely divided	22
	Nasal scale incompletely divided	*Psammophis condanarus*
22.	Supralabials 4th and 5th or 5th and 6th in contact with eye	*Psammophis schokari*
	Supralabials fourth, fifth and sixth in contact with eye	*Psammophis lineolatus*
23.	Adult exceeds 1250 mm in total length; head scales with regular dark borders (Fig. 27A)	*Ptyas mucosus*
	Adult does not exceed 1100 mm in total body length; no borders around head scales	24
24.	Anterior temporals 2; eyes large	25
	Anterior temporal 1; eyes small (Fig. 27C)	*Pseudocyclophis persicus*
25.	Scales at midbody in 19 rows	26
	Scales at midbody in 21 rows	*Hemorrhois ravergieri*
26.	Dorsals smooth, without keels	27
	Dorsals keeled	29

Chapter 2. Keys to the Identification of Families, Species, and Subspecies   17

27. Single labial (5th) touching eye; body dorsum with vivid pattern of sooty
 black crossbars, scales of bars uniformly black (Fig. 28Biii) ....*Coluber karelini*
 A pair of supralabials touching eye; dorsum unicolor, or with a pattern of
 spots or rhombs..28
28. Ventrals 199–211, subcaudals 82–119; dorsal pattern of light brown rhombs,
 center of scales forming pattern lighter (Fig. 28Bii) ....*Platyceps ventromaculatus*
 Ventrals 205–244, subcaudals 110–144; dorsal pattern of dark spots or
 unicolor, scales of the pattern unicolor (Fig. 28Bi) ....*Platyceps rhodorachis*
29. 3 supralabials in eye 30
 Less than 3 supralabials in eye 31
30. Ventrals 160 or fewer ....*Amphiesma stolatum*
 Ventrals more than 160 ....*Amphiesma platyceps*
31. A pair of supralabials touching eye 32
 1 supralabial touching eye (Fig. 28A) ....*Xenochrophis cerasogaster*
32. Anterior temporals 1 pair; ventrals 160 or less..33
 Anterior temporal single; ventrals 160 or more ....*Xenochrophis tessellata*
33. A pair of oblique orbitolabial dark stripes; dorsum light olive, with pattern
 of dark spots (Fig. 28C) ....*Xenochrophis piscator*
 Orbitolabial stripes indistinct; dorsum yellowish green, unicolor ....*Xenochrophis sanctijohannis*

Family Elapidae
1. Vertebral row of dorsals distinctly enlarged, subcaudals not divided (Fig. 30Aii) 2
 Vertebral row not enlarged; subcaudals divided (Fig. 30Bii)..3
2. Scales at midbody 15 ....*Bungarus caeruleus*
 Scales at midbody 17 ....*Bungarus sindanus*
3. Dorsum jet black to light black; usually a cuneate scale present (Fig. 30Bi)......................*Naja naja*
 Dorsum yellow brown or variegated; the cuneate scale usually absent ....*Naja oxiana*

Family Hydrophiidae
1. Ventrals distinct throughout, normally undivided 2
 Ventrals, except anterior, either divided by a median longitudinal fissure or indistinct 3
2. Mental scale elongate, partially hidden in a groove between first infralabial;
 ventrals uniform in size ....*Enhydrina schistosa*
 Mental scale normal 5
3. Head very small; neck long, slender; body thick; ventrals divided by a
 longitudinal fissure 4
 Head and body not as above 12
4. Prefrontal touches 3rd supralabial; ventrals 404–468 ....*Microcephalophis cantoris*
 Prefrontal not touching 3rd supralabial; ventrals 320–350 ....*Microcephalophis gracilis*
5. Ventrals broad anteriorly, narrower posteriorly ....*Praescutata viperina*
 Ventrals uniform in size 6
6. Adult with very small head and long, slender neck 7
 Adult with large head; neck not distinct from body 8
7. Ventrals 390 or more ....*Hydrophis fasciatus*
 Ventrals fewer than 390..*Hydrophis mamillaris*
8. Scales on thickest part of body with round or blunt tips; imbricate; adult
 length more than one meter 9

	Scales on thickest part of body hexagonal or quadrangular; imbricate or juxtaposed; length rarely exceeds 1 meter 10
9.	Anterior temporal normally 1 ....*Hydrophis spiralis*
	Anterior temporals normally 1 pair ....*Hydrophis cyanocinctus*
10.	Adult with bluish gray bands on body; maxillary teeth more than 13 ....*Hydrophis caerulescens*
	Adult with dark gray or greenish bands on body; maxillary teeth less than 13 11
11.	Head top with curved yellow mark; ventrals 314–322 ....*Hydrophis lapemoides*
	Head top mark absent; ventrals 209–312 ....*Hydrophis ornatus*
12.	Dorsals juxtaposed 13
	Dorsals pointed, strongly imbricate; ventrals indistinct ....*Astrotia stokesii*
13.	Ventrals distinct anteriorly, indistinct posteriorly ....*Lapemis curtus*
	Ventrals with a longitudinal fissure; dorsals subquadrangular ....*Pelamis platurus*

Family Viperidae

1.	Snout flanked with enlarged butterfly-like scales jutting out laterally (Fig. 31A) ....*Eristicophis macmahonii*
	Snout flanked with small scales 2
2.	Laterals small, strongly oblique, with serrated keels (Fig. 31Bi, ii) ....*Echis carinatus*
	Laterals with entire keels 3
3.	A group of elevated supraoculars forming "horn"; keels ending in knobs (Fig. 31C) ....*Pseudocerastes persicus*
	No supraorbital horn; keels normal 4
4.	Pattern of indistinct blotches on body; midbody scale rows 23–27 ....*Macrovipera lebetina*
	Pattern of 3 rows of vivid dark brown ocelli; ocelli of middorsum, often fused to form a chain; 27–33 scale rows at midbody (Fig. 31D) ....*Daboia russelii*

Family Crotalidae

	Head with symmetrically arranged large scales; a pit between eye and naris (Fig. 32) ....*Gloydius himalayanus*

Figures

(2 through 32)

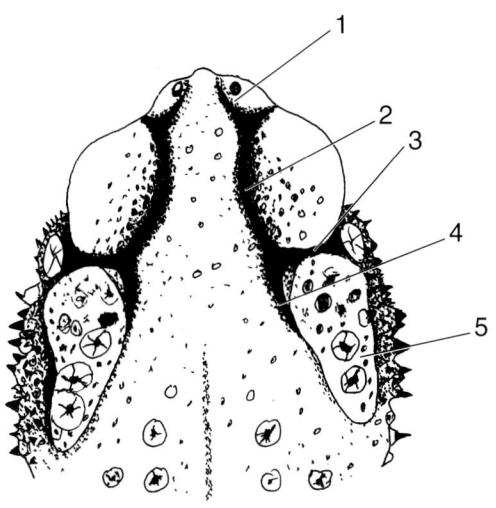

Figure 2A. *Bufo melanostictus:* Head, cranial crests, 1. Canthal; 2. Supraorbital; 3. Postorbital; 4. Temporal. 5. Parotid gland

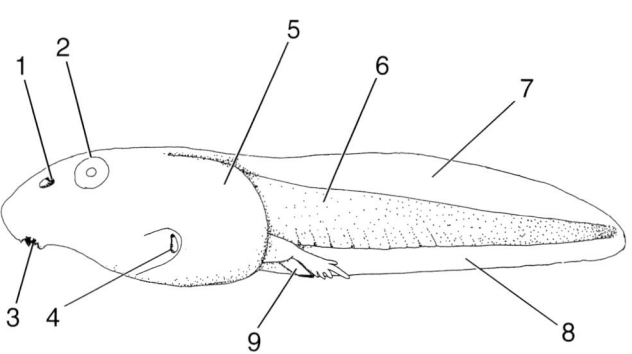

Figure 3A. Amphibian tadpole morphology, *Bufo stomaticus:* Tadpole, 1. Naris; 2. Eye; 3. Oral disc; 4. Spiracle; 5. Body; 6. Tail; 7. Dorsal fin; 8. Ventral fin; 9. Cloacal tube.

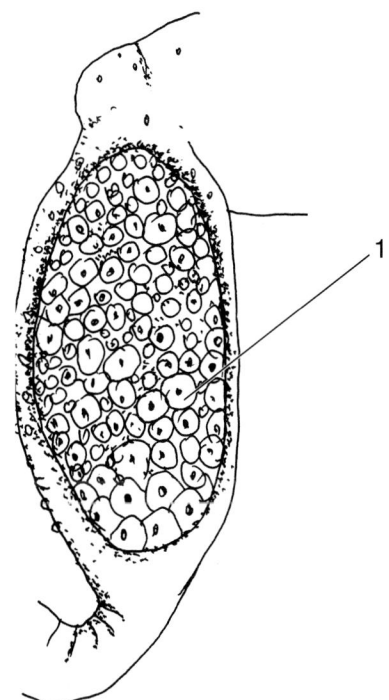

Figure 2B. *Bufo stomaticus*, tibia, 1. Tibial gland

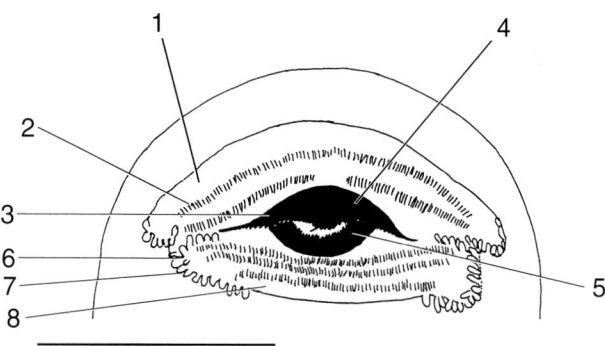

Figure 3B. Amphibian tadpole morphology, *Bufo stomaticus:* Oral Disc, 1. Anterior labium; 2. Dental row; 3. Anterior beak; 4. Oral opening; 5. Posterior beak; 6 and 7. Oral papillae; 8. Posterior labium (scale = 1 mm).

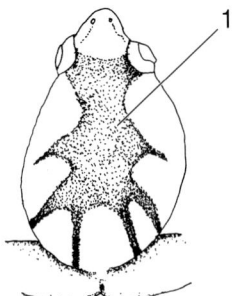

Figure 4A. *Microhyla ornata:* Dorsal pattern.

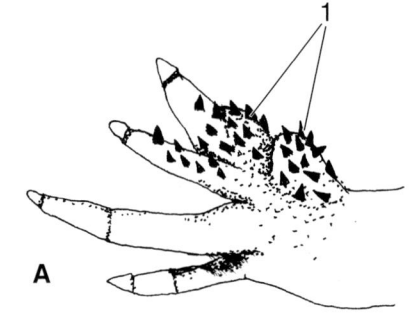

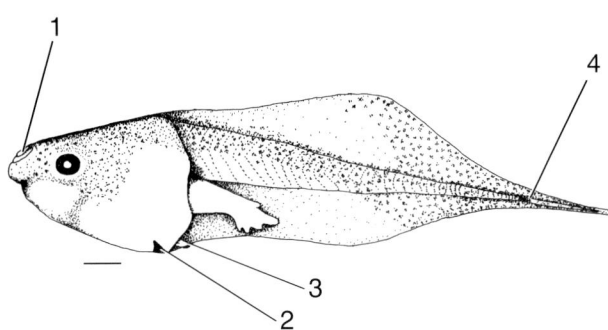

Figure 4B. *Microhyla ornata:* Tadpole, 1. Mouth; 2. Spiracle; 3. Anal tube; 4. Tail flagellum.

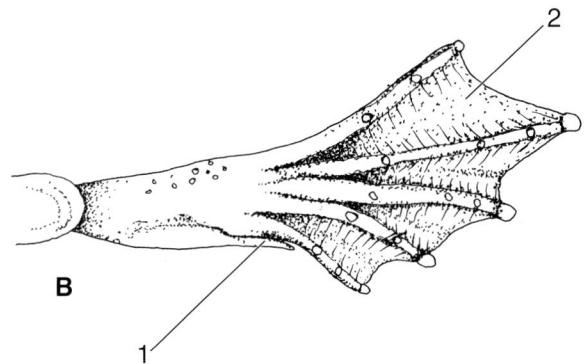

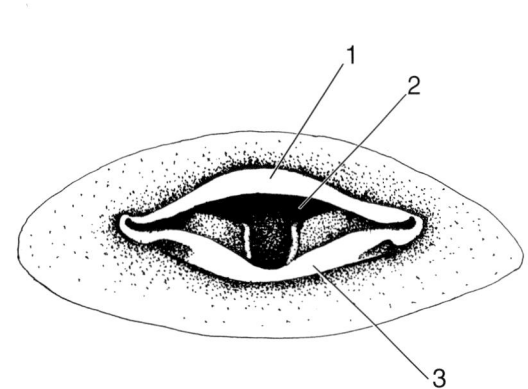

Figure 4C. *Microhyla ornata:* Oral disc: 1. Anterior labium; 2. Oral opening; 3. Posterior labium, (scale = 1 mm).

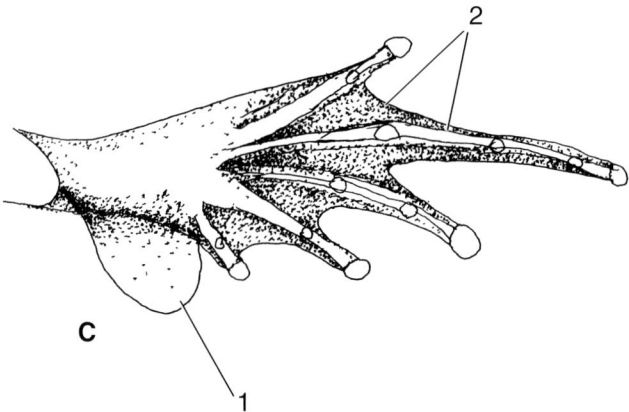

Figure 5. Frog morphology:
A. *Paa hazarensis,* hand, 1. Nuptial spines.
B. *Euphlyctis cyanophlyctis*, foot, 1. Inner metatarsal tubercle; 2. Web.
C. *Sphaeroteca breviceps*, foot, 1. Inner metatarsal tubercle; 2. Web.

Figures

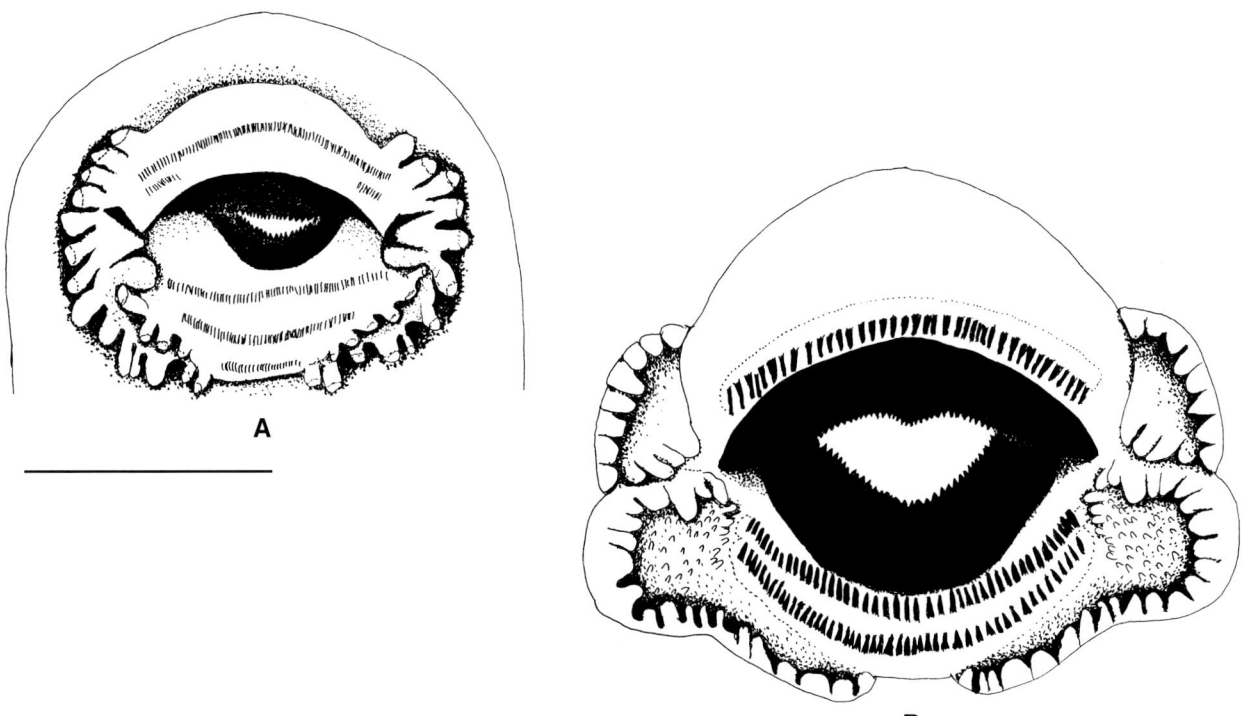

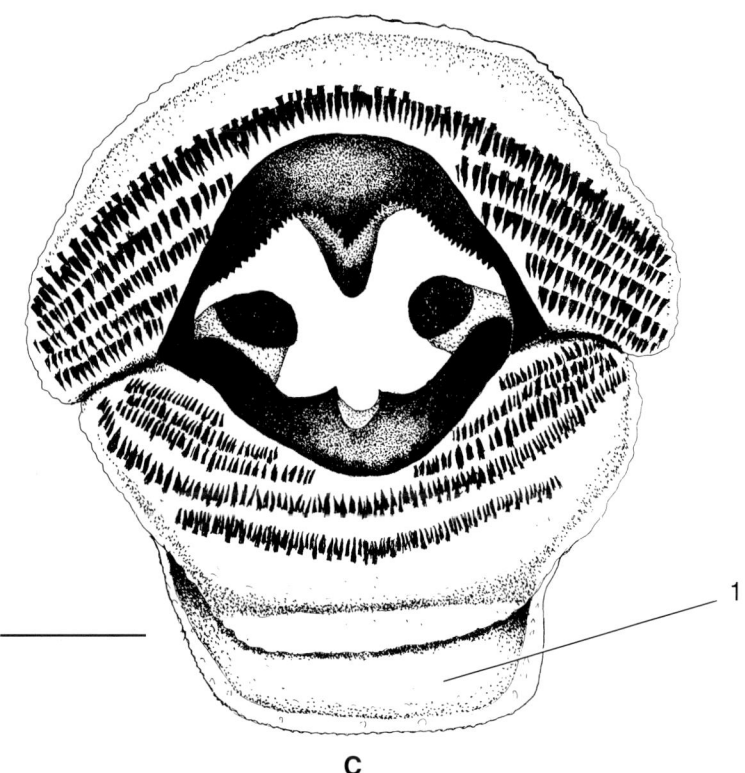

Figure 6. Oral discs of some amphibian tadpoles (scale = 1 mm):
A. *Fejervarya syhadrensis.*
B. *Euphlyctis cyanophlyctis.*
C. *Hoplobatrachus tigerinus*, 1. Postoral disc.

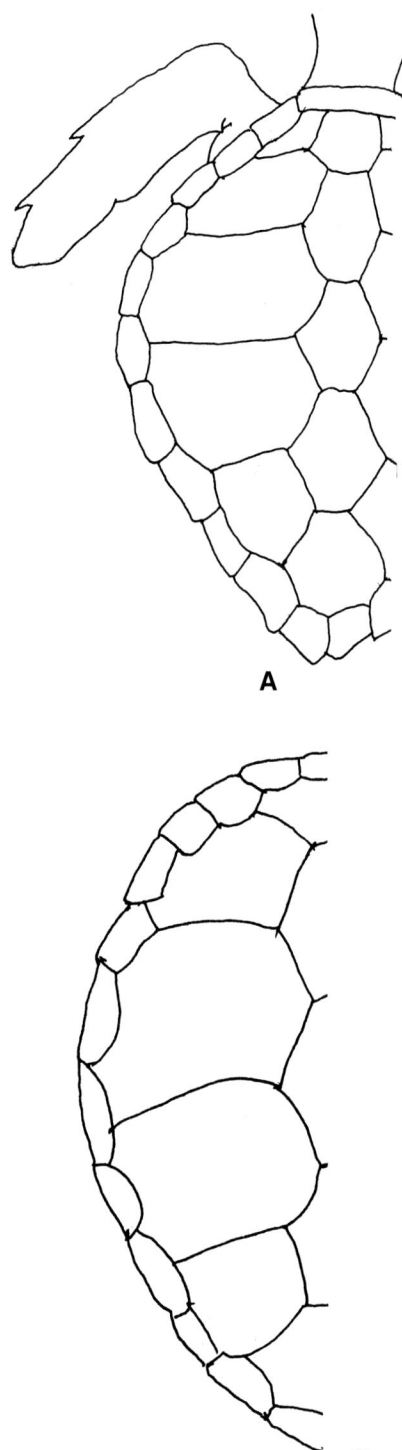
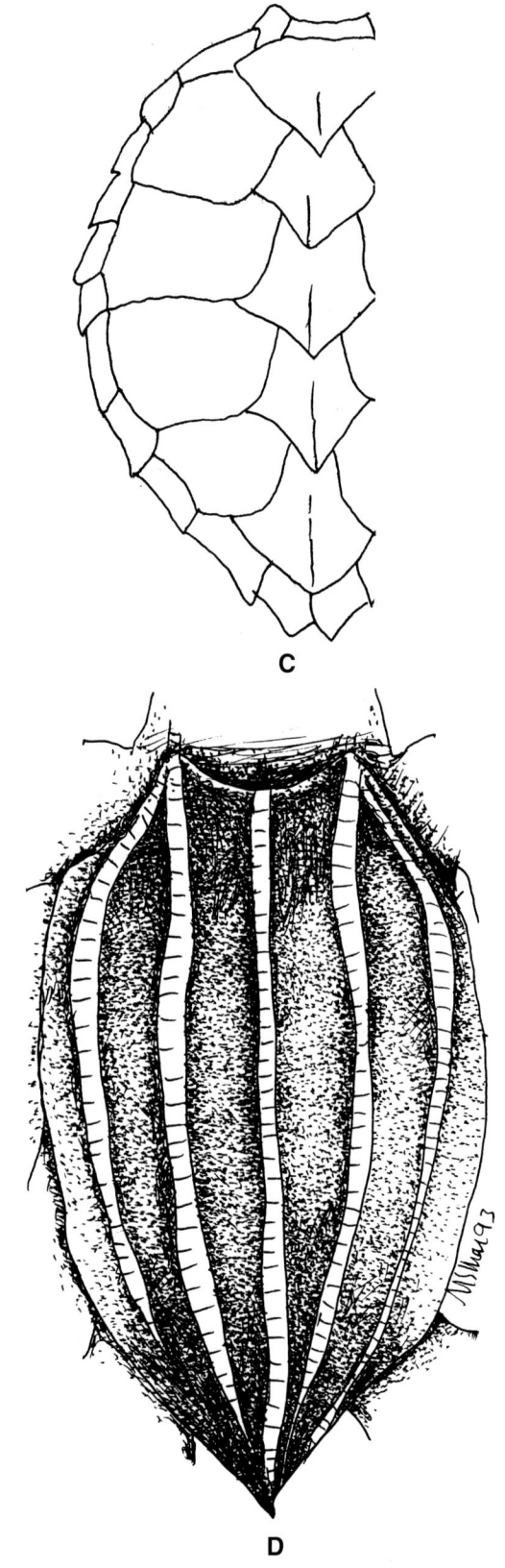

Figure 7. Turtle carapaces:
A. *Caretta caretta.*
B. *Chelonia mydas.*
C. *Eretmochelys imbricata.*
D. *Dermochelys coriacea.*

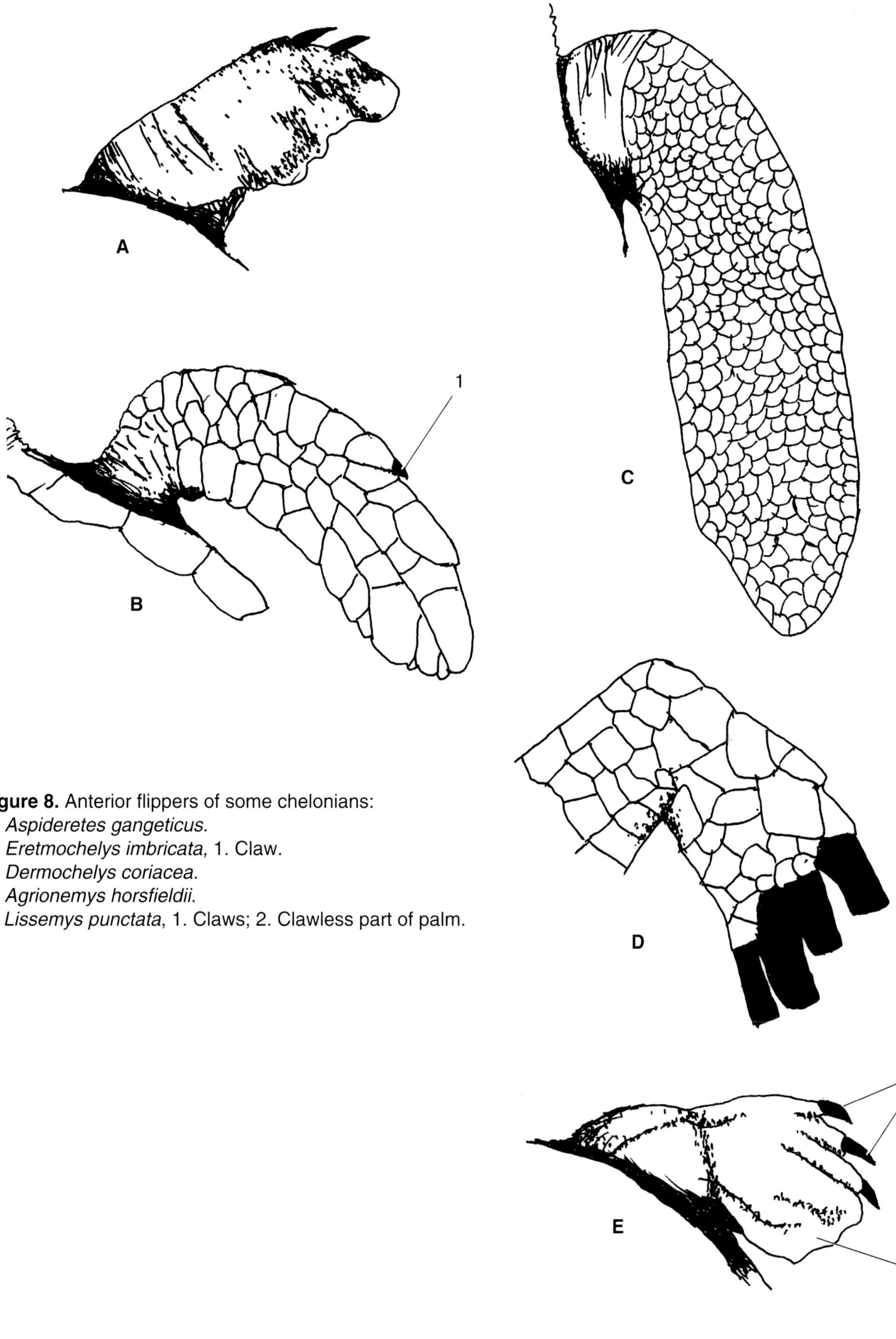

Figure 8. Anterior flippers of some chelonians:
A. *Aspideretes gangeticus.*
B. *Eretmochelys imbricata,* 1. Claw.
C. *Dermochelys coriacea.*
D. *Agrionemys horsfieldii.*
E. *Lissemys punctata,* 1. Claws; 2. Clawless part of palm.

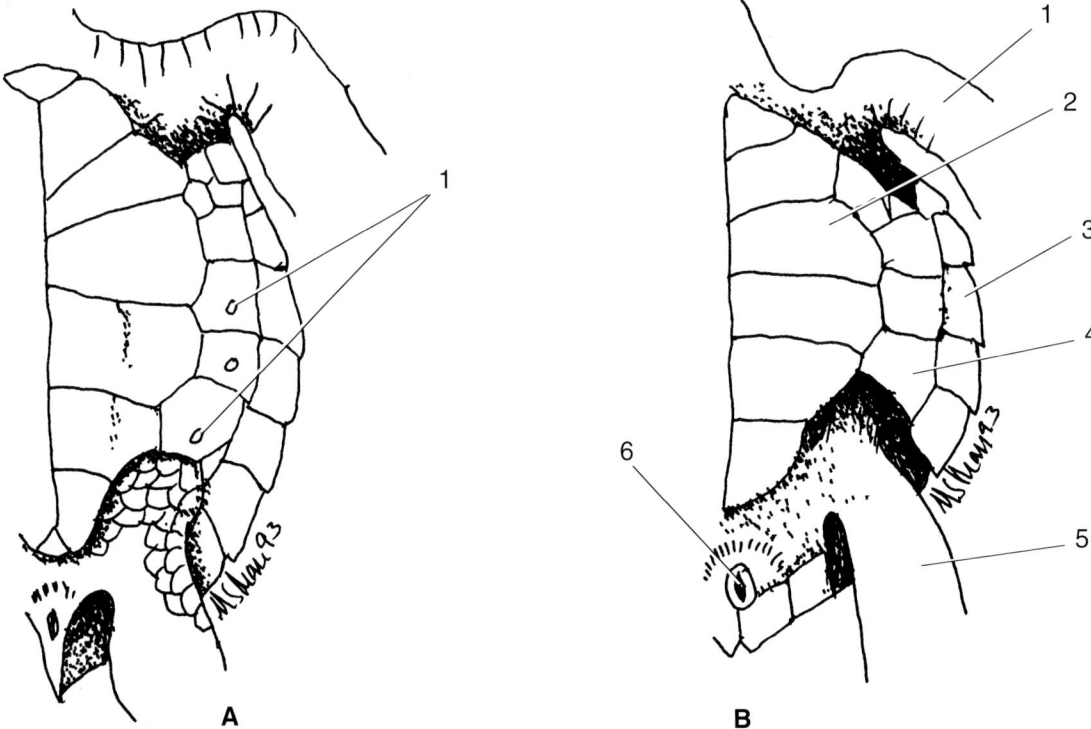

Figure 9. Plastrons of two turtles:
A. *Lepidochelys olivacea*, 1. Costal pores.
B. *Caretta caretta*, 1. Anterior flipper; 2. Plastron; 3. Carapace; 4. Costal bridge; 5. Posterior flipper; 6. Cloacal opening.

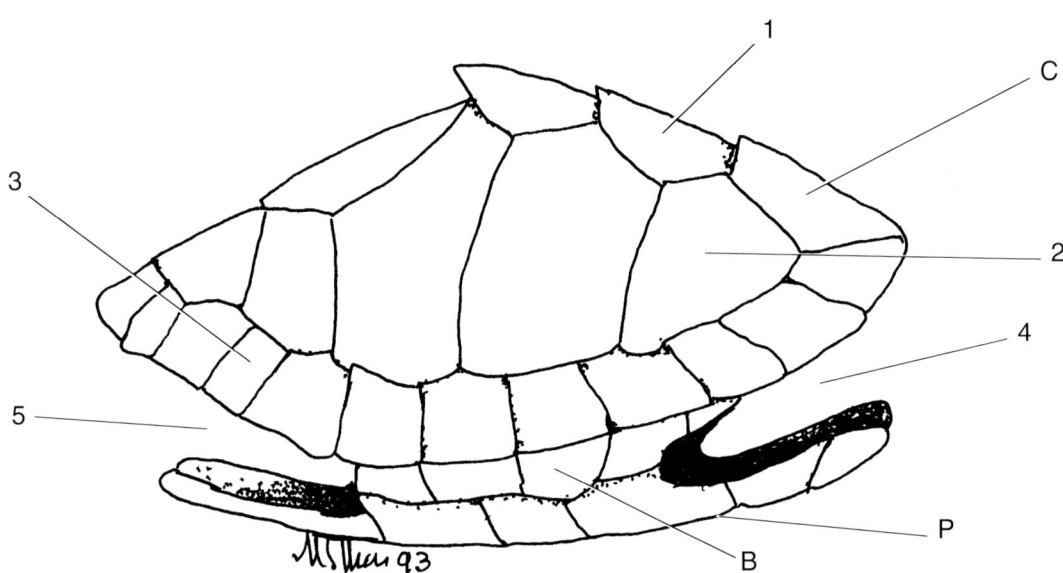

Figure 10. Turtle shell, *Kachuga tecta:* B. Bridge; C. Carapace; P. Plastron; 1. Neurals; 2. Pleurals (costals); 3. Marginals; 4. Cephalic opening of shell; 5. Caudal opening of shell.

Figures

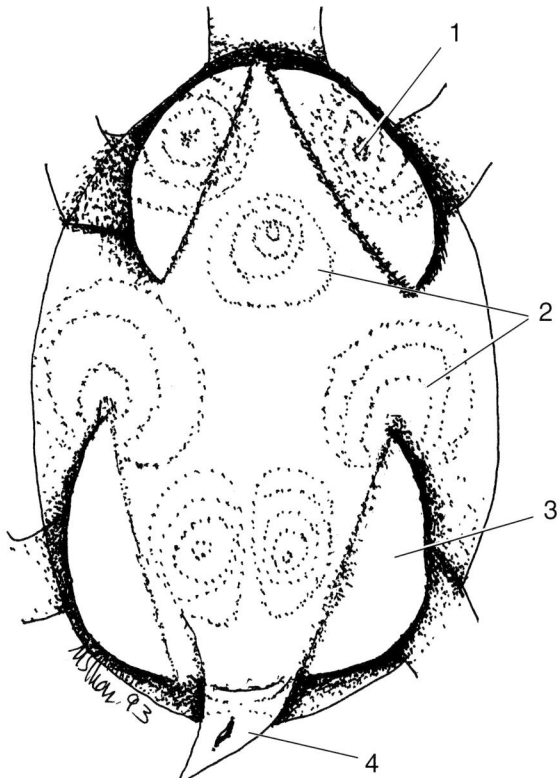
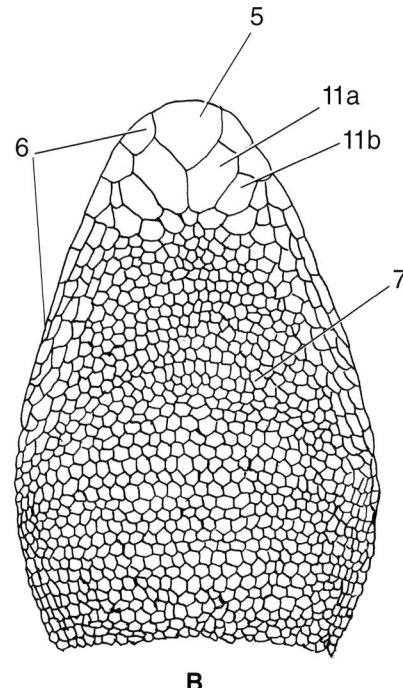

Figure 11. *Lissemys punctata*, plastron: 1. Pectoral valve; 2. Callosities; 3. Femoral valve; 4. Tail.

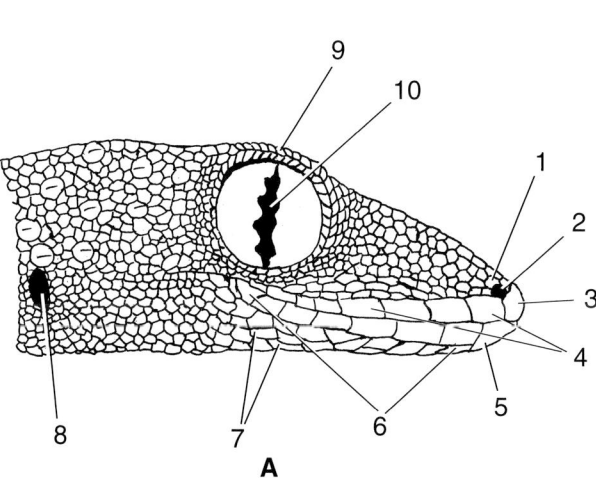
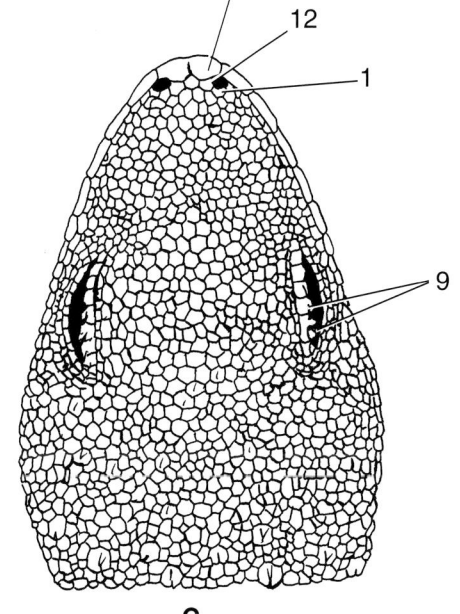

Figure 12. Family Gekkonidae, head morphology:
A. Head, lateral side. 1. Nasal; 2. Naris; 3. Rostral; 4. Supralabials; 5. Mental; 6. Infralabials; 7. Gulars; 8. Ear opening; 9. Supraciliary scales; 10. Pupil.
B. Head, ventral side, 11a. First pair of submentals; 11b. Second pair of submentals (other indicators as in Figure A.)
C. Dorsal side: 12. Internasals (other indicators as in Figure 12.A).

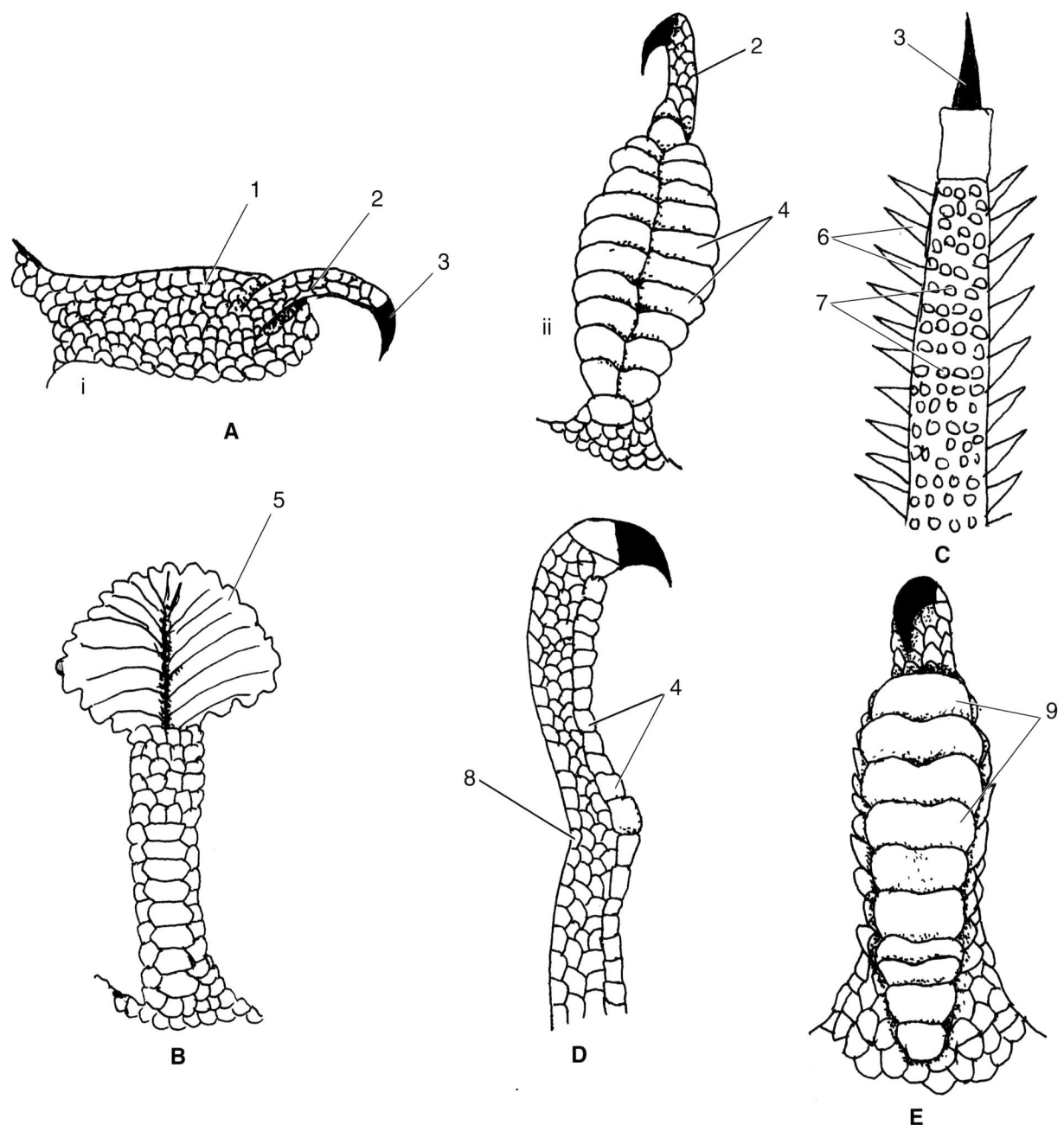

Figure 13. Family Gekkonidae, foot morphology:
A. Genus *Hemidactylus*, a toe, i. Dorsal side; ii. Ventral side: 1. basal expanded part; 2. Terminal angular part; 3. Claw; 4. Divided subdigital lamellae.
B. *Ptyodactylus homolepis*, a Finger, 5. Expanded part.
C. *Teratoscincus scincus*, fourth toe, ventral side; 6. Fringe of lateral pointed scales; 7. Subdigital tubercles (other indicators as in Figure 13A).
D. *Cyrtopodion montiumsalsorum*, angularly bent finger, 8. Angle (other indicators as in Figure 13A.)
E. *Teratolepis fasciata*, 9. Undivided subdigital lamellae.

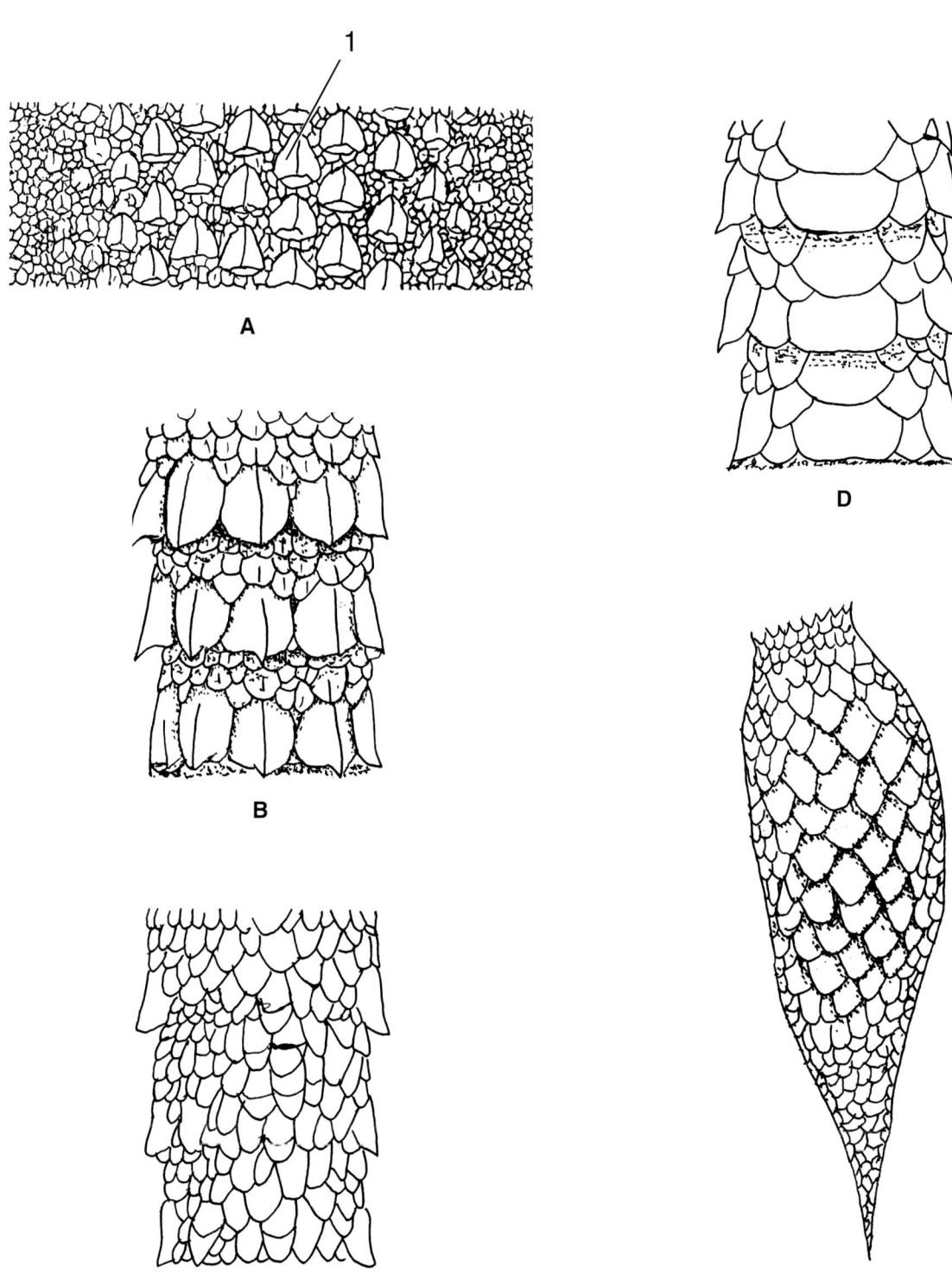

Figure 14. Family Gekkonidae, scale morphology:
A. *Cyrtopodion kachhense* and *Cyrtopodion scabrum*, dorsal scalation of the body, 1. Trihedral tubercles.
B. *Cyrtopodion scabrum* and *Cyrtopodion kachhense,* tail dorsum, trihedral tubercles.
C. *Cyrtopodion kachhense*, subcaudals.
D. *Cyrtopodion scabrum*, subcaudals.
E. *Teratoscincus scincus*, tail dorsum.

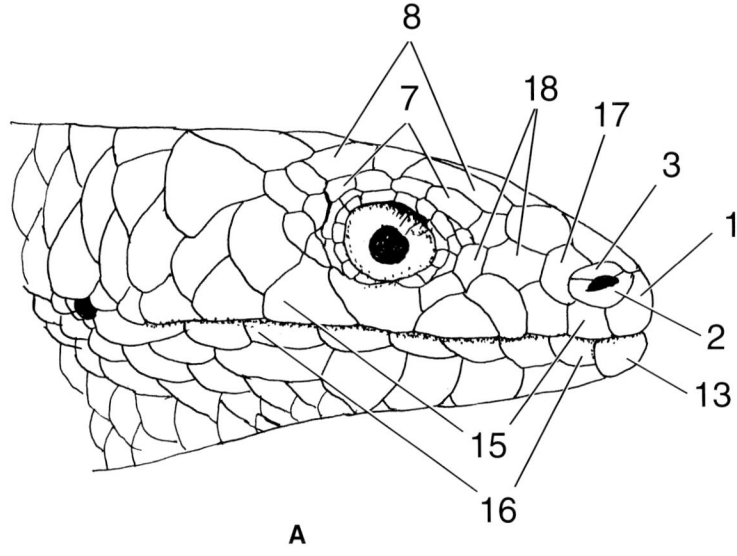
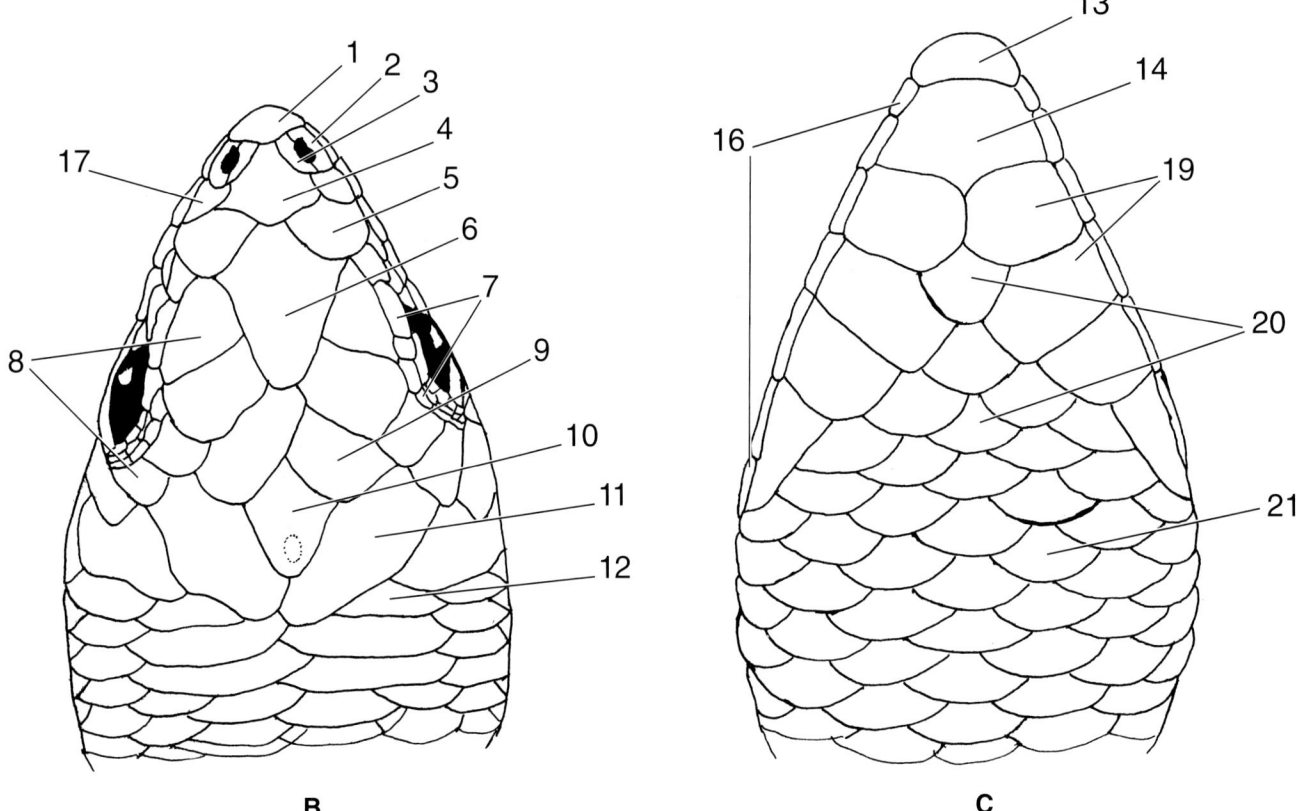

Figure 15. Family Scincidae, head scales: Lateral side; **B.** Dorsal side; **C.** Ventral side.
Legend for Figures A-C: 1. Rostral; 2. Infranasal; 3. Supranasal; 4. Frontonasal; 5. Prefrontal; 6. Frontal; 7. Supraciliaries; 8. Supraoculars; 9. Frontoparietal; 10. Interparietal; 11. Parietal; 12. Nuchal; 13. Mental; 14. Sub-mental; 15. Supralabials; 16. Infralabials; 17. Postnasal; 18. Loreals; 19. Submaxillary; 20. Genials; 21. Gulars.

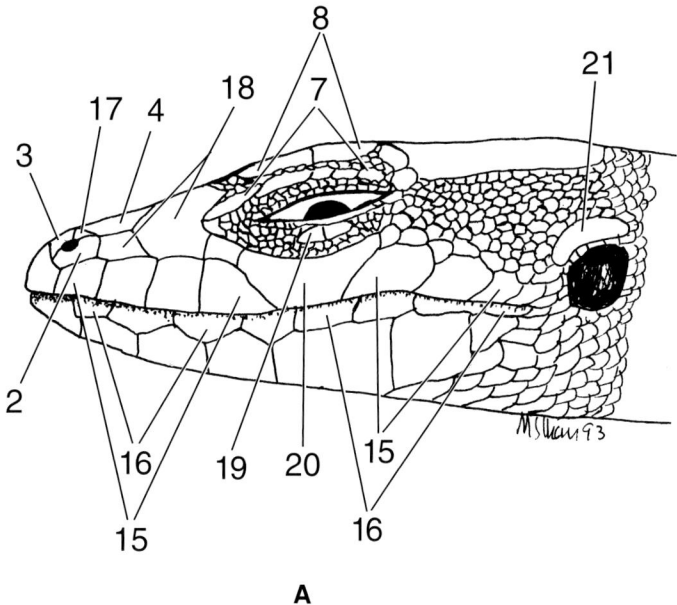
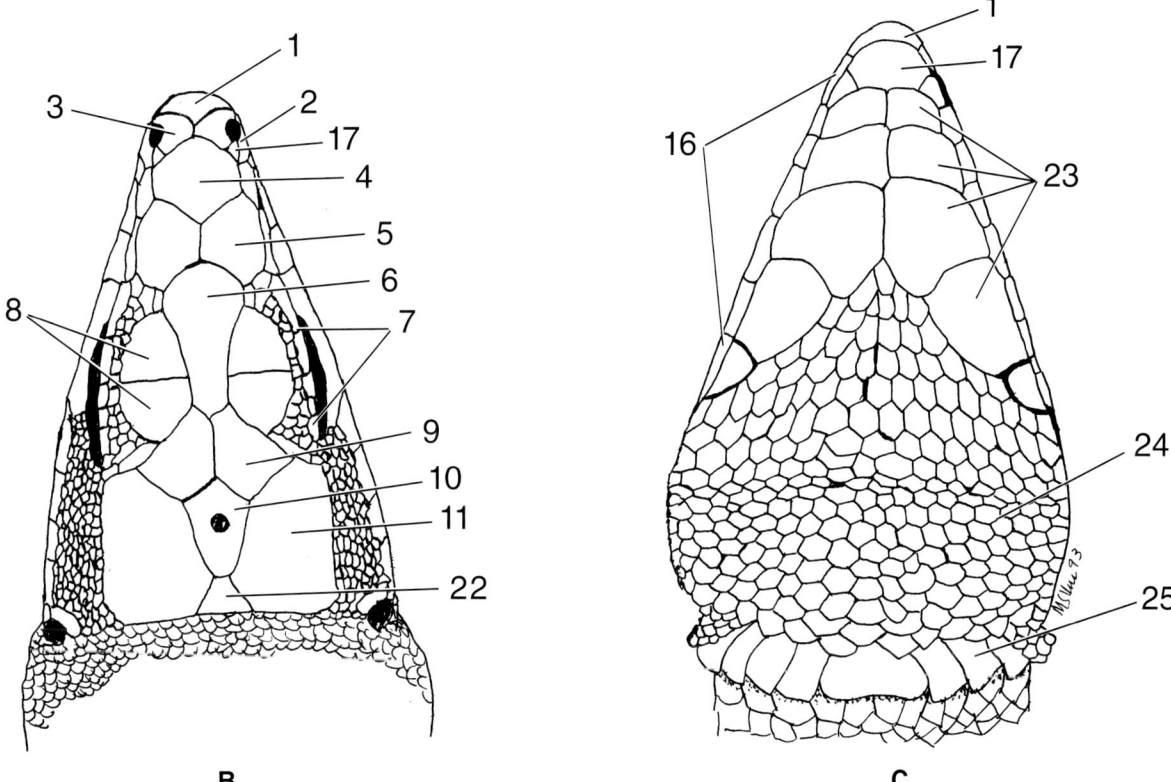

Figure 16. Family Lacertidae, head scales: 1. Rostral; 2. Infranasal; 3. Supranasal; 4. Frontonasal; 5. Prefrontal; 6. Frontal; 7. Supraciliaries; 8. Supraoculars; 9. Frontoparietal; 10. Interparietal; 11. Parietal; 15. Supralabials; 16. Infralabials; 17. Postnasal; 18. Loreals; 19. Subocular transparent disc; 20. Suborbital; 21. Supra-ocular; 22. Post-parietal; 23. Submaxillaries; 24. Gulars; 25. Gular fold.
A. Lateral side.
B. Dorsal side.
C. Ventral side.

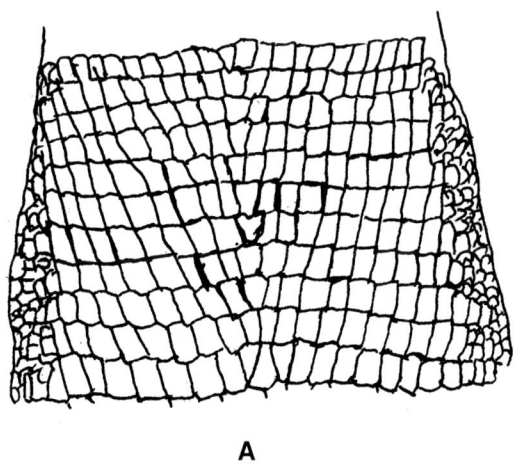
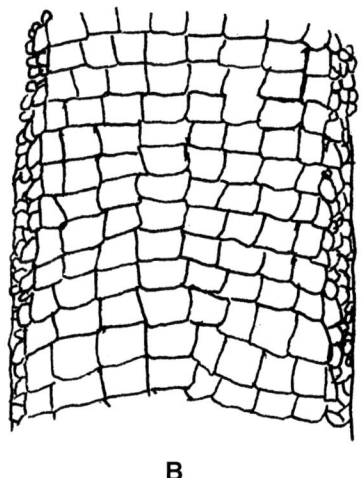
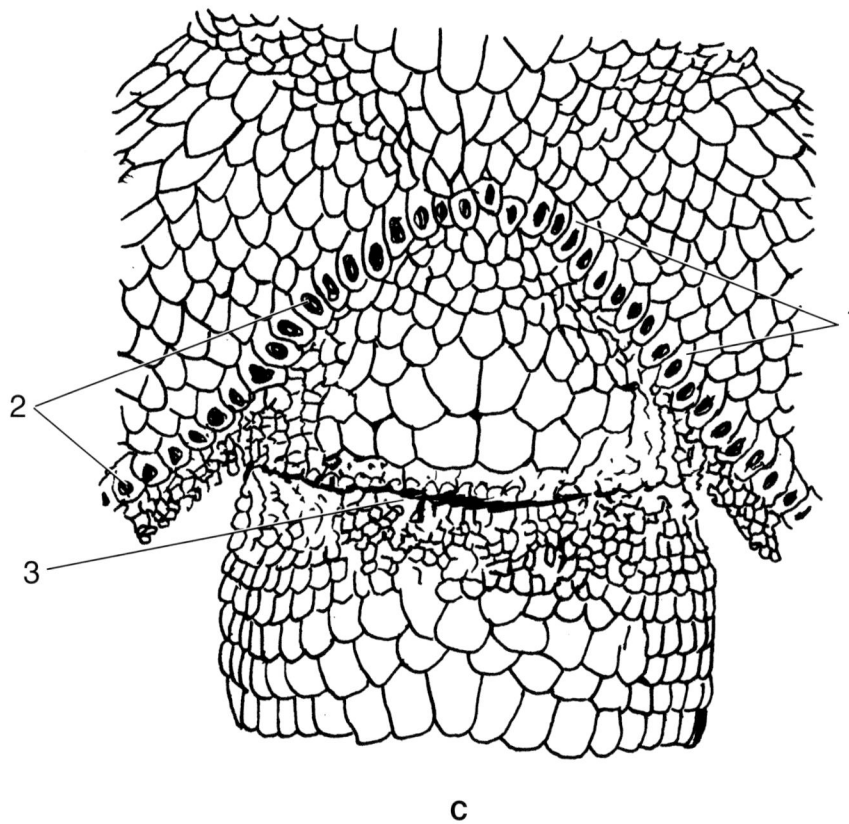

Figure 17. Family Lacertidae, arrangement of abdominal scales:
A. Oblique series, genus *Eremias*.
B. Longitudinal series, genus *Mesalina*.
C. Anal region: 1. Precloacal pores; 2. Femoral pores (in a continuous series); 3. Cloacal aperture.

Figures

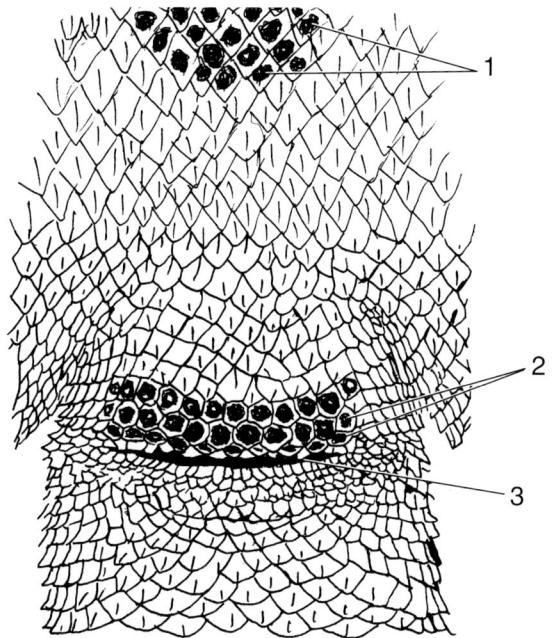

Figure 18. Family Agamidae, body ventrum:
1. Abdominal callous scales; 2. Cloacal callous scale;
3. Cloacal opening.

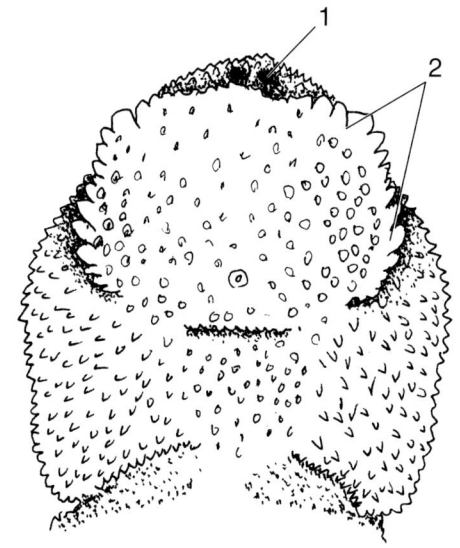

Figure 20. Genus *Phrynocephalus*, head dorsum:
1. Naris; 2. Supraorbital ridge.

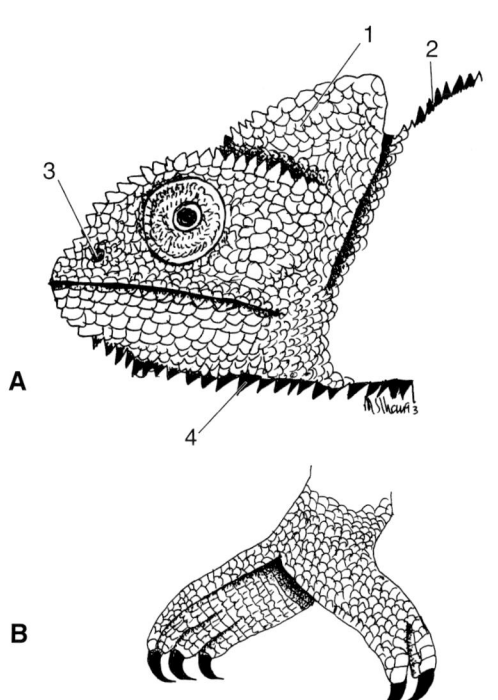

Figure 19. Family Chamaeleonidae:
A. Head, 1. Casque; 2. Middorsal row of pointed scales; 3. Naris; 4. Midventral row of pointed scales.
B. Foot, fused toes.

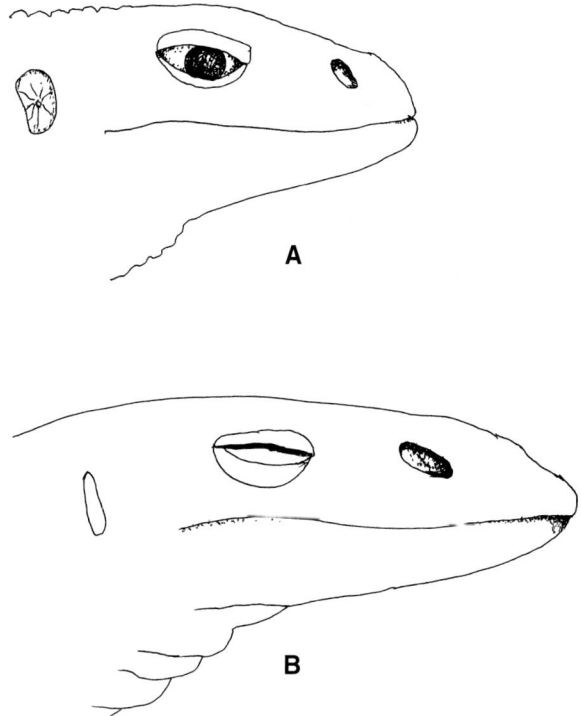

Figure 21. Family Varanidae, position of head and naris:
A. *Varanus flavescens.*
B. *Varanus bengalensis.*

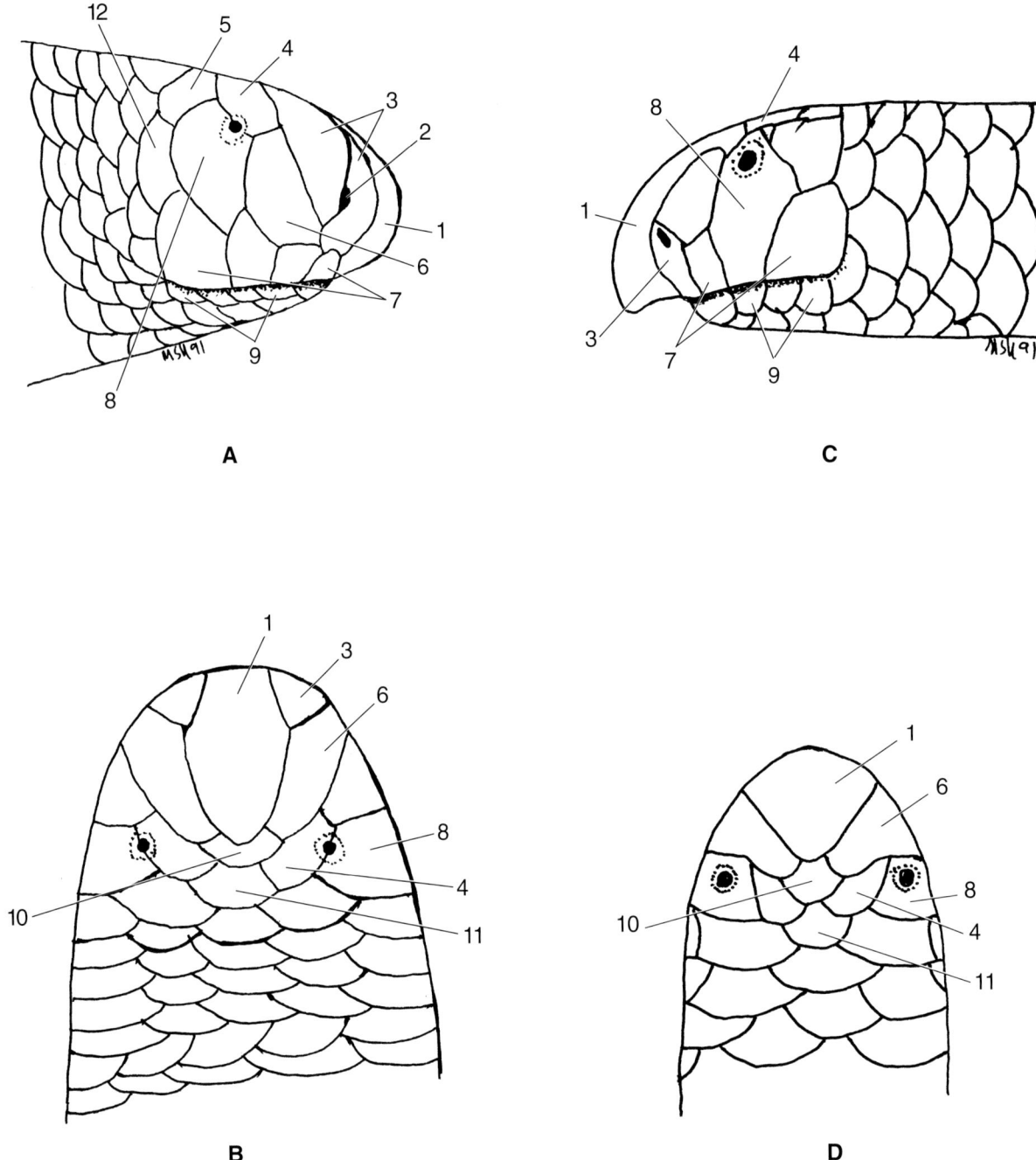

Figure 22. Family Typhlopidae, nomenclature:
A. *Ramphotyphlops braminus*, head, lateral side, 1. Rostral; 2. Naris; 3. Nasals; 4. Supraorbital; 5. Preparietal; 6. Preocular; 7. Supralabials; 8. Ocular; 9. Infralabials; 12. Temporal.
B. Head, dorsal side, 10. Prefrontal; 11. Frontal.
C. Family Leptotyphlopidae: *Leptotyphlops macrorhynchus:* head, lateral side (indicators as in Figure 22A, B).
D. Head, dorsal side (indicators as in Figure 22A, B). [After Smith, 1943]

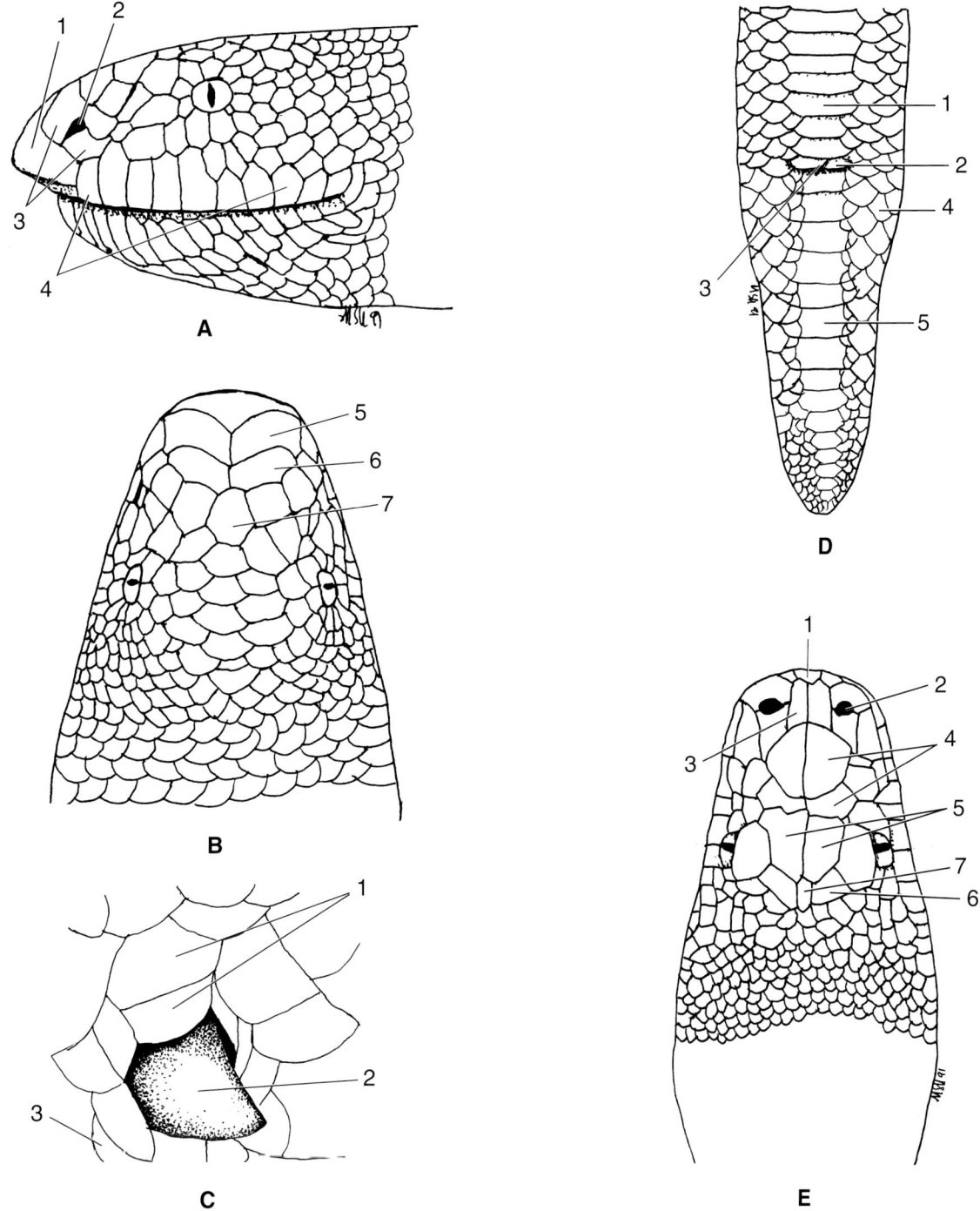

Figure 23. Family Boidae, nomenclature, *Eryx johnii:*
A. Head, lateral side, 1. Rostral; 2. Naris; 3. Nasals; 4. Supralabials.
B. Head, dorsal side, 5. Prefrontal; 6. Frontal; 7. Postrostral.
C. Circumcloacal region, showing spur, 1. Preanal scales; 2. Spur; 3. Postanal scales.
D. Posterior part of body and tail, ventral side, 1. Ventrals; 2. Cloacal scale; 3. Cloacal aperture; 4. Laterals; 5. Subcaudals.
E. *Python molurus*, head dorsum, 1. Rostral; 2. Naris; 3. Internasals; 4. Prefrontals; 5. Frontals; 6. Parietal, 7. Interparietal. [After Smith, 1943]

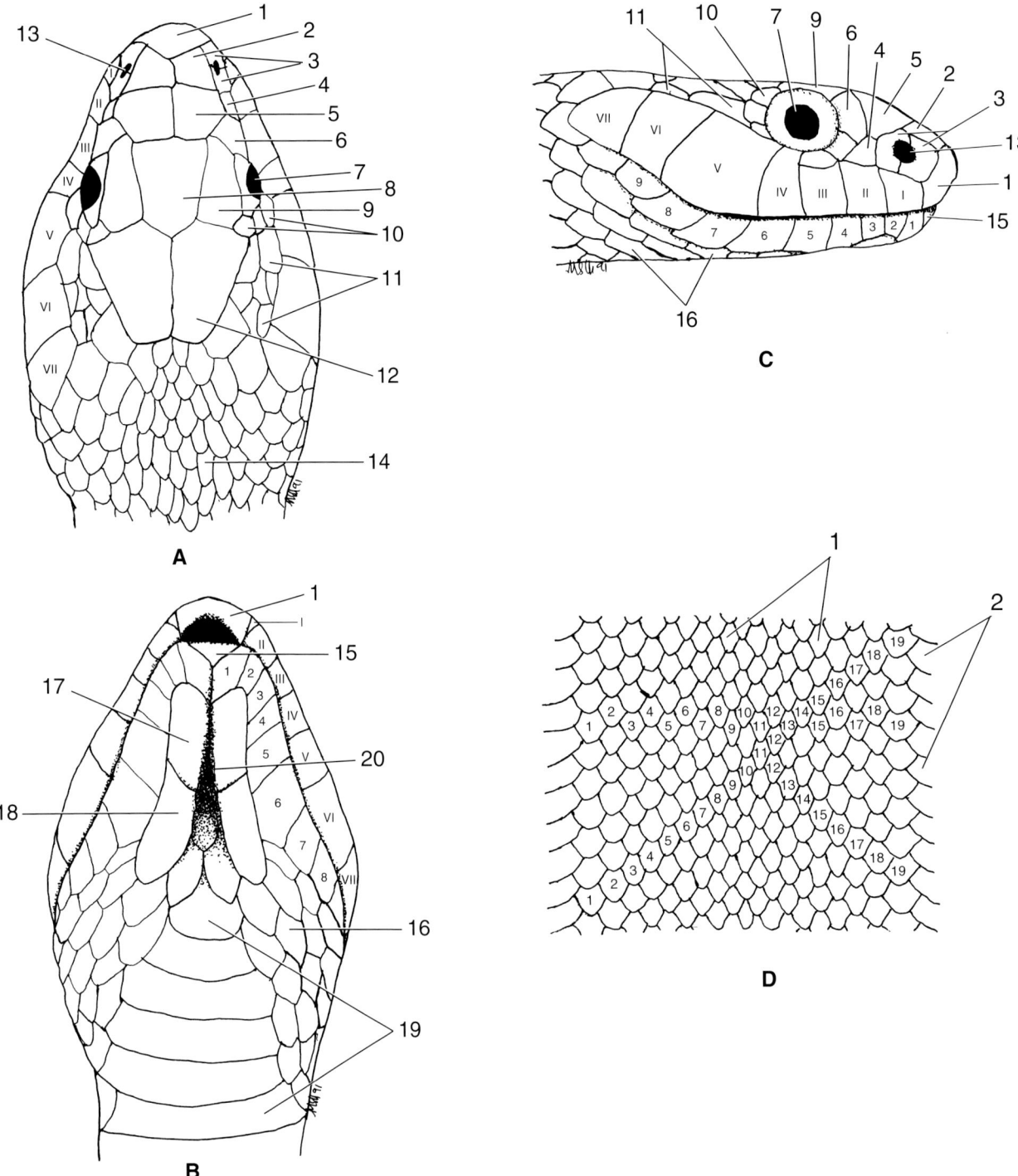

Figure 24. Family Colubridae, nomenclature:
A. Head dorsum, 1. Rostral; 2. Internasal; 3. Nasal; 4. Loreal; 5. Prefrontal; 6. Preocular; 7. Eye; 8. Frontal; 9. Supraorbital; 10. Postorbitals; 11. Temporals; 12. Parietal; 13. Naris; 14. Nuchals.
B. Head ventrum, 15. Mental; 16. Gulars; 17. Anterior genials; 18. Posterior genial; 19. Ventrals; 20. Mental groove.
C. Head, lateral side, 4. Loreal; 5. Prefrontals; 6. Preoculars; 8. Frontal; 9. Supraorbital; 10. Postoculars; 11. Temporals; 12. Parietal; I-VII (within key). Supralabials; 1–9 (within key). Infralabials (other indicators as in Figure 24A, B).
D. Dorsal scale counting, 1. Dorsals; 2. Ventrals.

Figures

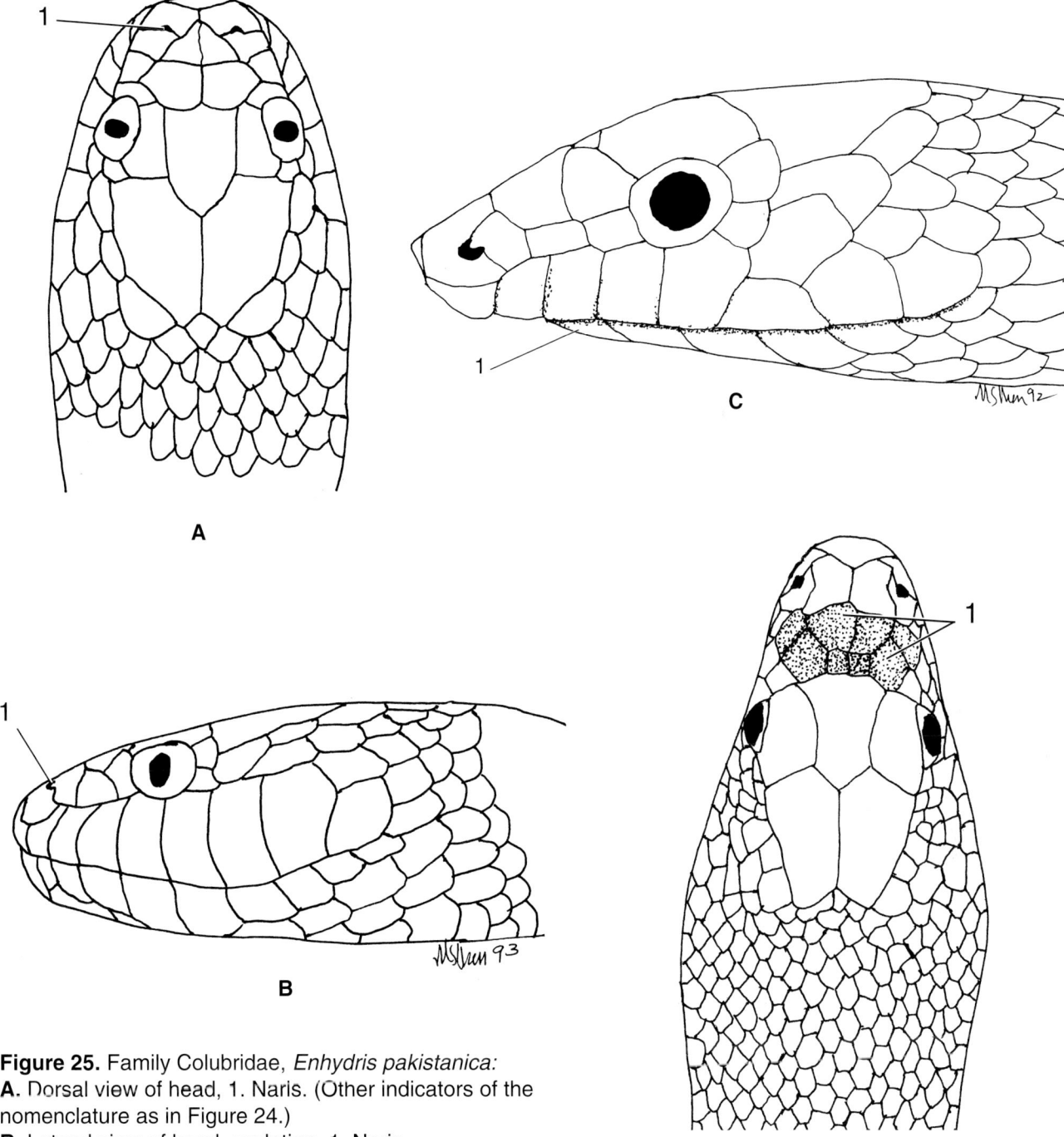

Figure 25. Family Colubridae, *Enhydris pakistanica*:
A. Dorsal view of head, 1. Naris. (Other indicators of the nomenclature as in Figure 24.)
B. Lateral view of head, scalation, 1. Naris.
C. *Argyrogena fasciolata*, lateral side of head, 1. Lower jaw.
D. *Spalerosophis diadema*, dorsal side of head, 1. Divided prefrontal. [After Smith, 1943]

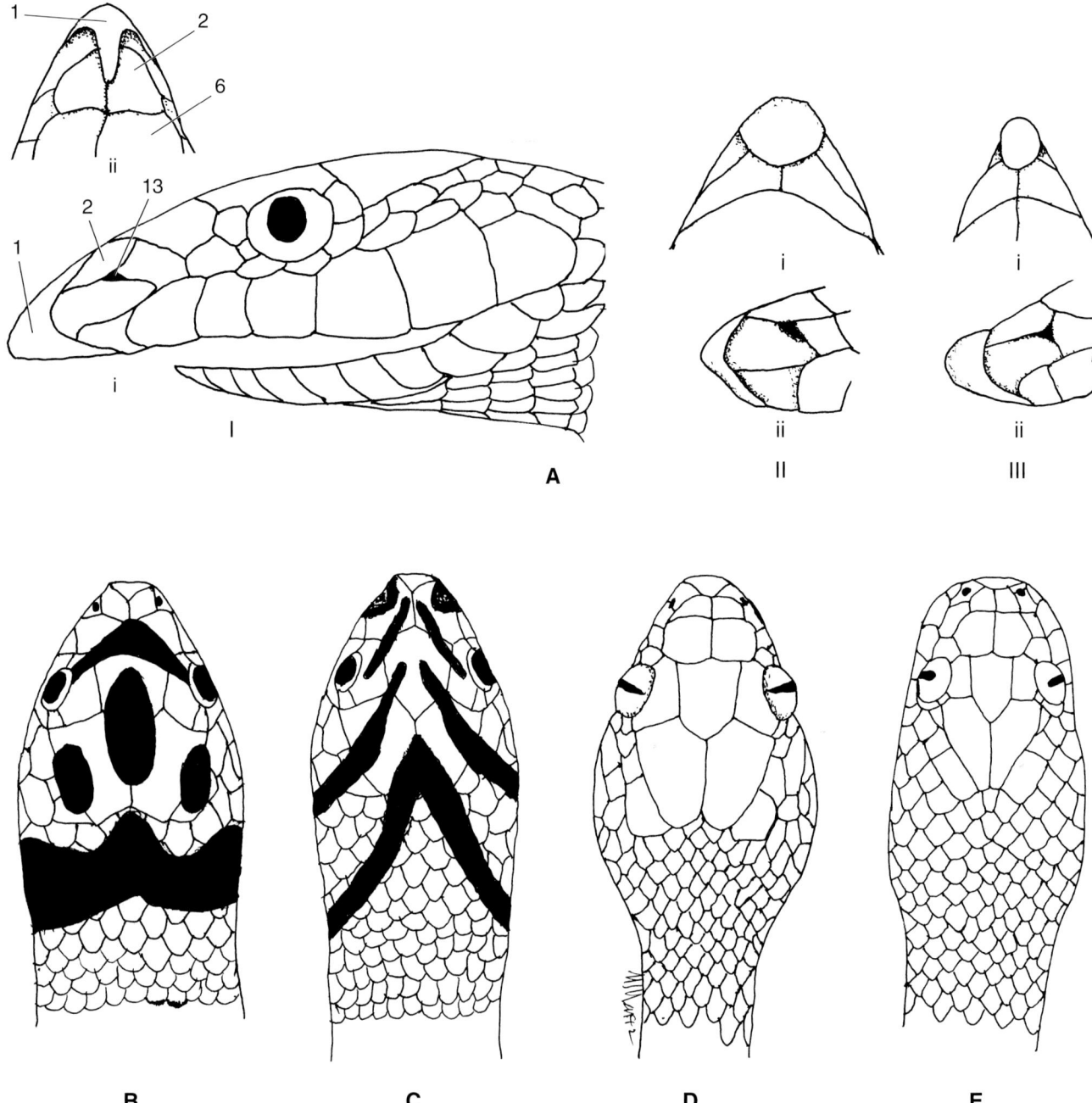

Figure 26. Family Colubridae (indicators as in Figure 24).
A. Genus *Lytorhynchus*, note rostral scale,
I. i. *Lytorhynchus maynardi*, lateral side of head; ii. Snout dorsum.
II. i. *Lytorhynchus ridgewayi*, dorsal side of snout; ii. Lateral side of snout;
III. i. *Lytorhynchus paradoxus*, dorsal side of snout; ii. Lateral side of snout.
B. *Oligodon taeniolatus*, head dorsum, scales, and pattern.
C. *Oligodon arnensis*, head dorsum, scales, and pattern.
D. *Boiga trigonata*, head dorsum.
E. *Lycodon aulicus,* head dorsum. [After Smith, 1943]

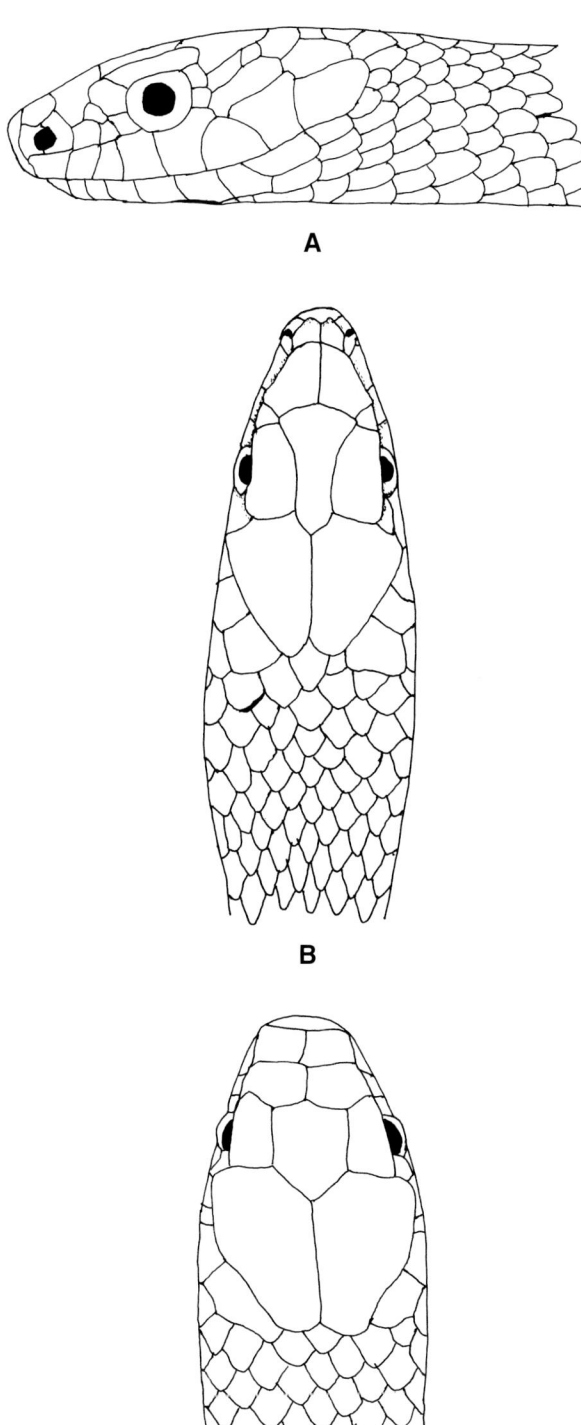
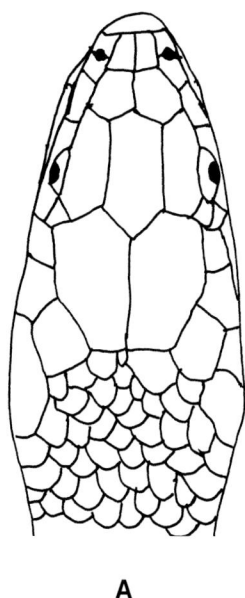

Figure 27. Family Colubridae (indicators as in Figure 24):
A. *Ptyas mucosus,* head lateral side.
B. *Psammophis leithii,* head dorsal side.
C. *Pseudocyclophis persicus,* head dorsal side.

Figure 28. Family Colubridae (indicators as in Figure 24):
A. *Xenochrophis cerasogaster,* head dorsum.
B. Pigmentation of scales in dorsal pattern of *Coluber* snakes: i. *Platyceps rhodorachis;* ii. *Platyceps ventromaculatus;* iii. *Coluber karelini.*
C. *Xenochrophis piscator,* head dorsum.

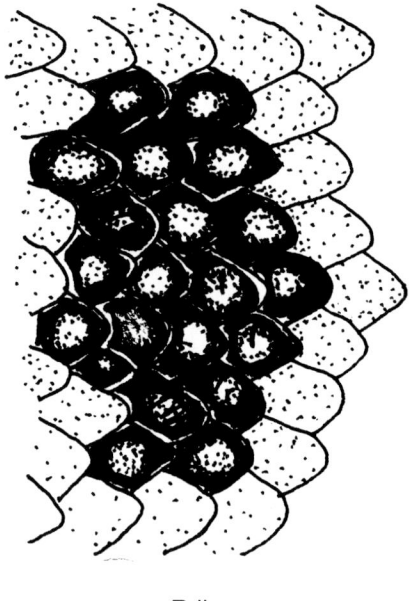
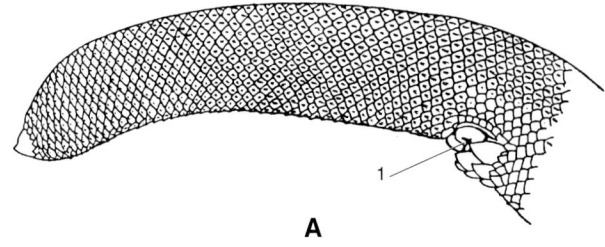
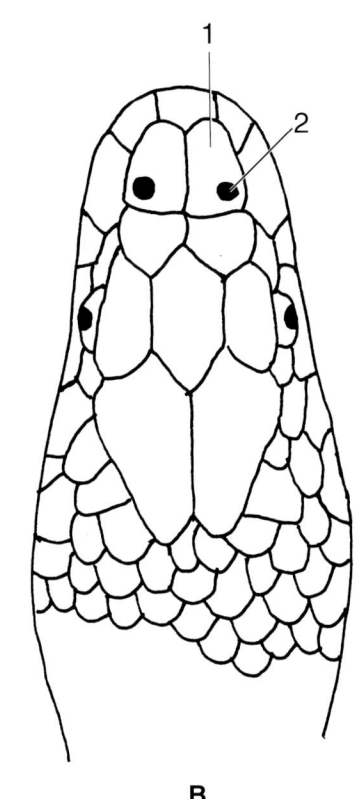

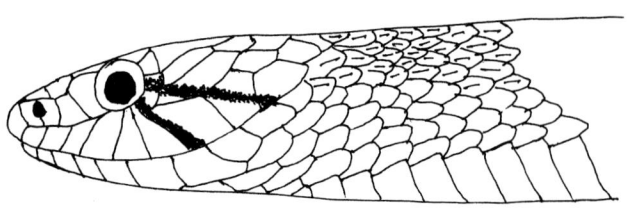

Figure 29. Family Hydrophiidae (indicators as in Figure 24). [After Smith, 1943]
A. Tail, lateral side, 1. Cloacal scales.
B. Head, dorsal side, 1. Nasal; 2. Nasal aperture.

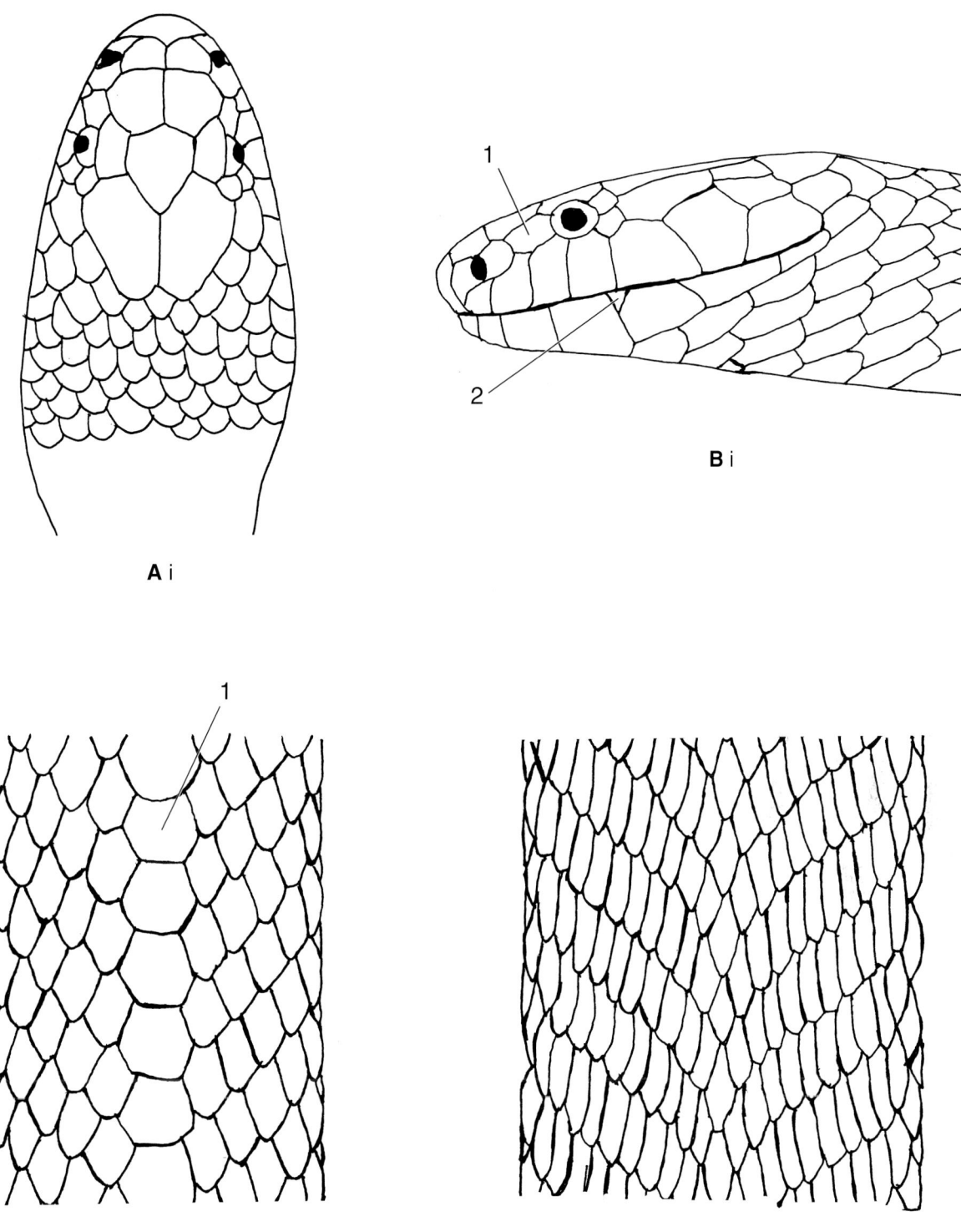

Figure 30. Family Elapidae (indicators as in Figure 24).
A. i. *Bungarus caeruleus*, head dorsum; ii. A part of body showing enlarged middorsal row of enlarged scales (1).
B. i. *Naja naja*, head, lateral side, 1. Preocular; 2. "cuneate" scale. ii. Dorsal scales.

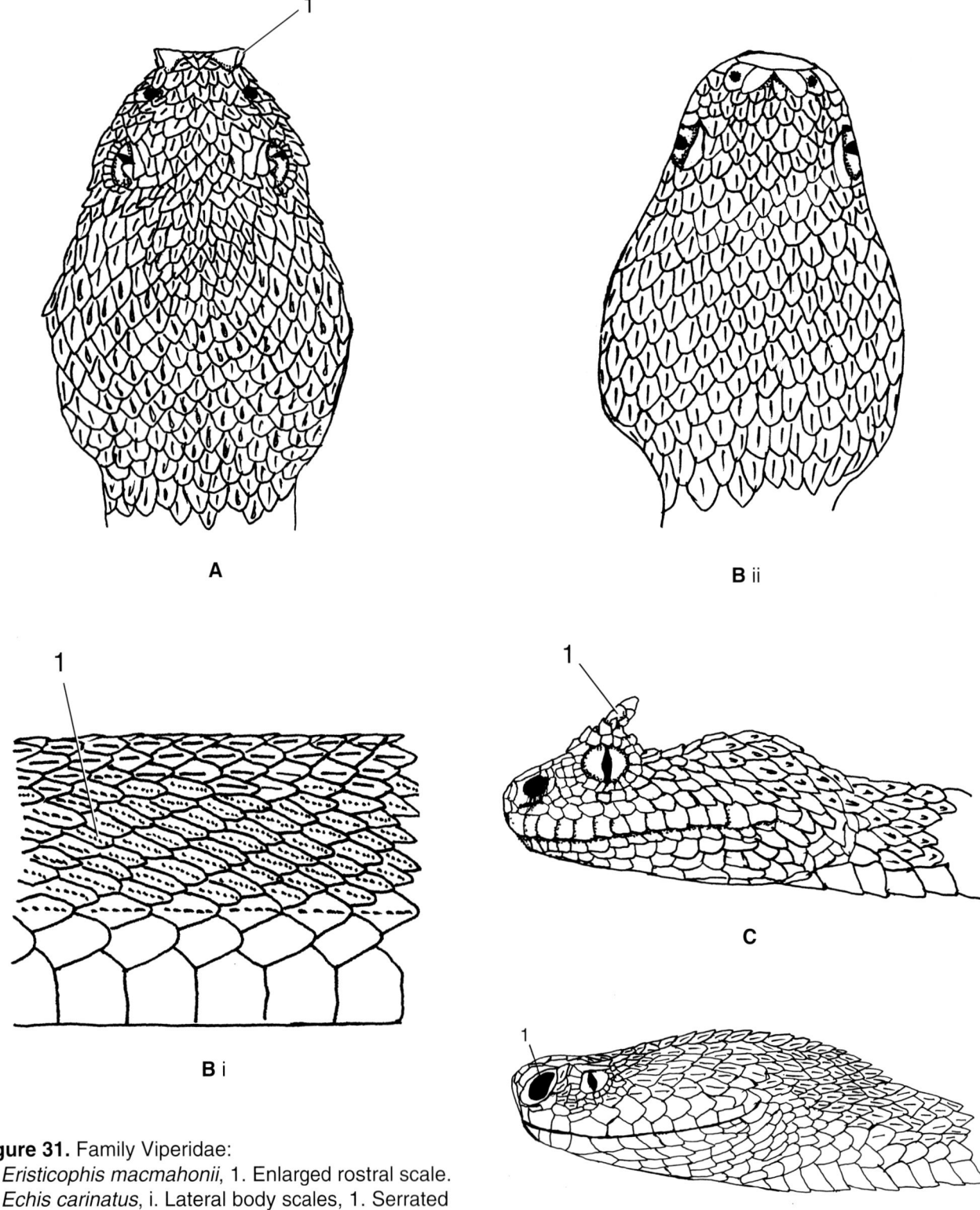

Figure 31. Family Viperidae:
A. *Eristicophis macmahonii*, 1. Enlarged rostral scale.
B. *Echis carinatus*, i. Lateral body scales, 1. Serrated scales; ii. Head dorsum.
C. *Pseudocerastes persicus*, 1. Supraorbital "horn."
D. *Daboia russelii*, 1. Naris.

Figures

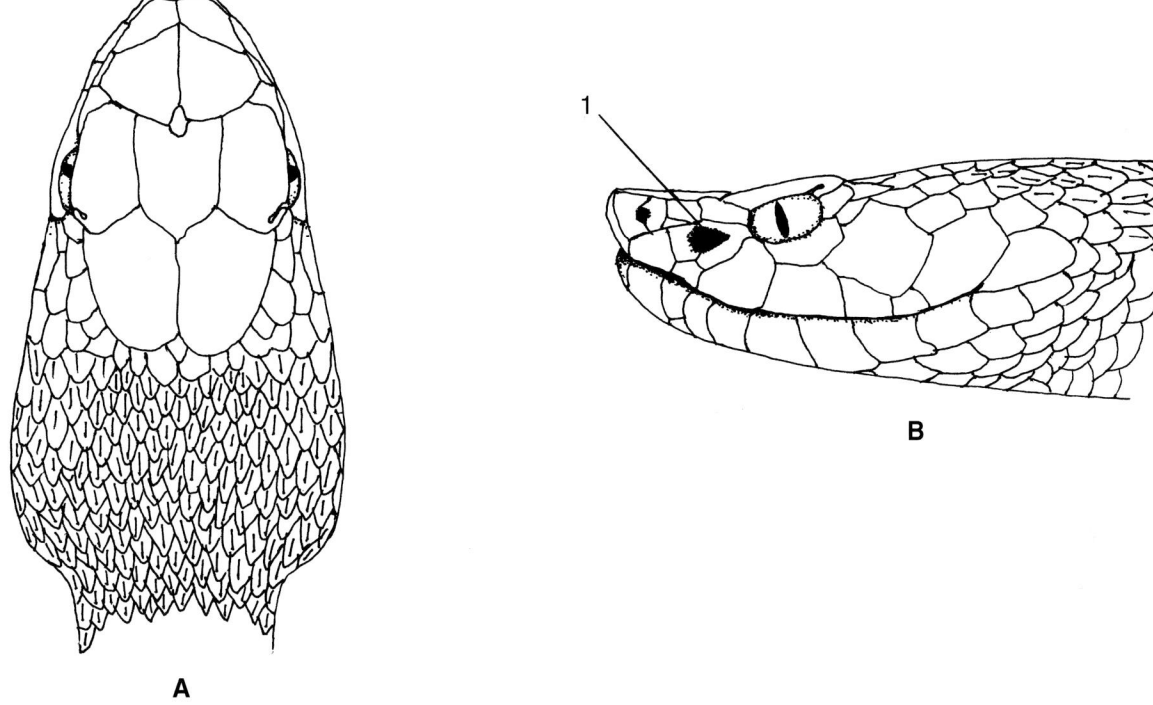

Figure 32. Family Crotalidae, *Gloydius himalayanus:*
A. Head, dorsal side.
B. Head, lateral side, 1. Loreal pit.

Chapter 3
Amphibians: Toads and Frogs

Family Bufonidae

Toads in Pakistan belong to genus *Bufo,* family Bufonidae. Toads belong to several species groups (Inger, 1972; Dubois and Ohler, 1999). Pakistani toads are in three species groups: *viridis*, *stomaticus*, and *melanostictus* (Khan, 1976).

Genus *Bufo* Laurenti, 1768

Parotoid gland present; body covered with horny tubercles; sometimes head has bony ridges; toes have weak web.

Bufo himalayanus Günther, 1864 (Plate 1)

(Himalayan toad: Hamalayai gauk)
1864 *Bufo melanostictus* var. *himalayanus* Günther, Rept. Brit. India: 422.
Type locality: Sikkim and Nepal.
Diagnosis: First described as a subspecies of *B. melanostictus*, from which it differs by the following characteristics:

1. Head deeply concave with only supraorbital ridges.
2. Interorbital space broader than the upper eyelid.
3. Tympanum very small or indistinct.
4. First finger does not extend beyond second.
5. Toes with single subarticular tubercle, no tarsal fold.
6. Parotoid is as long as the head.
7. Body with irregular porous tubercles.
Snout-vent length 130–132 mm.

Color: Uniformly brown. Cranial crest and tips of digits dark brown.
Tadpole: Head flat, body darker, belly bulging, tail weak, low fins; naris slightly nearer to eye than snout; eyes small and sunken; oral disc anteroventral, labial tooth row formula typically bufonid: 2(2)/3, beak serrated, oral papillae lateral; color uniformly black, ventrum lighter. The tadpoles are found at a high elevation in the Himalayan range, in small calm pools along streams with algal vegetation (Bhaduri, 1944).

Total length of tadpole 28–30 mm; tail 19–20 mm.

Natural history notes: *Bufo himalayanus* is a mountain species. It is primarily nocturnal; however, it is often seen moving about in broad daylight among rocks and vegetation feeding on grasshoppers, moths, ants, and other invertebrate animals. It rests during the day under stones or in fissures and holes in the ground.

During May, June, and July breeding activity starts after a downpour; males croak in a low tone, "curr, curr," repeated several times. Eggs are laid in a double string of jelly in shallow pools along streams. The toad hibernates during the winter under stones and in fissures in the ground from September to March.

The karyotype number recorded for this species is 22 (Chatterjii and Barik, 1970).
Distribution: It has been recorded from the Himalayas at 2000–3500 m of elevation. In Pakistan it has been recorded from Azad Kashmir, Hazara Division, NWFP.

Plate 1. *Bufo himalayanus.*

Chapter 3. Amphibians: Toads and Frogs

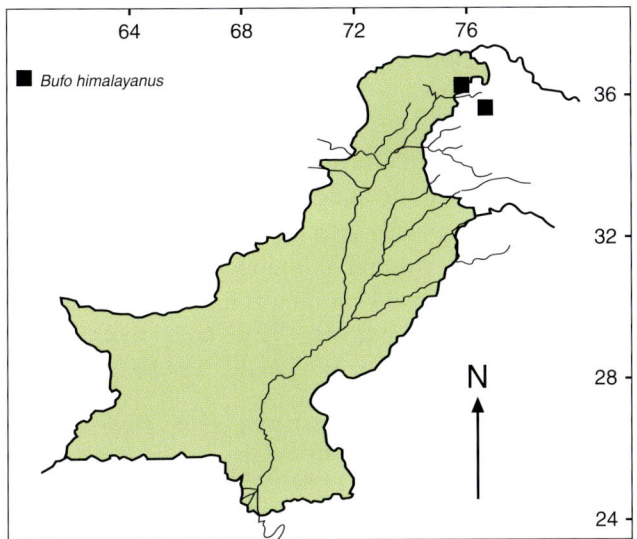

Bufo latastii Boulenger, 1882
(Ladakh toad: Ladakhi gauk)
1882 *Bufo latastii* Boulenger, Cat. Batr. Sal.:294.
Type locality: Ladakh, northeastern Pakistan.
Diagnosis:
1. Head flat without bony ridges.
2. Interorbital space narrower than the upper eyelid.
3. Tympanum distinct, half the diameter of the eye.
4. Toes with double subarticular tubercles.
5. Parotoid kidney-shaped.
   Snout-vent length 50–62 mm.

Color: Dorsum olive, spotted or marbled with black; a light vertebral band, ventrum more or less spotted and tips of digits black.

Tadpole: Dark brown, oval bulging body, weaker tail with low fins; inhabits small pools with vegetation. The tadpole has typical bufonid labial tooth row formula 2(2)/3, and feeds on pond vegetation.

Total length of tadpole 25–27 mm; tail 17–19 mm.

Natural history notes: This high altitude toad is found in shallow pools of water along streams in valleys. During the day it hides under stones, rocks, and vegetation. It is nocturnal; however, it is occasionally seen roaming about in vegetation and feeding on arthropods (Gruber, 1981).

After a downpour, from May to late July, breeding pairs gather in pools and ponds. Eggs are laid in double strands which are wound around submerged vegetation or rocks.

Distribution: Dubois and Martens (1977) extensively collected *B. latastii* from Nepal, Ladakh, Kashmir, and northern Indian Punjab. Earlier reporters identified toads from these regions as *B. viridis* (Steindachner, 1869; Stoliczka, 1872a; Anderson, 1872).

In Pakistan this toad has been recorded from Ladakh in Baltistan between 2600 and 3000 m of elevation.

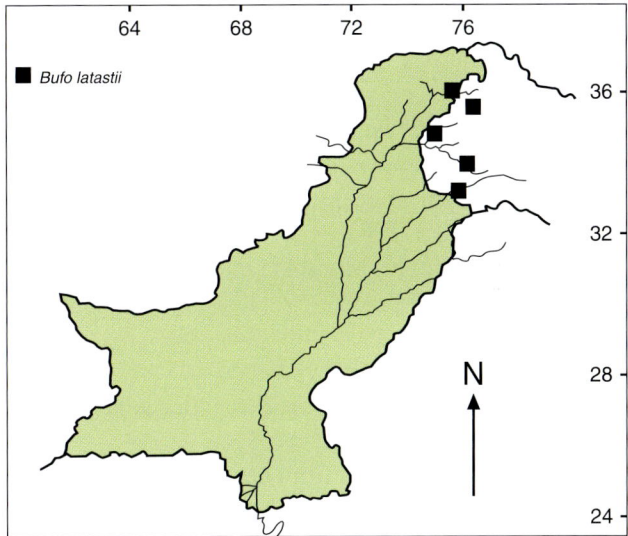

Bufo melanostictus Schneider, 1799
(Southeast Asian toad: Mashriqi gauk)
1799 *Bufo melanostictus* Schneider, Hist. Amph. l:216.
Type locality: India.
Diagnosis:
1. Head with distinct rostral, preorbital, supraorbital, postorbital and short orbitotympanic, cranial crests. No temporal ridge (Figure 2A).
2. Interorbital space much broader than upper eyelid.
3. Tympanum very distinct, at least two-thirds the diameter of the eye.
4. First finger generally but not always extends beyond second.
5. Double subarticular tubercles only under third finger. Toes with single subarticular tubercle.
6. Parotoid elliptical, with dark brown scattered branching concretions.
7. Skin heavily tuberculated on flanks, tubercles usually tipped with dark brown spines. A

lateral dorsal staggered row of 8–9 enlarged tubercles.
8. Cranial crests, lips, digit tips, and metacarpal and metatarsal tubercles are cornified with dark brown, which tend to peel off in preserved specimens; head is almost smooth.

Distribution: It is the most common and widely distributed toad in Southeast Asia (Church, 1959, 1960) extending throughout northern and peninsular India (Daniel, 1963a).

Bufo melanostictus hazarensis Khan, 2000 (Plate 2)
(Hazara toad: Hazara Gauk)
2000c *Bufo melanostictus hazarensis* Khan, Pakistan J. Zool., 33(4):293–298.

Type locality: Ooghi, Manshera, and Datta, Hazara Division, NWFP, Pakistan.

Diagnosis:
1. Parotoid glands kidney-shaped.
2. Double subarticular tubercles under penultimate phalanx of all fingers.
3. Rostral ridge absent from head.
4. Temporal ridge present.
5. Dorsum light brown.

Pakistan's largest toad, the female exceeds 150 mm in snout-vent length.

Color: Dorsum uniformly gray of various shades, brown or reddish with dark spots, ventrum uniformly dirty white; speckled with light brown on chin and throat.

Plate 2. *Bufo melanostictus hazarensis.*

Throat of the breeding male is light orange or yellow. It develops cornified pads on inner side of first and second fingers.

Tadpole: The tadpoles are uniformly dark, inhabiting side pools along hilly, fast-flowing streams. Schools of them swarm along the marginal waters of ponds and puddles feeding on any type of algal material. The body is typically bufonid, globular with a weak tail; dorsal fin is broad while ventral is narrow. The oral disc is typically bufonid with 2(2)/3 labial tooth row formula, the oral papillae are lateral. The beak is finely serrated and sharp (Khan, 1991e).

Total length of tadpole 26–27 mm, tail 19–20 mm.

Natural history notes: A rare toad in Pakistan, it is mostly confined to the low northern hilly ranges and Azad Kashmir. Nocturnal, it appears soon after sunset; during the day it hides under stones, logs, piles of vegetation, in holes and crevices among stones, and in the ground. Once a suitable place is selected, it is permanently shared with several toads.

This toad is a lethargic timid animal. It moves about with deliberate hops from place to place in search of insects on which it feeds. In tropical Southeast Asia it is a most common amphibian, coming out after sunset in large numbers and frequenting human habitations where it congregates under street lamps to feed on photophilic insects (Church, 1960).

In the temperate environs of the western Himalayas, breeding is initiated by the monsoon rains from July to August. Males gather in shallow side pools along fast-flowing streams and ponds. The call, a low melodious, "curr, curr, curr," is repeated several times and ends in a whistling note. The calling males become quite aggressive, tugging and jumping over each other. Males far exceed females in number and are much smaller than females. It breeds in every available space containing some water from the first showers of the monsoon rains in southern India (McCann, 1938). However, in tropical Southeast Asia, the toad is known to breed throughout the year (Church, 1960).

Calling males occasionally jump over each other and try to secure a nuptial holds on each other; however, kicks and zestful wriggling dislodge them from each other and they soon resume calling. The females lurch around and as soon one comes close, the nearest male jumps over it and quickly tightens its nuptial clasp. The other suitors are shaken off as the nuptial pair moves to a quieter place away from the site.

The eggs are laid in a double jelly string, generally in deep quieter water where the egg string is entangled in the vegetation; or the female moves around the submerged vegetation to wind the egg string around it. Eggs are enclosed in double gelatinous capsules (Khan, 1982b).

The swarms of recently metamorphosed toadlets from synchronized pairings leave the water, many falling prey to several kinds of predators, while some are crushed underfoot and by passing traffic.

Karyotype number recorded for this species is 22 (Nataranjan, 1953).

Distribution: Until recently *Bufo melanostictus* was reported to be the common toad of the Indo-Pakistan subcontinent (Günther, 1864; Murray, 1884; Boulenger, 1890; Annandale and Rao, 1918). However, in Pakistan, this toad is confined to District Hazara, NWFP, Alpine Punjab, and Azad Kashmir (Mertens, 1969; Khan, 1972b).

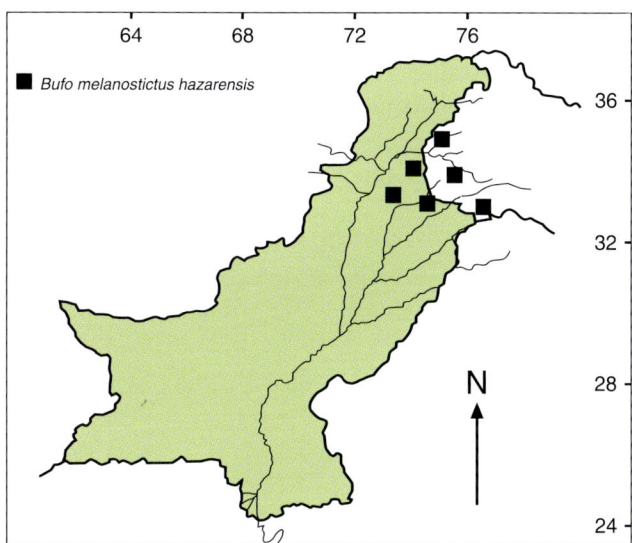

***Bufo olivaceus* Blanford, 1874**
(Olive toad: Zatooni gauk)
1874 *Bufo olivaceus* Blanford, Ann. Mag. Nat. Hist., London 14(4): 35.
Type locality: Dasht, Baluchistan.
Diagnosis:
1. No cranial crest.
2. The interorbital space slightly concave, a little broader than the upper eyelid.
3. Tympanum very distinct.
4. First finger longer than the second.
5. Subarticular tubercles of toes single, no tarsal fold.
6. Parotoids depressed, elongated, reaching to sacral region.
7. Tarsometatarsal articulation reaches in front of the eye.
8. Dorsum is smooth.
Snout-vent length 52–65 mm.

Color: Uniformly gray dorsum, with darkish spotting on limbs; ventrum whitish.

Tadpole: Typically bufonid, with oval, bulging body and weak tail. It inhabits ponds and puddles in oases and date palm groves in Dalbandin and Kharan, southwestern Baluchistan. Body is light brown, with dark specks on tail and fins, ventrum darkish white. The oral disc is typically bufonid, labial tooth row formula 2(2)/3, with lateral oral papillae. The tadpole feeds on algal vegetation and other concretions deposited on the surfaces of submerged objects.

Total length of tadpole 24–26 mm, tail 20–22 mm.

Natural history notes: The olive toad is confined to the oases and pools of karez water (underground streams, typical for Baluchistan). Nocturnal, after sunset it roams about in the meager vegetation, feeding on arthropods. It often ventures into human habitations to feed on photophilic insects. Anderson (1963) collected it from irrigation channels and groves near Minab, southern Iran.

Summer rains or the flow of irrigation water in karez streams trigger breeding activity. The breeding males gather in shallow ponds and puddles along water channels and call in their continuous guttural croak, "kreen, kreen, kreen," repeating it several times. Fights among males for possession of females are common. Pairs move to the side pools. The eggs are laid embedded in a long double string of gelatinous material.

Distribution: *Bufo olivaceus* is recorded from the extreme western parts of Baluchistan and adjoining Iran. Minton (1966) found little difference between *Bufo olivaceus* and *Bufo stomaticus*, while Eiselt and Schmidtler (1973) regard it as a subspecies of *Bufo stomaticus*. However, in a recent study Auffenberg and Rehman (1977) have shown *Bufo olivaceus* to be morphologically distinct from *Bufo stomaticus*.

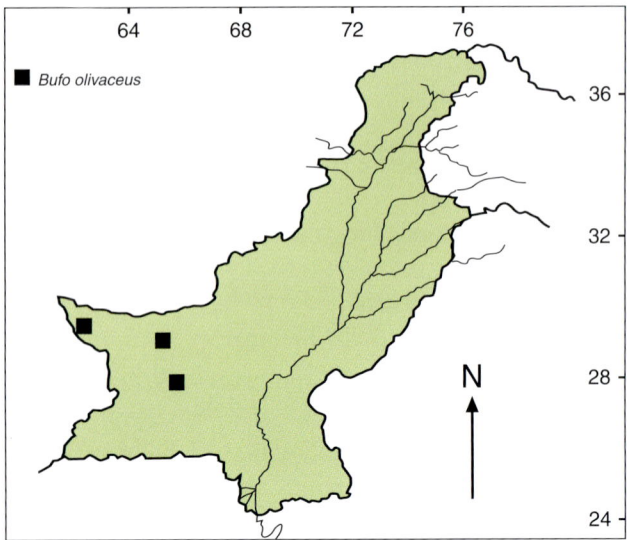

Bufo pseudoraddei pseudoraddei Mertens, 1971
(Plate 3)
(Swat green toad: Swati gauk)
1971 *Bufo viridis pseudoraddei* Mertens, Senckenb. Biol. 52(1–2):7–15.
1999 *Bufo pseudoraddei baturae* Stöck, Schmid, Steinlein, Grosse, Ital. J. Zool. 66:221–226.
Type locality: Mingora, Swat, Pakistan.

Diagnosis: Recognized from *B. viridis* by its dorsal body color and pattern.
Color: The dorsum is dark green with a light vertebral stripe and lighter spots on flanks. Limbs are with large dark blotches. Dorsal tubercles are not so prominent, rather they are flat.
Tadpole: Dark brown tadpoles swarm the water-land interface in side pools along streams. It feeds on pond vegetation and other concretions deposited on submerged objects. The tadpole has a typical bufonid globular body, a short tail with low fins, and is a typical bufonid tadpole in morphology of its oral disc and tooth row formula.

Total length of tadpole 23 mm, tail 16 mm.
Natural history notes: This toad inhabits rocky areas around Mingora, Swat. During the day, it hides under stones and retreats into fissures among rocks, swarming out at dusk. It feeds on insects and other arthropods. It is attracted under light posts to feed on insects and worms gathering under the light.

Breeding is triggered by summer rains from May to July, when temperature and water conditions are ideal. Males form calling groups. The call is typical *Bufo viridis* type. Amplectic pairs move to shallow puddles and small pools along fast-flowing streams. Eggs are laid in a double string.

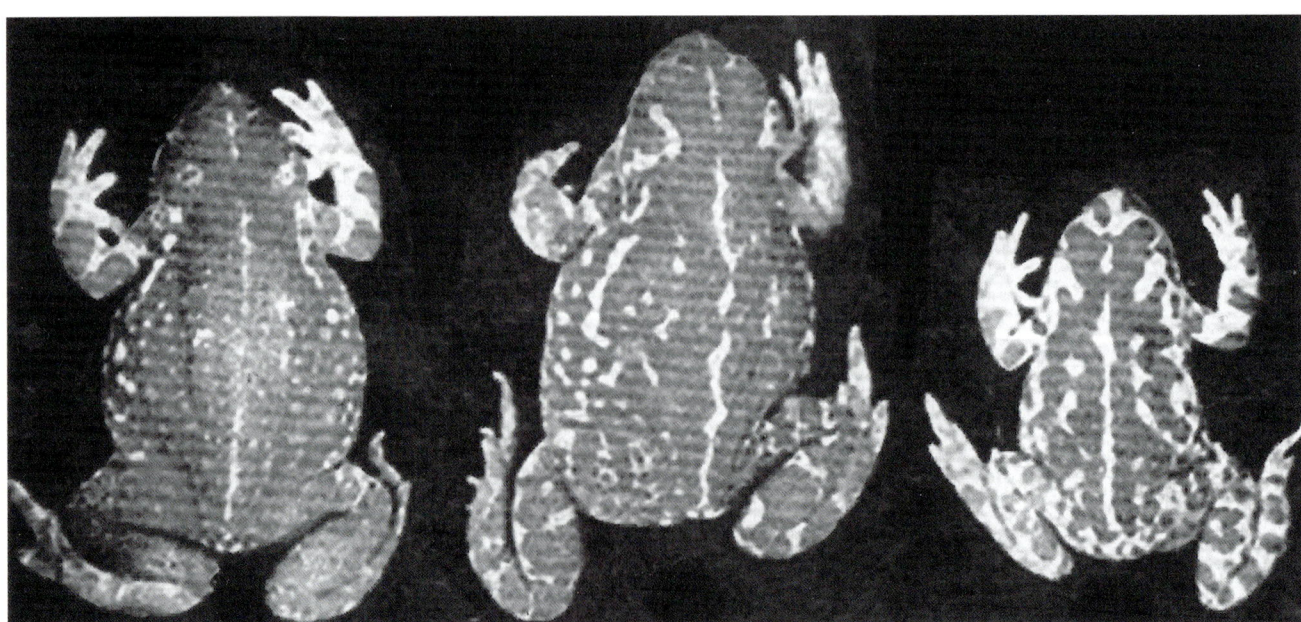

Plate 3. *Bufo pseudoraddei pseudoraddei.* (Mertens, 1971)

Chapter 3. Amphibians: Toads and Frogs

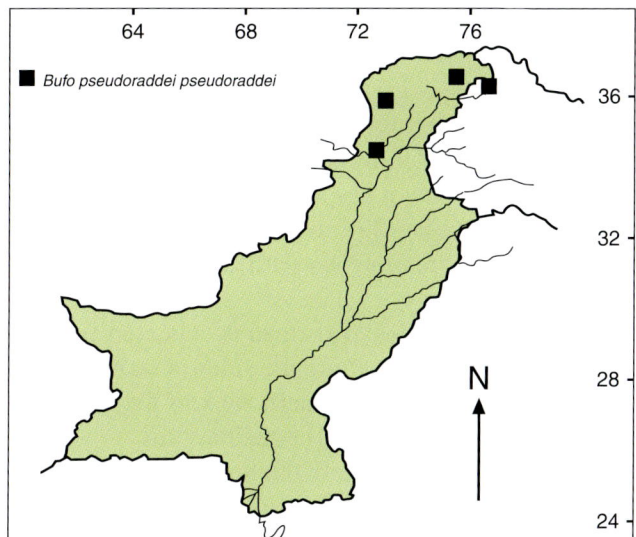

Distribution: So far, *Bufo pseudoraddei pseudoraddei* has been collected from Mingora, Swat. Eiselt and Schmidtler (1973) include the high mountains of eastern Afghanistan and the northwestern mountains of Pakistan in its range.

Taxonomic notes: Recently Stöck et al. (1999) have found it to be a distinct species, widely distributed in northern and northwestern Pakistan. They have partitioned it into two subspecies. It differs from *Bufo viridis* and *Bufo latastii* in its shorter parotoids, color pattern, and open habitat ecology within the zone of Himalayan coniferous forests and moist temperate forests.

Remarks: In Pakistan this toad is reported from northwestern NWFP, between 1500 and 2000 m of elevation. The toad ranging in North Western Frontier Province was described as *Bufo viridis pseudoraddei* by Mertens (1971).

Recently, the following new taxon, *B. p. baturae*, has been described from Karakoram, northern Pakistan by Stock et al. (1999).

Bufo pseudoraddei baturae Stöck, Schmid, Steinlein, and Grosse, 1999

(Batura glacier toad: Batura gauk)
1999 *Bufo pseudoraddei baturae* Stöck, Schmid, Steinlein, Grosse, Ital. J. Zool. 66:221–226.
Type locality: Hunza River, north of Passu River, Gilgit Agency, Baltistan, northeastern Pakistan.

Diagnosis:
1. Parotoids are inconspicuous.
2. Interorbital space smaller or nearly equals the internarial space.
3. Subarticular tubercles single under toes; often double on first, second and, in some cases, third finger.
Snout-vent length 60–65 mm.

Color: Dorsum grayish green or grayish brown, with irregular dark green spots, mostly smaller or about the size of the eye. Spots partially connected to form a marbled pattern with indented margins. In male brownish or grayish irregular spots at extremities.

Natural history notes: A mountain species, it inhabits marginal vegetation along fast-flowing streams in the glacier-ridden northeastern part of Pakistan.

Distribution: Passu, Swat, Karakoram Range, Gilgit Agency, Baltistan, Pakistan.

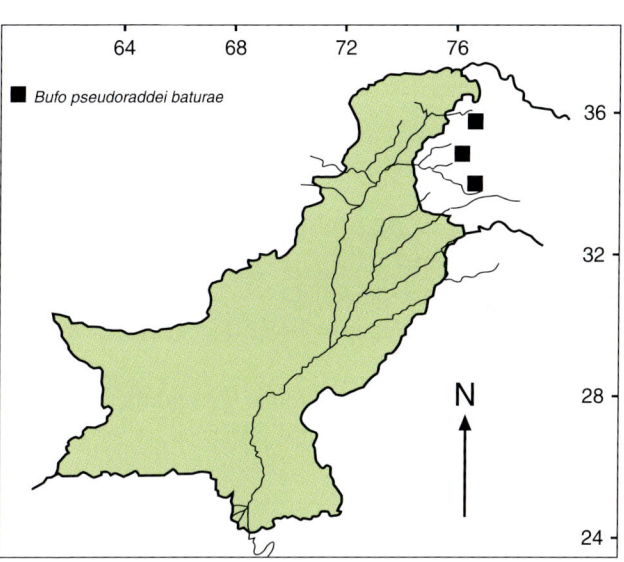

Bufo siacheninsis Khan, 1997 (Plate 4)

(Siachine toad: Siachin gauk)
1997 *Bufo siacheninsis* Khan, Pakistan J. Zool. 29(1):43–48.
Type locality: Shinu village, on the left bank of the River Shyok, 140 km east of Skardu, at the foot of Siachen Glacier, Baltistan, northeastern Pakistan.

Plate 4. *Bufo siacheninsis.*

Diagnosis:
1. Medium size *Bufo* with 6 dark green stripes on light dorsum.
2. First 2 fingers subequal.
3. Single subarticular tubercle under fingers and toes.
4. Tip of fourth finger extends beyond the base of terminal phalanx of third finger.
5. No tarsal fold.
6. Parotoids narrow, about twice as long as broad.
7. Distinct ulnar, tibial, and tarsal glands.
8. Dorsum smooth with low smooth nonspinulated tubercles.
    Snout-vent length 59–62 mm.

Color: Body leathery brown (in formalin-preserved material) which is visible in interstripe areas. Six dark green stripes form the dorsal pattern: an epiparotid stripe on each side, running from snout, upper eyelid, along upper border of parotoids to the coccal sides. These stripes are joined by a transverse midparotid band; a subparotid stripe on each side, starting from snout, running along lower border of parotoid to inguinal region. Third is a broad abdominal stripe, running from tympanum, on sides to shoulder, lateroventrum of the abdomen, where it is broken in dark specks, which are more concentrated under thighs and posterior part of the abdomen, continuing mesolaterally to the chest, chin, and under hindlimbs. The crown has a few elongated green blotches on the upper lip, snout, and upper eyelid. Dorsally, limbs are blotched with green. The tips of digits are white.

Natural history notes: This toad is collected from small seepage puddles along the sides of terraced cultivated fields on the southern bank of the Shyok River, near Shinu village. The puddles are surrounded with moderately thick marginal grass with no emergent vegetation. No other amphibian or reptile was collected from the site; however, several specimens of *Laudakia himalayana* were collected from the nearest rocks, and *Altigekko baturensis* Khan and Baig, 1992 were collected from houses in Shinu village.

Baltistan is a high-altitude (2300 m) cold desert lying in the north of the state of Jammu and Kashmir. The Karakoram glaciers, Baltoro, Concordia, Kaberi, Kandus, and Siachen, lie in the east, demarcating the border with China, with the desolate Deosai Plateau forming its western border. Several fast-flowing streams drain glaciers into the River Shyok which ultimately joins the River Indus (Adamson and Shaw 1981).

High altitude snowy desert conditions and hard stony soil, low temperatures (summer 13–29°C, winter 0–2° C) and low annual precipitation (Ahmad, 1989) support little natural vegetation (sparse grass and thorny bushes); however, fruit orchards and other cash crops are grown wherever water is available, mostly along sides of waterways in carefully tended terraced fields.

During the short-lived summer (May-August), several species of arthropods are common in the area, attracting the toad, the gecko *Altigekko baturensis,* and the local agamid, *Laudakia himalayana* which is collected from the nearby settlements and rocks. The short-lived summer is utilized by these animals for reproductive activity.

Remarks: Stock et al. (1999) have identified this toad as *Bufo latastii.*

Distribution: It is as yet known only from its type locality.

Chapter 3. Amphibians: Toads and Frogs

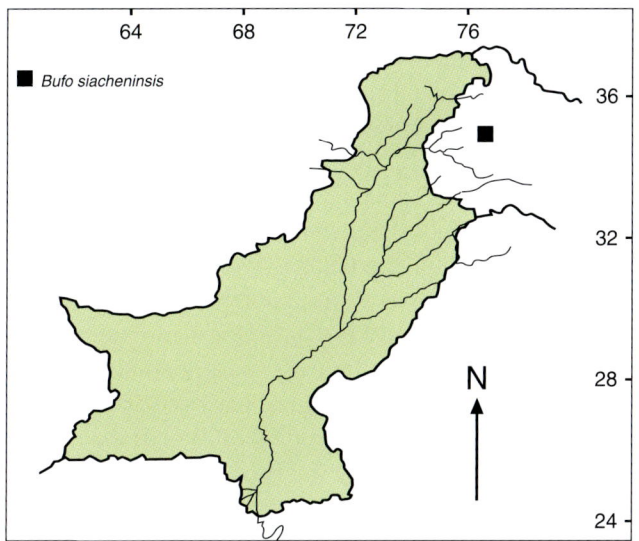

Bufo stomaticus Lütkin, 1862 (Plate 5)
(Indus Valley toad: Maidani gauk)
1863 *Bufo stomaticus* Lütkin, Vidensk. Meddel. Naturhist. Foren. Kjobenhavn 14:305.
Type locality: Assam, western Himalayas, India.
Diagnosis:
1. No cranial crests.
2. Interorbital space a little broader than the upper eyelid.
3. Tympanum distinct, round, its diameter two thirds that of eye.
4. First and second fingers subequal.
5. Toes with single subarticular tubercles.
6. A tarsal spinulated ridge.
7. The parotoid gland is longer than broad.
8. A distinct tibial gland is present (Figure 2B).
Snout-vent length 67–70 mm.

Color: Dorsum light gray or olive to almost black, with dark mottling or gray to dark reticulation; upper lip cream. Ventrum dirty white, dark mottling on throat, 3 dark transverse bands on anterior aspect of forearm. Tips of digits dark brown.

Tadpole: Schools of dark tadpoles of this species are a common sight in ponds and puddles in the plains of Punjab and Sind during monsoons. In the hills the tadpoles are confined to the side pools of fast-flowing streams. Daniel (1963a) recorded this toad as breeding in pooled water about 90 m away from the sea in Mumbai, India.

The tadpole is typically bufonid, with an oval bulging body, weak tail, high dorsal, and narrower ventral fin (see Figure 3A). The oral disc is anteroventral with a 2(2)/3 labial tooth row formula; the beak is finely serrated with lateral oral papillae (see Figure 3B). The body and fins of the tadpole are speckled with light brown (Khan 1968b; Khan and Mufti, 1994a,b).

Total length of tadpole 30–31 mm, tail 20 mm.

The gregarious habit of this tadpole continues up to the 36th stage of development (Khan, 1965), later the tadpole becomes solitary and when disturbed takes to the deep water (Khan, 1991e; Khan and Mufti, 1994a).

Natural history notes: *Bufo stomaticus* is the most common toad in the plains of Punjab and Sind. It extends to an elevation over 2000 m in the northern and western hilly areas (Khan, 1972b). It occurs in varying climatic conditions and is especially common in dry semidesertic conditions (Auffenberg and Rehman, 1997). It is essentially nocturnal; however, in breeding season it is active during the day. It emerges at sunset from holes and crevices among stones or brick walls and roams about widely in vegetation, feeding on insects and worms. It is readily attracted to areas under light posts to feed on photophilic arthropods. It ventures freely into inhabited houses where it usually hides under household articles, occasionally emerging to feed on houseflies. Under drought conditions several toads will aggregate at damp places, especially around hand pumps, to allow water to splash on their bodies. The toads squat by pushing out

Plate 5. *Bufo stomaticus.*

their hind legs on the sides of the body and pressing their hindquarters against the moist soil.

Young toads (snout-vent length 30–40 mm) often climb 1.5–2 m high compound walls, taking advantage of the irregularities between the bricks, to drop to the other side of the wall.

An evening temperature of 13°C, a downpour, and flow of rainwater are sufficient elements to trigger breeding activity in *Bufo stomaticus* (Khan and Malik, 1987b). It is one of the first amphibians to arrive at the flooded areas, where males soon form very noisy choruses. The monsoon rains of July and August are utilized mainly for breeding, however, if rains are late, water in the irrigation channels and recently watered fields is freely utilized for the purpose.

The croak consists mainly of guttural notes, "cree, cree, cree, cree, cree" repeated several times. There are fewer females than males. Several groups of calling males are established in different water bodies, each body holding one group. The females lurch around the calling males in vegetation (Khan, 1965). Calling males jump over each other every time to secure a nuptial hold. There is much kicking and tugging between them to dislodge the assailants; others continue croaking. When a female comes close to a calling male, it at once jumps onto her to secure a firm nuptial hold. Soon other males in the group try to do the same, so a tug-and-push game starts. The pair is clasped from all sides by aggressive males. These fighting groups may roll on the ground or float in the water for a long time until, perhaps due to muscular fatigue, the males thin out. The ability of the first male to hold on, together with the female quickly moving away from the spot saves their pairing. The female withdraws to a quieter place, carrying the male on her back. Eggs are laid in a double gelatinous string. While laying, the female circles around submerged vegetation to wind the egg string around it (Khan, 1965).

During summer large numbers of toads are crushed under passing traffic while crossing roads to reach their potential breeding sites. Crows and other birds feed on the exposed viscera, leaving the skin of the dead toad behind (Khan, 1982c, 1990b). The toad is included in the diet of several sympatric animals: *Hoplobatrachus tigerinus, Varanus bengalensis, V. griseus, Ptyas mucosus, Amphiesma stolatum, Bungarus caeruleus,* and *Echis carinatus*. To avoid attack from a potential predator, the toad escapes into a burrow or fissure. If no hiding place is available, it buries itself in the loose soil by pushing aside the soil particles (Khan and Tasnim, 1992).

Karyotype number recorded for *Bufo stomaticus* is 22 (Asana and Khardi, 1937).

Distribution: *Bufo stomaticus* is widely distributed throughout the Indo-Pakistan subcontinent, from Bangladesh through the Ganges Plain, peninsular India, upper and lower Indus Valleys, Baluchistan, Afghanistan, Iran, and Muscat. It does not extend beyond Mumbai along the west coast of southern India (Daniel, 1963a). *Bufo stomaticus* has been collected from the plains to an elevation of 1800 m in the northern and western hilly tracts of Pakistan, where it occurs sympatricly with *Bufo melanostictus* and *B. viridis*. Annandale (1907) expresses the possibility of occurrence of this toad at 2330 m of elevation in the Himalayas.

Taxonomic note: A recent taxonomic study on *Bufo stomaticus* suggests the partition of the Indus Valley population of this toad into five separate populations, though none are worthy of subspecific recognition (Auffenberg and Rehman, 1977).

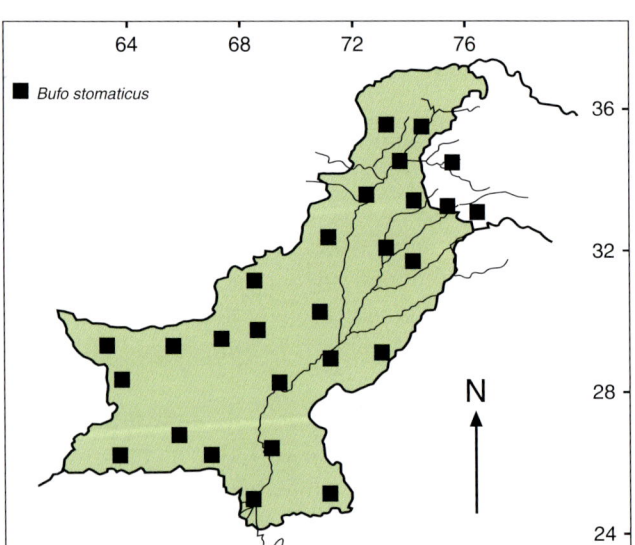

Bufo surdus **Boulenger, 1891**
(Iranian toad: Irani gauk)
1891 *Bufo surdus* Boulenger, Ann. Mag. Nat. Hist. London, (6)7:282.
Type locality: Baluchistan., Pakistan.

Diagnosis:
1. Head without cranial crests.
2. Interorbital space narrower than the upper eyelid.
3. Tympanum distinct, about half the diameter of the eye.
4. Fourth toe with single subarticular tubercles.
5. Parotoid squarish, as long as upper eyelid.
6. Dorsal tubercles unispinulate.
   Snout-vent length 45–65 mm.

Color: Dorsum gray, with greenish spotting; a dark blotch on the upper eyelid.

Tadpole: A typical light to dark brown bufonid tadpole, with bulging globular body, comparatively weaker tail, and narrow low dark spotted fins. The oral disc is typical bufonid, with finely serrated beak; labial tooth row formula 2(2)/3, oral papillae are lateral. The tadpole feeds on pond vegetation.
   Total length 21–22 mm, tail 16 mm.

Natural history notes: *Bufo surdus* is nocturnal, spends the day hiding under stones, vegetation, and falling logs, retreating into holes and crevices in the ground. It breeds during monsoons. Males gather in temporal pools and are very noisy. Eggs are laid in double strings which are wound around the pond vegetation.

During the hot season when there is a general dearth of water, the toad seeks dampness under stones and in vegetation; such sites may hold several toads.

Distribution: *Bufo surdus* is a little known species. It is reported along western Baluchistan and around Quetta.

However, it is widely distributed in Iran (Schmidtler and Schmidtler, 1969; Eiselt and Schmidtler, 1973).

Bufo viridis Laurenti, 1768
(Green toad: Hari gauk)
1768 *Bufo viridis* Laurenti, Synops. Rept.: 27.
Type locality: Arabian peninsula, Sinai.
Diagnosis:
1. No cranial crests.
2. The interorbital space narrower than the upper eyelid.
3. Tympanum is distinct, about half the diameter of the eye.
4. First finger extends a little beyond second.
5. Toes with single or slightly divided subarticular tubercles.
6. A distinct tarsal fold present.
7. Parotoids variable in shape and size, usually elongated, kidney-shaped.
8. Dorsum with distinct depressed irregular porous tubercles.
   Snout-vent length 65–70 mm.

Color: Dorsum olive to dark gray or buff, with large irregular dark green scattered spots; ventrum white.

Tadpole: Schools of sooty dark tadpoles of *Bufo viridis* swarm along margins of side pools of streams in summer. The body of the tadpole is typically bufonid as it is globular with weak low finned tail. The anteroventral oral disc has finely serrated thin beak, with typical bufonid labial tooth row formula 2(2)/3.
   Total length of the tadpole 25–26 mm, tail 18–19 mm.

Natural history notes: The toad is found in relatively more humid areas, around ponds, puddles, karez water channels, and lakes. It excavates holes in the moist marginal high ground to retreat into during the day. At sunset swarms of toads swim ashore to roam about in search of food, returning at dawn. The toad is a voracious feeder—it almost clogs its stomach with insects. In the suburbs of Quetta toads swarm under light posts feeding on the photo-attracted arthropods. They also readily feed on worms, etc.

May–July rains trigger breeding activity. Males gather in pools and temporal water catchment sites and call. The call is a melodious weak trill, "crr, crr, crr," repeated several times. There is no competition for mates; usually males and females are equal in number. Amplectic pairs move to deeper puddles. Eggs are laid in a double string which is entangled in the vegetation at the site.

Distribution: *Bufo viridis* has an extensive range in Europe and central Asia. It extends from Germany to Mongolia, central Siberia, Tibet, Middle Eastern countries, Israel, Egypt, Morocco, Arabia, Iraq, Iran, and Afghanistan. Pakistani populations of *Bufo viridis* represent the easternmost border of its range.

Bufo viridis zugmayeri Eiselt and Schmidtler, 1973 (Plate 6)

Baloch green toad: (Baloch gauk)
1973 *Bufo viridis zugmayeri* Eiselt and Schmidtler, Ann. Naturhist. Mus. Wien. 77:206–207.
Type locality: Peshin, southeastern Baluchistan, Pakistan.
Diagnosis:
1. The post-tympanic part of the parotoids extends on side to lower tympanic fold.
2. The interorbital space wider than the upper eyelid.
3. First finger longer than second.
4. Double subarticular tubercles under fourth toe.
5. Smooth tarsal fold.
6. Body dorsum with numerous weakly indicated warts, which are arranged in double rows on flanks.
Snout-vent length 50–70 mm.

Color: Light brown, head and body with numerous small dark green spots, smaller than eye. Finger and toe tips light.
Tadpole: Similar to that of *Bufo viridis*.
Natural history notes: This toad breeds in side pools along fast-flowing streams. Its habit is similar to *Bufo viridis*.
Distribution: The range of this toad includes the mountain ranges around Quetta, extending southward to Chagai, Baluchistan. Mertens (1969a) and Khan (1987) identified this toad as *B. v. arabicus*, while Hemmer et al. (1978) consider *Bufo viridis zugmayeri* a subspecies of *Bufo latastii*.

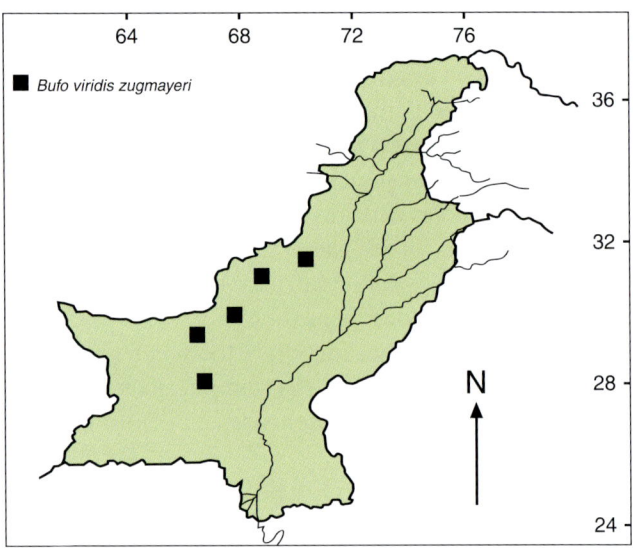

Family Megophryidae

Genus *Stutiger* Theobald, 1868
Medium sized frogs; tympanum indistinct; body dorsum with thick broken longitudinal folds and tubercles.

Scutiger nyingchiensis Fei, 1977
(Tibitin lazy toad: Tibti sust ghauk)
1977 *Scutiger nyingchiensis* Fei, Animal Journal, 23 (1):54–55.
Type locality: Linzhi, Tibet

Plate 6. *Bufo viridis zugmayeri.*

Diagnosis:
1. Body long and narrow.
2. Head flat, wider than long; snout obtusely pointed.
3. Canthus prominent, nostril between snout and eye, below canthus..
4. Eyes large, pupil vertical.
5. Tympanum indistinct; a prominent elevated tuberculated postorbital ridge.
6. No vomerine teeth.
7. Fingers short, cylindrical, first and second equal in length; toes flattish, half webbed; digit tips colorless, obtuse; shin has heterogeneous spiny warts on its front and back.
8. Tibiotarsal articulation reaches shoulder or temple.
9. Skin very rough with heterogeneous warts, those on lateral sides mostly arranged in rows, otherwise haphazardly arranged. The warts around the mid dorsal line of the back short and round with 1 to 7 big black spines, mostly in a straight line. The back of head smoother or with small flat simple warts. These are small spines on the sides of the head and lower lip.
10. Sexually dimorphic: female longer in snouth vent length (68–69), than male (51–64) mm; male with strong forelimbs, first three fingers with nuptial spines, a pair of spiny chest glands, and a large spiny axillary gland.

Color: Dorsum grey olive, with dark brown triangular spots; sides of body light yellowish; limbs spotted; chest and abdomen greenish yellow.

Tadpole: Eggs are laid in egg masses, which floats on water surface, get attached under stones or vegetation. The tadpole is medium in size, black to greenish brown. Dorsal fin brown with darker spots, lower fin lighter, with spotted tip. At hind limb bud stage tadpole measures 65 mm in length. The snout is round, eyes located well back on the head. The spiracle opening on left side, a bit closer to the rear end of the body. The cloacal tube long and open by an oblique aperture under the right side of the lower tail fin, tail tip obtuse.

The oral disc anteroventral; dental formula I:4-4/I:4-4, jaws are strong, long finger like oral papillae are present.

Natural history notes: The frog is found sitting under stones along slow running brooks, mostly in well shaded part, at elevation between 2730–4500 m. The habitat is moist with abundant vegetation of trees, bushes and clumps of vegetation. The reproductive season extends from June to August. The breeding toads collect in large numbers at such secluded places and breed. The couples stay in shallow water, the floating egg mass comes to rest under moist vegetation and stones etc., along the sides of the brook. After breeding season nuptial and chest spines of the male fall off. Toads rest during the day under rocks and vegetation, normally they do not venture far from water. During winter they hibernate under the stones along frozen brooks.

When approached it becomes motionless or it may escape into water, immediately climbing back, as it is a bad swimmer. Its movements on land are slow and deliberate, hence its name "lazy toad". It is easily captured. At night, toads come out of their hideouts and go foraging, normally they move about on the rocks waiting for the prey, which mostly are insects, insect larvae, millipedes and centipedes.

Distribution: *Scutiger nyingchiensis* has been collected from the Himalayas between 3000 and 3500 m. Mertens (1969a) reported a specimen as *Nanorana pleskei* from the border between Pakistan and Azad Kashmir. Subsequently several specimen have been collected from Gilgit, Dosai Plains (Khan, 2005). The frog ranges through Western Nepal, Kashmir, Ladakh, and southeastern Xizang (= Tibet), China, 2730–4560 m elevation (Khan, 2005).

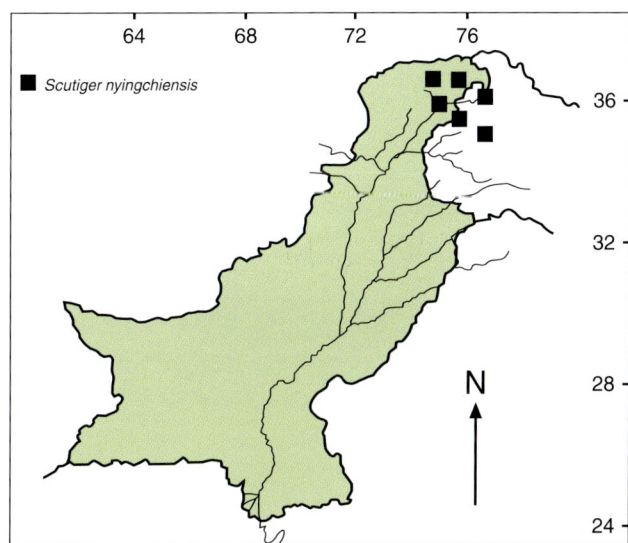

Family Microhylidae

Family Microhylidae is represented in Pakistan by 2 species belonging to genera *Microhyla* and *Uperodon*.

Genus *Microhyla* Tschudi, 1828

Tongue elliptical; a dermal ridge between internal naris; adult does not exceed 30 mm in body length.
Single species represented in Pakistan.

Microhyla ornata (Duméril and Bibron) (Plate 7)
(Ant frog: Bauna maindak)
1841 *Engystoma ornatum* Duméril and Bibron, Erpet. Gen. 8:745.
Type locality: Malabar Coast, India.
Diagnosis:
1. A small, slender-bodied, narrow mouth frog, snout-vent length does not exceed 30 mm.
2. Interorbital space broader than the upper eyelid.
3. Fingers slender, first much shorter than second.
4. Toes long with slight rudiment of web.
5. Subarticular tubercles distinct, tips of digits slightly swollen.
6. Tibiotarsal articulation reaches to eye, or between eye and shoulder.
7. Dorsum smooth or slightly tuberculated.
8. No tarsal fold.
Snout-vent length 27–29 mm.

Color: Dorsum reddish olive, with an elongated dark brown mark along midback, extending from between eyes backward, narrowing on nape, widening above shoulders, narrowing then widening at back, sends a stripe to groin and another to thigh on each side (see Figure 4A). A dark streak from the eye to the shoulder, the limbs are barred. Ventrum white, throat and chest dark brown in breeding male.

Tadpole: Transparent, with streamlined body; head dorsoventrally depressed and body laterally compressed (Figure 4B). The snout is countersunk, with anterodorsal displacement of U-shaped oral opening. A typical oral disc with beak and denticles is absent (Figure 4C). Tail long, more than twice the length of the body, produced posteriorly into a long vibratile flagellum. Tail fins are transparent and widest at midtail.

The body is widest at the level of laterally placed eyes. Position of imperforate naris is marked by anterolateral narial pits lying just anterior to the level of the eyes. The spiracle is medioventral with a prespiracular flap. Belly has a silver shine due to the presence of iridiocytes, which are soon lost upon preservation (Rao, 1917; Khan, 1982c, 1991e, 2001a).

The tadpole is midstream microphagus. Schools of tadpoles swim at midstream feeding on planktonic bloom at midstream of deep water pools. Water is continuously engulfed to filter the required amount of food material. The tadpole never settles at the bottom of the pond. Sensing danger, the whole school dips deep into the water, reappearing sometime after at midstream (Khan, 1991e, 2003g).

The tadpoles are best collected from large permanent ponds in the plains of Punjab, Pakistan, from mid-May to mid-August (Khan, 1982c, 2001a).

Total length of the tadpole 22–24, tail 13–15 mm.

Natural history notes: *Microhyla ornata* is a small active and elegant frog. During the day, it hides under stones, logs, and heaps of vegetation, and in fissures and holes in ground, along the sides of streams and ponds. It emerges at dusk, feeds on ants and other soft-bodied insects of small size.

Breeding is triggered by midmonsoons, when ponds already have enough water and are rich in planktonic bloom. Males gather in ponds after a heavy downpour and call via their characteristic "brrrat, brrrat, brrrat," outbursts of astonishing volume considering the small size of the frog. The call is ventriloquistic, making it difficult to locate a calling frog. The males are well placed from each other, generally sitting in shallow

Plate 7. *Microhyla ornata*: adult males.

depressions at the roots of vegetation formed by the hooves of water-visiting cattle (Khan and Malik, 1987b).

Small groups of greenish brown tiny eggs of *Microhyla ornata,* which are embedded in gel-like material, float on the water surface, and soon sink into the water as development proceeds. They are the smallest amphibian eggs recorded for an amphibian in Punjab, Pakistan, measuring 0.5–0.7 mm in diameter (Khan, 1982c).

Karyotype number recorded for *Microhyla ornata* is 26 (Sharma et al., 1977).

Distribution: *Microhyla ornata* is a wide-ranging Southeast Asian species. It extends from the Malay Peninsula, Siam, southern China, Cambodia, Myanmar, Nepal, Kashmir, Sri Lanka, and throughout India. It has recently shown to have wide distribution in Punjab, Pakistan (Khan, 1974, 1976, 1991e; Khan and Tasnim, 1987a).

Diagnosis:
1. Rotund, stout body, kept puffed round like a ball.
2. Snout scarcely longer than diameter of the orbit.
3. No canthus, snout round.
4. Fingers moderate, first a little shorter than second.
5. Toes short webbed at the base.
6. A pair of strong shovel-shaped metatarsal tubercles, inner larger.
7. Tibiotarsal articulation does not reach to the shoulder joint.
8. Dorsum smoothly tubercular, supratympanic fold present, throat and chest smooth, belly and anal region granular (Kirtisinghe, 1957).
9. A medial buccal ridge ending in 2 papillae lying between the internal nares, a similar papilla below each internal naris.
Snout-vent length 50–60 mm.

Color: Body dorsum with more or less symmetrical dark brown pattern on a pinkish brown background. Ventrum whitish immaculate.

Tadpole: The head and body of the tadpole is oval, with long pointed tail, broader ventral fin. The tail is about twice the length of head and body. The mouth is terminal and lacks a typical denticulate oral disc. Naris is nearer to the eye than the tip of the snout. The spiracle is a median ventral tube, opening in front of cloacal aperture.

Total length of the tadpole 27–27.5 mm, tail 16–17 mm.

Natural history notes: A completely fossorial species, it surfaces only during summer monsoons.

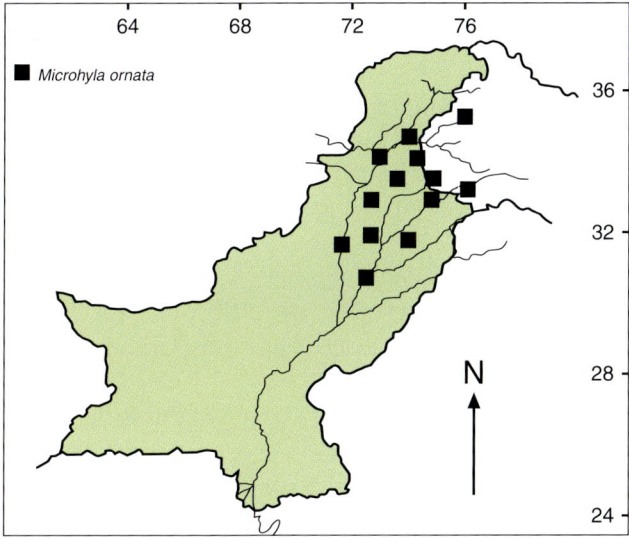

Genus *Uperodon* Duméril and Bibron, 1841
Tongue oval; a pair of tubercles between internal naris; adult size 50–60 mm.

A single species represented in Pakistan.

Uperodon systoma (Schneider, 1799) (Plate 8)
(Marbled balloon frog: Marmareen maindak)
1799 *Rana systoma* Schneider, Hist. Amphib. I:144.
Type locality: Carnatic, Biligiriranga Hills, Mysore, Madras, India.

Plate 8. *Uperodon systoma.*

The frog moves by small hops or slowly walks on the ground. It is a weak swimmer that usually floats in water; however, it is known to be an excellent burrower, quickly burrowing into the loose moist soil with the help of its well-developed and powerful metatarsal tubercles. During burrowing the loosened soil is thrown sideways by its powerful hind legs and the animal sinks into the cavity so formed, with the eyes the last to disappear underground. Moist soil is essential for this frog. During dry months the frog retreats into the moist environs of termite nests, termites being its main food item. Mukerji (1931) collected a specimen of *Uperodon globulosum* from a depth of 1–1.5 m. It had lived for 13 months without food. The body of the frog is globular due to the enormously developed inflated lungs.

The breeding season extends from May to July, during monsoon rains (Ferguson 1904). The males call from the banks of streams or paddy fields. The call is like a bleating goat, and during calling the subgular vocal sac is distended so enormously it looks like a float. The eggs are laid in masses which float on the surface of the water (Ferguson, 1904; Annandale and Rao, 1918).

Karyotype number recorded for this species is 26 (Bole Gowda, 1948).

Distribution: *Uperodon systoma* is a very rare frog in Pakistan. Recently a specimen was collected from the foot of Shakarparian Hills, Islamabad, from the side of a stream during a wet May night (Gvozdik and Radek, 1997; Baig and Gvozdik, 1998). Abdulali (1962) reported large numbers of this frog in Khanapur, Mysore, southern India, in the month of May. This species is widely distributed in southern and eastern India, extending into northern Sri Lanka (Kirtisinghe, 1957).

Family Ranidae

The heterogeneous assemblage of ranid frogs of the Indo-Pakistan subcontinent, has long posed problems for taxonomists (Boulenger, 1920). Recently, Dubois has partitioned the family Ranidae into several genera (1975, 1981, 1983a, b, 1986a, b, 1992). I adopt Dubois's generic partition in reallocation of Pakistani ranid frogs.

Family Ranidae is represented by 6 genera in Pakistan (Dubois, 1992).

Genus *Euphlyctis* Fitzinger, 1843

Tympanum distinct; toes extensively webbed; body dorsum pustulate; inner metatarsal tubercle elongate, toe-like; no nuptial pads in males; vocal slits under the sides of lower jaw.

Euphlyctis cyanophlyctis cyanophlyctis (Schneider, 1799) (Plate 9)

(Skittering frog: Tapakta maindak)
1799 *Rana cyanophlyctis* Schneider, Hist. Amphib. I:137.

Type locality: Eastern India.
Diagnosis:
1. The interorbital space is narrower than the upper eyelid.
2. Tympanum is distinct, about two thirds the size of the eye.
3. Fingers slender, pointed or slightly swollen at the tips, first not extending beyond second.
4. Toes are completely webbed (Figure 5B).
5. Inner metatarsal tubercle long, conical much like a rudimentary toe.
6. Male with vocal slits under the lower jaw.
7. Dorsum with numerous scattered small smooth tubercles, sides of body rugose, ventrum smooth.

Male (43–46 mm) is smaller than female (55–67 mm) in snout-vent length.

Color: Dorsum light gray, olive-green or light brown, sometimes black, with irregular black spots. Thighs posteriorly dark with one or two yellow or white irregular longitudinal stripes; ventrum white, immaculate or with dark speckling or reticulation; vocal sacs light brown.

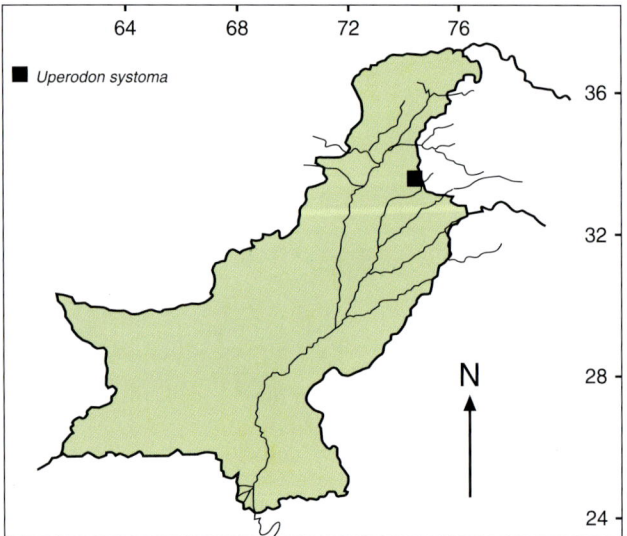

Chapter 3. Amphibians: Toads and Frogs

Plate 9. *Euphlyctis cyanophlyctis cyanophlyctis.*

Tadpole: Tadpole large, with oval bulging body, broadest at midbody, venter flat. The eyes are large and lateral. Tail is long, muscular, with wider dorsal and narrower ventral fins, tail tip is obtuse.

The anteroventral oral disc has broad anterior labium with a single tooth row, posterior labium is narrower with two rows of teeth. The labial tooth row formula is ½ (Figure 6B). The teeth are arranged in a single row. A tooth is a squarish, medially curved, 0.13–0.34 mm long, blunt-tipped rod. The beak is broad, finely serrated. The posterior labial palp extends well beyond posterior labium, is narrowly interrupted medially, while its anterior half forms an outpocket to include a patch of smaller papillae. Dorsum of tadpole blackish with dark black blotches and spots extending onto tail and fins (Khan, 1982c, 1991e).

This tadpole remains solitary, stays most of the time at the bottom, feeds mostly on debris, almost clogging its digestive tract. Usually no fresh vegetation is detected in its digestive tract. It also feeds on dead tadpoles, drowned animals like earthworms, etc. It attacks sympatric tadpoles and feeds on them (Khan and Mufti, 1995; Khan, 2003c).

Euphlyctis cyanophlyctis tadpoles are most common in water bodies throughout the plains of Punjab and Sind, from late February to mid-September.

Total length of the tadpole 42–44 mm, tail 23–24 mm.

Natural history notes: *Euphlyctis cyanophlyctis* is a highly aquatic and littoral frog. It remains permanently resident in different types of habitats with pooled water, in the plains and submountainous parts of Pakistan. The frog is remarkably capable of adjusting itself to the uncertain aquatic conditions in temperate arid parts of Pakistan.

Its peculiar unique habit of skittering over the water surface, is reported by the Mogul Emperor Babar in his autobiography (Beveridge, 1979; Khan and Tasnim, 1987a).

The frog either floats or remains squatting in the vegetation along marginal water. An intruder initiates the frog's skipping behavior during which the flattened and inflated ventral surface of the body rests on the water surface while the push comes from the completely distended webbed feet which steer the body forward so that the frog is speedily carried to the center of the pond. When further provoked, it plunges into the depths.

The frog can tolerate a wide range of pH variations, from fresh water to considerably brackish and polluted refuse water; it thrives equally well in sewer systems of towns and cities.

Individual frogs call from permanent water bodies almost throughout the year. However, active breeding activity is initiated as early summer water temperature rises to 10°C –12°C (Khan and Malik, 1987b). The calling males usually gather in a corner of a pond with some marginal vegetation. Some sit on the moist margin, others float. The tone of the call is very variable, depending on water and atmospheric temperatures, and the age and breeding state of the frogs. It is just "chuutt, chuutt, chuutt" repeated several times. Calling males are very active, calling and squeaking and continuously jumping over each other, causing commotion in the pond water. They actively assault each other in reproductive frenzy. When a pair is formed it does not leave the site. A female may pair with several males, laying eggs with each.

In Baluchistan *Euphlyctis cyanophlyctis* breeds sympatrically with *Paa sternosignata*, and in northern hilly tracts with *Sphaerotheca breviceps*. Often its relatively active males pair with relatively docile frogs of these species. No eggs are known to result from such pairings (Khan, 1987; Khan and Ahmed, 1987).

Euphlyctis cyanophlyctis is a voracious feeder, feeding mostly on aquatic insects, beetles, tadpoles, dragonflies, grasshoppers, fry, etc. It is known to come out of the water during the night and go foraging in the surrounding grass, returning to the pond at dawn.

Karyotype number recorded for this species is 26 (Yadav and Pillai, 1975).

Distribution: The skittering frog is one of the most widely distributed Oriental frogs. It extends from Thailand to Nepal, throughout India, Sri Lanka, almost throughout Pakistan below 1800 m (Khan, 1997c). It extends westward to Iran and Afghanistan. Its several races have been described from Pakistan. Its Saudi Arabian population was described as a distinct species, *Euphlyctis ehrenbergii* Peters (Balletto et al., 1985).

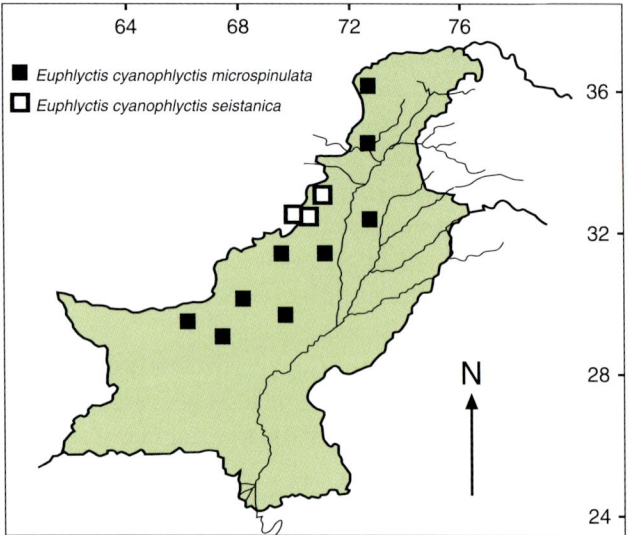

***Euphlyctis cyanophlyctis microspinulata* Khan, 1997**
(Spiny skittering frog: Khardar tapakta maindak)
1997 *Euphlyctis cyanophlyctis microspinulata* M.S. Khan, Pakistan J. Zool., 29:107–112.
Type locality: Khuzdar, southeast Kalat Division, Baluchistan, Pakistan.
Diagnosis:
1. Naris transversally enlarged. Its raised rim posteriorly produced in a distinct brown-tipped post-narial papilla.
2. Dorsum smooth, with scattered longitudinally enlarged tubercles, interspersed with minute spinules.
3. Fingers and toes with minute spinules.
4. Inner metatarsal tubercle curved inward, its length equals its breadth at base, not digitiform, with thickened outer border.
Snout-vent length 41–63 mm.

Color: Dorsum light grayish brown, with broken light brown oblique blotches, breaking on flanks into heterogeneous irregular spots. Similar large blotches on thigh and shank breaking on sides in spots. Ventrum whitish, with faded dark reticulum.

Tadpole: Essentially similar to that of *Euphlyctis cyanophlyctis cyanophlyctis*, except that the oral papillae are broadly interrupted anteriorly as well as posteriorly.

Natural history notes: Habits and breeding activity essentially as described for *Euphlyctis cyanophlyctis cyanophlyctis*. The frog floats on the surface of side pools, or spends most of its time sitting in crevices in the marginal rocks of karez channels.

Distribution: Widely distributed in Baluchistan, Afghanistan. Specimens have been collected from Waziristan and the Mianwali region, and around Jhelum city in Punjab, Pakistan.

***Euphlyctis cyanophlyctis seistanica* (Nikolsky, 1900)**
(Seistan skittering frog: Sestan tapakta maindak)
1900 *Rana seistanica* Nikolsky, Ann. Mus. Zool. Acad. Sci. St. Petersburg, iv, 1899:375–418.
Type locality: Seistan.
Diagnosis:
1. The snout is longer than length of the orbit.
Distribution: Seistan, along the Pakistan-Afghanistan-Iran border (Khan, 1997c).

Genus *Fejervarya* Bolkay, 1915
Small frogs; tympanum distinct; toes ½ webbed; inner metatarsal tubercle elongate; breeding male with loose dark gular skin.

Two species in Pakistan.

***Fejervarya limnocharis* (Boie, 1834) (Plate 10)**
(Alpine cricket frog: Paharri tidda maindak)
1835 *Rana limnocharis* Boie, *in:* Wiegmann, Nova Acta Acad. Leop. Carol., 17, Pt.1:255.
Type locality: Java.
Diagnosis:
1. Snout pointed, projecting beyond mouth.
2. Canthus obtuse, loreal oblique, more or less concave.
3. Internarial space is longer than interorbital width, which is much less than width of the upper eyelid.
4. Tympanum distinct, half to two-thirds the diameter of eye.
5. Fingers obtusely pointed, first longer than second, subarticular tubercles very prominent.

Chapter 3. Amphibians: Toads and Frogs

Plate 10. *Fejervarya limnocharis.*

6. Tibiotarsal articulation reaches tympanum or naris.
7. Toes obtuse or with slightly swollen tips, half webbed, subarticular tubercles small and prominent.
8. Body with small tubercles, sometimes small longitudinal folds are present, ventrum smooth except belly and thighs which are granular posteriorly.
9. Male with loose gular region, with brown or blackish W-shaped mark, forelimbs stronger, with pad-like subdigital tubercles under first finger.

Snout-vent length 39–43 mm.

Color: Gray brown or olive above, sometimes suffused with bright carmine; a V-shaped dark mark between eyes, a yellow vertebral stripe mostly present; lips and limbs barred, a light line along calf, thighs laterally yellow, marbled with black, ventrum white, throat is mottled with brown in male.

Tadpole: Delicate, has long oval body which is broadest and deepest at the middle. Ventrum convex, anterior half of the body flexed forward, upward. The eyes are dorsolateral in position, nearer to snout than vent. Tail is long, about twice the length of body, gradually tapering, acutely pointed, dorsal fin is broadest at middle; ventral fin runs parallel to the tail.

The anteroventral oral disc has anterior labium broader than posterior; the papillae are lateral, short, and thick. A complete preoral denticle row is followed by medially widely interrupted second row. Of the three postoral rows, the outermost is the smallest.

Labial tooth row formula is 2(2)/3. A tooth consists of three similar 0.4–0.5 mm long dental pieces, lying on each other; the crown of each piece has 5–6 sharp fine serrations. The beak is delicate, broadly arched, and finely serrated (Khan, 2003d).

Total length 26–27 mm, tail 11–11.5 mm.

Natural history notes: *Fejervarya limnocharis* is the most common and widely distributed species in the waters of neighboring Azad Kashmir, alpine Pakistan, and extends into the Potwar Tableland, descending into most of the riparian Punjab. The frog frequents marginal vegetation along canals, streams, fast-flowing streams, ponds, and puddles; when disturbed it leaps into the water, swimming back at once.

Breeding is triggered by monsoon rains. *Fejervarya limnocharis* is one of the first amphibians to arrive at the calling sites. Calling males are widely spaced from each other and generally stationed close to the stream of flowing water. The call is a characteristic "ta, ta, ta, ta" repeated rapidly several times.

Eggs are medium sized, enclosed in a double jelly capsule, and laid in batches; they adhere to the grass blades.

Karyotype number recorded for this species is 26 (Prakash, 1988).

Distribution: This species is mainly distributed in the sub-Himalayan parts of Pakistan, descending into the waters of Potwar Tableland to most of the Punjab plains and some of the lower Indus Valley where it is scarcer. It ranges from Japan to Pakistan.

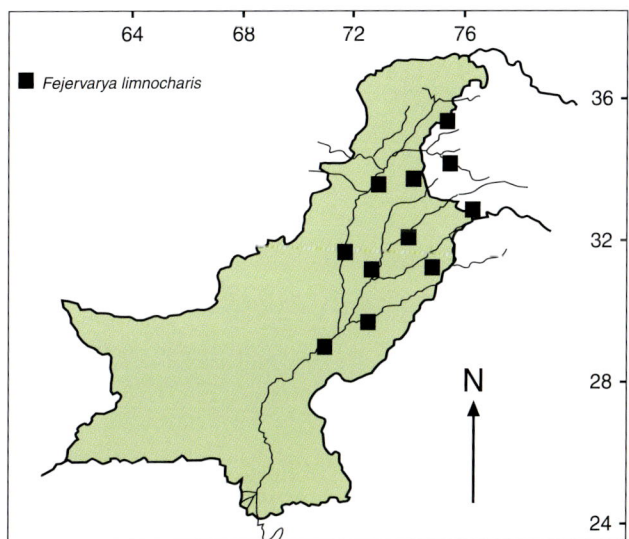

Fejervarya syhadrensis **(Annandale, 1919) (Plate 11)**
(Southern cricket frog: Dakhni tidda maindak)
1919 *Rana syhadrensis* Annandale, Rec. Ind. Mus. Calcutta 16:123.
Type locality: Bombay Presidency, between 300 and 500 m in southern India.
Diagnosis:
1. First finger hardly extends beyond second.
2. Tibiotarsal articulation reaches anterior border of eye or a point between it and the tip of snout. Snout-vent length 32–34 mm.

Color: Grayish dorsum, with dark spots, sometimes with reddish and orange suffusion, a light narrow middorsal line is often present. Ventrum white. In breeding male gular region is black.
Tadpole: Medium sized tadpole, Khan (1991e, 1996b) finds no apparent morphological difference from that of *Fejervarya limnocharis*.
Natural history notes: *Fejervarya syhadrensis* abounds in paddy fields, marginal vegetation of ponds, puddles, and streams in plains. It becomes rarer in northern hilly tracts along sub-Himalayan ranges. Its call is typical, like the clatter of a typewriter, a loud "trr, trr, trr, trr, trr", repeated several times.

The calling males sit quite apart from each other, away from the water, in the roots of marginal vegetation. Calling is triggered by the first monsoon downpour when water temperature reaches 20°C. Egg diameter ranges from 0.8 to 1.2 mm. Eggs are laid in small batches, embedded in gelatinous material, each enclosed in double jelly capsule. Eggs soon separate and adhere to the grass blades (Khan 1996b).
Distribution: *Fejervarya syhadrensis* occurs sympatric in most of its range in Pakistan with *Fejervarya limnocharis*, which becomes rarer in the lower Indus Valley. *Fejervarya syhadrensis* is widely distributed throughout southern India.

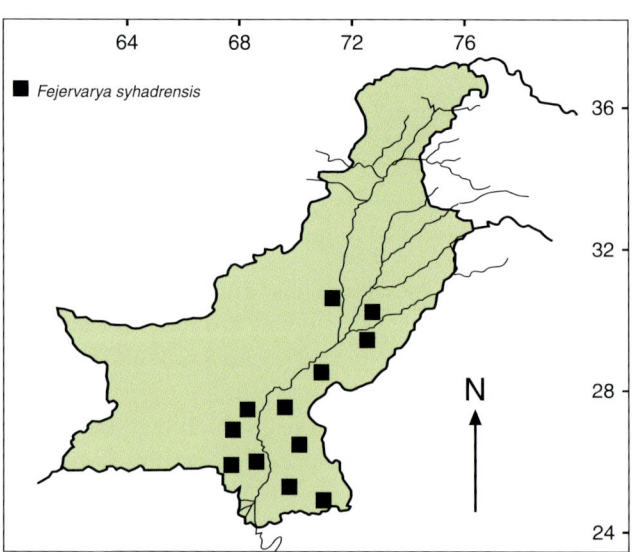

Genus *Hoplobatrachus* Peters, 1863
Large frogs; tympanum distinct; body dorsum smooth, with longitudinal broken folds; toes extensively webbed; male with nuptial pads, and vocal sacs on the sides of lower jaw.

Hoplobatrachus tigerinus **(Daudin, 1802) (Plate 12)**
(Bullfrog: Basanti maindak)
1803 *Rana tigerinus* Daudin, Hist. Nat. Rainettes: 64, Plate 20.
Type locality: Bengal.
Diagnosis:
1. The head is slightly longer than wide; in older specimens it is wider.
2. Snout pointed, projecting, canthus obtuse; loreal oblique, slightly concave.
3. Interorbital space much narrower than the upper eyelid.
4. Tympanum distinct, almost as large as the eye.
5. Fingers obtusely pointed, first longer than second.

Plate 11. *Fejervarya syhadrensis.*

Chapter 3. Amphibians: Toads and Frogs

Plate 12. *Hoplobatrachus tigerinus*.

6. Tibiotarsal articulation reaches eye or between eye and the naris.
7. Toes obtuse, with slightly swollen tips, entirely webbed, feebly emarginate.
8. Outer metatarsal tubercle separated nearly to its base.
9. Subarticular tubercles small, a dermal fold along outer border of the fifth toe, inner metatarsal tubercle small, blunt and compressed.
10. Dorsum smooth or granular, with 6–14 longitudinal broken folds, occasionally interspersed with smooth tubercles, ventrum smooth.
11. Forelimbs of breeding male are thick, first finger is swollen, with grayish brown velvety horny layer at its base, blue vocal sacs are located on sides of the throat.
Snout-vent length 130–145 mm.

Color: Dorsum olive green, olive or gray, with dark blotches, a light yellow vertebral streak, rarely absent; a dark canthal and a lighter labial streak often present; limbs with dark bars, which may break into dark blotches; thighs posteriorly marbled with black and yellow; a fine yellow line along upper surface of thigh, another on the inner side of calf. Ventrum white, sometimes feeble pigmentation on throat (Khan and Tasnim, 1987a).

Tadpole: Tadpole of *Hoplobatrachus tigerinus* has a cylindrical body, which does not bulge out; tail is muscular, almost as broad as body, fins are narrow, parallel, tail tip is acutely pointed. Anterior oral disc, with nonpapillate rim. Posterior labium extensible into an additional postdisc sucker. Beak strong, prebuccal half of it is strongly serrated, medially produced into a long serrated tooth, while postbuccal half is sharp, nonserrated with a median recess to receive the median tooth of the prebuccal half. The labial tooth row formula is 5(4)/5(3), teeth are biserial in arrangement (Figure 6C). A tooth is a 0.3–0.4 mm long cylindrical body, with a gradual taper toward acute tip (Khan, 1991e, 1996a).

The tadpole is predominantly carnivorous and feeds primarily on sympatric tadpoles and bodies of drowned animals (Khan, 1996a). It is benthic in habits, eyes and nostrils are dorsally placed. It stalks its prey, while lying at the bottom of water, darting to catch it in its powerful jaws. Melanophores are concentrated just below eyes, and along dorsolateral sides of body; tail and fins are speckled with black, tail tip heavily pigmented.

Total length of the tadpole 40–43 mm, tail 23–26 mm.

Natural history notes: *Hoplobatrachus tigerinus* is the largest frog in the Pakistani plains. It hibernates by burrowing in soil during winter as well as during drought.

Breeding activity is primarily confined to monsoons. The breeding males are lemon yellow in color, hence locally called "Basanti Dadoo," while females remain dull and drab-colored. The deep blue vocal sacs of male are prominent against the yellowish white color of the throat of the male. The call is a powerful nasal "cronk, cronk, cronk," which sometimes sounds like "oong wang, oong wang, oong wang," repeated several times. Calling males sit close to each other in shallow water, now and then jumping over each other. Females lurch around. One falling within the range of a male is grabbed by the male in an amplectic hold, with neighbors soon jumping on the pair and trying to dislodge them, which starts much fighting, pushing, and tugging. The pair somehow moves to a quieter place where large eggs (2.5–2.8 mm diameter) are laid in several groups, each egg enclosed in a double coat of jelly. Eggs are soon attached to grass blades and often sink into the water (Khan, 1969, 1996a).

Hoplobatrachus tigerinus is a voracious feeder; anything that is moving is pounced upon and swallowed. If needed, it uses its anterior limbs to thrust larger food into its mouth. In addition to a great variety of insects, it feeds on a variety of items: mice, shrew, young frogs, earthworms, roundworms, juvenile snakes, and small birds. Vegetable matter and several odd objects are recorded from its stomach (Khan, 1973). Lizards like *Uromastyx* (Daniel, 1975), snakes: *Lycodon aulicus, Ramphotyphlops braminus, Leptotyphlops sp.,* and young *Ptyas mucosus* (pers. obs.) have also been recorded from its diet.

The frog does not stay in water for a long time; it spends most of its time hiding and feeding in surrounding vegetation. On approach of danger, it plunges into deep water, stays underwater for 2–3 minutes, then returns quietly to the marginal vegetation undetected. In clear pools of water it hides under bottom gravel.

Karyotype number recorded for this species is 26 (Natarajan, 1958).

Distribution: The bullfrog is the most common frog of the Indo-Gangetic plains. It frequents mostly cultivated areas and swampy wastelands. In Pakistan it does not extend into Baluchistan, however it has been reported from Afghanistan close to the Khyber Pass (Kullmann, 1974).

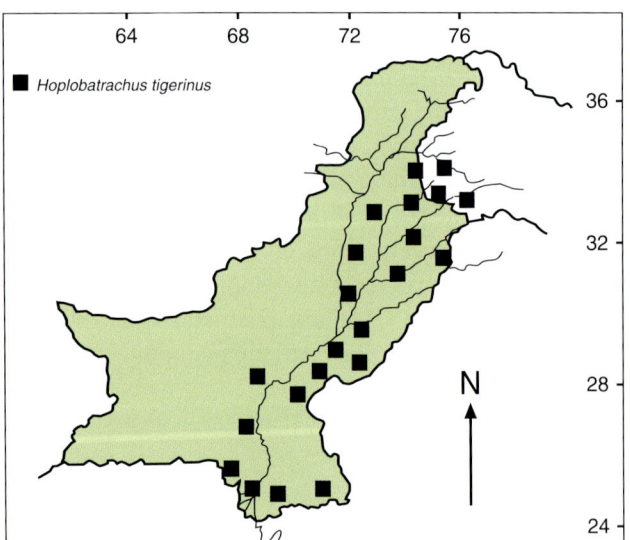

Genus *Paa* Dubois, 1975

Large frogs; indistinct tympanum; body dorsum smooth, few tubercles on sides of body; toes extensively webbed, with distinctly swollen flat tips.

Four species in Pakistan.

Paa barmoachensis (Khan and Tasnim, 1989) (Plate 13)

(Kashmir Torrent frog: Kashmir nadi maindak)
1989 *Rana barmoachensis* Khan and Tasnim, J. Herpetol. 23 (4):419–423.

Type locality: Barmoach, Goi Madan, Azad Kashmir.

Diagnosis:
1. Head is wider than long.
2. Naris is below canthus, distance between naris greater than width of upper eyelid.
3. First finger longer than second.
4. Tips of digits swollen, with distinct large subarticular tubercles. A distinct groove between penultimate and last phalanx.
5. Body dorsum with dark unispinulate small tubercles.
6. Interorbital region tuberculated.
7. Male with nuptial spines on first finger and inner metacarpal tubercle (Figure 5A).
8. Tympanum is distinct.
   Snout-vent length 56–61 mm.

Color: Dorsum gray olive in life with scattered dark spots more numerous on flanks than in middorsum, a distinct pair of dark spots between eyes; dark mottling on forehead, jaws, supratympanic fold, canthus dark; fore- and hindlimbs with crossbars breaking into a network of small brownish spots on hind of thighs. Ventrum dirty white, with light suffusion under hand and foot; throat and gular region mottled with gray; web between fingers olive gray.

Tadpole: Tadpole large, robust, with bulging body, strong muscular tail with moderately broad fins. It is a slow but powerful swimmer, swims about in pools of

Plate 13. *Paa barmoachensis.*

clear water in the bed of the fast-flowing stream; feeds on algal and other concretions deposited on the surface of the submerged rocks. Most of the time it remains settled on stones at the bottom and keeps feeding.

Eyes and naris are dorsal in position, while the oral disc is anteroventral; fins are spotted. The beak is strong, sharply serrated; labial tooth row formula is 6(4)/3. The labial papillae are in a double row, extending from anterolateral sides to posterior sides of the disc, narrowly interrupted medially along anterior labium (Khan and Tasnim, 1989).

Total length of the tadpole 78–80 mm, tail 50–55 mm.

Natural history notes: *Rana barmoachensis* is a torrenticolous frog. In the dry season a small stream of water continuously trickles from one pool of water to other in the stream bed. Overhanging branches of high broad leaved trees growing along banks partially cut off sunlight. The frog sits on the exposed side rocks, waiting for worms and insects to appear, on which they feed. When frogs are disturbed, they dive into the depth of the clear pool of water and stay at the bottom for considerable time.

In the strong current of water, the frog climbs to the rocks not in the way of the current. The tadpoles retire to crevices in the rock and get firm hold on rocks with their heavily papillate and multidental oral discs. Adult frogs also migrate under stones which are not under the direct force of the current.

The upstream pools with *Rana barmoachensis* are invaded by *Euphlyctis cyanophlyctis* from the larger pools at the lower stream. Both frogs breed together in these pools. Larvae of *Euphlyctis cyanophlyctis* often attack those of *Rana barmoachensis* and feed on them.

The breeding activity in this frog is observed twice a year: premonsoon activity is initiated during April and May, followed by a short pause during torrential monsoon rains during July and August. A second breeding phase occurs from mid-August to early September. I collected tadpoles of *Paa barmoachensis* during late November from pools at 4°C. These tadpoles were apparently from late amplectic pairs. The empty intestines of the tadpoles indicate they were not feeding; they probably derive energy for winter from their heavily laden fatbodies. These tadpoles probably survive the winter to metamorphose early the next summer.

Distribution: The Barmoach frog has been collected from Aram Bari, Tatta Pani, Charnali, and Kotli, all in District Kotli, Azad Kashmir.

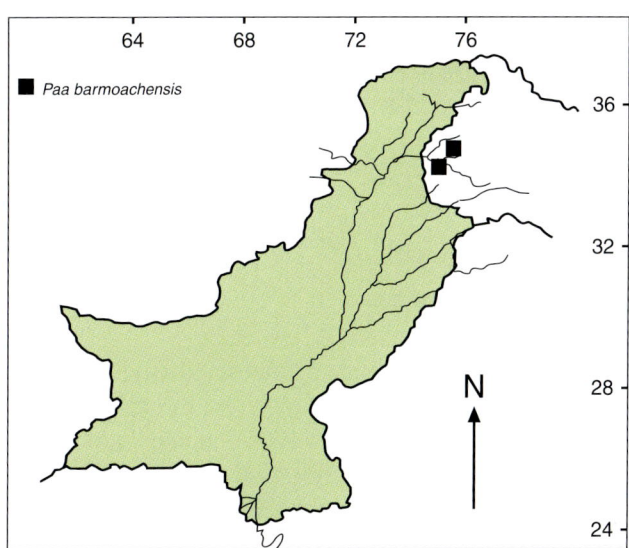

Paa hazarensis **(Dubois and Khan, 1979) (Plate 14)**
(Hazara fast-flowing stream frog: Hazara nadi maindak)
1979 *Rana hazarensis* Dubois and Khan, J. Herpetol. 13:403–410.
Type locality: Datta, District Manshera, Pakistan.
Diagnosis:
1. Head longer than wide.
2. Dorsal tubercles on short longitudinal folds.
3. Naris above the canthus.
4. Fewer or no tubercles in the interorbital region.
5. Forelimbs enlarged in breeding males, with nuptial spines on inner finger and metacarpal tubercle (Figure 8).
Snout-vent length 55–62 mm.

Color: Body dorsum grayish, superimposed with a network of darker color (dark olive green in life). A clear graey transverse band on the back of the head joining the posterior borders of the eyelids. Later edge of sypratympanic fold and canthus dark. Upper parts of the hindlimbs with dark cross-bars; hinder parts of thighs with a network of small blackish spots. Lower parts of the body and limbs whitish; throat mottled with graey. Webbing of feet graeyish.

Tadpole: The tadpole lives in pools of clear water in the course of fast-flowing streams. The anteroventral oral disc is bordered with two rows of long papillae

Plate 14. *Paa hazarensis*.

which are widely interrupted anteromedially; posteriorly it is uninterrupted and has 3 rows of papillae. Anterior labium has 8 tooth rows of which 7 are medially interrupted. The posterior labium has 3 rows of teeth of which 2 are interrupted. The labial tooth row formula is 8(7)/3(2). The beak is large, with preoral half strongly arched and finely serrated, overhanging similar postoral half (Khan and Malik, 1987a).

The tadpole is a typical Himalayan torrenticole in habits. It feeds on algal growths on the surface of submerged stones. Tadpoles of *Euphlyctis cyanophlyctis* also occur in these pools. During rainy season, to avoid the fast flow of water, the tadpoles either migrate into crevices under stones where the force of flow is minimum, or hold on to the surface of rocks with the oral disc which acts as an effective sucker giving the tadpoles a very firm hold.

Total length of the tadpole 75 mm, tail 65 mm.

Natural history notes: This frog frequents quieter and clear water pools in the beds of fast-flowing streams, feeding on water-visiting insects. It breeds from March to May; call is low-pitched, barely heard away from the fast-flowing stream. Large eggs are laid singly and are enclosed in a double jelly capsule.

Distribution: This frog is known from streams in the Rush Valley in Hazara Division, NWFP, Pakistan.

Paa sternosignata (Murray, 1885) (Plate 15)

(Karez frog: Karez maindak)

1885 *Rana sternosignata* Murray, Ann. Mag. Nat. Hist., London (5)16:120.

Type locality: Quetta, Baluchistan, Pakistan.

Diagnosis:
1. Head much broader than long, much depressed, scarcely projecting.
2. No canthus, loreal oblique, slightly concave.
3. Internarial distance greater than interorbital space, which is less than the width of the upper eyelid.
4. Tympanum not very distinct, about half the diameter of the eye.
5. Tibiotarsal articulation reaches the temple or posterior border of the eye.
6. Digits obtuse or slightly swollen at tips, toes entirely webbed, subarticular tubercles small.
7. Skin on body very loose; when the frog is on land, it is thrown in folds. Dorsum is smooth or tuberculated; tubercles are tipped with dark spines.
8. Sexually dimorphic: flanks of body of male are heavily tuberculated. Dark nuptial spines are on the first 2 fingers and the inner carpal tubercle. There is a large patch of pustules on both ventral sides of the chest. Venter in male is tuberculated all over, that of female is smooth. The anterior limbs of the breeding male are considerably thickened. Snout-vent length 88–90 mm.

Color: Olive brown or dark green above, with small irregular yellow, orange, or reddish spots; ventrum white with profuse fine dark brown or black mottling. Young dark brown or olive, with blackish blotches. Pupil reduced to a narrow slit. A transverse dark streak runs across the eye, a similar vertical streak runs downward across the golden brown iris, giving the eye a peculiar appearance. This feature is also present in tadpole.

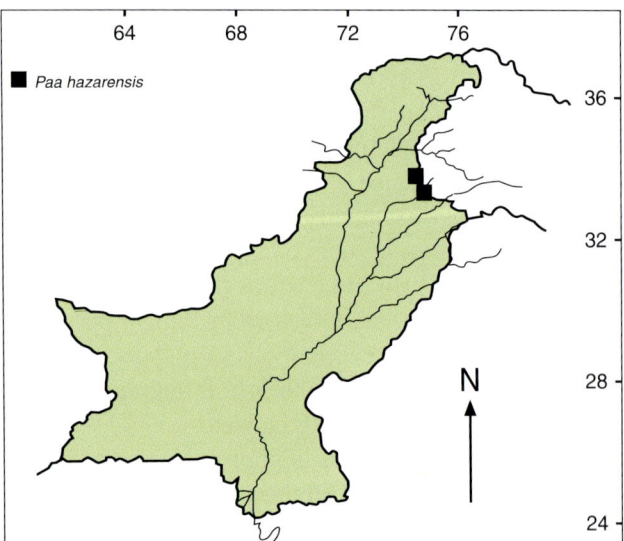

Plate 15. *Paa sternosignata*.

Tadpole: The tadpole is the largest and stoutest of all Pakistani tadpoles. It coexists with tadpoles of *Euphlyctis cyanophlyctis* and *Bufo viridis* in karez water, later mostly confined to side pools. The tadpole feeds on algal vegetation amply available in the karez channel.

The body is elongated, bulging, eyes are anterodorsal, tail very muscular, with broad dorsal and narrower ventral fins, acutely pointed tip.

Anteroventral oral disc is surprisingly small for this heavy and broad bodied large tadpole. Fringe of oral papillae does not extend along anterior labium, however it is entire along posterior labium. Beak narrow, finely serrated, labial tooth row formula is 5(4)/3. Total length of the tadpole 78–80 mm, tail 50–52 mm.

Natural history notes: *Paa sternosignata* is a thoroughly aquatic species, inhabiting clear pools with flowing karez water. When disturbed it jumps into the depths of the water and hides under gravel at the bottom or under dense marginal vegetation and thick floating algal cover. The frog never leaves the water; even in freezing winter when the upper water surface is frozen, it remains sluggishly active beneath in the unfrozen water. In summer it usually sits in the marginal vegetation or under undercut rocks along sides of streams (Khan and Ahmed, 1987).

Reproductive activity is at its peak from April to June. Males call from the margin of flowing water at sunset. The call is a low-pitched melodious "taroon, taroon, taroon" uttered three to four times rapidly. One has to wait for 5–10 minutes to hear the next outburst of calls. Calling males are solitary. Eggs are large (2.6–3 mm) and laid in groups enclosed in jelly coats, which soon are attached to the submerged vegetation.

Abortive interspecific amplectic pairing is common between *Paa sternosignata* and *Euphlyctis cyanophlyctis* (Khan and Ahmed, 1987).

The frog feeds on insects, crabs, small fishes, dragonflies, and a host of other water-visiting arthropods.

Distribution: The karez frog abounds around Quetta and Mastung in karez channels. It has also been reported from Afghanistan up to 1800–2000 m of elevation (Kullmann, 1974).

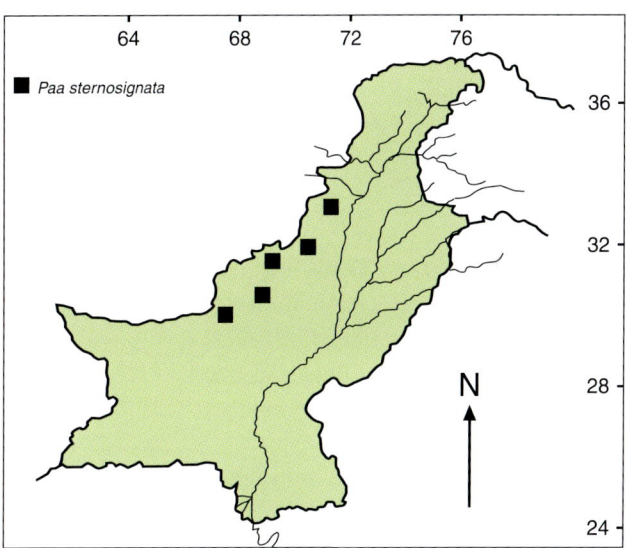

Paa vicina (Stoliczka, 1872)
(Murree hills frog: Maree maindak)
1872 *Rana vicina* Stoliczka, Proc. Asiatic Soc. Bengal, Calcutta 1872:124–131.
Type locality: Murree, alpine Punjab, Pakistan.
Diagnosis:
1. Naris a little nearer to the eye.
2. Tympanum indistinct.
3. First finger as long as second.
4. Toes half-webbed.
5. Outer metatarsal tubercle reduced, inner narrow, feebly developed.
6. Body dorsum tuberculate.

Distribution: Stoliczka (1872b) reported this frog from fountain water in Murree, alpine Punjab, Pakistan. A single specimen was collected by Mertens (1969a) from the area bordering Azad Kashmir and assigned to *Paa pleskei*, which was later identified as *Paa vicina* by Dubois (1976).

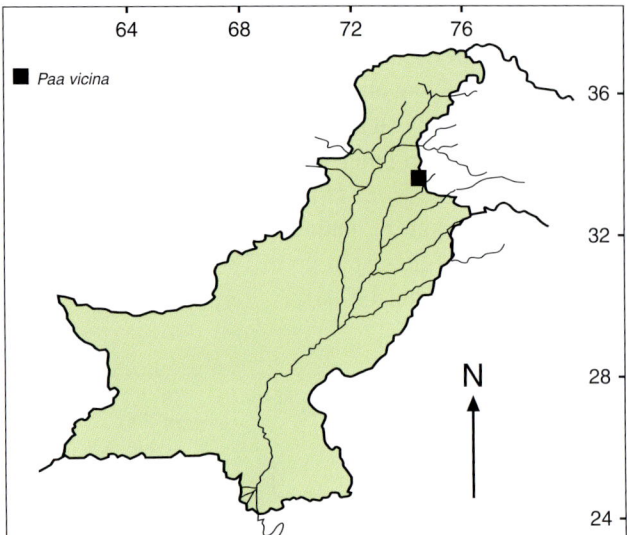

Genus *Sphaeroteca* Duméril and Bibron, 1841

Medium sized frogs; habitus toad like; large eyes and tympanum; inner metatarsal tubercles flat shovel-shaped; toes half to one-fourth webbed.

A single species.

Sphaeroteca breviceps (Schneider, 1799) (Plate 16)
(Burrowing frog: Gauk-maindak)
1799 *Rana breviceps* Schneider, Hist. Amphib.:140.
Type locality: India.
Diagnosis:
1. Habitus short, stocky, toad-like.
2. Head broader than long, top flat.
3. First finger longer than second, tips swollen.
4. Hindlimbs short, tibiotarsal articulation reaching the axilla.
5. Toes half to one fourth webbed. Inner metatarsal tubercle long, compressed, flat, shovel-shaped (Figure 5C).
6. Skin smooth, slightly varicose, with feeble longitudinal dorsal folds.
7. Tympanum distinct, as large as eye.
Snout-vent length 49–52 mm.

Color: Dorsum olive golden, with a light yellow median line; a well marked large golden ocellus on shoulder on each side, continuing obliquely along flanks. Indistinct yellowish bars on limbs and jaws. Sometimes dorsum is faintly mottled. Ventrum cream, gular region black in breeding males.

The eyes of juvenile *Sphaeroteca breviceps* are surrounded by a dark circle. Its broad head is reddish.

Tadpole: The tadpoles are pinkish brown with dark brown spots. The tadpoles have stout habitus; the snout is high broad and rounded. The oral disc is anterior; the fringe of oral papillae does not extend on anterior and posterior labium. The posterior labium double row of papillae. Labial tooth row formula is 2(2)/3. The beak is narrow and finely serrated.

The lanceolate tail has dorsal fin broader than the ventral. Body is light brown with large dark brown blotches, ventrum grayish white with fine brown dots.

The tadpole is found in small ponds and side pools along streams.

Total length of the tadpole 34–36 mm, tail 25–26 mm.

Natural history notes: This burrowing frog inhabits relatively humid parts of Pakistan and is found in abundance along the Himalayan foothills in the northwest. It is essentially nocturnal, emerging at dusk from of its burrow which it excavates in soft sandy soil with the help of its broad shovel-shaped inner metatarsal tubercle. At dusk these frogs swarm fields around Manshera city. It is insectivorous; however, it devours centipedes and millipedes which are common in its habitat.

Breeding season starts with the first showers of summer monsoons, when males gather in the shallow marginal waters of large ponds, spaced well apart from each other. The call is a quickly repeated "awang, awang, awang" uttered 5 to 7 times in a burst. The same habitat is a potential breeding site for other

Plate 16. *Sphaeroteca breviceps.*

local amphibians, e.g., *Bufo stomaticus*, *Euphlyctis cyanophlyctis*, *Hoplobatrachus tigerinus,* and *Microhyla ornata*. Occasionally, a relatively agile male of *B. stomaticus* or *E. cyanophlyctis* pairs with a *Sphaeroteca breviceps* (regardless of sex) which being docile, offers little or no resistance; the pair soon separates.

Eggs of *Sphaeroteca breviceps* are large, enclosed in two envelopes of jelly, and laid in small batches which float and adhere to the grass blades.

Sphaeroteca breviceps has a remarkable resemblance to the burrowing frogs of the genus *Uperodon*.

Distribution: The burrowing frog is distributed throughout India, Bangladesh, Nepal, Myanmar, and Sri Lanka. In Pakistan it is reported from the Himalayan foothills, and extends into the Potwar Tableland. It has spotty distribution in the riparian system of Punjab (Khan, 1976). Its colonies have been reported in desolate parts of the Cholistan Desert

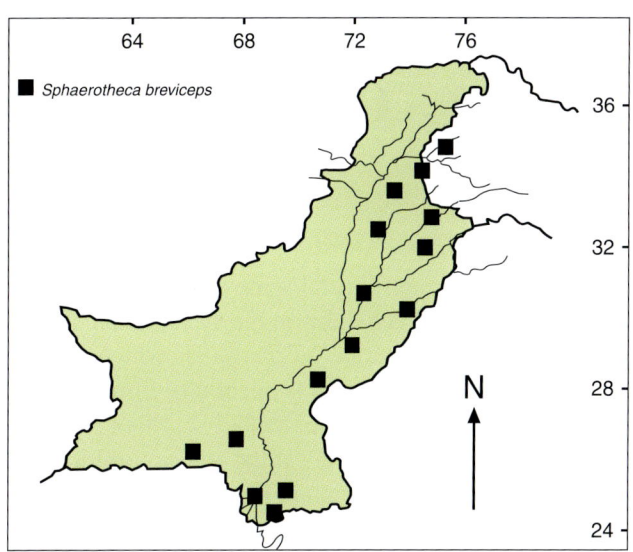

(Khan, 1985a). Around Karachi, it is collected from the Hab and Malir River Valleys (Minton, 1962, 1966).

Chapter 4
Chelonians: Turtles and Tortoises

The study of the interplastral scute of the carapace of the chelonian shell is helpful in the systematics of these animals. The number of marginals which contact the pleural scute is recorded sequentially from anterior to posterior, followed by the symbol <, M, or >, indicating whether the contact is at the anterior, middle, or posterior third of the marginal, respectively. Thus in carapace seam contact formula, 3<6<8<10<12<, the first costal seam is toward the anterior of the third marginal scute; the second costal seam is toward the anterior of the sixth marginal scute; the third costal seam is toward the anterior of the tenth marginal scute; and the fourth costal seam is toward the anterior of the twelfth marginal scute.

Similarly the plastral contact formulae measure all plastral seams (on one side) deciding whether it is to be the shorter or longer, then writing the seams in descending order. For example in plastral formula, an<abd><fem>hum>pec>gul>intergular, the anal is shorter than abdominal, which is greater or less than the femorals, which is greater than humerals, which is greater than pectoral, which in turn is greater than gular, which is larger than the intergulars.

Turtles and tortoises are represented by five families in Pakistan.

Family Cheloniidae

This family of sea turtles is represented in Pakistan by 4 genera.

Genus *Caretta* Rafinesque, 1814
Five or more pairs of costals; inframarginals three, without pores.

A single species in Pakistan.

Caretta caretta (Linnaeus, 1758)
(Loggerhead sea turtle: Paan kachhuwa)

1758 *Testudo caretta* Linnaeus, Syst. Nat. Ed. 10(1):197.
Type locality: Isles of America.
Diagnosis:
1. Carapace elongate, tapers posteriorly.
2. Inframarginals 3, without pores (Figure 9).
3. First costal touching nuchal.
4. Prefrontals four.
5. Coastals 5 or more pairs, first touches nuchal, usually 26 marginals.
6. Seam contact formula: 3<6<8<10<12<.
7. Plastral formula: an<abd><fem>hum>pec>gul>intergular.

Sexually dimorphic, male has longer tail; long recurved claws, shorter plastron, depressed, and wider carapace. The hatchling has 3 strong keels on the carapace.

Carapace 75–150 cm long, weight 100–110 kg (Das, 1991).
Color: Carapace reddish brown, plastron yellowish brown or yellowish orange. Hatchlings uniformly dark brown or blackish, paler below.

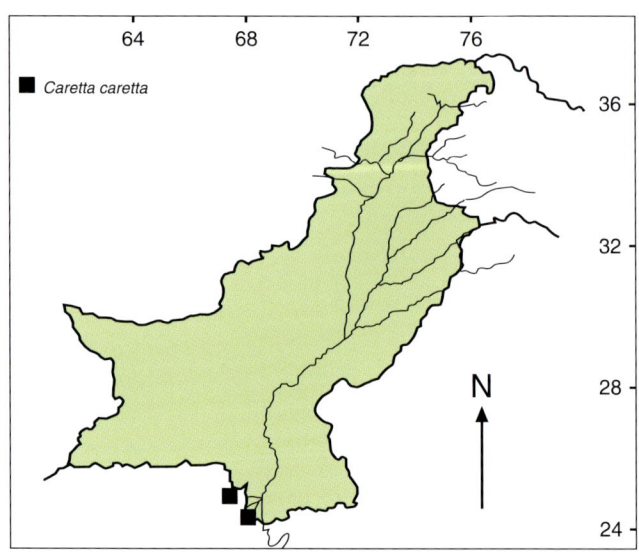

Chapter 4. Chelonians: Turtles and Tortoises

Natural history notes: *Caretta caretta* is global in distribution. It inhabits temperate, subtropical, and tropical seas, invading continental shelves, bays, lagoons, and estuaries. Nesting grounds are located in warm temperate and subtropical regions. It is known to hibernate in the muddy parts of the undersea channels (Carr et al., 1980).

It feeds on hard-shelled molluscs and crustaceans, a host of marine animals, and algae.

Mating involves courtship, which occurs in temperate waters. Gravid females visit beaches for nesting. Eggs are spherical, enclosed in leathery shells, and their diameter ranges from 3.4 to 5.5 cm; clutch size varies from 23 to 178 eggs, with an incubation period of 49–80 days, depending on nest temperature (Das 1991).

Distribution: The loggerhead sea turtle is global in distribution. It is very rare in Pakistani coastal waters and does not nest in Pakistan (Ghalib and Zaidi, 1976).

Genus *Chelonia* Brongniart, 1800

Four pairs of costals; single pair of prefrontals; dorsal plates juxtaposed; jaws not hooked.

A single species in Pakistan.

Chelonia mydas (Linnaeus, 1758) (Plate 17)

(Green sea turtle: Hara samundri kachhuwa)
1758 *Testudo mydas* Linnaeus, Syst. Nat. Ed. 10(1):197.
Type locality: Ascension Island.
Diagnosis:
1. Carapace cordiform, oval.
2. Shell naked, dorsal shields juxtaposed, marginals sloping, vertebrals broader than long.

Plate 17. *Chelonia mydas.* (Photo courtesy of S.A. Minton).

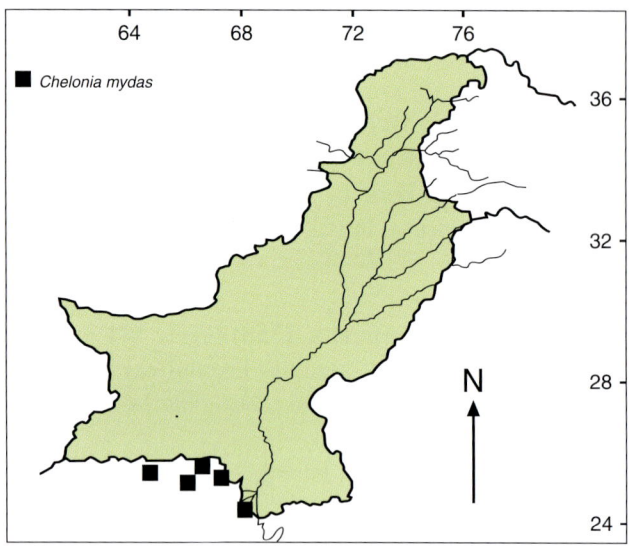

3. Single pair of prefrontals, 4 postoculars.
4. Marginals 25, 4 pairs of costals, first not touching nuchal.
5. Inframarginals 4.
6. Limbs paddle-shaped flippers, with large scales, digits not distinct, a single claw in adult, 2 in young.
7. Tomium of upper jaw feeble, while lower sharply denticulate.
8. Seam-contact formula: 2<5<7>9<11>.
9. Plastral formula: abd><fem>an><gul>pect>hum>intergul.

Sexually dimorphic: male smaller, more tapering carapace; foreclaws longer and curved; tail longer. Carapace length 70–140 cm, weight 155 kg (Daniel, 1983). Carapace feebly unicarinate in the young, sometimes with slight indication of lateral keel.

Color: Carapace olive or brown, usually with radiating pattern, soft parts brown, head scales outlined with yellow.

Juveniles with carapace plates gray to black with cream edges, limbs outlined yellow; ventrum yellow; large dark blotches on limbs.

Natural history notes: It is a tropical species, concentrating around wide sandy beaches along islands and continental coasts. In Pakistan its nesting areas are scattered along the coastal strip: Hawks Bay, Buleji, Paradise Point, Ormara, Somiani, and Ras Jiunri (Minton, 1966; Sternberg, 1981). The juveniles are carnivorous, while adults are herbivores feeding on marine vegetation of different types. The nesting female excavates a deep pit on a sandy beach, laying her eggs at night. Clutch size varies from 9 to 173 eggs, with diameter ranging from 41 to 52 mm; hatchlings appear after 22–166 days, depending on the temperature of the nest (Das, 1991). Minton (1966) observed hatchlings from July to October, with the greatest number during September and early October. Large numbers of nests are destroyed by jackals and dogs, and are later plundered by crows, kites, vultures, and varanid lizards. Crabs, birds, and different carnivorous mammals also take a heavy toll on the hatchlings.

Distribution: *Chelonia mydas* is a pantropic species; it centers on breeding and foraging grounds along the seacoasts of Pakistan, Kutch, western and eastern ghats of peninsular India, Maldives, Sri Lanka, the Andamans, and the coast along Orissa (Das, 1991).

Genus *Eretmochelys* Fitzinger, 1843

Costals 4 pairs; 2 pairs of prefrontals; dorsal plates imbricate; jaws hooked downward.

Single species in Pakistan.

Eretmochelys imbricata (Linnaeus, 1766) (Plate 18)

(Hawksbill turtle: Cheel-sar samundri kachhuwa)
1766 *Testudo imbricata* Linnaeus, Syst. Nat. Ed. 12:350.

Type locality: American and Asiatic seas.
Diagnosis:
1. Carapace roundish-oval, naked, with strong imbricate shields (Figure 7).
2. Prefrontals 2 pairs.
3. Neurals and costals 5 each, first neural touches nuchal.
4. Marginal laminae 26, a distinct notch between twelfth marginals.
5. Inframarginals 3 without pores.
6. Jaw hooked, feebly denticulate or without denticulation.
7. A pair of claws on forelimbs (Figure 8B).
8. Neurals wider than long, costals and neurals with slight ridges converging at an elevated point at the posterior border of the plates, due to which shell appears tricarinate.
9. Seam contact formula: 2<5<7>9<11>.

Chapter 4. Chelonians: Turtles and Tortoises

Plate 18. *Eretmochelys imbricata*.

10. Plastral formula:
    an>fem>abd><hum><pect>gul>intergul.
    Adult size 1.2 m, weight 140 kg.
    Sexually dimorphic: male has concave plastron, long, thick tail, and long forelimb claws.
    Juveniles with tricarinate carapace, with strongly imbricate shields, vertebrals rhomboidal.
Color: Adult with olive brown carapace and yellow plastron, carapace of juvenile blotched with dark and plastron is blackish; head scales dark brown to black with light edges; jaws yellow with brown marks.
Natural history notes: *Eretmochelys imbricata* abounds around sandy beaches along oceanic islands. It is a specialized eater of sponges, however, other reef organisms, corals, algae, and gastropods have been recorded from its stomach. Clutch size varies from 115 to 138 eggs, measuring 35–38 mm in diameter; incubation period varies from 60 to 65 days.
Distribution: The hawksbill turtle inhabits warm seas, preferring islands and coastal sandy beaches which are its foraging as well as nesting sites. Sternberg (1981) gives 7 nesting grounds in the subcontinent for this turtle, all confined to the Indian coastal line and

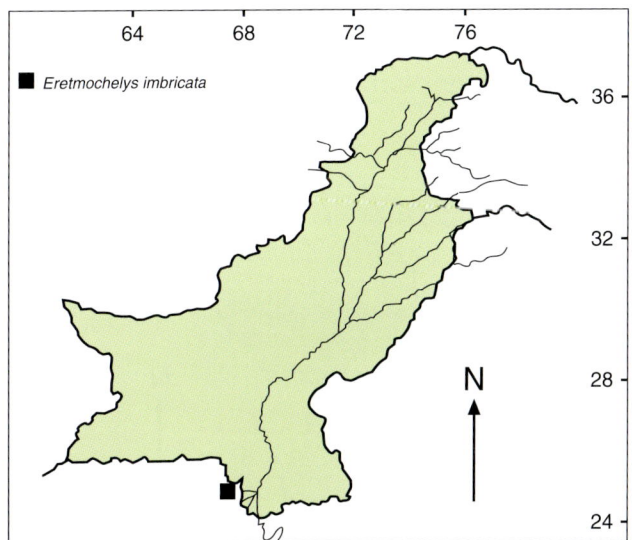

oceanic islands. The only report from Pakistan is by Mertens (1969a) of seeing small *Eretmochelys imbricata* shells in a shop in Karachi.

Genus *Lepidochelys* Fitzinger, 1843

Five or more pairs of costals; inframarginals four, some pierced with pores.

A single species in Pakistan.

Lepidochelys olivacea (Eschscholtz, 1824)
(Olive Ridley: Zatooni samundri kachhuwa)
1829 *Chelonia olivacea* Eschscholtz, Zool. Atlas: 3.
Type locality: Manila Bay.
Diagnosis:
1. Carapace broad, cordiform, tactiform with flattish top.
2. Pleurals 5–9 pairs, 12–14 pairs of marginals, posterior marginals serrated.
3. A broad cervical, touching first pleural, 1–2 small intergulars and a single internasal.
4. First and fifth vertebrals broader than long, second, third and fourth longer than broad.
5. The inframarginals are with costal pores (Figure 9A).
6. Head small, triangular, 4 prefrontals, upper jaw hooked.
7. Seam contact formula very variable, plastral formula: an>fem>gul>abd>pect>hum. Maximum length 80 cm, weight 50 kg.

Color: Dorsum olive green, ventrum greenish yellow.
Natural history notes: The Ridley turtle is less plentiful than green turtle along the Karachi coast. It prefers mainland nesting beaches, especially with mangrove vegetation, in tropical and subtropical waters. It is omnivorous taking fishes, snails, crabs, jellyfish, lobsters, shrimps, tunicates, oysters, and echinoderms; algae has also been recovered from its stomach (Das, 1991).

To nest, it migrates to shallow coastal waters in large numbers known as *flotillas* close to nesting grounds. Mating occurs in the surface water. This turtle nests in mass which is often referred to as "arribada." Nests are covered with beach vegetation. In Hawks Bay and Sandspit beaches, Karsachi, Pakistan, nesting is between June and October, and peaks in September. Clutch size varies from 80 to 160 eggs measuring 29–39.4 mm, weighing 31–35 g, and hatching within 42–69 days (Das, 1991).
Distribution: It is widely distributed on tropical and subtropical beaches with shallow muddy bottoms, high in detritus and low in salinity. It is known from the mangrove vegetation the along the Sind coastal strip.

Family Dermochelyidae

Represented by a single genus along the sea coast of Pakistan.

Genus *Dermochelys* Blainville, 1816

Body enclosed in a bony shell, no epidermal plates on the shell; limbs without external evidence of digits, oar-shaped, clawless.

Single species.

Dermochelys coriacea (Vandelli, 1761)
(Leatherback sea turtle: Saat-ubhar Khaal-pusht kachhuwa)
1761 *Testudo coriacea* Vandelli, pistola de Holothurio,et *Testidinae coriacea*
Type locality: Palermo, Sicily.
Diagnosis:
1. Elongate leathery shell, with 7 ridges, tapering on sides to a posterior point, 5 ridges on plastron (Figure 7D).

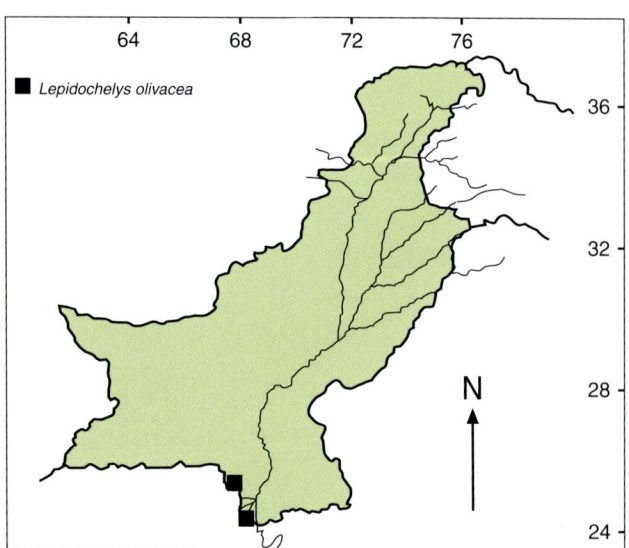

2. Snout blunt, upper jaw with a median W-shaped cusp.
3. Skull roof complete, neck short, nonretractable.
4. No scales on limbs in adult.
5. Entoplastron absent, the bridge is long.
6. Limbs paddle-shaped, clawless (Figure 8C).
7. A thick membrane includes hindlimbs and the tail.
8. Anterior limbs much longer than the posterior.

Male with depressed shell, concave plastron, and longer tail.

Largest of all chelonians, attaining a total length of more than 2 meters and an estimated weight of slightly above 450 kg (Smith, 1931). Eckert and Luginbuhl (1988) reported a specimen with carapace length of 256.5 cm, weighing 916 kg. Daniel (1983) records length of carapace 150 cm, weight 400 kg.
Color: Dorsum black with white spots. Neck and base of flippers with blue or pinkish spots, ventrum pale pink or white.
Natural history notes: The most widely distributed sea turtle of temperate freezing waters, it is truly pelagic in habits, coming ashore only to nest. It can dive to a depth of 1200 m (Eckert et al., 1986). It forages widely feeding exclusively on jellyfish (Das, 1991). It nests at night; clutch size varies from 90 to 130 eggs, ranging from 48 to 51 mm in diameter. A considerable number of yolkless undersized eggs are also laid. Incubation period recorded is 50–56 days.

The body and limbs of the hatchlings are covered with small irregular polygonal shields, the largest on the carapace and plastron. There are 7 dorsal and 5 ventral rows of raised enlarged squarish shields. Top and sides of head are with symmetrical plates.
Distribution: The leatherback forages very widely, even extending into cold waters in search of jellyfish which it relishes. It nests on wide exposed beaches along the coasts of India, Sri Lanka, and islands in the Indian Ocean. It is rare along coastal Pakistan. There are occasional reports of it nesting on islands near the mouth of the Indus River (Das, 1991), while stranded individuals have been reported from along the Karachi coastal strip (Firdous, 1989).

Family Emydidae

Family of mud-turtles, represented by three genera in Pakistan.

Genus *Geoclemys* (Gray, 1821)
Alveolar surface of jaws broad with a median ridge.
A single species.

Geoclemys hamiltonii (Gray, 1821) (Plate 19)
(Yellow-spotted mud turtle: Chitra kachhuwa)
1831 *Emys hamiltonii* Gray Synops. Rept. 1:21.
Type locality: India.
Diagnosis:
1. Carapace elongate, strongly convex with sloping sides, with 3 interrupted keels, formed by a series of node-like prominences on neural and costal plates.
2. First vertebral longer than broad, second and third broader than long.
3. Plastron deeply notched posteriorly, truncate anteriorly; large auxiliary and inguinal plates.
4. Head massive, snout short, posterior part of head with symmetrical large scales.
5. Digits webbed to the base of claws, limbs with broad scales.
6. Seam contact formula: 1>5<7<9<11<.
7. Plastral formula: abd>fem>pect>gul>an>hum.

Male smaller than female with comparatively longer tail, arched plastron, tips of the anal plates are pointed outward in female.

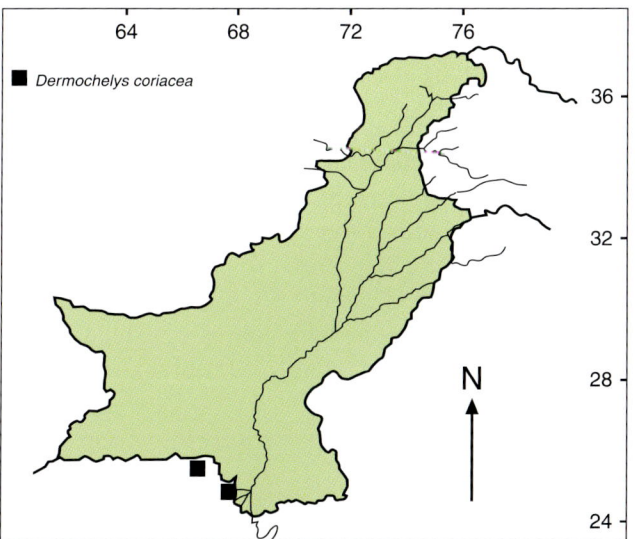

Plate 19. *Geoclemys hamiltonii.*

Length 25–36 cm, weight 2–4.6 kg.
Color: Dark shell with yellow streaks and triangular marks. Dark head spotted with yellow, neck gray with white spots; dark forelimbs spotted with white.
Natural history notes: It inhabits shallow ponds and pools, however, in drier areas of Rajasthan is known to occur in rivers. This turtle is reported to dig in riverbeds (Prakash, 1982). It is shy and crepuscular in habits; when cornered it opens its mouth and utters a low croak while retracting its neck. It rarely attempts to bite. It inhabits permanent ponds and lakes with ample vegetation. During drought it migrates to seepage pools along canals and rivers. Its massive head and strong jaws are an adaptation to crush the shells of molluscs, which form a major part of its diet. Larvae of dragonflies are also recorded from its diet. In captivity it accepts meat, fish, snails, crustaceans, and several types of insects; however, it does not feed on vegetable matter (Minton, 1966). It basks in winter; in drought it digs deep in the moist beds of ponds and rivers (Prakash, 1982).

It is a poorly known species. Most probably this species lays eggs twice a year, before and after monsoons (Das, 1991).
Distribution: The spotted pond turtle inhabits the flood plains of the Indus-Ganges-Brahmaputra. It has been reported from Nepal, Assam, Bihar, Jammu, Rajasthan, Uttar Pradesh, Western Bengal, and Jessore in Bangladesh. In Pakistan it has been recorded from Tunsa Barrage, Saidabad, Balloki Headworks in Punjab; Jacobabad, Tharparker, and Sehwan in Sind.

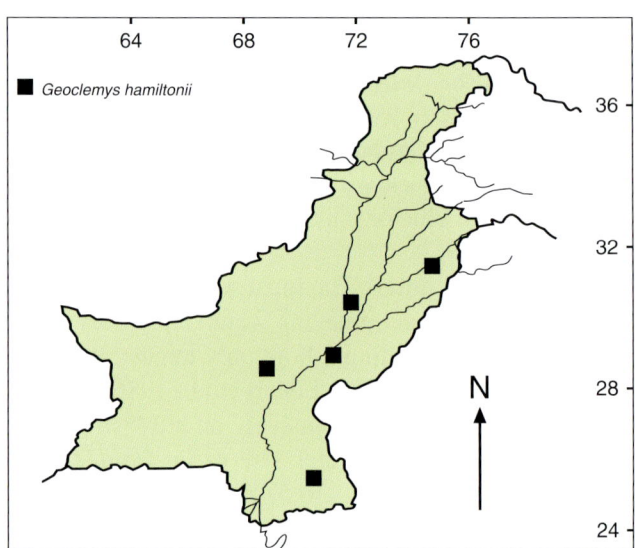

Genus *Hardella* Gray, 1870
Alveolar surface with 1–2 ridges; fourth vertebral not longer than broad, not longer than third.

A single species represented.

Hardella thurjii (Gray, 1870) (Plate 20)
(Common river turtle: Daryai kachhuwa)
1831 *Emys thurjii* Gray, Synops. Rept. 1:22.
Type locality: Indus River system.
Diagnosis:
1. The carapace is extensively in contact with plastron through strong long buttresses extending on to the neurals. An interrupted neural keel is present.
3. Plastron is produced posteriorly into a narrow lobe.
4. Marginals are serrated in adult.
5. Neurals are broader than long.

Chapter 4. Chelonians: Turtles and Tortoises

Plate 20. *Hardella thurjii.* (Photo courtesy of S.A. Minton).

6. Snout projecting, skin of its posterior half smooth.
7. Jaws serrated with broad triturating surface.
8. Limbs with narrow transverse rows of scales, digits are fully webbed.
9. Seam contact formula: 1<4>6>8>11<.
10. Plastral formula: abd>pect><fem>hum>an>gul.

Female three times bigger than male, reaches 15–23 cm in carapace length, while male is 10–15 cm long, with tail thick at its base. Female weighs 1.5–3 kg.

Color: Carapace dark brown with a grayish black vertebral keel; plastron yellow, scutes with large blackish blotches. Four yellowish orange stripes on sides of head and a similar stripe often present on forehead. Limbs brownish, edged with yellow.

Natural history notes: This turtle is thoroughly aquatic, frequenting pools, ponds, canals, rivers with slow currents, and lakes (Das, 1991). It has also been recorded from salt water. Often leeches and egg masses of aquatic animals are attached to its shell and other parts of its body. It is seldom seen out of water, and it does not bask in winter; however, during drought it wanders in search of water and buries itself in moist debris at the bottom of drying ponds.

The female matures at 320–350 mm carapace length. Follicular development occurs from July to September. Courtship occurs from April to July; eggs are laid from September to January. The clutch size is 8–13, eggs are ellipsoidal, measuring 40–56 by 28–36 mm, weighing 19–41 g; incubation period is 273 days (Basu, 1998).

The turtle is vegetarian, taking aquatic vegetation and fruits.

Distribution: It is a widely distributed turtle in the flood plains of the Brahmaputra, Ganges, and Indus Rivers. In Pakistan it has been recorded from localities in Sind and around Karachi.

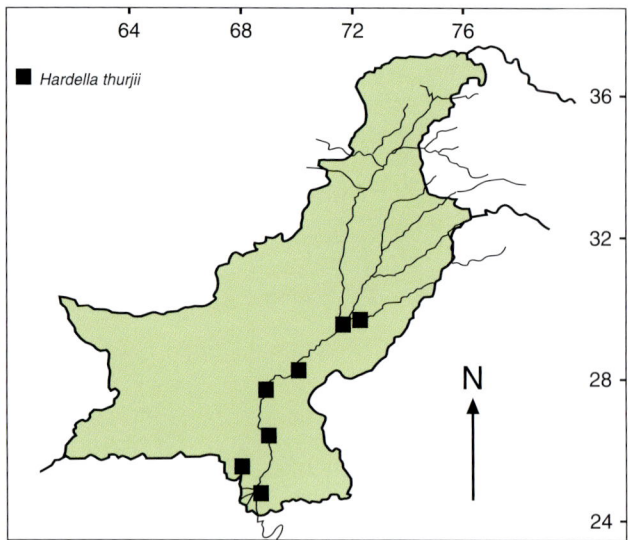

Genus *Kachuga* Gray, 1856

Alveolar surface of jaws with 1–2 ridges; fourth vertebral much longer than broad, longer than third.

Two species represented.

Kachuga smithii (Gray, 1863) (Plate 21)

(Brown river turtle: Bhoora daryai kachhuwa)
1863 *Batagur smithii* Gray, Proc. Zool. Soc. London 1863:253.

Type locality: Chenab River, Punjab, Pakistan.
Diagnosis:
1. Carapace depressed, oval, strongly arched, highest at midpoint, margin flared, spine feeble.
2. Vertebrals longer than wide, except second and fifth which are wider than long, third quadrangular to pentagonal, 24 marginals.
3. Plastron truncated anteriorly but notched posteriorly.
4. Forehead with large irregular scales.
5. Snout projected anterior to lower jaw, upper jaw serrated, anteriorly notched.
6. Seam contact formula: 1M4>6M8<10>.
7. Plastral formula:
   abd>fem>hum>pect>an>gul.
   Shell length 23–24 cm.

Male smaller than female, male has a longer tail with thick base.

Color: Iris pale green to blue. Carapace brown olive with black or dark brown vertebral stripe; a single, large black blotch on each plastral scute. Head olive, with a reddish brown blotch behind the eyes. Neck and limbs with yellowish blotches.

Natural history notes: It is the characteristic turtle of flowing waters; however, it has usually been found in lentic waters (Moll, 1987). It is a shy species, never attempting to bite. In winter, from early December to March, it becomes inactive, while in dry season it hibernates by burrowing into the drying mud (Minton, 1966).

Stomach contents contain remains of crayfish, crabs, and macrophytes (Das, 1991), rotten flesh, and plant matter (Parshad, 1914b), while Moll (1987) records plant matter and remains of prawn. Nesting season extends from April to mid-September (Gupta, 1987; Duda and Gupta, 1982). Seven to eight eggs are

Plate 21. *Kachuga smithii.* (Photo courtesy of S.A. Minton).

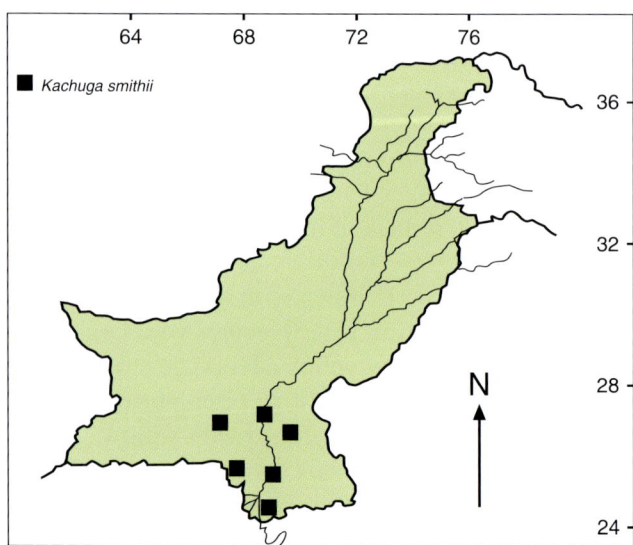

Chapter 4. Chelonians: Turtles and Tortoises

laid measuring 43–45 by 22–24 mm (Minton, 1966; Auffenberg and Ahmad, 1991a).
Distribution: Lower Sind and Indus Delta, Pakistan.

Kachuga tecta (Gray, 1831) (Plate 22)
(Sawback turtle: Ari-pusht daryai kachhuwa)
1831 *Emys tecta* Gray, Synops. Rept., 1:23.
Type locality: India.
Diagnosis:
1. The carapace is dome-shaped, oval with a distinct neural spiked keel, which is more prominent on third vertebral scute; fourth vertebral is longer than wide, flask-shaped (Figure 10).
2. Plastron is truncate anteriorly, notched posteriorly.
3. Snout pointed, posterior half of head with irregular scales.
4. Upper jaw serrated with concave alveolar surface.
5. Seam-contact formula: 1>4M 6M 8M 10M.
6. Plastral formula:
   abd>fem>an><hum>an>gul.
   Maximum length 23 cm, weight 1.35 kg.

Male is smaller, has longer tail which is thicker at its base, carapace is darker, iris red; while in female the carapace is paler and iris is pink.
Color: Carapace brownish, with a light brown, red, or orange stripe along first 3 vertebrals, marginals with narrow yellow border; plastron yellow or pink, with 2–4 black marks on plastral scutes; head with an orange or reddish crescent-shaped postocular mark,

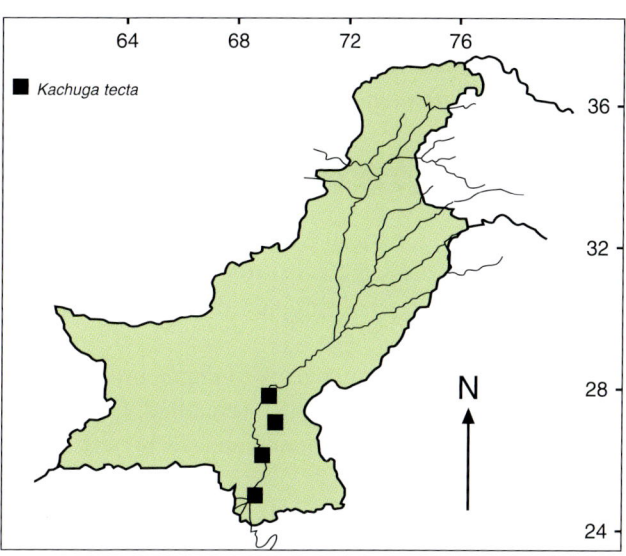

curves up from eyes to meet on forehead; neck dark with narrow yellow stripes (Mertens, 1969b).
Natural history notes: The sawback turtle frequents standing, slow moving water bodies and may invade large rivers. Basking turtles are a common sight during winter. It is non aggressive, and when handled refuses to bite (Minton, 1966). It is primarily vegetarian; however, it might occasionally be carnivorous (Das, 1991).

It breeds from March to May; 8–12 eggs are laid in a nest excavated in soft clayey soil, along banks of rivers, canals, large ponds, and lakes. Eggs measure to 37 by 21 mm, and are 10–11 grams in weight. Incubation period extends from 125 to 144 days.
Distribution: It is one of the most common turtles in the flood plains of the Ganges, Narmada, Brahmaputra, and Indus Rivers. It is widely distributed from Nepal, Bangladesh, and India throughout the Indus Valley in Pakistan. Mertens (1969a) and Minton (1966) report it from the River Indus.

Family Testudinidae

Tortoises are represented by two genera in Pakistan.

Genus *Agrionemys* Khozatsky and Mlynarsky, 1966
Forelimb with 4 claws; carapace with flat plates. Single species.

Plate 22. *Kachuga tecta.* (Photo courtesy of S.A. Minton).

Agrionemys horsfieldii **(Gray, 1844) (Plate 23)**
(Central Asian tortoise: Pahari kachoor)
1844 *Testudo horsfieldii* Gray, Cat. Tort. Croc. Brit. Mus.: 7.
Type locality: India, Afghanistan.
Diagnosis:
1. Shell oblong, widest at the level of hindlimbs, strongly arched, flattish at top.
2. Posterior margin of carapace distinctly flared and serrated.
3. Cervical scute long, narrow; neurals broader than long, 23 marginals, twelfth separated.
4. Plastron large, emarginated anteriorly, notched posteriorly.
5. Head moderately large, with scales on top and sides, single frontal and a large divided prefrontal.
6. Jaws not serrated, upper with 3 anterior cusps.
7. Fore- and hindlimbs with 4 large claws (Figure 8D).
8. Seam contact formula: 1>5<7<9<11<.
9. Plastral formula: Abd>gul><an>fem>pect.
Female larger, with smaller tail. Male with concave plastron.
Maximum size 16–22 cm, weight 1.5–3.0 kg.
Color: Carapace dark to light brown, without any definite pattern; plastron more or less diffusely and heavily clouded with dark brown; head and limbs dark gray to light yellowish; iris dark brown.

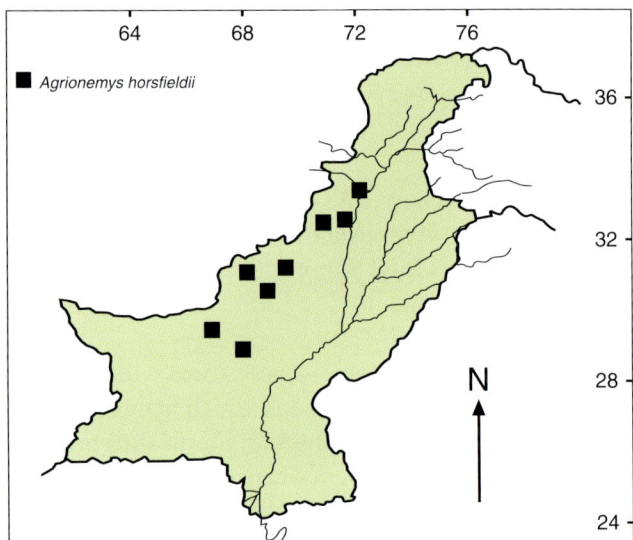

Natural history notes: This tortoise frequents hilly and rocky terrain with clayey and sandy deserts, steppes, ephemeral communities at foothills, and semi-steppes near oases. It is plentiful among grasses near fast-flowing streams and springs. It climbs very steep hillsides to reach rodent holes or caverns under overhanging stones (Roberts, 1975). It ranges in moderately arid rocky deserts, avoiding sandy and clay deserts. It is found between 1500 and 2000 m of elevation (Minton, 1966).

It comes out of hibernation by mid-March and follows a set schedule of activity during summer: it is active by midmorning till noon, becomes inactive at noon, when it rests in its den. In the afternoon, it is active for 2–3 hours, retiring early in the evening till the next day. This schedule is strictly followed throughout summer. A resting place is usually a shallow burrow beneath boulders, rocks, piles of bricks, or the roots of some plant. The turtle returns repeatedly to the site for rest, and shares it with several individuals. Winter hibernation lasts from late September to mid-March, while the hottest months, May and June, are spent mostly resting.

Diet includes leaves, fruits, and flowers. The tortoise is especially attracted to the red color of the tomato, and watermelon, and red flowers; however, it readily eats cucumber, cabbage, apple, and grasses.

Fights among breeding males are frequent, each trying to topple its adversary. Breeding season extends

Plate 23. *Agrionemys horsfieldii.*

from March to May, with eggs being laid from April to June. Two to three clutches are laid during one breeding season. Three to five eggs are laid in a clutch, eggs measure 41–50 by 26–35 mm, and weigh 21–22 g. Eggs have a hard brittle shell. The hatching period is from 126 to 128 days.

This tortoise has a gentle disposition. Sensing danger it withdraws its limbs and neck, and hisses loudly, but never attempts to bite.

Distribution: A wide-ranging species, it is found from the Caspian Sea eastward through Khazakistan, through Iran, Afghanistan, and western China. In Pakistan it occurs throughout northern and western Baluchistan and Waziristan.

Khozatsky and Mlynarski (1966) have placed this turtle in genus *Agrionemys*.

Genus *Geochelone* Fitzinger, 1835

Forelimb with 5 claws; carapace with umbovate plates, with star-shaped marks.

A single species.

Geochelone elegans Schopff, 1792 (Plate 24)

(Star tortoise: Satara kachoor)
1792 *Testudo elegans* Schopff, Hist. Test.:111
Type locality: India.

Diagnosis:
1. The shell is elongate, dome-shaped, flattened on sides, widest at the insertion of hindlimbs. Its posterior sides are not flared.
2. Neurals and costal laminae umbovate, nuchal absent.
3. Marginals 23, the twelfth about twice the size of others.
4. Plastron with a small shallow anterior and a broad posterior notch.
5. Edges of jaws finely serrated, lower jaw with a central cusp.
6. Head with small scales, tympanum distinct.
7. Forelimbs with 5 claws, hind with 4.
8. Seam-contact formula: 1>5<6>8>11<.
9. Plastral formula: abd>hum>gul>fem>pect><an.

Female exceeds male in body mass, male has longer tail, flattened shell top, concave plastron, and narrower postanal gap.

Maximum carapace length 20–28 cm, weight 8 kg.

Color: Carapace dark brown to black; umbo on neural and costal laminae yellow with a series of 5–10 yellow stripes radiating from it; plastron plates with similar pattern; head and limbs dull yellow, spotted, mottled with black.

Natural history notes: The star tortoise is mainly associated with dry scrub land, along the desert edge, tilled fields, grasslands, and thorn scrub wasteland. Diurnal in habits, it is active mostly early in the morning and late afternoon, resting during the warm hours; however, it remains active throughout the day during the rainy season. The star tortoise is primarily herbivorous, taking different fruits and vegetables. It likes red fruits, vegetables, and flowers, with a special appetite for tomato, watermelon, cabbage, grasses, and cucumber. In captivity it accepts dead rats, fresh water mussels, dead garden lizards, and snails. It drinks water frequently during summer (Das, 1991).

Breeding follows monsoons. Male fights are a common sight during the breeding period, each attempting to topple the other; mating pairs are a common sight during rains. Eggs are laid within 90 days of mating; clutch size ranges from 2 to 10 eggs, with diameters varying from 44 to 52 mm and weighing from 30 to 34 g. A female may lay 3 to 4 clutches in a year. Incubation period varies from 111 to 127 days.

Plate 24. *Geochelone elegans.* (Photo courtesy of S.A. Minton).

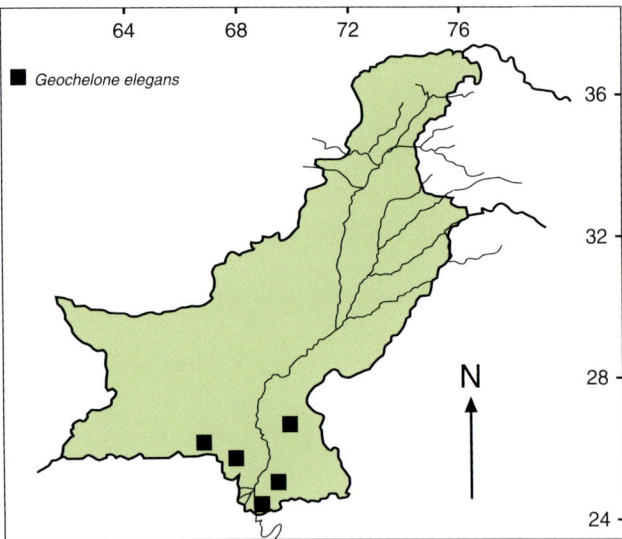

The star tortoise makes a desirable pet due to its vividly attractive pattern and gentle disposition. It is easy to feed with common fruits and vegetables (Minton, 1966).

Distribution: The star tortoise is widely distributed in peninsular India and Sri Lanka. In Pakistan it is found around Karachi, where it lives in semidomesticated conditions. It has also been recorded from Kutch and Nagar Parker, southeastern Sind.

Family Trionychidae

Family of turtles of rivers and lakes, represented by 3 genera in Pakistan.

Genus *Aspideretes* Hay, 1835

Head broad, massive, dorsally convex; nasal septum with lateral ridges; plastron without femoral valves; no marginal bones; 4 plastral callosities.

Two species are represented.

Aspideretes gangeticus (Cuvier, 1825) (Plate 25)
(Indian soft-shell: Prait)
1825 *Trionyx gangeticus* Cuvier, Rech. Ossem. Foss. Ed.3, 5, 2: [186] 203.
Type locality: Ganges River, India.
Diagnosis:
1. Carapace low, oval, vertebral region depressed in adult.
2. Costals 8 pairs, the eighth meeting at midline, first separated by preneural and first neural plate.
3. Plastron much shorter than carapace, with 5 callosities.
4. Plastral bones strongly rugose in adult.
5. Head broad, massive, eyes dorsolateral, placed well anterior on snout.
6. Nasal septum with lateral ridges.
7. Few transversally enlarged scales on forelimbs, which have 2 claws (Figure 8A).

Sexually dimorphic, male with longer and thicker tail, vent near the tail tip.

Maximum length 42–75 cm, weight 17 kg.

Color: Carapace dull olive to green, unicolor or with dark reticulation, plastron ivory, slightly grayish callosities; head and limbs sage green with dark mottling, head sometimes with black oblique stripes; iris greenish yellow flecked with black. Carapace of young bright green, with fine intricate black reticulation, 4 fairly distinct ocelli.

Hatchling has flat oval body of olive green color, with 3–6 ocelli on the carapace, forehead markings of 3–5 black lines persist; however, after 3 years markings on carapace fade away.

Natural history notes: This huge soft-shell turtle inhabits rivers, lakes, and large permanent ponds where it remains buried in the bottom gravel. It is extremely aggressive, shooting out its neck to give a very painful bite. Once it gets a firm grip, it holds on and retracts its neck in an attempt to tear off what is in its jaws (Das, 1991). Basking turtles are a common

Plate 25. *Aspideretes gangeticus.*

Chapter 4. Chelonians: Turtles and Tortoises

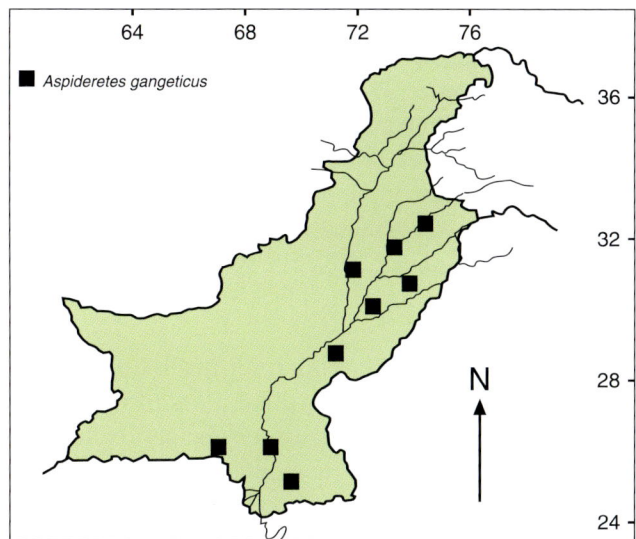

Plate 26. *Aspideretes hurum*.

sight along quite sandy banks of water bodies in Punjab, Pakistan, during winter.

This soft-shell is omnivorous and is known to attack waterbirds, fishes, amphibians, and even small mammals. It is a notorious carrion eater, feeding on dead human bodies thrown into the Ganges and Gumna Rivers. It is known to attack swimming people and drown them by pulling them to the bottom (Das, 1991). It attacks turtles of its own species and eats its own eggs. Due to its large size, it is a voracious feeder. Apart from animal diet, it eats pond vegetation; arthropods have also been recovered from its stomach (Das, 1991; Vyas and Patel, 1992).

Breeding coincides with monsoons (April–May). Breeding males and females migrate to comparatively shallower water, along rivers, lakes, and ponds. Breeding males become aggressive. Eggs are spherical with brittle hard shells. Clutch size varies from 20 to 40 eggs, diameter ranges from 33 to 34 mm, weight 24–28 g.

Distribution: This turtle is widespread in rivers, canals, and large lakes throughout the Indo-Pakistan subcontinent.

Aspideretes hurum (Gray, 1831) (Plate 26)
(Peacock soft-shell: Peeli prait)
1831 *Trionyx hurum* Gray, Synops. Rept. 1:47.
Type locality: Fatehgarh, Ganges, India.
Diagnosis:
1. Carapace low, oval, with several longitudinal rows of tubercles on its posterior side.
2. A preneural and neural between first pair of costals.
3. Eighth pair of costals meet at carapace midline.
4. Plastral callosities 5, large.
5. Head large, with downward-bent snout.

Sexually dimorphic, male with longer and thicker tail, anal aperture close to the tail tip.

Maximum carapace length 40–60 cm.

Color: Carapace olive with yellow rim. Head and limbs olive, head with dark reticulation, and yellow blotches on snout, postocular, and tympanic regions. Plastron light gray.

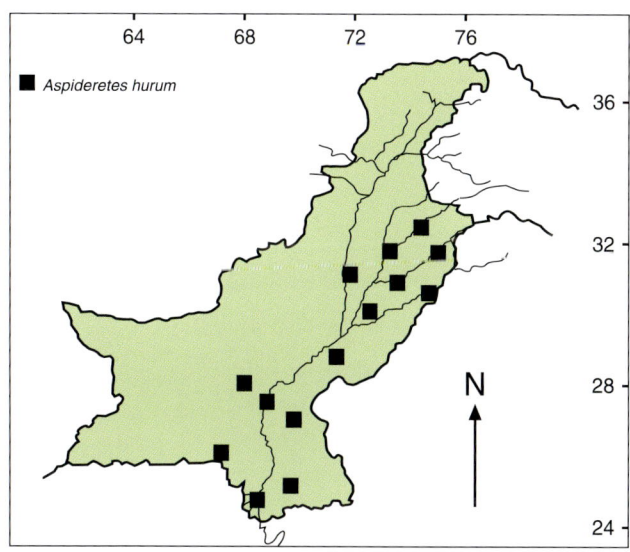

The juveniles are brightly colored, with 4–6 dark-centered, yellow-bordered ocelli on the disc.
Natural history notes: Little is known about the habits of this turtle. It inhabits rivers, lakes, and ponds. It is aggressive and bites with force; it is nocturnal and omnivorous. Nesting takes place in winter. Eggs are round with brittle shells (Das, 1991).
Distribution: This turtle has been recorded from throughout India and Bangladesh, and in Pakistan it is known from the Indus and its tributaries.

Genus *Chitra* Gray, 1844

Head long, narrow, flat above; nasal septum without lateral ridges; plastron without femoral valves; no marginal bones; 4 plastral callosities.

A single species is represented.

Chitra indica (Gray, 1831) (Plate 27)
(Narrow-head soft-shell: Tang-sar prait)
1831 *Trionyx indicus* Gray, Synops. Rept. 1:47.
Type locality: Fatehgarh, Ganges, India.
Diagnosis:
1. Depressed oval shell, with prominently pitted scutes.
2. Costals 8 pairs; a single neural between first pair of costals.
3. Plastral callosities 4.
4. Head extremely long and narrow, eyes situated close to the nostrils, proboscis short.
   Maximum size 35–115 cm, weight 120 kg.
   Male with longer tail which is thick at its base.

Color: Carapace dull olive or bluish gray, with a pattern of wavy reticulations, neck, and forelimbs with similar pattern. A V-shaped mark starts from nape and extends to the carapace. Juvenile with 4 ocelli on carapace or numerous elongated black spots; plastron cream or pale pink.
Natural history notes: This river turtle hunts in the more sandy shallow marginal parts of rivers. Its food consists of fish and crustaceans; snails have also been reported from its stomach. It is known to bury in bottom sand, keeping only its eyes exposed; as a passing fish approaches, it lunges and devours it. The dorsal pattern is an effective camouflage so that its prey is unable to detect it in natural environments. It feeds on fishes and molluscs (Das, 1991).

This turtle is bad tempered; when disturbed, it lunges, when handled, it may exude a foul smell.

Clutch size ranges from 60 to 120 eggs, each having a diameter of 34 mm, and weighing 18–20 g.
Distribution: This turtle is widely distributed in the river systems of the Oriental region, right from Thailand to Pakistan, where it is common in the Indus and its tributaries.

Plate 27. *Chitra indica.*

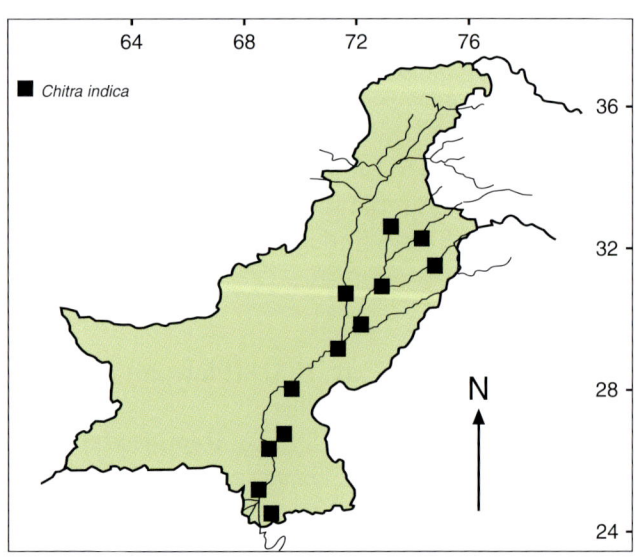

Genus *Lissemys* Smith, 1931

Plastron with cutaneous femoral valves; marginal bones present; 7 plastral callosities.

A single species is represented.

Lissemys punctata andersoni Webb, 1980 (Plate 28)
(Indian flapshell: Moonji kachhuwa)
1980 *Lissemys punctata andersoni* Webb, Bull. Mus. Hist. Nat. Paris 4, ser. 2, sec. A(2):547–557.
Type locality: Pondicherry, Coromandel Coast, India.
Diagnosis:
1. Shell low, oval, scutes with fine granulation and covered with soft skin.
2. Costals 8 pairs, last pair meet each other medially.
3. A small prenuchal present.
4. Callosities on plastron 7.
5. Plastron with hinged pectoral and femoral valves, under which hindlimbs are withdrawn (Figure 11).
6. Anterior half of the shell hinged, can close the shell completely from anterior side, with neck safely withdrawn inside shell.
7. Tail short, does not extend beyond shell.
8. Limbs with 3 claws, fingers webbed (Figure 8E).

Males smaller 14–20 cm, weight 4.5 kg than females 17–28 cm, weight 7 kg. Male has longer tail.
Color: Shell light olive brown, with scattered bright yellow round dashed spots, sometimes carapace is with a yellow reticulation and edged with pale yellow; plastron cream; head and limbs gray, with bright yellow spots on head and neck.

Natural history notes: This turtle is plentiful in the riparian system of the upper and lower Indus Valleys and frequents muddy ditches, lakes, and marshes with considerable marginal vegetation. It adapts to wide variations in environments and different habitats: salt marshes, rivers, ponds, lakes, streams, rice fields, seepage pools, streams, and often extends in sewer systems of metropolitan cities. This turtle is often seen moving from one drying pond to another in dry season. It is known to dig in moist soil to escape desiccation (Minton, 1966; Auffenberg, 1980a).

Individuals vary in temperament; some refuse to bite and allow free handling by withdrawing their limbs and neck into their shells, however, they occasionally flip their limbs backward injuring the careless holder with sharp claws, while others hiss loudly and lunge with open mouth, tearing off whatever comes into it, voiding foul-smelling secretions from gland openings on the sides of the tail and on the bridge between carapace and plastron.

It hibernates from November to February. In drought, May to June, it burrows into moist soil and hibernates. During monsoons it is a common sight in a wetland, moving from place to place.

Diet of the flapshell turtle includes adult frogs, tadpoles, fishes, crustaceans, fish larvae, carrion, water plants, bivalve molluscs, and snails. It is reported to

Plate 28. *Lissemys punctata andersoni.*

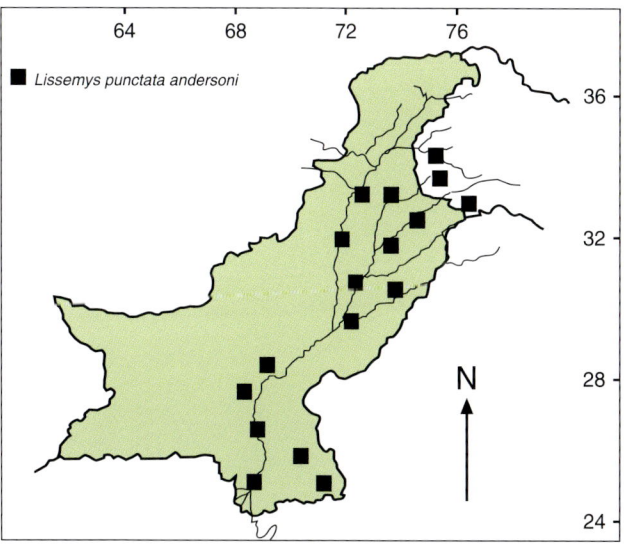

ambush fish. Seasonal changes in its dietary habits appear throughout the year. After emergence from hibernation in February, it is herbivorous, changing to carnivory throughout its active period, reverting to herbivory at the onset of hibernation in November. It is known to be cannibalistic also.

The male is smaller than the female. Mating is observed from May to June, preceded by complex courtship. The mating pair settles down at the bottom of the water for a long time. Eggs are laid in summer, however, in Myanmar it is reported to lay during fall (Smith, 1931). Clutch size varies from 6 to 14 eggs, which are spherical, measuring 24–30 mm in diameter. Incubation period varies from 30 to 40 days. During drought predation on young turtles is high by water birds, hawks, and jackals. Vultures are reported to feed on dead *Lissemys* turtles (Auffenberg, 1981).

Distribution: *Lissemys punctata andersoni* is widely distributed in Bangladesh, Nepal, India, and Pakistan.

Chapter 5
Crocodiles: Muggers and Gharials

Family Crocodylidae

Body large, heavy, often exceeds 2 m in length; hand with 5, foot with 4 digits, only inner 3 digits clawed; anal opening longitudinal; body covered with rows of sculptured scutes.

Snout broad, not more than twice as long as broad at the base; 19, 19 teeth in upper jaw. A single species of genus *Crocodylus* is represented.

Genus *Crocodylus* Laurenti, 1768
Snout long, flat, about 60% wider than long; postoccipital plates small, arranged in a transverse series, followed by 5 large nuchals.

Crocodylus palustris Lesson, 1831 (Plate 29)
(The mugger: Magar machh)
1831 *Crocodilus palustris* Lesson, Bull. Sci. Nat. Paris 25, 2:121.

Type locality: Plains of India.
Diagnosis:
1. Snout long, flat, about 60% wider than long.
2. Thecodont teeth in upper jaw 19, 15 in lower jaw, on each side, fourth and tenth are the largest.
3. Postoccipital plates small, arranged in a transverse series, followed by 5 large nuchals.
4. Neural plates from neck to the level of vent 18, each plate sculptured by several ridges.
5. Rectangular abdominal plates 16 across midbelly.
6. Caudal segments 33, each with a series of 4 ridged plates on dorsal side, distal half of tail with plates with a single high ridge.
7. Fingers webbed at base, toes completely webbed, emarginate; a serrated fringe along the outer side of limbs.

Maximum total size 3 m.

Color: Dorsum light brown, with darker mottling to almost uniformly black.
Natural history notes: The mugger inhabits rivers and lakes from plains up to 600 m of elevation. It is

Plate 29. *Crocodylus palustris.*

extremely shy; it does not like human company, and quickly takes to water. A basking mugger sits on the sandy bank of a river, during early and late winter, facing river with its wide open mouth. It is a silent swimmer, floating quietly on the surface of 2- to 3-m-deep water, with meager emergent vegetation. On land it rests on its belly, and drags it along; however, when running the belly is lifted clear off the ground. It excavates a 2- to 4-m-long and 60- to 90-cm-wide passage, like a burrow, with a spacious terminal chamber on the banks of lakes in the Sind Delta.

It is a silent hunter; all hunting is done in water. Once a prey is sighted in the shallows or at the edge of the water, the mugger nears it undetected from under water, suddenly lunging to catch a part of the prey's body in its strong jaws, and sinks with it into deep water. There the corpse is left to rot which makes its dismemberment easy. If there are several muggers around, the prey is torn into pieces on the spot. Usually the mugger feeds on fishes, frogs, turtles, varanids, birds, different mammals, snakes, and any other animal which can be easily crushed between its strong jaws and engulfed. The mugger also scavenges on kills made close to the water. It feeds on human corpses which are set afloat or are drowned. Often jewelry is recovered from the mugger's stomach from such feasts. The mugger usually swallows large stones which remain in its stomach, perhaps to facilitate crushing of hard parts of its prey (Simcox, 1906). There are reports of muggers attacking human beings and watering cattle, which is possible under unusual circumstances.

The mugger is bad tempered; when it is disturbed it hisses loudly, and lunges furiously, violently lashing its tail. Adults occasionally roar, repeating the call twice or thrice, which resembles the bellow of cattle. Short grunts are the typical form of communication between the hatchling and adult. The distress cry of the hatchling from within the egg compels assistance from nearby adults.

Muggers live in colonies. There is a dominant male to which other males of the locality submit by raising their heads to show the underside of their jaws. Breeding season extends from January to March. The breeding males and females locate each other by the secretions from their scent glands. Copulation (in water) lasts for about 10 minutes. The male half rides the female entwining his tail under hers;

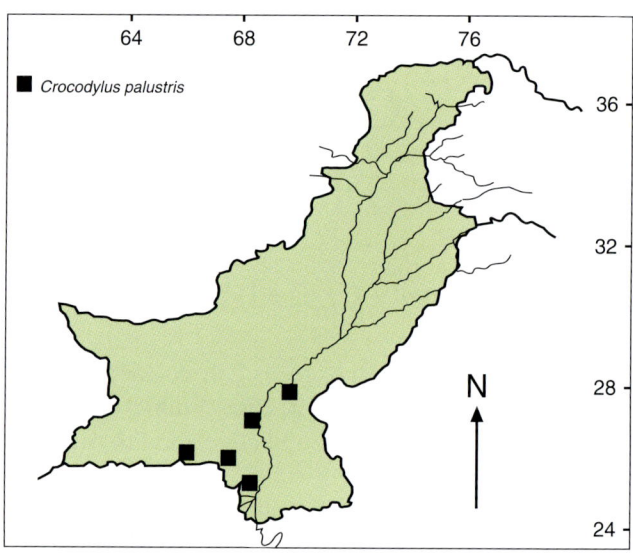

3–45 eggs, depending on the age and size of the female, measuring 71–74 by 46–50 mm, weighing 92–105 g, are laid in a 50-cm deep and 30-cm wide pitcher-shaped nest, which is excavated by the female, 2–500 m away from the water, on sandy earth or the gravel bank of the water body. The eggs are covered with vegetation and the female keeps a constant watch. Young are hatched within 4–6 weeks, depending on the warmth in the nest. At the time of hatching, young muggers utter characteristic calls. In

Table 5.1 Mugger Presence in Pakistan (Ahmed, 1985b)

Locality	Counted	Estimated
Sind		
Nara Canal	17	—
Lakes	19	120
Baluchistan		
Nari Nadi	3	10
Las Bela		
Titian Nadi	7	—
Hab River	14	—
Hab Dam	—	5
Makran Mirani Dam	11	?

Chapter 5. Crocodiles: Muggers and Gharials

answer, the mother, who is close by, removes the vegetation cover from the nest and helps the young in cracking the hard eggshells. Juveniles soon scramble out of the nest (Daniel, 1983).

Distribution: The mugger ranges from Assam through India, from Sri Lanka to westernmost Baluchistan and eastern Iran. In India it inhabits all river systems and their connecting streams and all sizable ponds. While in Pakistan it is reported to inhabit the lower Indus Valley (Mertens, 1969a, 1972). Small populations are on record in the Hub and Dasht Rivers, along the Makran coast. A small population of muggers is being maintained in a pond near Mangu (or mugger) Pir shrine, in the Karachi area (Moses, 1948; Minton, 1966).

Mugger presence in Pakistan, as reported by Ahmed (1985b), is given in Table 5.1.

Attempts by the Sind Wildlife Board to establish a breeding mugger colony in Haleji Lake, Sind have been successful. Similar attempts are being made to establish crocodile populations throughout the lower Sind Delta.

Family Gavialidae

Snout long, slender, at least 3 times as long as broad at base; 27–29 teeth on each side of the upper jaw. A single species is represented in Pakistan belonging to genus *Gavialis*.

Genus *Gavialis* Oppel, 1811

Snout long, narrow, its length 5 times its breadth; a pair of small postoccipital scutes.

Gavialis gangeticus (Gmelin, 1789) (Plate 30)
(Gharial: Gharial: Sassar)

1789 *Lacerta gangetica* Gmelin, partim, Linn. Syst. Nat. (13)1:1057.

Type locality: Senegal, Africa, and Ganges, India.

Diagnosis:
1. Snout long, narrow, its length 5 times its breadth.
2. Teeth 27–29 teeth in the upper, 25–26 in the lower jaw on each side, first 3 fitting into notches in upper jaw.

Plate 30. *Gavialis gangeticus.*

3. A continuous series of 21–22 transverse rows of plates from nuchal to level of vent.
4. A pair of small postoccipital scutes.
5. Fingers webbed.

Males longer than females. During breeding season the male develops a large hollow cartilaginous protuberance at tip of its snout, known as "ghara" from which the name "gharial" is derived.

Maximum length 6.5 m.

Color: Dark olive or brownish above, whitish or yellowish below. Young with spots or crossbars.

Natural history notes: Gharials are restricted to deep, fast-flowing rivers. About 100 years ago they were plentiful in the rivers of the Indo-Gangetic plain. Their range extends from throughout the Gangetic system to the Indus Delta. Decline in population of these big reptiles is due largely to uncontrolled all-season hunting for hides, meat, medicine, and sport. Moreover, loss of habitat due to human activity and encroachment has led to the present deteriorating conditions, which will quickly lead to the extinction of these fascinating reptiles.

The gharial is an agile swimmer; however, on land it is clumsy, propelling its body with laterally extended limbs and heavily dragging its body along. It rarely ventures far from water. Gharials groan when disturbed, and bellow and groan when in distress. Breeding males develop large lumpy excrescences known as *ghara* at the tips of their snouts, which are said to act as resonators for bellowing and hissing.

The gharial's food consists mainly of fishes and waterbirds; it rarely attacks dogs, goats, deer, etc., It is a known corpse-eater; humans are known to have been attacked, but gharials are not feared on this account by the local people. Like the mugger, it swallows stones. Fish caught are manipulated in long jaws then engulfed headfirst. In winter and summer, the gharial basks on sandy riverbanks, keeping its mouth wide open. At nightfall it slips into the water and goes hunting. The gharial uses its jaws to seize its prey, piercing the unfortunate animal with its long, sharp teeth, then with a flick of its head the prey is engulfed.

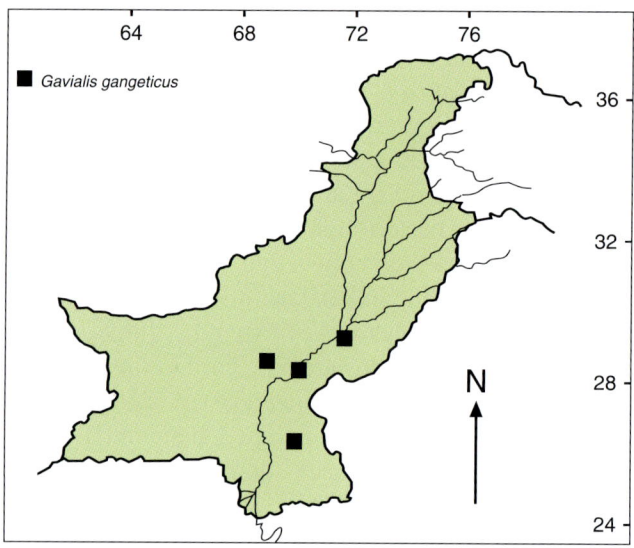

In the rainy season the gharial migrates into smaller streams, avoiding the main water channel. Usually, it remains afloat on the surface of the water, with the eyes and the tip of the snout exposed, observing activity along the bank. Mating season extends from December to January. Nesting occurs from late March to early April. Nests measuring 30–37 cm deep and 22 cm wide are excavated in sand along banks. Clutch size varies from 10 to 96 hard-shelled white eggs, measuring 85–90 by 65–70 mm. Incubation period 72–92 days. Hatchlings grunt from within the egg, and the mother, which is guarding the nest, helps the young in breaking shells of the eggs. Eggs and young are robbed by nest predators like rats, pigs, jackals, monitor lizards, large wading birds, etc.

Distribution: The gharial occurs in pockets throughout the Gangetic plains in India (Daniel, 1983). In Pakistan the gharial is extremely rare, if not absent. This is why nothing definite can be said about its distribution in Pakistan. Ahmed (1985b) reported 2–3 gharials between Sukkhur and Guddu Barrages in Sind. Similarly there is evidence in support of the existence of a population in the lower Indus, the East Nara in Sanghar District, Sind (Mertens, 1972; Ahmed, 1985a).

Chapter 6
Lizards

Lizards are represented by 8 families in Pakistan.

Family Agamidae

This family is represented by 6 genera.

Genus *Brachysaura* Blyth, 1856

Body dorsoventrally depressed; tympanum distinct, small, deeply sunk, caudal scales irregular; tail equals or is slightly less than body length; males without callous preanal scales.

A single species is represented.

Brachysaura minor (Hardwicke and Gray, 1827) (Plate 31)

(Short-tail ground agama: Dum-ktta kirla)
1827 *Agama minor* Hardwicke and Gray, Zool. Jour. 3:218.

Type locality: Chittagong, Bangladesh.
Diagnosis:
1. Habitus stout, head enlarged, body feebly depressed.
2. Dorsal head scales large, unequal, strongly keeled or tubercular.
3. Canthus and superciliary sharp edged.
4. Groups of spines on ear 2, diameter of tympanum almost half that of the orbit.
5. Supra- and infralabials 11 to 15.
6. Body with large, strongly imbricate keeled, often mucronate scales, pointing backward and upward.
7. Ventrals smaller than dorsals, more or less keeled.
8. Scalesaround midbody 48–58.
9. A denticulate nuchal and dorsal crest.
10. Tail shorter than body, round, with heterogeneous keeled scales.
    Snout-vent length 94–99 mm, tail 92–98 mm.

Plate 31. *Brachysaura minor.*

Color: Yellowish brown dorsum, with 3 middorsal rows of dark brown light-edged blotches, those of median row larger and rhomboidal. A white streak from nape, bifurcating behind and another from eye to the angle of mouth; limbs with dark crossbars; ventrum yellowish white, throat speckled with gray.

Juvenile pinkish brown dorsal blotches edged with light pink.

Natural history notes: The short-tail ground agama has been collected from hard barren desert and desolate areas. It is crepuscular and diurnal in habits. Its movements are sluggish; it does not attempt to escape, rather it flattens itself against the ground and does not move. In the badlands it lives in burrows close to the roots of thorny bushes. When caught it emits short squeaks. It feeds on leaves and flowers; moreover, insects are also included in its diet.

The female is larger than the male and more brilliantly colored during breeding season, which extends from April to June; 4–6 hard-shelled, white eggs are laid in burrows.

Distribution: The short-tail ground agama is rare. It is wide-ranging in the Indo-Gangetic plains. It extends from Bangladesh through the Central and United Provinces of India (Smith, 1935). Westward it extends into the upper and lower Indus Valleys in Pakistan, where it is rare and spotty in distribution. Its definite records are from Sind and Punjab, District Jhang (Mertens, 1974; Khan and Mirza, 1977; Perveen and Auffenberg, 1988).

Genus *Calotes* Cuvier, 1817

Body laterally compressed; a distinct median dorsal row of pointed elevated scales, extending to tail.

A single species is represented.

Calotes versicolor versicolor (Daudin, 1802)
(Plate 32)
(Common tree lizard: Girgit)
1802 *Agama versicolor* Daudin, Hist. Nat. Rept. 3:395.
Type locality: India.
Diagnosis:
1. Forehead concave. Head with unequal, smooth, or feebly keeled scales.
2. A pair of well separated supraorbital spines.
3. Diameter of tympanum half or less than half that of the orbit.
4. Body compressed.
5. Dorsals 35–52 around midbody, keeled, all pointing backward and upward, larger than ventrals. Ventrals keeled.
6. Median row of dorsals elongated, sharp tipped, forming dorsal crest which extends from nuchal to the level of vent, progressively decreasing in size backward.

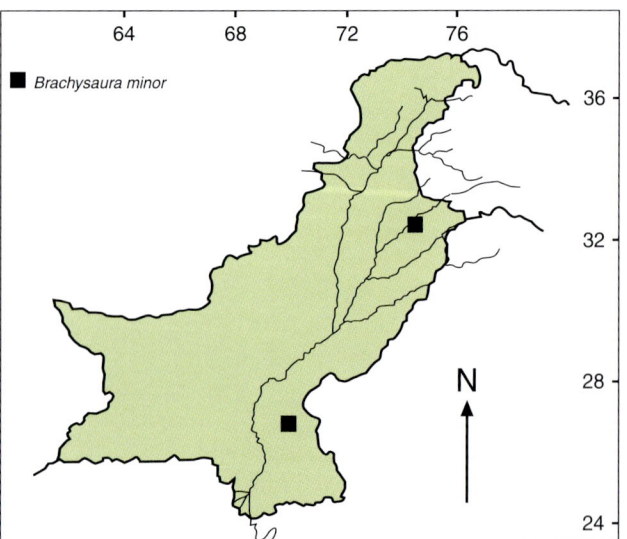

Plate 32. *Calotes versicolor*.

Chapter 6. Lizards

7. Third and fourth fingers nearly equal, fourth toe longer than third. Sexually dimorphic, males are longer in snout-vent and tail length, have a larger body mass, a prominent gular region, cheeks and longer dorsal spines; develop bright color pattern at the beginning of rainy season.

Snout-vent length 135–140 mm, tail 338–350 mm.

Color: Uniformly light brown or grayish dorsum, with distinct dark brown transverse spots or bars, or variegated with dark brown. Dark streaks radiate from eyes. Juveniles and females often with a pair of dorsolateral stripes on body dorsum, tail with light and dark annuli, ventrum dirty white. Adult male is usually more or less uniform in color with greenish tinge, throat is dark-barred (Smith, 1935).

Natural history notes: The tree lizard is arboreal and diurnal. It inhabits forests and any situation with trees and shrubs. It invades different biotopes from dry deserts to thick forests throughout the Indo-Pakistan subcontinent and Southeast Asia, from the plains to about 2000 m above sea level (Auffenberg and Rehman, 1993). It is common in gardens, often extending into backyard plantations in cities and towns, where it invades holes and crevices in compound walls and the undergrowth. In the desert it inhabits oases where it lives on trees and thorny bushes (Blanford, 1876a). It abounds in roadside trees of *Dalbergia sisso* and *Acacia sp.* It is an agile climber; it scrambles into branches, jumping from one branch to another. Its long toes and long prehensile tail help the lizard during its movements on trees. It dodges its predators by pressing its body against the tree trunk, so that it remains invisible. It changes its position by sliding around the tree trunk according to the movements of the predator. The lizard has the remarkable ability to change its body color to camouflage itself according to its surroundings, and it sometimes becomes immobile to deceive its predators. It basks during winter on exposed branches. It feeds on tree-visiting insects; however, small birds, nestlings, frogs, and small mammals are reported to be in its diet. It is usually observed feeding on flower-visiting insects in backyard gardens.

Breeding season extends from April to September. The head and neck of the breeding male becomes thick and brightly colored. Its sole territory is a particular tree, to which it remains confined. Several females are escorted into the branches above, while the male stays at the tree trunk, defending its harem of females from rival males who are on the lookout to seduce the females. The defending male continuously threatens other males nearby by head bobbing while slidingaround the tree trunk, often fighting off the rival males while trying to get the females to stay on the branches above. The restrained females scramble in the branches, with some escaping to the neighboring trees by climbing into the intertwining branches to the nearby harem. Mating takes place in the branches of the tree.

Breeding males often cross roads by clumsily walking in search of new territories and are crushed under passing traffic. Mortality on roads involves males exclusively which are greater in number during breeding season than otherwise, since non breeding males are fast runners and seldom cross roads (pers. obs.).

A gravid female uses her forelimbs to dig an 8- to 10-cm-deep pit in soft soil among vegetation; 11 to 23 soft leathery white-shelled eggs, which vary from 12 to 23 mm in size, are laid in the pit which is then covered with earth and levelled. Incubation period varies from 37 to 47 days. Hatchlings are observed from August to October. The juveniles scramble into low bushes, undergrowth, and the leaf litter. They feed on soft-bodied insects and worms; they reach the breeding size after 16–20 months (Minton, 1966; Daniel, 1983).

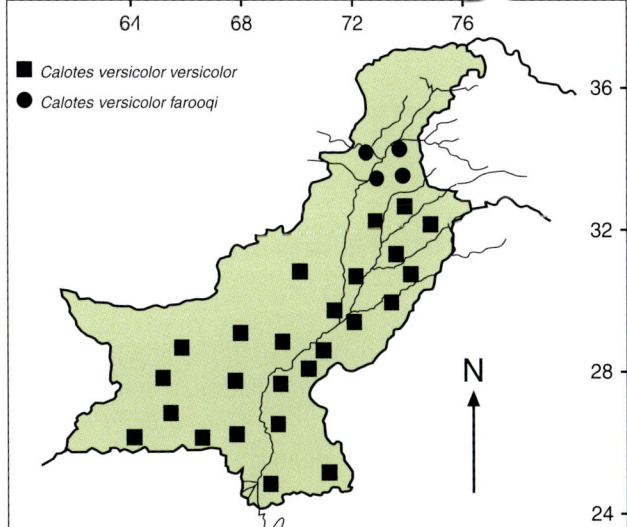

Taxonomic notes: Auffenberg and Rehman (1993) distinguished the alpine population of this lizard as a new race, *Calotes versicolor nigrigularis*, on the basis of its short head, lateral body scales pointing upward and backward, and absence of a prescapular fold or pit. Later they designated it as *Calotes versicolor farooqi* (1995), since the former name was preoccupied.

Distribution: The common tree lizard is most widely distributed in Southeast Asia, from Sumatra to southern China, throughout India, Sri Lanka, and Pakistan, and extends into Iran and southern Afghanistan. In deserticolous habitat the lizard is confined to oases and vegetation growing along the water courses.

Calotes versicolor farooqi Auffenberg and Rehman, 1995

1995 *Calotes versicolor farooqi* Auffenberg and Rehman, Asiat. Herpetol. Res. 6:27.
Type locality: Alpine Punjab, Pakistan
Diagnosis:
Distinguished on the basis of:
1. Comparatively short head.
2. Lateral body scales pointing upward and backward.
3. Prescapular fold or pit absent.

Distribution: Alpine Punjab, Pakistan.

Genus *Japalura* Gray, 1853

Body laterally compressed; an indistinct row of elevated scales from head to middorsum of body.

A single species is represented.

Japalura kumaonensis (Annandale, 1907) (Plate 33)
(Kumaon agama: Kumaon kirail)
1907 *Acanthosaura kumaonensis* Annandale, Rec. Ind. Mus. 1:152.
Type locality: Naini Tal, western Himalayas, India.
Diagnosis:
1. Body compressed.
2. Dorsals heterogeneous.
3. Dorsal crest a mere denticulation, reaches to the midbody.
4. No preanal or femoral large scales.
5. Supralabials 6.
6. Fourth toe as long as tibia.
    Snout-vent length 60–63 mm, tail 123–127 mm.

Plate 33. *Japalura kumaonensis*.

Color: Body gray to dark brown, with a median row of triangular dark markings edged with white.
Natural history notes: This agama is characteristic of low hilly tracts, with moderate vegetation in mesic habitat. It is diurnal and feeds on arthropods and their larvae. It lives under stony ledges and slabs with grasses and thorny bushes around.

Breeding period extends from late March to June, 4–6 eggs are laid under a stone. Juveniles are seen by late April.
Distribution: This highland agama has recently been collected from Hazara District in eastern NWFP,

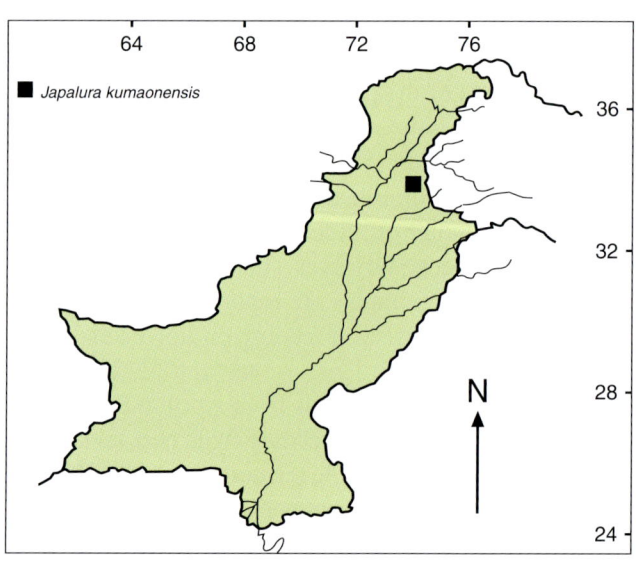

Chapter 6. Lizards

Pakistan. Its range extends to Kumaon District in western Himalayas, India.

Genus *Laudakia* Gray, 1845

Body dorsoventrally depressed; tympanum distinct, large, superficial; fifth toe extends beyond second; caudal scales in distinct annuli.

Thirteen species are represented.

Laudakia agrorensis (Stoliczka, 1872) (Plates 34, 35, 36)

(Agrore Valley agama: Agror wadi kirla)
1872 *Stellio agrorensis* Stoliczka, Proc. Asiatic Soc. Bengal, Calcutta July 1872:128.
Type locality: Sussel Pass, Agrore Valley, Hazara District, NWFP, Pakistan.
Diagnosis:
1. Head with keeled scales.
2. Numerous spinose scales on sides of head and neck.
3. Median dorsals distinctly larger than ventrals and strongly keeled, arranged in 8–12 longitudinal rows, an intervening vertebral series of small scales may be present, which extend to the occiput.
4. Flanks of body with numerous enlarged strongly keeled scales.
5. In addition to the callose preanal scales an oblong patch of large callose scales is present along the flanks of the body.
6. Hindlimb extends to eye or tip of snout.
7. Scales around the base of tail 30–40.
Snout-vent length 105–110 mm, tail 245–250 mm.

Color: Dorsum dark olive, spotted and reticulated with dark spots usually arranged in longitudinal series; head paler, ventrum uniformly whitish, throat and chest heavily marbled with dark blue. Juveniles have variegated dorsum spotted with black and pale yellow, or there are three longitudinal yellow dorsal stripes, the median extending onto the tail; throat is reticulated with dark while rest of the ventrum is dirty white.

Natural history notes: The Agrore Valley agama is very common around Ooghi Fort, in District Manshera, NWFP, Pakistan. It lives in crevices

Plate 34. *Laudakia agrorensis:* male.

Plate 35. *Laudakia agrorensis:* female.

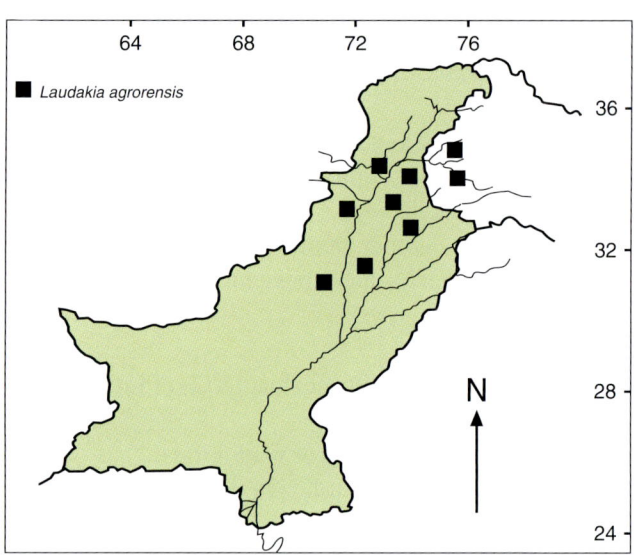

among rock blocks. During March and April it is active, feeding on insects and caterpillars. Mating takes place during April and May. Juveniles are seen by early May. Chasing males are observed till August, indicating that its breeding period is quite extended. Clutch size varies from 6 to 8 eggs, and most probably more than one clutch is laid in a breeding season.

The lizard hibernates from November to late February.

Distribution: The Agrore Valley agama is widely distributed in Ooghi Valley, around Manshera and Abbottabad, extending down into the Jhelum Valley, and northward into Chitral.

Plate 36. *Laudakia agrorensis:* juveniles.

Laudakia badakhshana **(Anderson and Leviton, 1969)**
(Badkhshan rock agama: Badkhshan kirla)
1969 *Agama badakhshana* Anderson and Leviton, Proc. Calif. Acad. Sci., 4th Ser. 37(2):33.
Type locality: Mazar-i-Sharif, northern Afghanistan.
Diagnosis:
1. Head scales smooth.
2. Scales around the base of the tail 19–25, caudals arranged in annuli.
3. Middorsals enlarged, smooth or weakly keeled, distinctly larger than ventrals.
4. A patch of large mucronate scales on flanks.
5. Scales on thigh large, strongly keeled.
6. A large patch of callose abdominal scales. Snout-vent length 65–70 mm, tail 99–108 mm.

Color: Dorsum olive gray, enlarged dorsals uniformly gray, bordered by longitudinal rows of irregular dark spots; dark-edged white ocelli are arranged in oblique rows on back, flanks mottled with dark gray. Tail with small dark spots arranged as cross bars. Ventrum grayish white, limbs marbled with gray, digits barred.
Natural history notes: The lizard inhabits rocks between 450 and 2400 m of elevation. It has been collected from clay loess, slopes, and plateau rock structures, dry water courses, and in dry mountain habitat.
Distribution: This species was originally described from northern Afghanistan. In Pakistan it has been recorded from Sost and Gulmit, near Khunjrab Pass, in northeastern Pakistan.

Chapter 6. Lizards

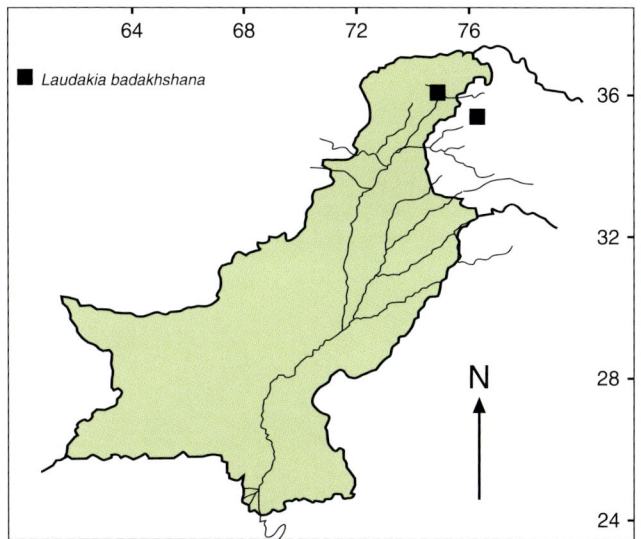

Laudakia caucasia (Eichwald, 1831) (Plate 37)
(Caucasian rock agama: Kohkaf ka kirla)
1831 *Stellio caucasia* Eichwald, Zool. Spec. 3:20.
Type locality: Tiflis and Baku, Caucasus.
Diagnosis:
1. Top of head with heterogeneous scales, smallest on the upper eyelid. Posterior sides of head and neck with numerous groups of large spinose tubercles, rest of neck scales very small.
2. Supralabials 12 to 16.
3. Median dorsals more or less hexagonal, imbricate, smooth or obtusely keeled, 6–10 across midback, abruptly separated from very small dorsolaterals.
4. Lateral scales of body intermixed with large squarish spinose scales and a large patch on flanks is usually present.
5. Ventrals smooth, smaller than median dorsals.
6. Gulars smaller than ventrals, no gular sac.
7. Limbs strong, hind reaches to the ear or eye.
8. Tail depressed at base, oval in cross-section, covered with large spinose scales arranged in whorls, 2 whorls to a segment.

Male with 4–5 rows of callose preanals, and an elongated midbelly patch.

Snout-vent length 143–145 mm, tail 185–205 mm.

Color: Olivaceus or yellowish brown above, with yellow or black markings; body dorsum with pale yellow spots bordered with black, spots may coalesce into a median row of crossbars, more vivid in anterior half of body. Head uniformly yellow or spotted with black. Tail yellowish or olivaceus, with or without dark annuli, yellowish below. Throat heavily marbled with dark blue or black; belly and under surface of limbs dark blue.

Plate 37. *Laudakia caucasia.* (Photo courtesy of S.A. Minton).

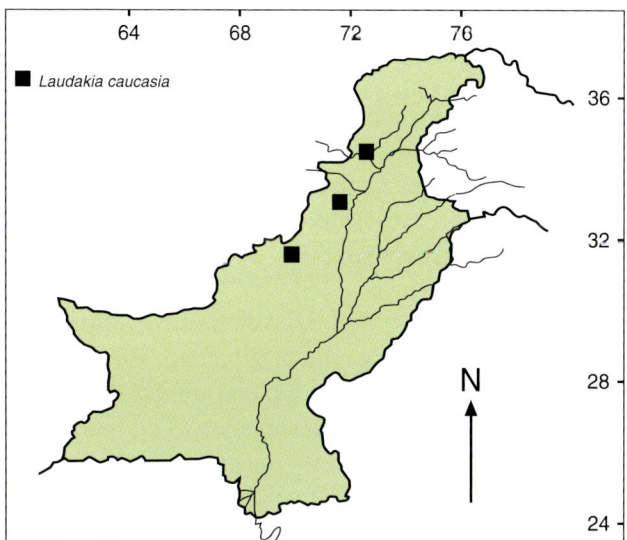

Young grayish or olivaceus above with jet black spots, which may join each other to form narrow crossbars.

Natural history notes: The Caucasian agama is rock-dwelling. It frequents boulders in stream beds, living in crevices among rock blocks. It comes out of hibernation by early March. As environments get warmer, its midday activity period is progressively increased. By June at midday, the rocks are very warm so the lizard takes to cooler parts of the rocks and is rarely seen.

It is vegetarian, occasionally feeding on insects and small lizards.

Breeding period extends from May to June, when 6–14 eggs are laid under rocks. Young are seen from late May to August.

Distribution: This agama has been recorded from Caucasus extending eastward to Waziristan and northern Baluchistan, between 1800 and 3000 m.

Laudakia fusca (Blanford, 1876) (Plate 38)

(Yellow-head rock agama: Pela kirla)
1876 *Stellio nupta fusca* Blanford, East. Pers. Bound. Commis. 1870–1872. London, Vol. 2:319.

Type locality: Kalagan and Jalk, Baluchistan, Iran.
Diagnosis:
1. Circumauditory spiny excrescences are numerous and prominent.
2. Supralabials 13 to 16, infralabials 14 to 16.
3. Enlarged mid-dorsals in 13–16 rows are strongly mucronate.
4. Tail annuli composed of 13–22 scales.
Snout-vent length 159–162 mm, tail 289–300 mm.

Color: Head of male bright yellow, body, limbs and tail dark brown to sooty black above and below, with few yellow speckles on back. Callose abdominal and preanal scales are reddish brown. Females light brown to reddish, with slight speckles of yellow. Tail distally dark, ventrum yellowish.

Juvenile grayish yellow, with narrow yellow irregular crossbars, or a reticulate pattern. Head black with a large yellow mark on top and similar spots on temporal, occiput, neck, and shoulders. Tail banded dark.

Natural history notes: The yellow-head agama is shy, frequenting vertical rocks along ravines with a lot of holes and crevices. When disturbed it takes to the

Plate 38. *Laudakia fusca.* (Photo courtesy of S.A. Minton).

rocks, waits for some time observing the intruder from behind the rocks, then scrambles about freely. It forages far and wide in nearby low-lying strips of land with grassy growths and occasional bushes.

They are active by early March, with males chasing females on the rocks. The males display by head bobbing on rock peaks, while guarding their territory for several days. Trespassers are chased, often resulting in fights. Breeding activity lasts till July. Juveniles are observed from April to early August.

This agama remains active almost throughout the year. Its activity period is curtailed as winter sets in, becoming active in the warmer hours of day and retreating earlier.

Distribution: The yellow-head agama is widely distributed in southern Iran, Baluchistan, and southwestern Sind, Pakistan, up to an elevation of 1800 m. I have seen it in the Kalabag area in northwestern Punjab, Pakistan.

Chapter 6. Lizards

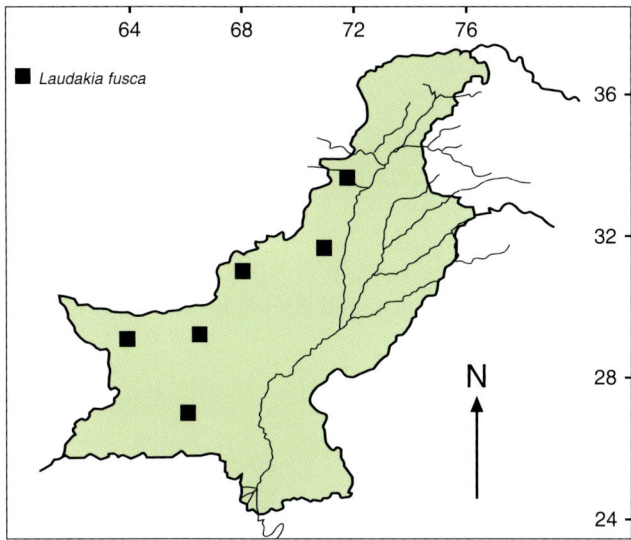

***Laudakia himalayana* (Steindachner, 1869) (Plate 39)**
(Himalayan agama: Himalayaie kirla)
1869 *Stellio himalayanus* Steindachner, Reise Novara, Rept. 1867: 22.

Type locality: Lei (Leh) and Kargil, Ladakh Province, Kashmir.

Diagnosis:
1. Head scales heterogeneous, convex, smooth or keeled, largest on snout.
2. Posterior side of head and neck with small spinose scales, periauricular spines arranged in rows while those on neck in groups, neck scales granular.
3. Supralabials 10–12.
4. Median dorsals subequal, round, imbricate, feebly keeled, 8–14 across middorsum.
5. Laterals small, with no enlarged scales, adult male may have a patch of large scales at midventrum of body.
6. The skin is loose on sides of neck and body, ventrals smooth, smaller than median dorsals. Gulars smaller than ventrals, no gular sac.
7. Hindlimb reaches to the ear or eye.
8. Tail depressed at base, oval in cross-section, 40 or more strongly keeled subequal scales around its thickest part.

Plate 39. *Laudakia himalayana:* subadults and juveniles.

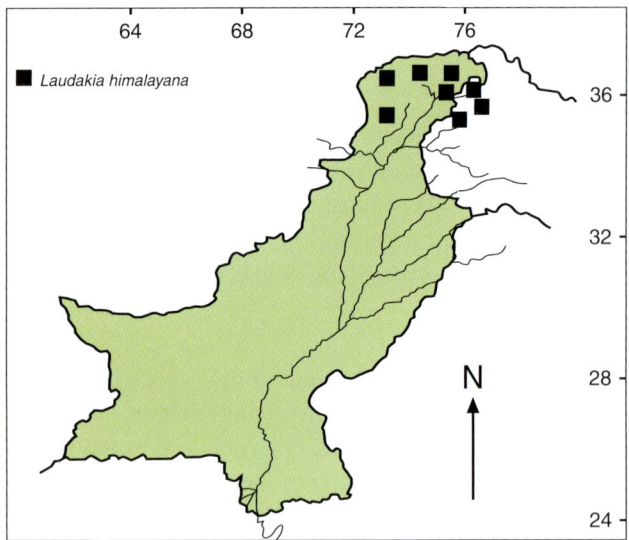

9. Male and female have 2–3 rows of callose preanal scales, no abdominal patch.
Snout-vent length 100–105 mm, tail 139–148 mm.

Color: Dorsum olive, marbled with black, round light spots may form a network; sometimes a dorsolateral series of black spots or spots may join to form sinuous groups. Tail barred with black, greenish white below, throat of male spotted or marbled with gray.

Natural history notes: The Himalayan agama has been recorded from remote areas of northern Pakistan, between 3000 and 3200 m of elevation. It is active by April, and becomes common on rocks by May-June. It feeds on vegetation and occasionally on lizards and arthropods.

It breeds during May and June; 12–16 eggs with diameters ranging from 19 to 22 mm, are laid under rocks which are largely exposed to sunlight. Juveniles are seen by July–August.

Distribution: The Himalayan agama extends from the western Himalayas to Tajikistan. In Pakistan it has been recorded from Gilgit and Chitral.

Laudakia lirata (Blanford, 1874)
(Yellow head spotted rock agama)
1874 *Stellio liratus* Blanford, Ann. Mag. Nat. Hist., London, (4) 13:453.
Type locality: Gedrosia, Baluchistan (=Saman, Dasht Province, Baluchistan).

Diagnosis:
Pholidosis like *Laudakia melanura*, however, distinguished from it by:
1. Bright, sandy head and yellow-spotted anterior part of body.
2. Dorsals are strongly keeled
Snout-vent length 121–135, tail 225–285 mm.

Distribution: Collected from Kirthar range, Hab Chauki, Mekran Coast, and Astola Island.

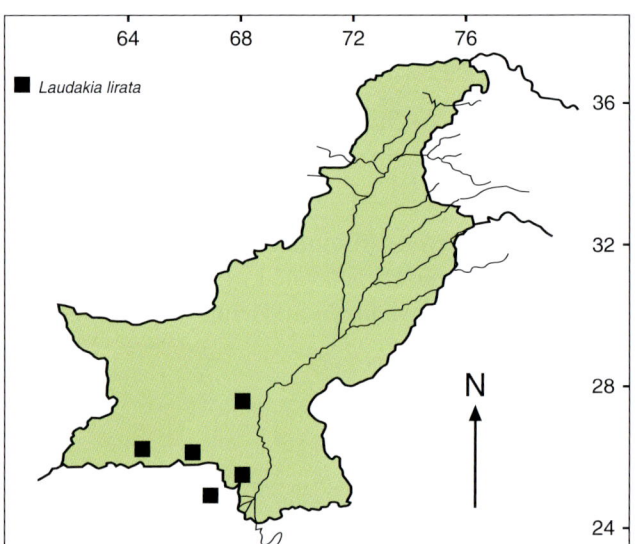

Laudakia melanura Blyth, 1854 (Plates 40A, B)
(Black rock agama: Siaah kirla)
1854 *Laudakia melanura* Blyth, J. Asiat. Soc. Bengal, Calcutta 23:738.
Type locality: Salt Range, Punjab, Pakistan.
Diagnosis:
1. Naris piercing canthus.
2. Head with large, subequal, smooth, or obtusely keeled scales.
3. Supralabials 12–16.
4. Posterior lateral sides of head with groups of short spines.
5. Neck scales granular, except those of median line which form a low crest.
6. Ventrals smooth, not half as large as median dorsals.
7. Gulars smaller than ventrals. A gular sac is absent.

Chapter 6. Lizards

Plate 40A. *Laudakia melanura:* dorsal view.

Plate 40B. *Laudakia melanura:* ventral view.

8. Tail depressed, oval, covered at base with large subequal scales, dorsal and lateral caudal scales strongly mucronate, ventral caudals are flat. Snout-vent length 140–145 mm, tail 280–295 mm.

Color: In life, dorsum brownish olive with smaller, more-or-less confluent dark spots, interspersed with large, bright-yellow spots; a dark reticulation on head; tail pale yellowish at base, but for the greater part of its length entirely black. Body dorsum in young is olive, with a yellowish white ventrum, head, chin, chest reticulated with black; neck, body, limbs, and base of tail with numerous small black and yellow spots; eyelids and superciliary ridges yellow; tail dusky black at its tip.

Natural history notes: The black rock agama lives in rock crevices. It feeds on vegetation, occasionally taking insects. It comes out of hibernation by March; as temperatures go up its activity period is increased. Among males, breeding scrambles on the rocks are quite common by late March and continue to May; 10–15 eggs are laid under rock and juveniles are active by June.

Distribution: The black rock agama has been recorded from rocks of moderate elevation from eastern Iran, Baluchistan, extending into the Salt Range, Punjab, and western hilly tracts of Sind, Pakistan.

Taxonomic Notes: Blyth (1854) described a melanistic agama, *Laudakia melanura,* from the Salt Range, Punjab, while Blanford's (1874a) yellow headed *Stellio liratus* is from Baluchistan. Apart from color differences both agamids are inseparable on pholidosic counts (Minton, 1966: 92). However, Mertens (1969a:33–34) regarded these agamids as separate races, noting that during mating season males develop yellow color on head and neck, which extends on body dorsum as light spottings (Fig. 8), however, females remain melanistic with dorsal yellow spottings (Fig. 9).

The morphology of *Agama m. nasiri* almost fits the description of *Laudakia m. lirata* (Blanford, 1874), restricting *Laudakia melanura melanura* to the Salt Range, Punjab, and placing Sind and Baluchistan

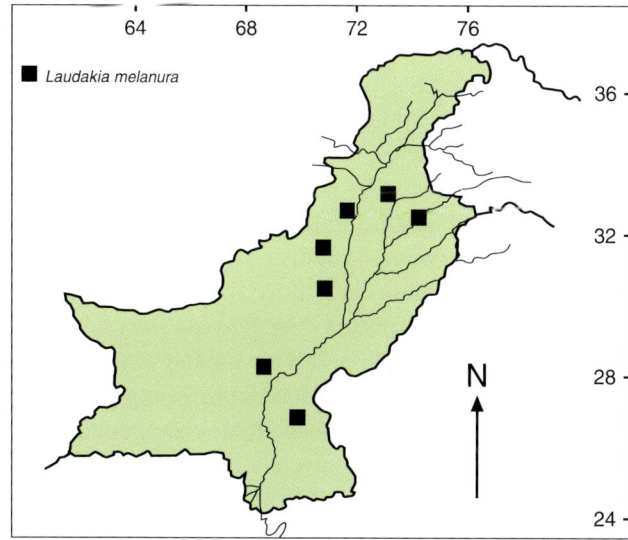

material in *L. m. nasiri*. Further collection and examination of whole series will solve the issue of validity of these races.

Scales around midbody 120–150. Male with a large patch (4–5 rows) of callose preanal and another at mid-abdomen .... *Laudakia melanura melanura*

Number of scales around midbody more than 150. No patch of spinose scales on flanks ....*Laudakia melanura nasiri*

Laudakia melanura melanura Blyth, 1854

(Black rock agama: Siaah kirla)
1854 *Laudakia melanura* Blyth, J. Asiat. Soc. Bengal, Calcutta 23:738.
Type locality: Salt Range, Punjab, Pakistan.
Diagnosis:
1. Naris piercing canthus.
2. Head with large, subequal, smooth, or obtusely keeled scales.
3. Supralabials 12–16.
4. Posterior lateral sides of head with groups of short spines.
5. Scales around midbody 120–150.
6. Neck scales granular, except those of median line which form a low crest.
7. Median dorsals 8–10 across midbody, broader than long, smooth or keeled, sometimes mucronate, in straight longitudinal series, laterals are much smaller, pointed backward and downward, without enlarged intervening scales.
8. Ventrals smooth, not half as large as median dorsals.
9. Gulars smaller than ventrals. A gular sac is absent.
10. Tail depressed, oval, covered at base with large subequal scales, dorsal and lateral caudal scales strongly mucronate, ventral caudals are flat.
11. Male with a large patch (4–5 rows) of callose preanal and another at mid-abdomen.

Snout-vent length 140–145 mm, tail 280–295 mm.
Color: In life, dorsum brownish olive with smaller more or less confluent dark spots, interspersed with large bright yellow spots; a dark reticulation on head; tail pale yellowish at base, but for greater part of its length entirely black. Body dorsum in young olive, yellowish white ventrum, head, chin, chest reticulated with black; neck, body, limbs, and base of tail with numerous small black and yellow spots; eyelids and superciliary ridges yellow; tail dusky black at its tip.

Natural history notes: The black rock agama lives in rock crevices. It feeds on vegetation, occasionally taking insects. It comes out of hibernation by March; as temperatures go up its activity period is increased. Among males, breeding scrambles on the rocks are quite common by late March and continue to May; 10–15 eggs are laid under rocks and juveniles are active by June.
Distribution: The black rock agama has been recorded from rocks of moderate elevation from the Salt Range, Punjab, and western hilly tracts of Sind, Pakistan.

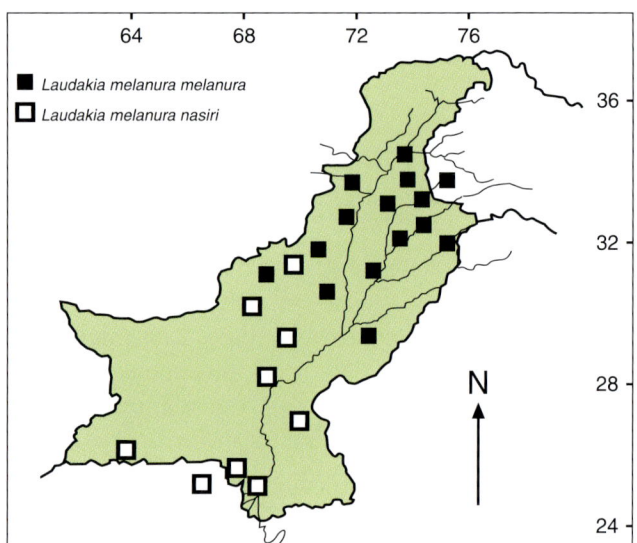

Laudakia melanura nasiri Baig, 1999

1999 *Laudakia melanura nasiri* Baig, Russian J. Herpetol., 6(2): 81–86.
Type locality: Tanishpa, Killa Saifullah, Toba Kakar Range, Baluchistan, 31°12′N, 68°28′E.
Diagnosis:
1. Head and body very depressed.
2. Group of spinose scales around tympanum.
3. Enlarged vertebrals in 8 rows of smooth scales.
4. Number of scalesaround midbody more than 150.
5. No patch of spinose scales on flanks.

Color: Grayish brown with yellow spots on head and body; no pattern on tail, except that its distal part more melanistic. Ventrum yellowish gray, with yellowish speckling distinct on chest, gular and chin.
Distribution: Recorded within the radius of 20 km of type locality.

Chapter 6. Lizards

Laudakia microlepis (Blanford, 1874)

Stellio microlepis Blanford, Ann. Mag. Nat. Hist. Ser.4, 13(78): 453.

Type locality: Khan-i-Surkh pass, north of Sarjan, Iran.
Diagnosis:
Close to *Laudakia caucasica* differing from it by having: Scales around the midbody 177–235 in male, 190–259 in female (Anderson, 1999).

A single male, collected from 5 miles N of Koh-a-Taftan, western Baluchistan, with 195 midbody scales, snout-vent length 135, tail 125 mm.
Color: Body dorsum darker than lighter tail
Natural history note: Collected from out crop of rocks. Sighted several lizards, caught one.
Distribution: Widely distributed in southern, central, and eastern parts of Iranian Plateau (Anderson, 1999), rare in Pakistan.

Plate 41. *Laudakia nupta*.

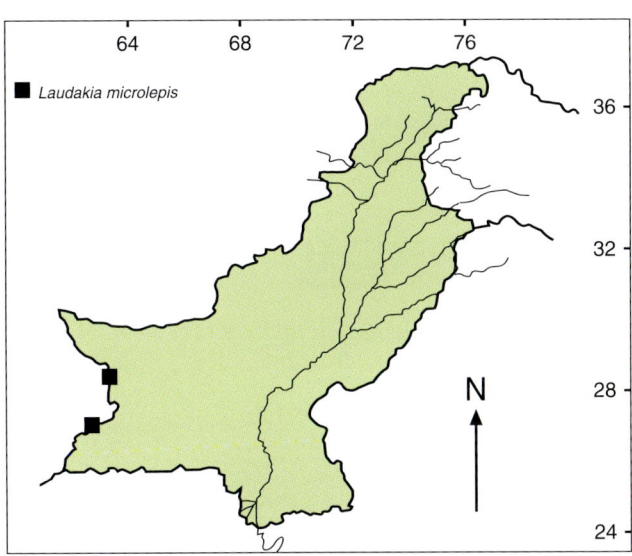

Laudakia nupta (de Filippi, 1843) (Plate 41)
(Spiny-head rock agama: Khaar-sar kirla)
1843 *Agama nupta* de Filippi, Jorn. Ist. Lomb. e Bib. Ital. 6:407.
Type locality: Persepolis, Iran.
Diagnosis:
1. Head scales subequal, smooth, or obtusely keeled.
2. Supralabials 14 to 18.
3. Posterior sides of head and neck with numerous groups of long spines, while rest of scales of the region are granular, except those of median dorsal line which form a low crest.
4. Median dorsals homogeneous, imbricate, broader than long, keeled, mucronate, arranged in 16–20 oblique series with a middorsal convergence.
5. Flanks of the body with very small scales pointed backward and downward, without enlarged scales.
6. Ventrals smooth, smaller than dorsals, gulars smaller than ventrals.
7. Tail depressed, oval, segmented, each segment with more than 2 whorls, or segments are indistinct. The caudal scales are large, subequal, mucronate; subcaudals are flat.
8. Males with 3 or 4 rows of callose preanal scales and a midabdominal patch of similar scales.

Snout-vent length 170–172 mm, tail 289–305 mm.

Color: Dorsum olivaceus, thickly speckled or mottled, limbs with yellow, tail similar or uniformly brownish, or banded with black, or its posterior half may be dark. Flanks marbled with gray and yellow. Ventrum yellowish brown, throat with large dark-blue spots.

Natural history notes: This agama is shy and inhabits mountain peaks, taking shelter in crevices among rock blocks at an elevation of 1800–2000 m.

Its natural habitat supports sparse vegetation of grasses and low thorny bushes. Breeding is observed from March to April. The male stations itself conspicuously on the peaks of rocks to guard his territory. The presence of another male is strongly resented; it is pursued vigorously, often resulting in a fight until the intruder is driven away. Eggs (6–8) are laid under stones and young appear by June. As breeding season ends, males become scarce and only the young roam about on the rocks.

This agama is mainly vegetarian, feeding on leaves, flowers, and fruits, with the young occasionally taking insects.

Distribution: Its range extends from eastern Iraq, Afghanistan, Iran, Baluchistan to southwestern Sind, extending on to the Waziristan Hills in Northwest Frontier Province, and the Kalabag area in northwestern Punjab, along the western bank of the Indus.

Laudakia nuristanica (Anderson and Leviton, 1969) (Plate 42A, B)

(Nuristan agama: Nuristan kirla)

1969 *Agama nuristanica* Anderson and Leviton, Proc. Calif. Acad. Sci. 4th Ser. 37:39.

Type locality: Kamdesh, eastern Afghanistan.

Diagnosis:

1. Head, body, and tail strongly depressed, tail longer than body, segmented, each segment of 4 whorls; 45 scales around base of the tail.

Plate 42A. *Laudakia nuristanica:* dorsal view.

Plate 42B. *Laudakia nuristanica:* ventral view.

Chapter 6. Lizards

2. Tympanum large, superficial. Three clusters of enlarged spinose scales around tympanum, and a pair on neck.
3. Median dorsals heterogeneous, not arranged in regular rows, with few strongly keeled scales which are about twice as large as largest ventrals, dorsolaterals granular, flanks with scattered enlarged scales, not in groups.
4. Limbs with strongly heterogeneous scales.
5. Naris in a large scale below canthus.
6. Supralabials 13, infralabials 11–12.
7. A preanal patch of 7 rows of callose scales, another at midabdomen of 17 rows.
Snout-vent length 130–135 mm, tail 238–250 mm.

Color: In formalin, olive brown above, head and limbs dark gray, corners of eyelids cream, 3 short dark bars on supracilliaries, median descending across eye; enlarged middorsals light gray interspersed with dark flecks; enlarged tubercles on flanks, light gray, contrasting sharply with olive-brown ground color; ventrum dirty white mottled with gray, throat with cream spots; posterior two-thirds of tail very dark brown, nearly black; fingers and toes barred with dark.

Natural history notes: Inhabits crevices among wet rocks and boulders along water courses in a conifer biotope of evergreen forests at 500–600 m of elevation.

Distribution: This species has recently been recorded from Ziarat, NWFP, Pakistan.

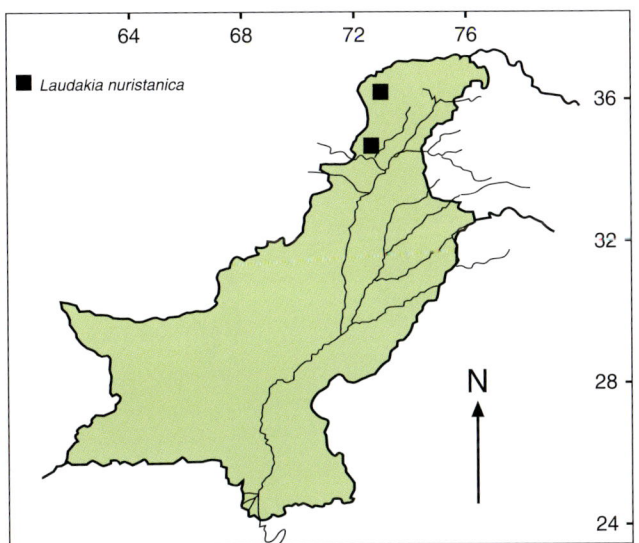

Laudakia pakistanica **(Baig, 1989)**
(North-Pakistan agama: Pakistan kirla)
1989 *Agama pakistanica* Baig, Bull. Kitakyushu Mus. Nat. Hist., 9:117–122.
Type locality: Ganglotgah, Gilgit, northern western Pakistan.
Diagnosis:
1. Head scales smooth.
2. Supralabials 11–13.
3. Distinct middorsal rows of spinose scales 8.
4. Vertical rows of enlarged spinose scales on flanks.
5. In male 6 and in female 3 preanal callose scales.
6. Number of scales around the base of tail 30–40.
7. Lamellae under 4th toe 26–32.
8. A distinct patch of large spines on flanks.
9. Tail segments with 3, rarely 4 whorls.
Snout-vent length 150 mm, tail 240+ mm.

Color: In formalin, dorsum jet black, chest, neck, and gular region speckled with orange; ventrum dark gray; groin speckled with yellow.

Natural history notes: The rock agama inhabits dry barren rocky mountains away from human settlements. The agama either scrambles about among rocks or is seen basking on cliff tops, taking refuge under rocks and boulders.

It feeds on twigs and leaves as well as beetles and other insects. Its probable breeding season extends from June to September (Baig, 1989).

Distribution: Widely distributed around Gilgit to Manshera, northeastern Pakistan, along the River Indus.

Taxonomic notes: Baig and Böhme (1996) distinguished three subspecies of *L. pakistanica* on the basis of color and number of scales around the midbelly:
1. Midbody scales 150–166; dorsum dull, dark brown-gray, with white bars, head and distal part of tail black. Gular region bluish black with irregular pale yellow blotches, ventrum pale yellow with thick dark reticulation ....*Laudakia pakistanica auffenbergi*
2. Midbody scales 146–158; dorsum silver gray with black mosaic, head black with scattered light scales; limbs barred with black................................*Laudakia pakistanica khani*
3. Midbody scales 168–178; uniform jet black, head speckled below with yellow orange ....*Laudakia pakistanica pakistanica*

Laudakia pakistanica pakistanica Baig, 1989
(Plates 46A, B)

1989 *Laudakia pakistanica* Baig, Bull. Kitakyushu Mus. Nat. Hist., 9:117–122.

Type locality: Ganglotgah, Gilgit, northern western Pakistan.

Diagnosis:
1. Scale rows midbody 168–178.
2. Lamellae under 4th toe 26–31.
3. Pericaudal scales around tail base 33–40.

Distribution: Known from type locality.

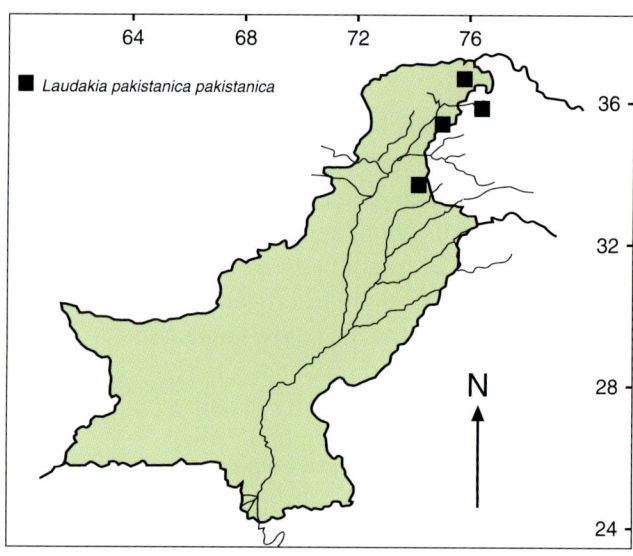

Plate 46A. *Laudakia pakistanica pakistanica:* dorsal view of holotype.

Laudakia pakistanica auffenbergi Baig and Böhme, 1996 (Plates 43, 44)

1996 *Laudakia pakistanica auffenbergi* Baig and Böhme, Russian J. Herpetol. 3: 3–5.

Type locality: Besham, District Swat, NWFP, Pakistan, elevation 700 mm.

Diagnosis:
1. Midbody scale rows 150–166.
2. Pericaudal scales at the base of tail 31–35.
Snout-vent length 115–132 mm, tail length 10–13 mm.

Distribution: Known from its type locality.

Plate 46B. *Laudakia pakistanica pakistanica:* ventral view of holotype.

Plate 43. *Laudakia pakistanica auffenbergi:* dorsal view.

Chapter 6. Lizards

Plate 44. *Laudakia pakistanica auffenbergi:* ventral view.

Plate 45A. *Laudakia pakistanica khani:* dorsal view.

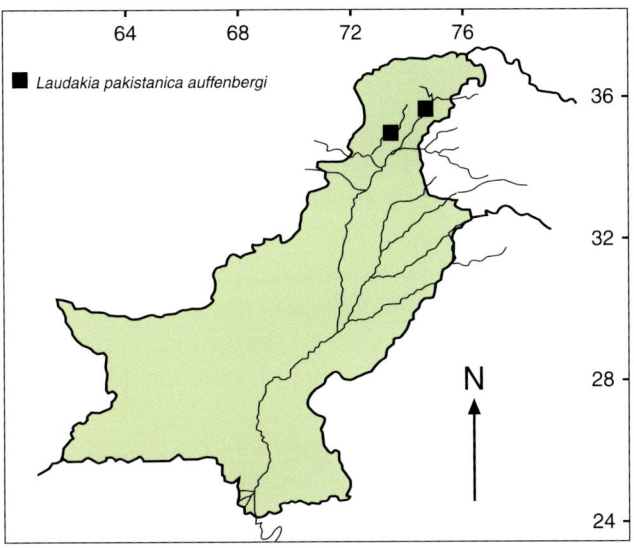

Plate 45B. *Laudakia pakistanica khani:* ventral view.

***Laudakia pakistanica khani* Baig and Böhme, 1996**
(Plates 45A, B)
1996 *Laudakia pakistanica khani* Baig and Böhme, Russian J. Herpetol. 3:1–3.
Type locality: Hadar, Chilas, North Western Frontier Province, Pakistan.
Diagnosis:
1. Number of scale rows around midbody 146–158.
2. Pericaudal scales around the base of tail 33–36.
3. Subdigital lamellae under 4th toe 30–32.
  Snout-vent length 81–125 mm, tail length 195–234 mm.

Distribution: Known from its type locality.

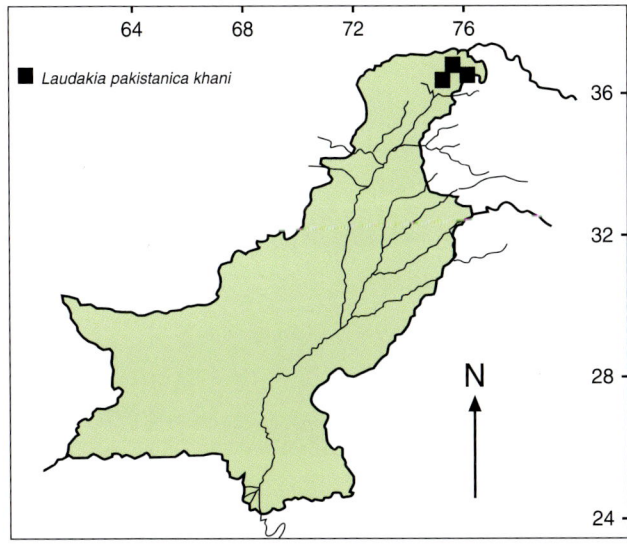

Laudakia tuberculata (Hardwicke and Gray, 1827) (Plate 47)

(Blue rock agama: Neela kirla)
1827 *Agama tuberculata* Gray, Zool. Journ. 3:218.
Type locality: Bengal.
Diagnosis:
1. Head with heterogeneous, convex, smooth, or keeled scales.
2. Supralabials 10 to 12.
3. Posterior sides of head and neck with small spinose scales; those around ear arranged in series, those on neck in groups which are sometimes absent; rest of the neck scales are granular.
4. Median dorsals subequal, roundish, imbricate, keeled; 10–15 across the midback.
5. Flanks with small scales, and few scattered enlarged keeled scales.
6. Ventrals smooth, as large as the dorsals.
7. Gulars much smaller than ventrals, no gular sac.
8. Skin of neck loose.
9. Tail depressed, with subequal strongly keeled scales, more than 40 at its thickest part.
10. Male with 6–7 rows of callose preanal scales and an elongated midabdominal patch.
    Snout-vent length 145–150 mm, tail 150–160 mm.

Color: In juvenile, dorsum is dark olive brown, with numerous dark spots arranged in longitudinal series on sides of a light vertebral line. In adult, dark spots tend to break and are replaced by a mixture of dark brown and yellowish specks. Head pale, ventrum whitish or brown, throat and often chest heavily marbled with dark blue.

In life, male is gorgeously colored; shoulders, chest, and flanks are spotted with bright yellow or orange. In adult male, ventrum is of bright bluish black tinge with purple on throat, sides of neck, shoulders, and belly.

Natural history notes: It is the commonest agamid species in the western Himalayas. It inhabits rock blocks at elevations of 1500–2500 m. The lizard is predominantly herbivorous, occasionally resorting to insectivory.

The agama comes out of hibernation in March and is seen basking on sun-exposed rocks. Diurnal in habit, it lives in holes and crevices between rocks. It is usually seen agilely scrambling on the rocks. Feeds on arthropods, leaves, and flowers, and breeds from April to July, during which time the male becomes strongly territorial and pugnacious. Males chase females and antagonistic fights among males are common. Six to nine hard-shelled eggs are laid under rock in more than one clutch. Juveniles appear from late May to August.

Distribution: It has been reported from eastern Afghanistan, northern Pakistan, and Kashmir to Nepal.

Plate 47. *Laudakia tuberculata*. (Photo courtesy of S.A. Minton).

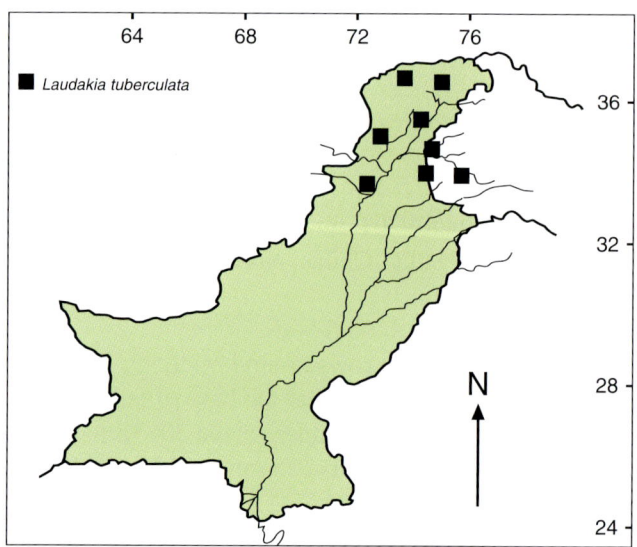

Chapter 6. Lizards

Genus *Phrynocephalus* Kaup, 1825
Body dorsoventrally depressed; tympanum concealed. Six species represented.

Phrynocephalus clarkorum Anderson and Leviton, 1967
(Afghan toad agama: Afghani gauk–sar)
1967 *Phrynocephalus clarkorum* Anderson and Leviton, Proc. Calif. Acad. Sci., San Francisco (4)35:228.
Type locality: 20 miles south of Kandahar, Afghanistan.
Diagnosis:
1. Nasals in contact with each other.
2. A single elongated suborbital scale.
3. Dorsals subequal in size.
4. No spiny scales on head and neck.
5. Both sides of fourth toe and lateral aspect of third, fringed with pointed scales.
Snout-vent length 40–45 mm, tail 50–54 mm.
Color: Dorsum sandy gray with two rows of dark spots along middorsum. A distinct, dark-lined dorsolateral light stripe, from posterior angle of eye, extending along body, onto tail. A series of elongated oval brown spots down the vertebral line, linked by a lighter chain. Ventrum white, underside of tail yellow-green in life, with 5 dark crossbars, tip white.
Natural history notes: The striped toad agama is adapted to varying ecological conditions: large active sand dunes, strips of firm sand, narrow ridges of hard soil, gravel plains and stony wastes, interspersed in the general scenario of a continuous sea of sand. Intervening strips of hard soil support grass tussocks, stunted bushes, and low xerophytic shrubs. The agama quickly takes to a nearby shelter where it remains motionless and undetected; however, where no shelter is available it quickly buries itself in sand (Khan, 1999a).
  Diurnal, it becomes active 2 hours after dawn until sunset, constantly moving and resting under meager shade provided by the stunted bushes, feeding on various psammophilous arthropods. Breeding season extends from March to May; 2–6 eggs are laid in burrows in the roots of vegetation (Khan, 1999a).

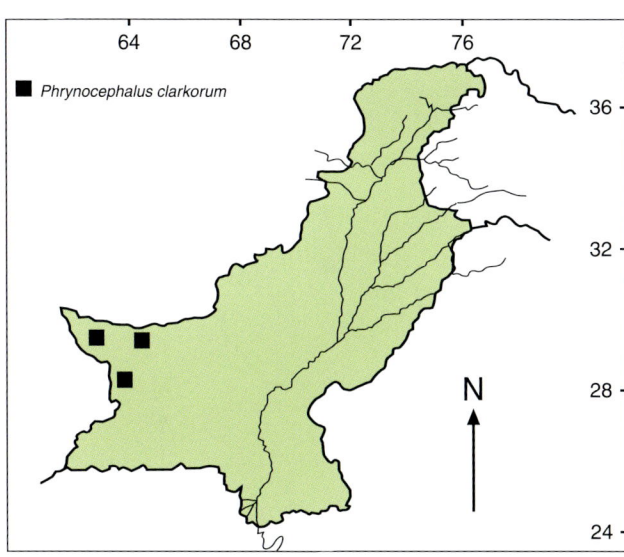

Distribution: This agama has been recorded from the sandy deserts of southern Afghanistan and western Baluchistan.

Phrynocephalus euptilopus Alcock and Finn, 1896
(Spotted toad agama: Chittr gauk-sar)
1896 *Phrynocephalus euptilopus* Alcock and Finn, J. Asiat. Soc. Bengal, Calcutta 65: 556.
Type locality: Darband, 900 m, northern Baluchistan.
Diagnosis:
1. Dorsals subequal.
2. Spinose scales on head and neck.
3. Nasals in contact with each other or partially separated.
4. Hindlimb reaches to the eye.
5. Digits long, with well developed lateral denticulations, length of which exceeds digit's breadth.
Snout-vent length 60–63 mm, tail 63–65 mm.
Color: Dorsum sandy, with black speckles. Head with irregular blotches. Five rounded spots on nape and shoulder. Ventrum whitish, tail tip black.
Natural history notes: The spotted agama is known only by its type locality; nothing is known about its natural history.
Distribution: It is reported from Darband, Baluchistan, at an elevation of 900 m.

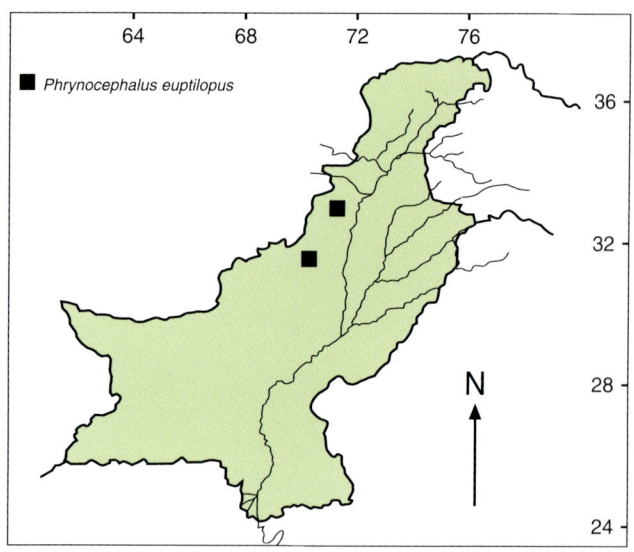

Plate 48. *Phrynocephalus luteoguttatus*.

Phrynocephalus luteoguttatus **Boulenger, 1887**
(Plates 48, 52B)
(Yellow speckled toad agama: Peela goak-sar)
1887 *Phrynocephalus luteoguttatus* Boulenger, Cat. Liz. Brit. Mus. 3: 497.
Type locality: Between Nushki and Helmand and along the Afghanistan-Pakistan border.
Diagnosis:
1. Dorsals markedly heterogeneous.
2. Sides of head and neck with long spinose scales.
3. Nasals in contact with each other or partially separated.
4. A series of elongated scales bordering head crown from front.
5. Hindlimb reaches to eye.
6. Digits strongly denticulate, length of denticulation equals breadth of toe.
 Snout-vent length 44–47 mm, tail 43–45 mm.

Color: Dorsum gray to blue-black, with numerous irregularly scattered small round yellow spots. Limbs sandy to pinkish. Anterior half of tail pinkish, while posterior black. Ventrum white.

Natural history notes: This toad agama is characteristic of fine windblown sand dunes with sparse stunted bushy vegetation. It is sluggish, moving rather slowly on the sand dunes. It usually it keeps its short tail parallel to its body; when excited it repeatedly curls its tail up and straightens it. It buries itself in sand by displacing it with rapid lateral shivering movements of

Plate 52B. *Phrynocephalus luteoguttatus*.

Chapter 6. Lizards

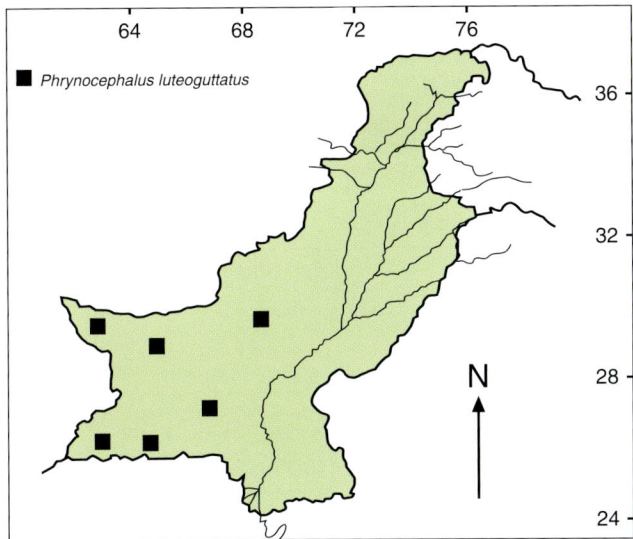

its body, practically sinking itself into the sand, leaving a characteristic impression of its body on the surface, indicating from where it can be picked up. Collected animals exhibit characteristic shivering movements of their bodies, even in the collection bag (Khan, 1999a).

It is diurnal, becoming active 2–3 hours after dawn, then moves about under bushes feeding and resting under their shade. Its food includes mainly deserticolous arthropods. At sunset it rests for the night by burying itself in sand.

Reproductive activity extends from late May to July; 2–4 hard-shelled eggs are laid in pits excavated at the roots of bushes.

Distribution: The yellow speckled toad agama is known from western Baluchistan around Nushki, southward to Las Bela, northward it extends into southern Afghanistan and western Iran.

Phrynocephalus maculatus Anderson, 1872
(Plates 49, 50)
(Whip-tail toad agama: Lambi-dum gauk-sar)
1872 *Phrynocephalus maculatus* Anderson, Proc. Zool. Soc. London 1872:389.
Type locality: Awada, Shiraz, Iran.
Diagnosis:
1. Dorsals subequal.
2. No spinose scales on sides of head and neck.
3. Nasals separated by 1–2 scales.
4. Mental larger than adjacent labials.

Plate 49. *Phrynocephalus maculatus.* (Photo courtesy of S.A. Minton).

5. Outer borders of third and fourth toe feebly denticulate.
Snout-vent length 70–74 mm, tail 115–120 mm.
Color: Dorsum sandy brown or grayish, speckled with black and white or dull yellow. Limbs with faint dark bars, sometimes with black spots which may unite to

Plate 50. *Phrynocephalus maculatus.* (Photo courtesy of S.C. Anderson).

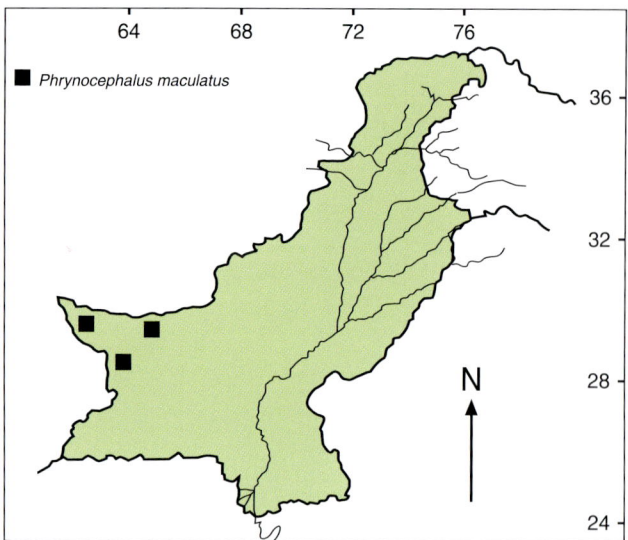

form crossbars on body and limbs. Distal one-third of tail black. Ventrum dirty white.

Natural history notes: The whip-tail toad agama prefers tracts of hard soil among sand dunes, close to scrubby growth. This vegetation supports various types of deserticolous arthropod populations which form the primary diet of this agama; it also feeds on leaves and twigs.

It is diurnal, becoming active 2–3 hours after dawn. Occasionally it perches on some elevated object to keep surveillance over its surroundings. It reacts to a disturbance by swaying its tail, occasionally curling it upward (Khan, 1999a).

Distribution: This species has a wide range in the Middle East, from Arabia, Iraq, Afghanistan, and Iran to western Baluchistan, Pakistan.

Phrynocephalus ornatus **Boulenger, 1887**
(Plates 51, 52A)
(Striped toad agama: Daharidar gauk-sar)
1887 *Phrynocephalus* ornatus Boulenger, Cat. Liz. Brit. Mus. 3:496.
Type locality: Between Nushki and Helmand, at the Baluchistan and Afghanistan border.
Diagnosis:
1. Dorsals subequal.
2. No spinose scales on sides of head and neck.
3. Nasals in contact with each other or partially separated.

Plate 51. *Phrynocephalus ornatus*.

4. Suborbital scales 2–3.
5. Hindlimb reaches to naris or beyond.
6. Outer borders of third and fourth toes denticulate, denticulation size equals the breadth of toe.

Snout-vent length 42–47 mm, tail 50–54 mm.

Color: The striped toad agama is light gray to sandy brown, with 1–3 pairs of reddish spots and scattered dark dots. A dorsolateral white to light yellow stripe, bordered with black. Tail ventrum pale yellow, with 4 or 5 dark crossbars, body ventrum white.

Natural history notes: This agama inhabits wind-blown sand dunes with sparse vegetation of stunted bushes. At midday when it is very hot, it crouches under shade of the bushes. It usually keeps its body as high as possible from the hot sand. It is quite speedy, running in zigzags between the bushes. During the night it buries itself in sand.

It feeds mainly on arthropods. Breeding takes place from March to May; 2–5 hard shell eggs are laid in pits formed in the roots of vegetation.

Distribution: The striped agama is known from the Chagai Desert, extends to Las Bela in the south and to southern Afghanistan in the north and west to south-eastern Iran.

Chapter 6. Lizards

Plate 52A. *Phrynocephalus ornatus.*

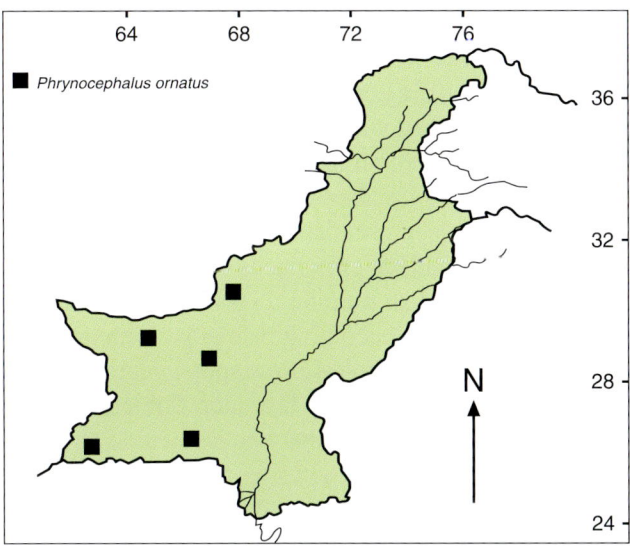

***Phrynocephalus scutellatus* Olivier, 1807 (Plate 53)**
(Banded toad agama: Pattay-dar gauk-sar)
1807 *Phrynocephalus scutellatus* Olivier, Voy. Emp. Othom. 3:110.

Type locality: Ispahan, Iran.
Diagnosis:
1. Dorsals unequal.
2. Head and neck sides without spinose scales.
3. Nasals in contact with each other, rarely separated by scales.
4. Mental larger than adjacent labials.
5. Outer borders of third and fourth toes denticulate, length of which equals the breadth of toe. Snout-vent length 50–56 mm, tail 72–78 mm.

Color: Grayish or brownish above, with light and dark markings, flanks darker. A median dorsal broad crimson stripe in adults. Limbs and tail barred. Ventrum white, with 6–7 dark bands under tail.

Natural history notes: It is characteristic of hard stony terrain with very sparse vegetation. It is very active and runs from one shelter to another. Unlike its congeners it does not bury itself in sand, but escapes into burrows made in the roots of bushes and under

Plate 53. *Phrynocephalus scutellatus.* (Photo courtesy of S.A. Minton).

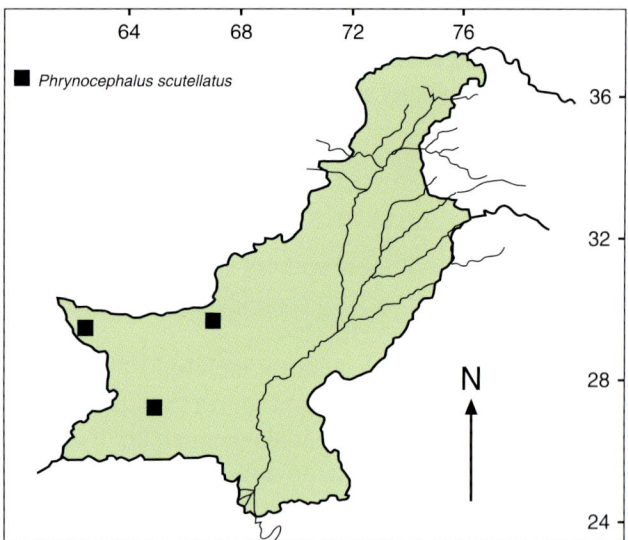

boulders. When cornered, it takes an offensive stance by raising its body and attacking with open mouth and lashing tail.

Distribution: This toad agama ranges in desert basins of Iran, Afghanistan, and Baluchistan.

Genus *Trapelus* Cuvier, 1816

Body dorsoventrally depressed; tympanum distinct, small, deeply sunk; caudal scales irregular; tail exceeds body in length; males with callous preanal scales.

Five species represented.

Trapelus agilis (Oliver, 1804) (Plate 54)

(Common field agama: Maidani Korrh-kirla)
1804 *Agama agilis* Oliver, voy. l'Empire Othm. l'Egypte et la Perse, vol.4.H. Agasse, Paris.
Type locality: Neighborhood of Baghdad, Iraq.
Diagnosis:
1. Head with heterogeneous, convex, smooth or keeled scales, largest on the snout.
2. Short spines on occiput and tympanic region.
3. Supralabials 15–18, their free margins form denticulated upper lip.
4. Tympanum deeply sunk, smaller than eye.
5. Dorsum with uniform rhomboidal, imbricate, keeled, often mucronate scales, spines of the median rows point straight backward and downward.

Plate 54. *Trapelus agilis pakistanensis*. (Photo courtesy of S.A. Minton).

6. Ventrals are as large as laterals, smooth or feebly keeled.
7. Male with short gular sac and has 1–3 rows of callose preanal scales.
   Snout-vent length 90–108 mm, tail 150–164 mm.

Color: *Trapelus agilis* displays rapid metachromatic changes. Usually the male is brown, gray, or dull yellow with dark crossbands on body and light spots on flanks, which may change to more or less completely suffused with bright cobalt blue; tail bright yellow with dark brown bands; throat, chest, and belly normally whitish, streaked with gray and lavender, may change to deep ultramarine blue. Female usually dull brown, with black crossbands enclosing a median dorsal row of reddish rhomboid spots and 2 rows of lateral lighter spots, becoming pale gray, with orange crossbands enclosing pale yellow spots; ventrum white lightly striped with gray and lavender. Sometimes color changes are so rapid that one is not able to follow them.

Natural history notes: *Trapelus agilis* is typical of open deserts, with clay, gravel and moderately sandy soil. Vegetation comprises of scattered shrubs or mounds with grass and herbs, or it is dry grassland with marginal thorny bushes. The male usually climbs into the bushes and keeps surveillance over its surroundings. Burrows are inaccessible, as they are located in the roots of thorny bushes. In rocky terrain the agama takes refuge under rocks.

It is diurnal, and when encountered in a field it runs to a nearby object and flattens itself against the ground to avoid detection. Here it is able to be caught by hand; upon handling, it turns around to bite.

In summer its daily routine has two foraging sessions: the morning period of activity starts 3 hours after dawn; at noon it rests in its burrow, afternoon session lasts till sunset. It feeds on grasshoppers, beetles, insect larvae, butterflies, moths, and field ants (pers. obs.).

Breeding activity takes place from March to May, with the peak period from April to mid-May. Young are seen from June to August. In the warmer parts of its range (Sind and Baluchistan), egg laying continues to September, with a female laying twice a year. Clutch size varies from 4 to 8 eggs, depending on the age of the female.

The field agama falls prey to hawks, owls, foxes, mongoose, and cats. Often their dried skins are found hanging from the branches of trees, left there by birds of prey after consuming the viscera.

Distribution: The field agama extends from western Iran through Afghanistan and Central Asia (to southern Kazakhstan), Pakistan and Rajasthan, India. In Pakistan this agama occurs below 1800 m in the plains of Punjab and Sind. Records from Baluchistan and southern NWFP are confined to plain areas, it avoids hilly tracts.

Recently Pakistani populations of *Trapelus agilis* have been recognized as two subspecies by Rastegar-Pouyani (1999):
1a. Tail almost always round, two (or more) rows of callose preanal scales; 65–91 scales around body; dorsal scales subequal, weakly to moderately keeled, often strongly mucronate; ventral scales smooth or weakly keeled; usually 2, sometimes 3 (rarely 4–5) rows of callose preanals; background coloration variable; central Iranian Plateau, central and southern Afghanistan, south-western Pakistan ....*Trapelus agilis agilis*
1b. Tail often compressed, often only one row of callose preanal scales, in the case of two, the second

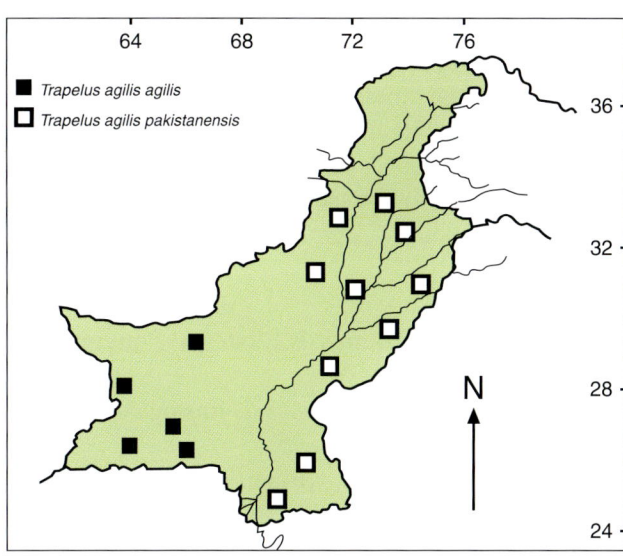

undeveloped (preanals absent or weakly developed in females) Body and limbs often distinctly slender, sometimes compressed in males; dorsal scales subequal to homogeneous, distinctly keeled and mucronate, usually clearly set off from small dorso-laterals; 67–83 scales around body; ventrals distinctly keeled in adult males; background coloration often sandy-grey; southeastern Pakistan and adjoining northwestern India..........*Trapelus agilis pakistanensis*

Trapelus agilis agilis (Oliver, 1804)
1804 *Agama agilis* Oliver, voy. l'Empire Othm. l'Egypte et la Perse, vol. 4.H. Agasse, Paris.
Type locality: Neighborhood of Baghdad, Iraq.
Diagnosis:
1. Tail round; usually 2 or 3 (rarely 4–5) rows of callose preanal scales.
2. 65–91 scales around the mid body.
3. Dorsal sales subequal, weekly to moderately keeled, often strongly mucronate.
4. Ventrals smooth or weakly keeled.

Color: Displays rapid metachromatic changes. Usually the male is brown, gray, or dull yellow with dark cross-bands on body and light spots on flanks, which may change to more or less completely suffused with bright cobalt blue; tail bright yellow with dark brown bands; throat, chest, and belly normally whitish, streaked with gray and lavender, may change to deep ultramarine blue. Female usually dull brown, with black crossbands

enclosing a median dorsal row of reddish rhomboid spots and 2 rows of lateral lighter spots, becoming pale gray, with orange crossbands enclosing pale yellow spots; ventrum white lightly striped with gray and lavender. Sometimes color changes are so rapid that one is not able to follow them.
Distribution: Southwestern, Baluchistan, Pakistan.

***Trapelus agilis pakistanensis* Rastegar-Pouyani, 1999 (Plates 55, 56)**
1999 *Trapelus agilis pakistanensis* Rastegar-Pouyani, Asiatic Herpetol. Res. 8:94–96.
Type locality: Gaj-River, Kirthar Range, southeastern Baluchistan, Pakistan.
Diagnosis:
1. Distinguished mainly on the basis of morphology of male, which has compressed head and body, head distinctly pointed; usually a single row of callose preanal scales (female without).
2. Dorsal scales relatively flat, subequal to homogeneous, distinctly keeled and mucronate, grading into small dorsolaterals rather abruptly (especially in males), 67–83 around body.
3. Ventrals distinctly keeled.

Distribution: *Trapelus agilis pakistanensis* is restricted in distribution to the lowland and semidesert regions of Sind province, southern Punjab, and some regions of eastern Baluchistan (southern and southeastern Pakistan), from around the Hab River in the west through Karachi and Thatta to the vicinity of Hyderabad and Mirpur Khas eastward into the Indian

Plate 56. *Trapelus agilis pakistanensis:* ventral view.

Desert. Biswas and Sanyal (1977) recorded this subspecies from inside the Indian territory (from Jaisalmir, Kolayat, Pugal, Phalodi, and some other localities in the Rajasthan Desert, northwestern India); to the north, it is distributed along the Kirthar Range up to the areas south of Khuzdar (south-central Baluchistan, Pakistan). It is parapatric with its *T. a. agilis* in the eastern regions of Baluchistan province.

***Trapelus megalonyx* Günther, 1864 (Plate 57)**
(Ocellate ground agama: Patta korrh-kirla)
1864 *Trapelus megalonyx* Günther, Rept. Brit. Ind.: 159.
Type locality: Afghanistan.

Plate 55. *Trapelus agilis pakistanensis:* dorsal view.

Plate 57. *Trapelus megalonyx.*

Chapter 6. Lizards

Diagnosis:
1. Dorsals heterogeneous, larger are arranged in groups, some scales smooth, other feebly or strongly keeled or mucronate, strongly imbricate and less rhomboidal.
2. Ventrals as large as small dorsals, feebly keeled. Snout-vent length 65–67 mm, tail 70–75 mm.

Color: Dorsum metallic bronze in male, with a vertebral series of 6 large light dark-edged crossbars, enclosing a reddish ocellus. Throat cobalt blue, a dark streak along nape. Ventrum whitish with light mottling. Female pale gray to brown, with dorsal row of cross-bands enclosing dull orange ocelli edged with black. Throat and ventrum white.

Natural history notes: This agama usually inhabits deserts with alluvial soil and little vegetation of grass and shrubs. Its burrows are located in the roots of shrubs and low thorny bushes. Diurnal, it becomes active 3–4 hours after dawn, feeding on insects and vegetation (Minton 1966).

Reproductive activity is observed from March to May; 4–6 eggs are laid in burrows or under rock. Juveniles are seen from April to August.

Distribution: It extends from southern Afghanistan (and possibly southeastern Iran), Baluchistan, the Sind and Cholistan Desert in Punjab, Pakistan. It is collected at elevations to 1800 m.

Trapelus rubrigularis Blanford, 1876 (Plate 58)
(Red-throat ground agama: Surakh-gani korrh-kirla)
1876 *Trapelus rubrigularis* Blanford, Proc. Asiatic Soc. Bengal Calcutta 1875:233.

Type locality: Kirthar Range, western Sind, Pakistan.

Diagnosis:
1. Head scales unequal, keeled or granular, largest on snout. No spine on occiput.
2. Supralabials 16 to 18, forming denticulate upper lip.
3. Dorsals rhomboidal, imbricate, subequal, feebly keeled, scales intermixed with large, strongly keeled scales, arranged in more or less regular transverse rows.
4. Gular and ventral scales smooth, as large as the small dorsals. Gular sac slightly indicated in male.
5. Limbs weak, hindlimb reaching occiput or ear.
6. Tail oval, covered with blunt tipped, subequal keeled scales.
7. Male with 1 or 2 rows of callose preanal scales. Snout-vent length 82–89 mm, tail 98–105 mm.

Color: Dorsum olive or grayish, with small golden yellow spots at the center of each enlarged dorsal scale; juveniles with 3–4 dorsal series of black spots, persisting in some individuals throughout life. A dark streak from nape, ventrum whitish. In life, a large red spot on throat in both sexes.

Natural history notes: The red-throat ground agama has been collected from flat desert with sparse grass and occasional thorny shrubs. When encountered in

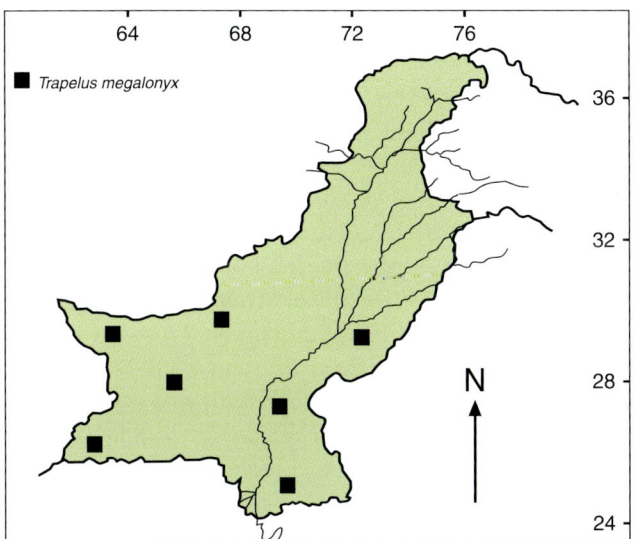

Plate 58. *Trapelus rubrigularis.*

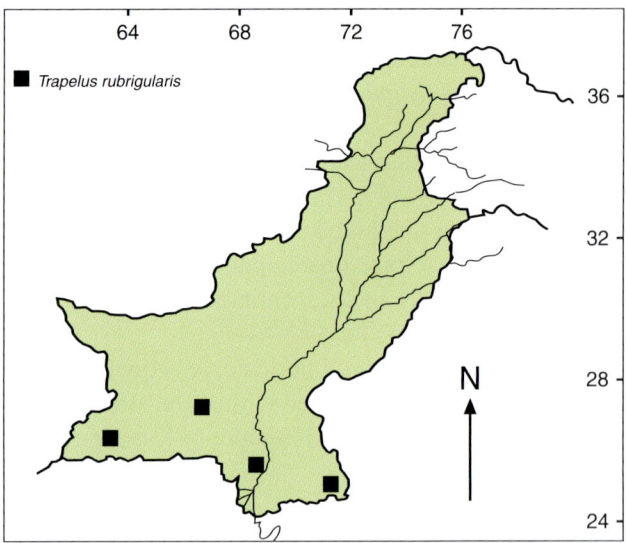

the open, it flattens its body, then sprints toward its burrow which lies at the roots of thorny bushes. Breeding season is quite extended from April to August, 4–6 eggs are laid in burrows. It is insectivorous and diurnal (Minton, 1966).

Distribution: The red-throat agama has been recorded from lower Sind and coastal Baluchistan.

Trapelus ruderatus baluchianus (Smith, 1935) (Plate 59)

(Spotted ground agama: Chittra korrh-kirla)
1935 *Agama ruderata baluchiana* Smith, Faun. Brit. Ind. 2:223.

Plate 59. *Trapelus ruderatus baluchianus.* (Photo courtesy of S.A. Minton).

Type locality: Quetta District, Baluchistan, Pakistan.
Diagnosis:
1. Head with heterogeneous convex keeled scales; those at the back of the head spinose.
2. Supralabials 19, orally dentate.
3. Dorsals heterogeneous, rhomboidal, imbricate, with numerous large strongly keeled scales which are irregularly arranged, sometimes partially erect, all pointing more or less backward.
4. Flanks with backward and downward-directed small pointed scales.
5. Ventrals as large as laterals, smooth or feebly keeled.
6. Gulars as large as ventrals, no gular sac.
7. Hindlimb reaches neck.
8. No strongly differentiated scales on neck and hindlimb.
9. Tail round in transverse section, with subequal strongly keeled scales.

Snout-vent length 75–80 mm, tail 100–106 mm.

Color: Dorsum grayish brown, ventrum pale, an indistinct dorsolateral series of paired dark brown spots.

Natural history notes: This ground agama has been collected from semiarid rocky highland at elevations of 1800–2000 m. It is shy and sluggish, and lives in dense clumps of thorny vegetation, or under rocks. Breeding season extends from March to June; 6–10 eggs are laid. It is carnivorous, feeding on arthropods (Minton, 1966).

Distribution: The spotted ground agama has been recorded from Quetta and Sibi Districts in Pakistan.

Taxonomic notes: Anderson (1999), and Rastegar-Pouyani (2000), both of whom examined the type specimens, decided that *Trapelus ruderatus baluchianus* (Smith, 1935) is a junior synonym of *T. megalonyx* (Günther, 1864). However, in Pakistan, there is a taxon of *Trapelus* distinct from the other species recognized here, (see key, diagnosis, and plates) and for the present, I follow Minton (1966) and Mertens (1969) in calling this taxon *Trapelus ruderatus baluchianus*. Rastegar-Pouyani (2002), after examining the type specimens of all related species of *Trapelus*, decided that *T. ruderatus* (and any of its subspecies) should be called *T. lessonae*. In light of the taxonomic uncertainties cited here, I continue to follow the conventional nomenclature.

Chapter 6. Lizards

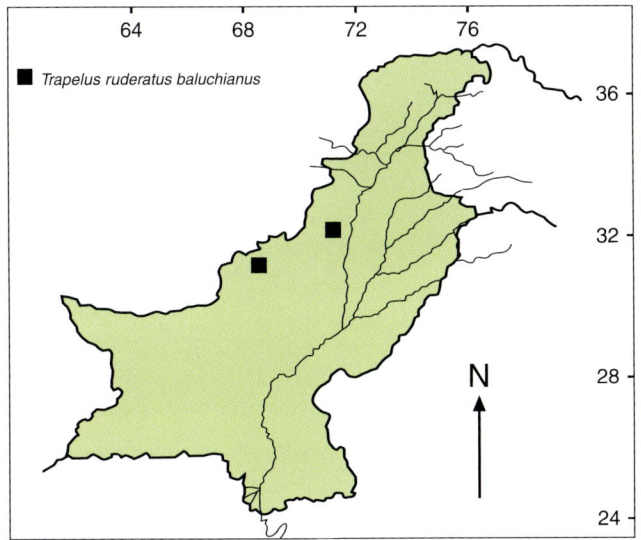

Family Chamaeleonidae

Represented in Pakistan by a single species.

Genus *Chamaeleo*
Head compressed, with an elevated median casque, covered with small, irregularly arranged scales; digits fused in two bundles.

Chamaeleo zeylanicus Laurenti, 1768
(Casque lizard: Taj-sar girgit)
1768 *Chamaeleo zeylanicus* Laurenti, 1768:46.
Type locality: By inference, Sri Lanka.
Diagnosis:
1. Head and body laterally compressed.
2. Prominent orbital and occipital ridges fusing posteriorly to form a casque (Figures 19A, B).
5. Tail coiled ventrally.
6. Eyes protuberant, capable of independent movement.
    Snout-vent length 218–235 mm, tail 241–265 mm.

Color: Color variable due to metachromatic changes. Usually dorsum pale yellow to dark olive, indistinct white or yellowish spots on sides, ventrum yellowish to greenish white, upper labials and enlarged midventrals ivory.
Natural history notes: The chameleon inhabits thick jungles, bushy country, with mesic to moderate xerophytic environments. It is diurnal and completely arboreal, very clumsy on the ground. Branches are held firmly between clasping digits of hands and feet; the prehensile tail also helps in movements by holding on to the branches. Movements of a chameleon in a tree are slow and deliberate. Similarly, on the ground, its walk is slow and stilted, waving each leg in a strange hesitant manner before setting it on the ground. The large revolving eyes of the lizard have a great range of movement independent from each other and permit binocular vision, enabling the animal to survey branches for insects—the primary constituents of its diet. The prey is caught by shooting out its sticky 30 cm long tongue with amazing speed and accuracy. The lizard is a voracious feeder and drinks water freely.

Breeding season lasts from April to July; 10–40 oval eggs are laid in a 30 cm deep pit dug by the female in soft soil under bushes. Average egg measurements are 19 by 12 mm, weighing 1.5–1.7 g. The pit with eggs is closed by pushing soil into it. Incubation period lasts almost to 90 days (Minton, 1966; Daniel, 1983).

Distribution: The chameleon is found in the wooded parts of peninsular India and Sri Lanka. It extends into southeastern Sind in Pakistan, where it is rare and local. Daniel's (1983) statement about extension of this species up to Peshawar is highly exaggerated, since in Pakistan it does not extend beyond southeastern Sind.

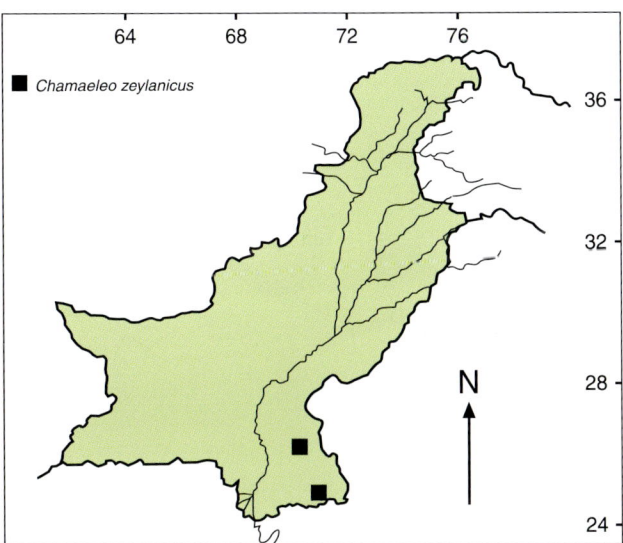

Family Eublepharidae

This family is represented by a single species of genus *Eublepharis* in Pakistan, which is characteristic in having moveable eyelids, digits are cylindrical and straight.

Genus *Eublepharis* Gray, 1827

Head with granular scales; pupil vertical with pinholes (Figure 12A); eyelids moveable.

A single species.

Eublepharis macularius (Blyth, 1854)
(Plates 60A, B)
(Fat-tail gecko: Khin-khin, Korrh kirly)
1854 *Cyrtodactylus macularius* Blyth, J. Asiatic Soc. Bengal, Calcutta 23: 737–38.
Type locality: Salt Range, Punjab, Pakistan.
Diagnosis:
1. Digits short, straight, cylindrical, with a single row of subdigital lamellae.
2. Eyelids well developed, movable.
3. Body dorsum with granular scales, with scattered round tubercles.
4. Tail plump, distinctly broader than body, tail unsegmented, small subcaudals.
5. Male with 9–14 preanal pores.

Plate 60A. *Eublepharis macularius:* juvenile (Photo courtesy of S.A. Minton).

Plate 60B. *Eublepharis macularius:* defense posture.

Snout-vent length 120–158 mm, tail 89–93 mm.
Color: Body dorsum in juvenile light brown, yellowish or pinkish with broad crossbars, sometimes with intervening dark brown spots. Head brown, tail ringed with light and dark; in older specimens the bars on body are broken into spots. Head spotted, sometimes a U-shaped mark on the nape.
Natural history notes: The fat-tail gecko is a common lizard in the stony countryside of Pakistan. It inhabits rocky stony terrain, mudflats with sparse grass and bushes, in mesic to xeric conditions. It is gregarious: several lizards live in colonies in holes in the ground, under stones, and in crevices among rocks, especially in stone walls demarcating fields or houses. The gecko climbs several feet to return to its selected crevice. It selects a permanent site which is shared with others. Nocturnal, it comes out just after sunset and goes foraging around, returning just before dawn. At some places it is semi-arboreal (Daniel, 1983). Its activity is restricted by environmental changes. Dry, cool, and very windy weather inhabits its activities, and it mostly stays home; hot, still, humid nights in the rainy season bring it out in large numbers. It hibernates from October to February, while in warmer parts of its range hibernation may be delayed until November.

Brandstaetter's (1992) vivarium observations show that a colony has a dominant male that reacts in the presence of other males in the colony by the following behavior: a) raising its body on stiffened legs in an intimidation display. Waxy secretions from its preanal glands are dropped on the ground by lowering the posterior part of its body in front of the rival male; b) licking the preanal secretions and smearing it all over its own body; and c) rubbing its body on stones and other objects in the territory. Moreover, a definite place is used by the dominant male for defecation, also used by all the animals in the colony. Feces also mark the territory.

The gecko is carnivorous, its diet including beetles, grasshoppers, spiders, and scorpions, etc., and other geckos (Matz, 1978). Prey is slowly stalked and ambushed, and is captured by a sudden lunge accompanied by a flick of the tail. The gecko is rather sluggish, with a deliberate slow walking style. When cornered, it raises and arches out its body by straightening its legs and flicks and twists its fat tail sideways. It snarls and lunges, its bite just a moderate pinch as the teeth are small.

It breeds from March to May; 2–6 oval eggs with smooth pliable shells are laid in more than two clutches in a season. The size of the egg is 31–35 by 13–16 mm. The eggs are laid in the humid environs of a crevice, and take almost a month to hatch.

Locally, this gecko is considered to be very venomous; its bite is believed to liquefy the body of the victim (Minton, 1966).

Taxonomic notes: There have been several attempts to distinguish the Indo-Pakistan population of *Eublepharis macularius* in different species and subspecies: Günther (1864): *Eublepharis fasciolatus* from Hyderabad, Sind, Pakistan; Börner (1974, 1976, 1981): *Eublepharis gracilis,* locality unknown; *Eublepharis afghanicus* from Kabul-Jalalabad, Afghanistan; *Eublepharis macularius fuscus* from Bombay, India; *Eublepharis macularius smithi,* from Delhi, India; *Eublepharis macularius montanus* from Karachi; an as ßßyet undescribed form from Nushki and Zhob region, Baluchistan, Pakistan, and *Eublepharis macularius fasciolatus* from Lahore, Punjab, Pakistan. All these forms have been found to be the color morphs of *Eublepharis macularius*, with which they are synonymized (Smith, 1935; Szczerbak and Golubev, 1986).

Distribution: The range of this gecko extends from Rajputana and Khandesh District of India. In Pakistan it has been recorded from Azad Kashmir, NWFP, northern Punjab, Baluchistan, and lower Sind.

Family Gekkonidae

Several assemblages of fragile gekkonid lizards inhabit deserts, scrublands, and alpine habitats throughout Pakistan. The taxonomy of these animals is yet little understood despite several efforts (Kluge, 1967, 1983, 1987, 2001; Bauer and Russell, 2002; Szczerbak and Golubev, 1977, 1986, 1996; Khan, 1997f, 2001d). Kluge (1983) placed genera *Agamura, Alsophylax, Bunopus, Crossobamon, Hemidactylus, Stenodactylus, Cyrtopodion* and *Tropiocolotes* in tribe "Gekkonini," and *Microgecko, Ptyodactylus,* and *Teratoscincus* in tribe "Ptyodactylini."

With recent descriptions of several new gekkonids from Pakistan (Sczerbak, 1991; Khan, 1980a, 1988, 1989, 1991a, 1993a, d, 2001d; 2003e, f; Khan and Baig, 1992; Khan and Tasnim, 1990b), my concept of the relationships of the angular-toed geckos which followed Szczerbak and Golubev (1986, 1996) for the generic definitions, has changed (Golubev, et al., 1995). Khan (1993a, 2001d, 2005b) and Khan and Rosler (1999) distinguished Pakistani angular-toed geckos in the following major morpho-ecological groups:

1. Circum-oceanic group

Circum-tropical, scattered along sub-continental coastal strips and oceanic islands, between 27–32° N and 75–105° E; dorsal pattern of vivid cross bars or spots; dorsal granular scales mixed with larger rounded, smooth, or slightly keeled tubercles; tail and body cylindrical, tail often longer than body. Species included are

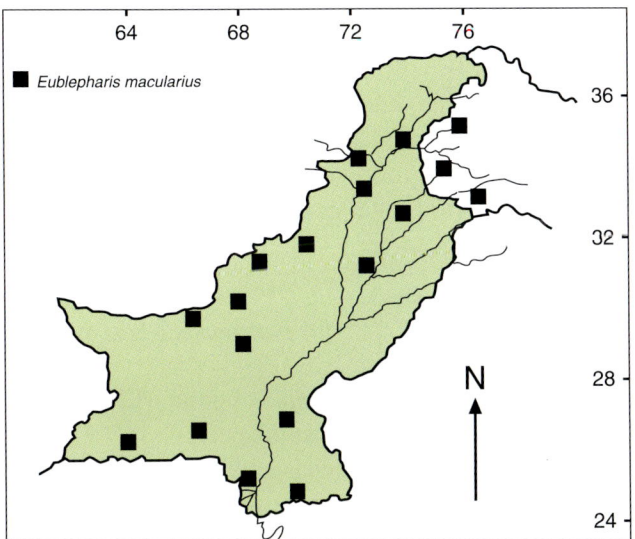

all extralimital southeast Asian: *Cyrtodactylus pulchellus, Cyrtodactylus intermedius, Cyrtodactylus consobrinoides, Cyrtodactylus frenatus, Cyrtodactylus condorensis, Cyrtodactylus oldhami, Cyrtodactylus peguensis, Cyrtodactylus irregularis, Cyrtodactylus angularis, Cyrtodactylus khasiensis, Cyrtodactylus rubidus, Cyrtodactylus triedrus, Cyrtodactylus nebulosus, Cyrtodactylus collegalensis, Cyrtodactylus dekkanensis, Cyrtodactylus albofasciatus,* and *Cyrtodactylus jayporensis* (Smith 1935). For recent groups of Philippine cyrtodactylid species see Inger (1957).

2. Circum-Himalayan group
Subtropical, highland forms, mainly extending between 34–40° N and 75–80° E, dorsal granular scales tubercular, bead-shaped, interspersed with 2–3 times larger oval keeled or keelless tubercles arranged in more or less 12–13 longitudinal rows at midbody; body and tail subequal and subcylindrical; subcaudals small in several rows; inconspicuous dorsal pattern of transverse bands, spots, or reticulations.

The circum-Himalayan group is further distinguished in three subgroups:

Stoliczkai subgroup
Body and tail rather flat; caudal tubercles feeble, flat, smooth; anterior half of the tail segmented, segments indicated by lateral lobulation, regenerated tail flattened and abnormally swollen, caudal tubercles feeble, subcaudals small, no preanal and femoral pores. Dorsal pattern of inconspicuous bands. Includes highland species: One Pakistani representative *Altigekko baturensis*. Extralimital species included are: *Altigekko stoliczkai* and *Altigekko yarkandensis*.

Walli subgroup
Body flatter; tail quadrangular in cross-section, distinctly segmented; caudal tubercles larger slightly keeled; median row of subcaudals transversally enlarged; inconspicuous dorsal banded pattern; 4–6 preanal pores. Species included are: *Mediodactylus walli* (=*chitralensis*) and *Tenuidactylus kirmanensis* (extralimital) (This last species is placed in *Cyrtopodion* by both Anderson (1999) and Kluge (2001). I (Khan, 2003a) tentatively placed them in Palearctic genus *Mediodactylus*.)

Highland geckos *tibetanus* subgroup
Body and tail round; tail segments indistinct; caudal tubercles feebly keeled, regenerated tail neither flattened nor swollen; 4–10 preanal pores. Dorsal pattern of vivid crossbars, spots, or reticulations. Includes Tibeto-Himalayan low altitude submontane geckos: *Tenuidactylus mintoni, Tenuidactylus dattanensis,* and *Tenuidactylus battalensis:* extralimital: *Tenuidactylus tibetanus* and *Tenuidactylus himalayanus*.

3. Palearctic group
Characterized by dorsoventrally depressed body and tail, which is longer than body; dorsum with large trihedral tubercles; large flat, sharply keeled caudal tubercles arranged in three latero-dorsal rows on each side; subcaudals in a single row of broader-than-long scales. The Palearctic group is further divided into the following two subgroups:

Ground geckos
Dorsum with trihedral tubercles; caudal tubercles broadly in contact laterally with each other; male with preanal, rarely with femoral pores; a single row of transversally enlarged subcaudals.

This subgroup includes Pakistani species: *Cyrtopodion montiumsalsorum, Cyrtopodion kohsulaimanai, Cyrtopodion agamuroides, Cyrtopodion scabrum, Cyrtopodion watsoni,* and *Cyrtopodion kachhense*.

Sandstone geckos
Body and tail much depressed, tail much longer than body; dorsum with flat, round, feebly keeled tubercles, arranged in longitudinal rows; in male 7–9 preanal pores, in some, 6–9 femoral pores also present. Tail distinctly segmented, with broad dorsal and subcaudal scales. Pakistani species included are: *Tenuidactylus indusoani, Tenuidactylus rohtasfortai, Tenuidactylus fortmunroi,* and *Tenuidactylus rhodocaudus*.

Now that the morpho-ecological groups of Pakistani angular-toed geckos are clearly defined, I placed them in the following three new genera (with exclusion of extralimital genus *Cyrtodactylus,* as defined above):

Altigekko Khan, 2003
Type species: *Tenuidactylus baturensis* Khan & Baig, 1992
Species included: *A. stoliczkai* subgroup (Khan, 2001d). See above for the list of species.
Etymology: The new genus is named because of high altitude distribution of its geckos in the Greater Himalayas (Karakorams), in northeastern Pakistan.

Chapter 6. Lizards

Indogekko **Khan, 2003**
Type species: *Cyrtodactylus indusoani* Khan, 1988
Species included: *S. tibetanus* (Khan, 2001d). See above for the list of species.
Etymology: The new genus is named after the Siwalik Mountain Range in the Lesser Himalayas where these geckos are widely distributed from Pakistan to Nepal.

Siwaligekko **Khan, 2003**
Type species: *Cyrtodactylus battalensis* Khan, 1993
Species included: Sandstone geckos (Khan, 2001d). See above for the list of species.
Etymology: The new genus is named after the Indus River, from the sandstone bank rocks in Soan Skaser Valley where the geckos were first collected (Khan, 1988).

Geckos are represented by 14 genera in Pakistan.

Genus *Agamura* Blanford, 1874

Characteristic geckos with long slender limbs; digits thin, cylindrical at base, the distal part of the digits angularly bent, slightly compressed; claws between two enlarged scales; subdigital lamellae transverse and smooth; no enlarged postmental shields; dorsal scales granular, intermixed with small round tubercles; tail cylindrical, suddenly reduced in diameter, just behind tail base, not longer than snout-vent length, the tail autotomizes only at the base; a row of more or less circular subcaudals.

Length of limbs, absence of distinct postmentals, and morphology of tail separate this genus from the rest of the geckos.

Agamura persica (Duméril, 1856) (Plate 61)
(Persian spider gecko: Irani makra-chapkali)
1856 *Gymnodactylus persicus* A. Duméril, Arch. Mus. Hist. Nat. Paris 8: 481.
Type locality: Persia.
Diagnosis:
1. No postmental.
2. No femoral or preanal pores. Anderson (1999) records two preanal pores in male.
3. Ventral scales in 23–26 rows across mid-abdomen.
4. Tail as long as body, tip blunt.
Snout-vent length 65 mm, tail 65 mm.

Plate 61. *Agamura persica*. (Photo courtesy of S.A. Minton).

Color: Dorsum light gray with yellowish tinge. Five dark gray crossbars on body, as broad as or slightly narrower than interspaces, 9–10 on tail. Belly dirty white, flecked with gray. In young, bands darker dotted with light.
Natural history notes: The blunt-tailed spider gecko has been collected from rocky and stony terrain close to sandy semidesert, and on the slopes of hills at

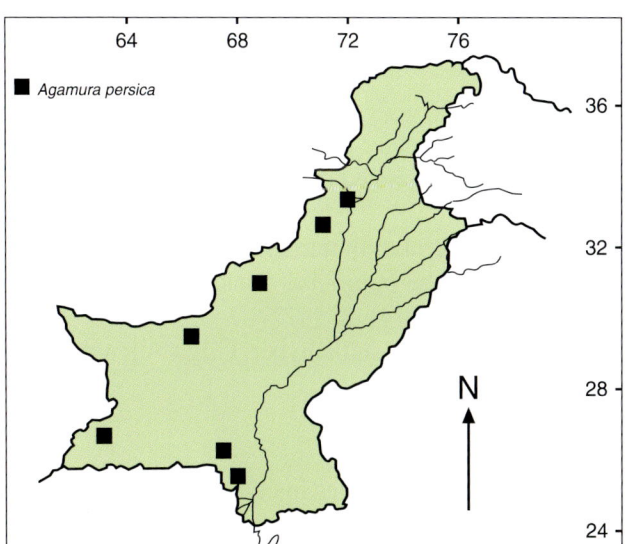

30–100 m of elevation. The sparse vegetation consists of undersized sagebrush (Minton, 1966). The gecko is sluggish, climbs rocks, and hides in crevices and under rocks. It does not climb vertical surfaces. It is essentially nocturnal; however, it may be active during the day. When disturbed, it sprints to nearby shelter where it remains motionless until picked up. It feeds on insects, spiders, and their larvae.

It breeds from March to May; the eggs are laid in June and juveniles appear at the end of September.

Distribution: The blunt-tailed spider gecko is reported from Iran, and extends eastward to near Karachi and northward to the Waziristan Hills, between 25 and 100 m of elevation.

Genus *Altigekko* Khan, 2003

Stoliczkai subgroup (Khan, 2001d). Body and tail depressed, tail a little longer than body, segments marked by deep lateral lobulations in the anterior half of un regenerated tail. Tail fragile at its base; regenerated tail much swollen; caudal tubercles as small, conical, protuberant tubercles given from the middle of segments, indistinct by midtail; subcaudals in several rows; postfemoral tubercles absent; dorsal granular scales convex, mostly juxtaposed, arranged in lateral transverse rows, interspersed with thrice large convex, smooth or feebly keeled oval tubercles arranged in more or less longitudinal rows, rare on head and limbs. no supracilliary spines on the posterior dorsal part of the upper eyelid; a distinct frontal and postnasal pit. Interorbital scales 16–20; scales across midabdomen 27–32, midventrals 117–150; preanal and femoral pores not indicated in both sexes (Gruber, 1981; Khan, 1992).

Three species in Pakistan.

Altigekko baturensis (Khan and Baig, 1992) (Plate 65)

(Batura glacier gecko: Batura chapkali)
1992 *Tenuidactylus baturensis* Khan and Baig, Pakistan J. Zool. 24(4): 273–277.

Type locality: Passu and Khyber, Gilgit Agency, northern Pakistan.

Diagnosis:
1. Interorbitals 16–20.
2. Dorsal granular scales interspersed with flat thin weakly keeled tubercles, in 11–12 longitudinal rows at midbody.

Plate 65. *Altigekko baturensis.*

3. Scales across midabdomen 26–30.
4. Scales along midventrum of the body 149–171.
5. No preanal or femoral pores.
6. Caudal tubercles flat, arising from the middle of tail segments.
7. Subcaudals small.

Snout-vent length 50–53 mm, tail 51–54 mm.

Color: Dorsum light gray, with wavy-margined faint transverse bars on nape, neck, 4 on body, one at the pelvis and the last one at the level of the vent. Tail with transverse dark bars.

Natural history notes: The gecko was picked from the stone wall of a thatched hut. During day the gecko was collected from under stones and other debris. The area has sparse grass and low bushes. The gecko forages in grass and feeds on the insects and their larvae. Those in buildings feed on photophilic insects.

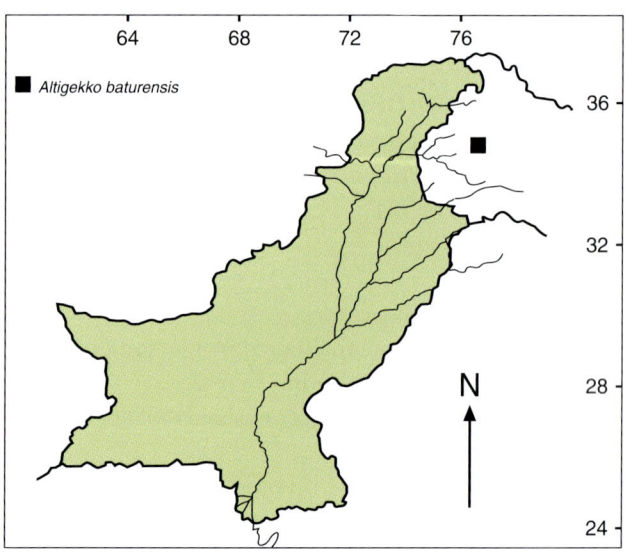

Chapter 6. Lizards

During winter the area remains covered with snow and the gecko hibernates (Khan and Baig, 1993).
Distribution: This gecko is known only from its type locality in Gilgit Agency, northeastern Pakistan.

Altigekko boehmei (Szczerbak, 1991)
(Karakorum gecko:koh karakram ki chipkali)
1991 *Alsophylax* (*Altiphylax*) *boehmei* Szczerbak, Salamandra 27:53–57.
Type locality: Skardu, Ladak, northeastern Pakistan, 2300 m.
Diagnosis: Supralabials 8–9; infralabials 6–7; interorbitals 20; nasals 3; postmentals in three pairs, scales of first pair broadly in contact with each other; midventral scales 109; finger tips of adpressed forelimbs reach to snout tip, toes of hind limbs to mid-neck; 22 subdigital lamellae under 4th toe.
  Snout-vent length 34–39 mm, tail length 35–50.
Color: Body light dark with 6–7 darker transverse bands, 11–12 similar bands on tail.
Ecological notes: A gecko of snowfields, inhabiting holes and crevices among walls of stone built huts and houses. Feeds on insects attracted to light. Breeding occurs May to July.
Distribution: Known from Skardu, Ladak, northeastern Pakistan, at 2300 m.
Taxonomic notes: Kluge (2001) followed Szczerbak (1991) in his original assignment of this species to *Alsophylax*. However, I regard it as a species of *Altigekko* (Khan, in preparation).

Altigekko stoliczkai (Steindachner, 1869)
(Baltistan gecko)
1869 *Gymnodactylus stoliczkai* Steindachner, Reise Novara, Zool., 1:15 Rept. 1, Plate 2, Fig. 2, 2a.
Cyrtodactylus stoliczkai Underwood, 1954, Proc. Zool. Soc. London, 124:475.
Type locality: Near Karoo, north of Dras, Kashmir, India.
Diagnosis:
1. Body and tail moderately depressed, tail a little longer than body, segments of unregenerated tail laterally deeply sected in anterior half of the tail, three dorsolateral rows of blunt, conical, short, smooth flat caudal tubercles. Regenerated tail much swollen.
2. Several series of small subcaudals arranged in several rows.
3. Body dorsum with flat, mostly juxtaposed granular scales, distinctly arranged in transverse rows, interspersed with large flat smooth oval tubercles, about three times larger than granular scales, more or less arranged in 9–10 transverse rows across middorsum and 19–20 along paravertebral line.
4. 16–20 interorbital tubercular scales.
5. 27–32 scales across midabdomen, 117–150 midventral scales from postmentals to anterior lip of cloaca.
6. Both preanal and femoral pores not indicated. Snout vent length 48 mm, tail length 52 mm.

Color: Body dorsum light blue or gray, with pink edged transverse bands with denser wavy posterior edges, broader than interspaces, three on nape, six on body and 13 on tail (in preserved specimens bands are dark. Head, labials and tail plates with fine gray dots, limbs and digits barred, ventrum light.
Ecological notes. Ladakh lies around 3000 m, above timberline. It is a completely dry snow desert, with sparse vegetation of herbs, shrubs and grasses. The area is highly arid with sub-tropical continental highlands cold climate. Heavily snowy winters, getting rain in winter and spring. Maximum summer July temperature is 24°C, minimum 10°C, while maximum winter temperature in January is −1°C, dropping to

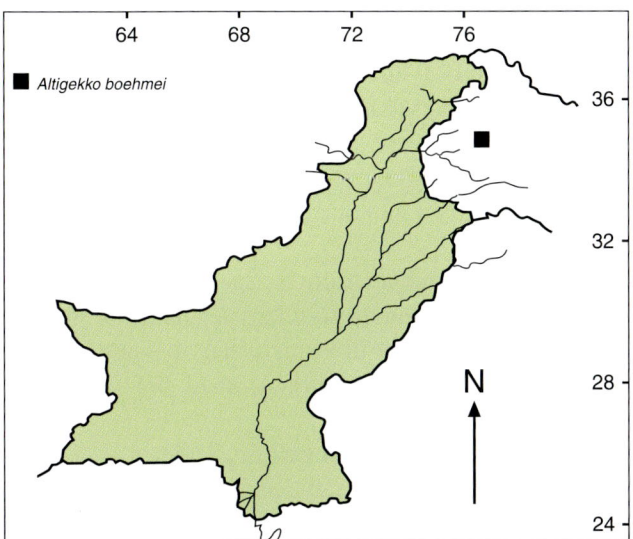

minimum −13°C. Maximum rainfall of 1.5 cm is received during August, minimum 0.1 cm during November (Ahmad, 1951).

Gruber (1981) collected *A. stoliczkai* from rocky habitat, where this gekko prefers desert, bare, dry situations in the non-irrigated areas without or with very sparse vegetation, apparently avoiding direct neighborhood of human settlements.

Distribution: Widely distributed in Baltistan, northeast Pakistan. Recently it has been recorded from upper Indus (K. Affenberg pers. comm.).

Genus *Bunopus* Blanford, 1874

Digits straight, without angular bend, clawed; subdigital lamellae transverse, tuberculated; body dorsum with juxtaposed small scales, intermixed with larger keeled, slightly trihedral flattish tubercles; tail cylindrical, round in cross-section, segmented, 2–3 whorls of caudal scales in a segment and one whorl of caudal tubercles, which are in close contact with each other; enlarged subcaudals may be present or absent; enlarged postmental may be present or absent; male with 4–6 preanal pores.

Anderson (1999) determined the holotype of *Crossobamon lumsdeni* as *Bunopus tuberculatus*.

A single species is represented in Pakistan.

Bunopus tuberculatus Blanford, 1874 (Plate 62)
(Tuberculated desert gecko: Khurdari reg chapkali)
1874 *Bunopus tuberculatus* Blanford, Ann. Mag. Nat. Hist. London (4) 13: 454.
Type locality: Persian Baluchistan.

Plate 62. *Bunopus tuberculatus*.

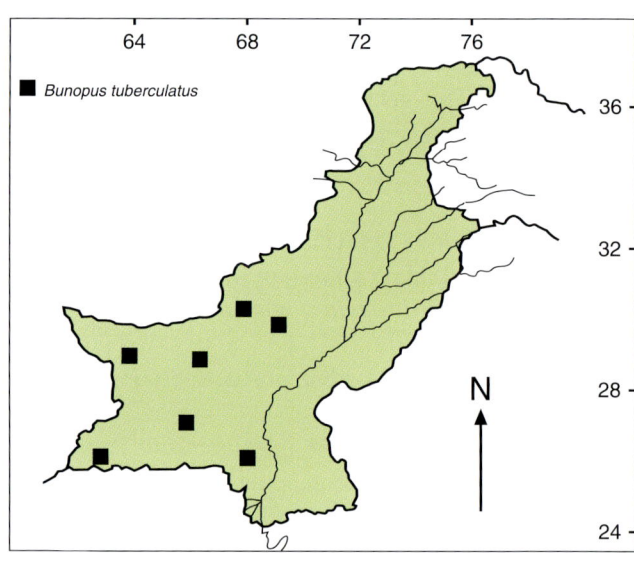

Diagnosis:
1. No postmental.
2. Digits straight, distal phalanx not compressed.
3. Subdigital lamellae with a single series of minute tubercles.
4. Male with 5–6 preanal pores.
   Snout-vent length 50–52 mm, tail 59–60 mm.

Color: Dorsum light grayish to pale brown, with 6 indistinct light gray crossbars, broader than interspaces. A brown crescentic curve around nape passes through eye. Labials barred. Limbs mottled or barred. Tail barred above, flecked below, belly white.

Natural history notes: The gecko is nocturnal, however, during day it can be extracted from its burrow, which is among the roots of shrubs into sand dunes. It also extends in the mountainous region with sandy soil (Zarudny, 1904). It inhabits crevices in the brick walls of dikes built around salt lakes known locally as "hamuns," crevices under bridges, and fissures in soil.

Breeding season extends from April to May; a pair of eggs is laid in a burrow or fissure in the roots of vegetation.

Distribution: It ranges from Syria, Iraq, eastern Arabia, southern Iran, to southern Afghanistan. In Pakistan it is common in Baluchistan, to southern Sind, Las Bela, and around Hyderabad, below 2000 m.

Genus *Crossobamon* Boettger, 1888
Digits straight, comb like scales projecting on each side; transverse multileveled subdigital lamellae;

dorsum with small imbricate keeled scales intermixed with larger flattish keeled tubercles; no distinct postmental scale; preanal pores in male present.

Three species are represented in Pakistan.

Crossobamon lumsdenii (Boulenger, 1887)
(Smooth-belled sand gecko: Naram kanghi-ungusht)
1887 *Stenodactylus lumsdenii* Boulenger, Cat. Liz. Brit. Mus. 3: 479.
Type locality: Between Nushki and Helmand, northern Baluchistan, Pakistan.
Diagnosis:
1. Length of snout greater than distance between eye and ear opening.
2. Hindlimb reaches beyond axilla.
3. Abdominals smooth.
4. Dorsal tubercles numerous.
5. Head with irregular scales.
Snout-vent length 33–38 mm, tail 40–42 mm.
Color: Dorsum sandy gray, with 7 distinct brownish crossbars, which are broader than interspaces. Body ventrum white.
Natural history notes: The smooth-belly sand gecko is truly psammophilous and essentially a nocturnal species. It lives in burrows at the roots of bushes and grass among packed and semi-packed sand dunes, and feeds on different arthropods and their larvae. The gecko is preyed upon by local snakes, *Platyceps rhodorachis*, *Platyceps ventromaculatus*, *Spalerosophis arenarius*, and *Echis carinatus*, and varanid lizards.

The breeding season extends from March to July. The courting male vocalizes. Multiple clutches, each of 2 white oval sticky eggs measuring 11.15–12.30 by 8.2–9.85 mm, are laid in a safe place. Incubation period lasts 45–53 days, averaging 48 days (Szczerbak and Golubev, 1996).
Taxonomic notes: Anderson (1999) determined the holotype of *Crossobamon lumsdeni* as *Bunopus tuberculatus*. Nonetheless, a fourth species of *Crossobamon* (i.e., in addition to *C. eversmanni*, *C. orientalis*, and *C. maynardi*), as diagnosed above, occurs in Pakistan. If Anderson (1999) is correct, this species will require a formal description, with a designation of a holotype and a new name. As this is not the place for a description of a new species, I retain the conventional taxonomy. Szczerbak and Golubev (1986) regarded this form as a subspecies of *C. eversmanni*.

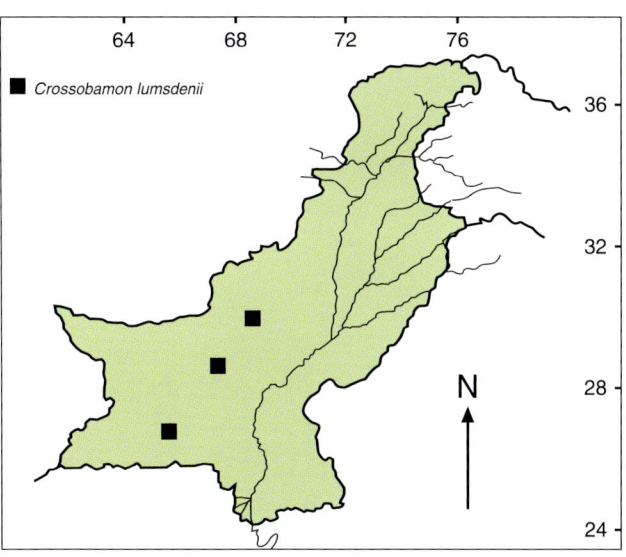

Distribution: Recorded in Pakistan between Nushki and Helmand in northern Baluchistan.

Crossobamon maynardi (Smith, 1933)
(Striped sand gecko: Dharidar kanghi-ungusht)
1933 *Stenodactylus maynardi* Smith, Rec. Ind. Mus. Calcutta 35: 18.
Type locality: Baluchistan, near the Afghanistan border.
Diagnosis:
1. Snout is longer than distance between eye and the ear opening.
2. Supralabials 13–15, infralabials 12–13.
3. Belly with small round keeled scales.
4. Hindlimb reaches to axilla.
5. Toes with marked lateral denticulations.
6. Male with 9 preanal pores, female with 9 enlarged pitted scales.
Snout-vent length 69–70 mm, tail 75–76 mm.
Color: Three yellow irregular stripes, with brownish black bands, from head to tail. Ventrum slightly pinkish.
Natural history notes: Like other members of the genus, the striped sand gecko is essentially nocturnal. It excavates burrows in the roots of vegetation and feeds on desert arthropods. When encountered it runs to the nearby shelter, spraying sand behind (Minton, 1966).

It breeds from April to early June. Clutch size is 2 eggs, which are laid in a burrow.

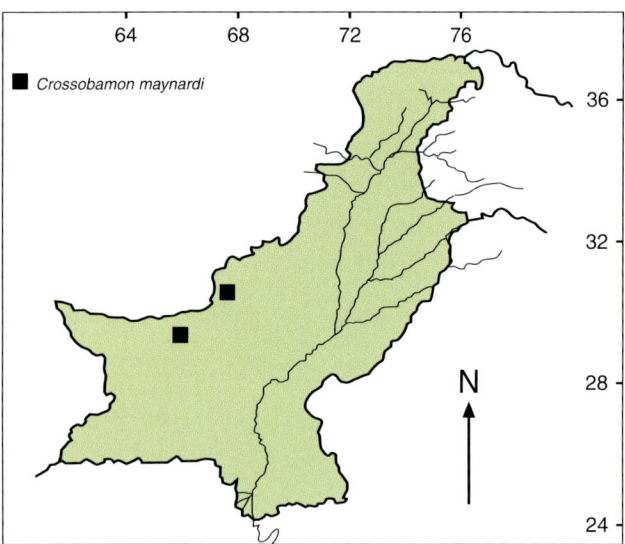

Distribution: The striped sand gecko has been collected in northwestern Baluchistan.

Szczerbak and Golubev (1996) regard this species as a synonym of *Crossobamon eversmann lumsdeni*.

Crossobamon orientalis (Blanford, 1876) (Plate 63)
(Yellow tailed sand gecko: Pelee-dum kanghi-ungusht)
1876 *Stenodactylus orientalis* Blanford, J. Asiatic Soc. Bengal, Calcutta (2)45: 21.
Type locality: Rohri and Shikarpur Districts, upper Sind, Pakistan.
Diagnosis:
1. Snout about as long as distance between eye and the ear opening.
2. Supralabials 11–13, infralabials 9–11.
3. Head scales fairly uniform, flat or feebly keeled.
4. Abdominal scales small, round and keeled.
5. Hindlimb reaches to the axilla.
6. Toes long, with well marked lateral denticulations.
7. Male with 1–4 preanal pores.
Snout-vent length 47–49 mm, tail 48–50 mm.

Color: Dorsum pale sandy, with indistinct dark crossbars. A dark line from eye to the sides of the body. Dorsal tubercles dark brown. Ventrum whitish.

Natural history notes: The gecko is nocturnal, feeding on different desert arthropods. Its burrow has a large round opening. Most of the day it stays close to the opening, catching passing insects. As the sun rises and the upper layers of sand become increasingly hot, it moves into the cool depth of its burrow.

When encountered in the open, it straightens its limbs, arches its body and twitches its tail, and sprints to escape to nearby shelter or into its burrow. While escaping it runs in short angular sprints, spraying sand behind (Khan, 1985a). While being handled it gives faint snarling squeaks and struggles by twitching its tail, which often breaks. In this effort it usually voids excrement.

It breeds from March to May. Normal clutch size is a pair of eggs, laid in a burrow. Juveniles are seen from April to late May.

Distribution: The yellow tailed sand gecko is widely distributed throughout the Thar, Cholistan, and Thal Deserts; moreover, it has been recorded from Sind Delta and Las Bela, southern Baluchistan (Minton 1966).

Plate 63. *Crossobamon orientalis*.

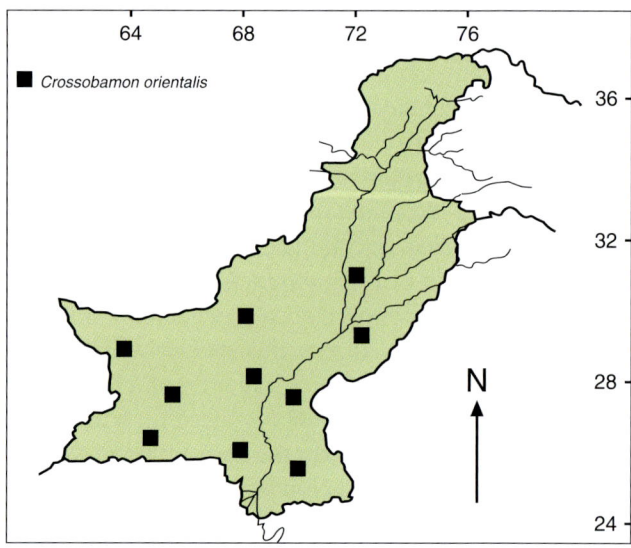

Chapter 6. Lizards

Genus *Cyrtopodion* Fitzinger, 1843

Dorsum with typical trihedral tubercles arranged in regular longitudinal rows, bigger than interspaces between them; caudal tubercles mucronate, laterally broadly in contact with each other; male with preanal, rarely with femoral pores; a single row of transversally enlarged subcaudals (Khan 2001a). There have been several attempts to group these geckos (Annandale, 1913; Smith, 1935; Szczerbak and Golubev, 1986; Khan, 2001a; Khan and Rosler, 1999; Anderson, 1999).

Cyrtopodion is the most advanced gekkonid genus (Szczerbak and Golubev, 1996). Evidence goes in favor of believing that Indus Valley *Cyrtopodion* group of species is monophylatic, evolved recently from a *C. scabrum* or *C. scabrum*-like ancestor, that probably invaded Pakistan along the Iran-Afghan border, and mostly through the Makran coastal strip (Minton, 1966; Mertens, 1969; Khan, 2003e). *Cyrtopodion scabrum* typifies the basic *Cyrtopodion* morphology (Annandale, 1913; Szczerbak and Golubev, 1986, 1996; Anderson, 1999; Khan, 2003e), while the rest of the Indus Valley *Cyrtopodion* geckos, have derived characters and are limited in distribution within the bounds of the valley. Moreover, their common ancestry from a *C. scabrum*-like ancestor is strongly indicated by their synapomorphic morphology and overlapping scutellation and occurrence of intergrades between them, suggesting that some level of inbreeding is still going on in the areas of their sympatry (Khan, 2003e).

This genus includes Pakistani ground geckos: *Cyrtopodion agamuroides, Cyrtopodion kachhense, Cyrtopodion kohsulaimanai, Cyrtopodion montiumsalsorum, Cyrtopodion potoharensis, Cyrtopodion scabrum,* and *Cyrtopodion watsoni.*

Cyrtopodion agamuroides (Nikolsky, 1900) (Plate 68)

(Makran spider gecko: Makrani makra chapkali)
1900 *Gymnodactylus agamuroides* Nikolsky, Ann. Mus. Zool. St. Petersburg, 4: 384.

Type locality: Pensarch (Pendzhsara), eastern Kirman, Iran.

Diagnosis:
1. Nasal scales 3.
2. Scales with preanal pores are much larger than surrounding scales.

Plate 68. *Cyrtopodion agamuroides.* (Photo courtesy of S.C. Anderson).

3. Transverse rows of subcaudal scales under a caudal segment 3.
4. Scales across midbelly 28.
5. Male with 2–4 preanal pores.

Snout-vent length 38–40.2 mm, caudal length 49–50 mm.

Color: Khaki brown dorsum with 7 cross-rows of 3–5 dark brown spots arranged between neck and level of vent. Tail with 13 thick bars; 4–5 bars on limbs; ventrum white.

Natural history notes: This gecko has been collected in different situations from barren scrubland to rocky

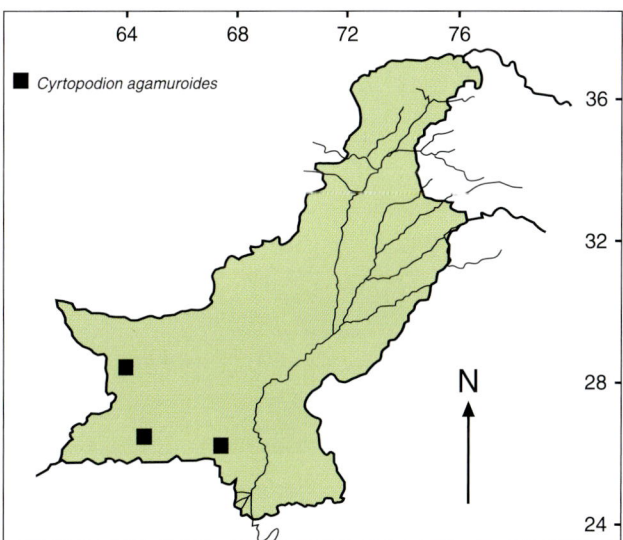

cliffs (Zarudny, 1904; Minton, 1966). It is nocturnal and very agile on rocks during the night, jumping from one rock to another. During the day it is found resting under stones or in crevices among rocks or bricks. Rare in the rainy season, it is found to be active under the shade during the day.

It lays 1–2 clutches of a pair of round-shelled eggs, from March to May, in crevices or under stones. Young are active by late May-July.

Distribution: From southeastern Iran to Las Bela.
Taxonomic notes: Minton (1966) and Mertens (1969a) identified this gecko as *Agamura agamuroides*. However, Szczerbak and Golubev (1984) placed this gecko in genus *Tenuidactylus*.

Cyrtopodion kachhense kachhense (Stoliczka, 1872) (Plate 69)

(Kachh spotted ground-gecko: Kachh chapkali)
1872 *Gymnodactylus kachhensis* Stoliczka, Proc. Asiatic Soc. Bengal, Calcutta (1): 79.
Type locality: Kutch, southwestern Sind, Pakistan.
Diagnosis:
1. Subcaudals small, not broader than adjacent scales (Figure 14C).
2. Scales across midabdomen 30.
3. Dorsal tubercles subtrihedral (Figure 14A). Snout-vent length 37–40 mm, tail 40–45 mm.

Color: Dorsum yellowish gray, with irregular dark brown spots, tail barred.
Natural history notes: The spotted ground gecko is plentiful in hard rocky terrain. It lives in cracks and holes in ground, readily colonizing crevices among rocks and nearby houses, and crevices under bridges and walls. It avoids competition with *Hemidactylus flaviviridis Hemidactylus persicus*, and *Hemidactylus brookii*, which are sympatric species (Khan and Tasnim, 1990b). At nightfall it is attracted to areas under light posts to feed on photophilic insects, and confines itself to the periphery of the feeding ring of the other resident geckos. Minton (1966) found it plentiful during the day after rains, however during dry conditions it seeks moist environs by retreating deep under rocks and in holes in the soil.

It feeds mainly on soft-bodied insects and their larvae. Breeds from late February to early June. One or two thin-shelled oval eggs, 9.5 by 5 mm, are deposited in a protected place on powdery soil or under decaying *Euphorbia* stalks (Minton, 1966). Juveniles are active by mid-March to late July.

Distribution: The spotted ground gecko has been collected from most of Kutch, coastal Sind, and Las Bela, Pakistan.

Gymnodactylus ingoldbyi described by Proctor (1923) from the hilly tracts along the western borders of Punjab Pakistan and southwestern NWFP, has recently been described as a northern race of *Cyrtopodion kachhense* by Khan (1997f).

Plate 69. *Cyrtopodion kachhense kachhense.* (Photo courtesy of S.A. Minton).

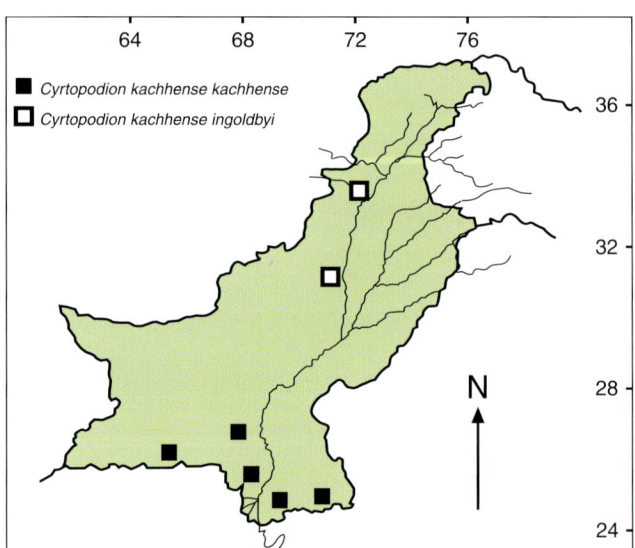

Chapter 6. Lizards

***Cyrtopodion kachhense ingoldbyi* Khan, 1997**
(Western rock gecko: Chattani chapkali)
1923 *Gymnodactylus ingoldbyi* Proctor, J. Bombay Nat. Hist. Soc. 29:121.
1997 *Cyrtopodion kachhense ingoldbyi*, Khan, Russian J. Herpetol. 4:83–88.
Type locality: Ladha, Dera Ismael Khan, southern NWFP, Pakistan.
Diagnosis:
1. Dorsum with trihedral tubercles, the sides of which are bowed in so that the tubercle has a sharp median longitudinal keel.
2. Interparietal scales 14–16.
3. Scales across midbelly 32–40.
4. Lateral cloacal tubercles 2–3.
5. Dorsolateral caudal keeled tubercles in 4 rows.
Snout-vent length 37–53 mm, tail 47–65 mm.

Color: Dorsum light gray with yellowish gray ventrum, with irregular dark brown bars, tail and limbs barred.

Natural history notes: Khar-Rakhni-Quetta road leading to the Punjab-Baluchistan border meanders though a series of low hills. Several bridges are constructed to allow fast-flowing stream water to flow under the road. The holes and crevices among the bricks and stones under the bridges are colonized by *Cyrtopodion kachhense ingoldbyi*. Tufts of meager grass are the only vegetation growing between boulders along the fast-flowing stream bed, supporting local arthropod fauna upon which these gecko feed. There is a flow of water only when there is rain.

Several geckos were also seen on boulders and rocks both near to and away from any bridges; moreover, they readily extend into nearby houses where they feed on photophilic insects. *Cyrtopodion kachhense ingoldbyi* appears to be the only gecko in these hills, however, Khan (1993b) reported a new gecko, *Indogekko fortmunroi* from the buildings around Fort Munro and Khar Garden, about 80 km northwest of Dera Ghazi Khan City, Punjab, Pakistan.

Distribution: This gecko is a mountain form, widely distributed in the hills along Khar-Rakhni-Quetta road, about 10 km west of Fort Munro.

Plate 70. *Cyrtopodion kohsulaimanai.*

***Cyrtopodion kohsulaimanai* (Khan, 1991) (Plate 70)**
(Sulaiman Range gecko: Koh-Sulaiman chapkali)
1991d *Tenuidactylus kohsulaimanai* Khan, J. Herpetol., 25:199–204.
Type locality: Sakhisarwar village, Dera Ghazi Khan District, northwestern Punjab, Pakistan.
Diagnosis:
1. Interorbitals 14–17.
2. Dorsum with large trihedral tubercles arranged in longitudinal rows, not in contact with each other.
3. Rows of scales across midabdomen 27–30.
4. Scales along midventrum of the body from postmentals to the anterior lip of the vent 120–138.
5. A continuous series of 30–40 preanofemoral pores.
Snout-vent length 54–59 mm, tail 70–73 mm.

Color: Dorsum light gray, with fine light brown granulations; head and supralabials mottled with brown. Brown and white tubercles alternate with each other. Juveniles with 6–7 indistinct transverse bands from neck to the vent. Limbs and tail barred.

Natural history notes: The Sulaiman Range gecko is essentially a ground gecko, inhabiting barren sparsely vegetated areas, in semi-sandy desert situations where it inhabits holes and fissures in the ground. However, where artificial structures in the form of buildings and bridges are available, the gecko readily invades them, inhabiting crevices and holes among bricks and stones. It is here the gecko is collected; otherwise, to collect it from its natural habitat is very difficult.

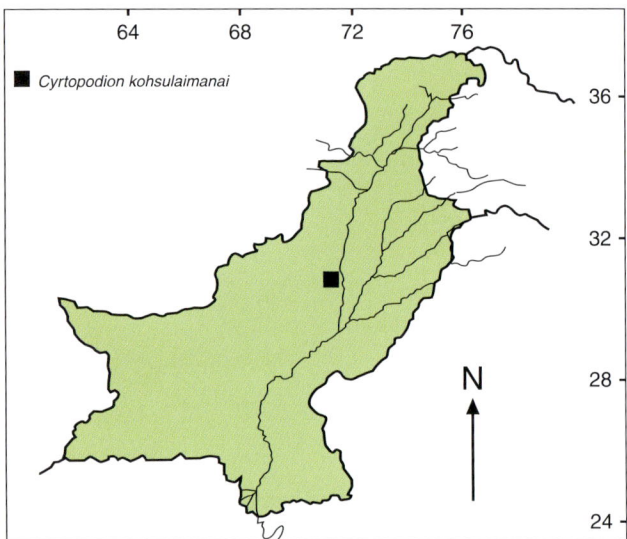

The gecko is nocturnal, foraging in stony terrain with low sparse vegetation, feeding predominantly on arthropods, grasshoppers, crickets, and moths. The gecko is very agile, quickly taking to deep crevices on approach of danger.

The Sulaiman Range gecko has quite an extended breeding season, from March to October. Most probably more than one clutch of 2–3 hard-shelled round eggs, 8.9–9.6 mm in diameter, laid in crevices. The juveniles are active by early April (Khan, 1990a).

Distribution: This gecko is known from two localities, Sakhisarwar and Rakhni Gorge, along Dera Ghazi Khan-Fort Munro road, in District Dera Ghazi Khan, northwestern Punjab, Pakistan.

Cyrtopodion montiumsalsorum (Annandale, 1913) (Plate 71)

(Salt Range ground-gecko: Koh-namak chapkali)
1913 *Gymnodactylus montiumsalsorum* Annandale, Rec. Indian Mus. Calcutta 9:309–326.
Type locality: Salt Range, Punjab, Pakistan.
Diagnosis:
1. Dorsal large trihedral tubercles are arranged in regular longitudinal rows. They are mostly in contact with each other.
2. Scales across midabdomen 21–23.
3. Interorbitals 11–13.

Plate 71. *Cyrtopodion montiumsalsorum.*

4. Scales along midventrum of body from postmentals to the anterior of vent 112–115.
5. A series of 26–32 preanofemoral pores.
Snout-vent length 43–46 mm, tail 58–60 mm.

Color: Dorsum light gray, dark brown and light tubercles are transversally disposed in alternate rows, giving the dorsum a variegated pattern. Tail with 13–14 crossbars. Head and labials mottled with dark brown to black. Belly light.

Natural history notes: The Salt Range gecko lives primarily in holes and crevices in badlands terrain along the southern Salt Range. It secondarily invades holes and crevices in the walls of buildings and rocks. It occurs sympatrically with *Cyrtopodion scabrum* and *Cyrtopodion watsoni*. In Rohtas Fort, it is sympatric with *Indogekko rohtasfortai*. Nocturnal, it feeds on grasshoppers, moths, larvae and other soft-bodied arthropods.

It breeds from March to July and the breeding male becomes aggressive. A male follows a female, coughing and grunting with inflated neck and half-open mouth. The female is led into a remote corner where copulation takes place. A pair of round hard-shelled eggs varying 7.7 to 8 mm in diameter is laid under a stone.

Distribution: This gecko has been collected from various localities in the southern Salt Range, Punjab, Pakistan (Khan, 1989).

Chapter 6. Lizards

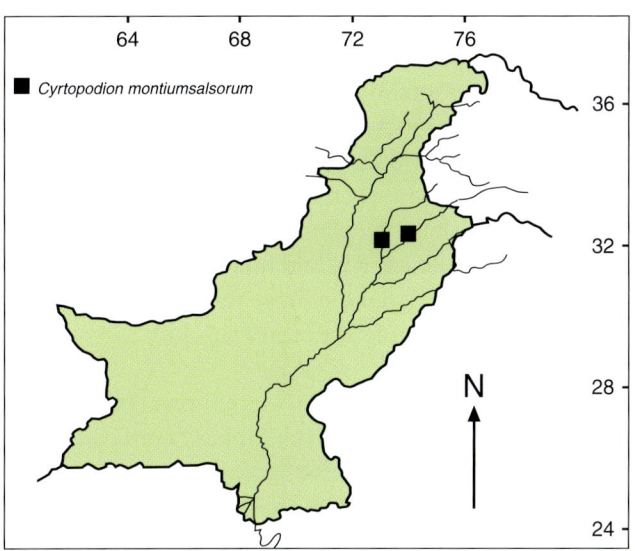

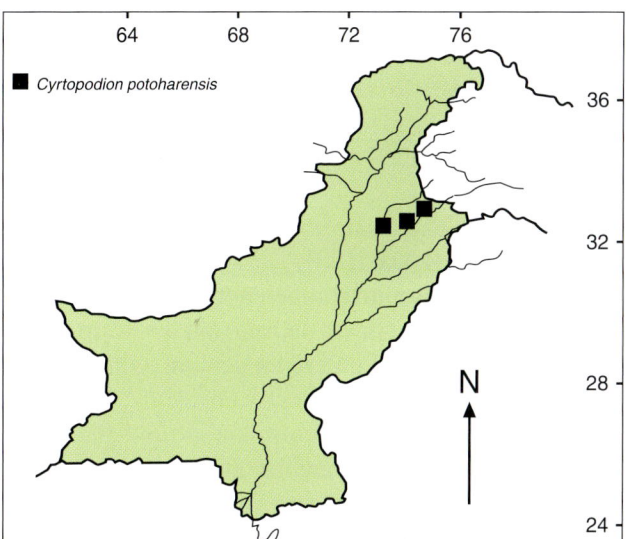

Cyrtopodion potoharensis **Khan, 2001 (Plate 72)**
(Potohar gecko: Potohar chapkali)
2001a *Cyrtopodion potoharensis* Khan, Pakistan J. Zool.33(1):15–16.
Type locality: Nazampur, District Attock, northern Punjab, Pakistan.
Diagnosis:
1. Supralabials 10–15.
2. Interorbital scales 12–17.
3. Abdominals squarish, 25–33 across midabdomen.
4. Midventrals 121–145, from the postmentals to the preanal lip.
5. Postfemoral tubercles 5–12.
6. Preanal pores 6–7, arranged in an angular arch, in distinctly enlarged scales.
7. Caudal tubercles distinctly mucronate.
Snout-vent length 37–52 mm, tail 47–64 mm.

Color: Dorsum light bluish with 3–4 heterogeneous squarish dark spots arranged in 5–8 transverse series, from nape to the level of vent, tail with 10–12 dark bands. Limbs and digits heavily barred with black.

Natural history notes: This gecko was collected during the day from holes and crevices under roads and rail bridges, and under from under stones and slabs in District Attock, northern Punjab. Some were collected from holes and crevices along sides of mudflats through which roads run. The geckos were extracted from the holes and crevices with the help of long forceps. The area is semiarid countryside, and its occasional thorny bushes, trees, and tufts of grasses give the impression of a scrubland. *Trapelus agilis* and *Uromastyx hardwickii* were collected from the area. In the central Potwar, several geckos were extracted from holes and crevices among rocks (Khan and Baig, 1988).

Range: The Potwar gecko was collected from different localities in central Potwar Plateau, Salt Range, Punjab.

Plate 72. *Cyrtopodion potoharensis.*

Cyrtopodion scabrum **(Heyden, 1827) (Plate 73)**
(Common tuberculate ground-gecko: Toor chapkali)
1827 *Stenodactylus scaber* Heyden, *in:* Ruppell, Atlas N. Afr. Rept.:15.
Type locality: Arabia.
Diagnosis:
1. Subcaudals in a single series of broad scales (Figure 14A, D).
2. Scales across midabdomen not more than 25. Snout-vent length 43–44 mm, tail 43–45 mm.

Color: Light gray to light brown dorsum, with regular brown spots, tail barred with dark brown.

Natural history notes: A common ground gecko, it inhabits dry scrublands, and dry stony hillsides with typical sparse xerophytic vegetation. Crevices, holes, and fissures in soil with sparse grass are inhabited by this gecko. Near human habitations, it lives in crevices in boundary walls of inhabited houses, occasionally venturing inside the buildings attracted by the photophilic insects: however, it avoids confrontation with common house geckos *Hemidactylus flaviviridis* and *H. brookii*.

It is very common around Quetta, Baluchistan, where it lives in crevices and under the plastering of stone and mud walls around houses and boundary walls built around fields. It comes out at dusk and swarms the surrounding vegetation to forage, returning to its own crevice before dawn.

In Punjab it breeds from March to July. Vocal breeding males chase each other to get hold of females. More than one clutch of a pair of spherical hard-shelled white eggs is laid in a protected place, a hole, under a slab, or in the loose mud plastering of boundary walls. Young are seen from July to September.

Distribution: This gecko is wide-ranging, from Egypt to Rajputana, India. In Pakistan, it has been reported from the upper and lower Indus Valleys, and along the eastern edge of the Thar Desert. It is widely distributed in Baluchistan and Waziristan.

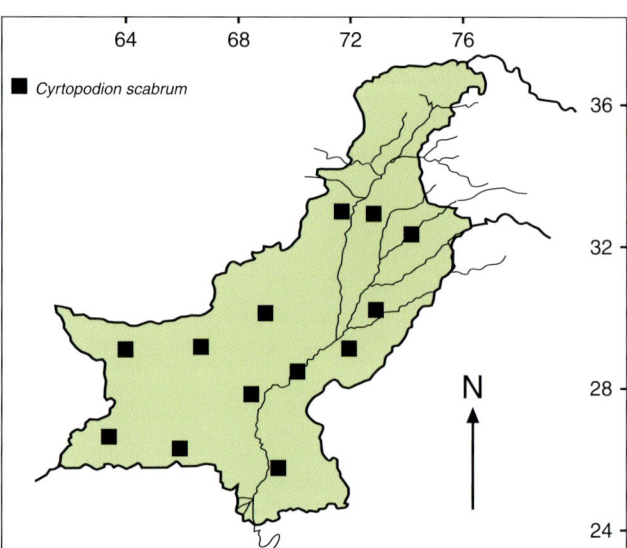

Plate 73. *Cyrtopodion scabrum.*

Cyrtopodion watsoni **(Murray, 1892)**
(Northern spotted ground-gecko: Shamali chapkali)
1892 *Gymnodactylus watsoni* Murray, Zool. Belooch.: 68.
Type locality: Quetta, Baluchistan, Pakistan.
Diagnosis:
1. More than 25 scales across midabdomen.
2. Dorsal granular scales intermixed with larger flat keeled scales and large trihedral tubercles. Snout-vent length 50–53 mm, tail 54–57 mm.

Color: Light brown dorsum, with scattered dark spots. Tail barred with dark.
Natural history notes: This gecko lives in holes and crevices in the ground, and stone or brick walls. It is nocturnal, coming out of its burrow just after sunset. It occurs sympatrically with *C. scabrum*; however, it avoids competition with other sympatric geckos, *Hemidactylus flaviviridis* and *H. brookii*.

It breeds from March to May, when more than one clutch of a pair of white hard-shelled eggs is laid in some safe humid place. The male vocalize to mark its territories and to attract females. Breeding males are seen chasing each other.
Distribution: Reported from western Salt Range, Punjab, and Quetta, Baluchistan, Pakistan.

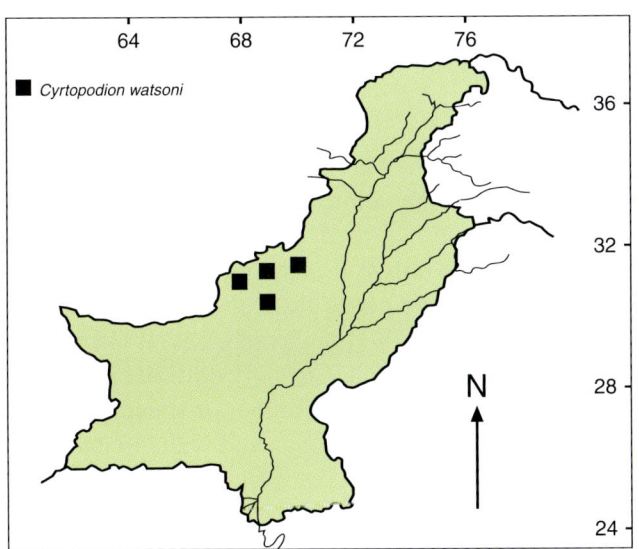

Plate 74. *Hemidactylus brookii*.

Type locality: Borneo, Southeast Asia.
Diagnosis:
1. The ear opening is oval, about one-third the size of the eye.
2. Dorsum granular, interspersed with small subtrihedral tubercles.
3. Supralabials 8–10, infralabials 7–9.
4. Lamellae under 4th toe 8–10.
5. Tail cylindrical, distinctly segmented, with 3 dorsolateral rows of small flat caudal tubercles, a single row of broad subcaudals.
6. Male with preanal and femoral pores, separated medially by 2–3 scales.

Color: Dorsum shows metachromatic changes, from dark brown to light gray, spotted with dark, a dark stripe from eye to the temple.
Natural history notes: The spotted barn gecko frequents piles of chopped vegetation, logs, crops, in crevices and holes in the ground, under tree bark, around potted plants, dark uninhabited huts, leaf litter, and piles of trash. It is known to frequent tilled areas, forests, oases where it is found under leaf litter, fallen trees, and anything which can provide shelter. The most common gecko in the countryside, they swarm out of their holes and crevices, foraging around just after sunset. They

Genus *Hemidactylus* Oken, 1817
From dilated basal part of digits distal phalanges arise angularly, digits covered with scales, not denticulate; all digits clawed; dorsum with uniform granular subimbricate scales, sometimes mixed with round, enlarged tubercles.
Seven species known from Pakistan.

Hemidactylus brookii Gray, 1845 (Plate 74)
(Spotted barn gecko: Barani chapkali)
1845 *Hemidactylus brookii* Gray, Cat. Liz. Brit. Mus.:153.

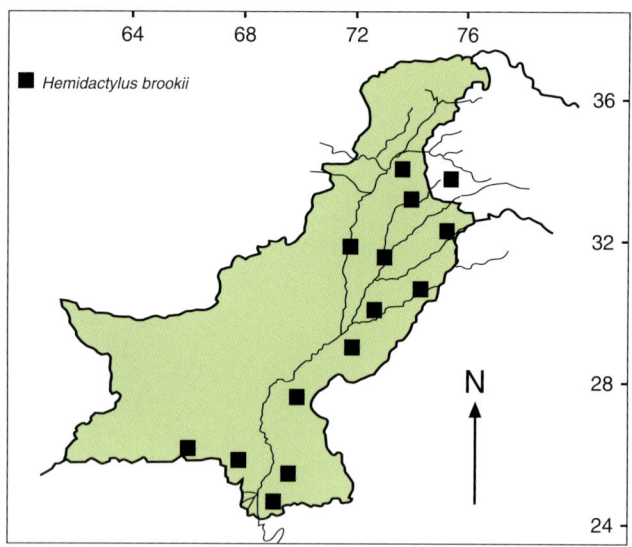

avoid competition with the common house geckos, *Hemidactylus flaviviridis*, which inhabits the interiors of buildings.

It is a sluggish animal, and gives little squeaks and voids excrement upon handling. Under slight pressure it breaks off its tail. It is nocturnal; however, it is active under leaf litter 3–4 hours before sunset. It feeds on soft-bodied arthropods which are in abundance under warm humid leaf litter.

Breeding season is extended, from March to October; more than one clutch of a pair of white hard-shelled eggs, is laid in a protected place during the breeding season.

It falls prey to local amphibians, larger lizards, and snakes.

Distribution: The spotted barn gecko has a wide range in Southeast Asia, from Borneo, China, through tropical and subtropical Asia, extending through India, Pakistan, and the Middle East to northern Africa. There are reports of it from the West Indies.

In Pakistan it is a common gecko in the plains, avoiding higher northern mountains, and extends into the peripheral humid areas around deserts and oases. In Rohtas Fort it occurs sympatrically with *Cyrtopodion montiumsalsorum*, *Indogekko rohtasfortai*, *Hemidactylus flaviviridis*, and *Hemidactylus persicus*. It is widely distributed in alpine Punjab and Azad Kashmir.

Hemidactylus flaviviridis **Rüppell, 1835 (Plate 75)**
(Yellow-belly common house-gecko: Ghar chapkali)
1835 *Hemidactylus flaviviridis* Rüppell, Neue Wirb. Faun. Abyss.:18.
Type locality: Massaua Islands, Eritrea.
Diagnosis:
1. Dorsum with granular scales, no distinct tubercles.
2. supralabials 12–15, infralabials 10–14.
3. Lamellae under first 7–10, under fourth toe 12–15.
4. Tail indistinctly segmented; caudal tubercles small and conical.
5. Preanofemoral pores 8–15.
   Snout-vent length 86–95 mm, tail 89–93 mm.

Color: Dorsum shows marked metachromatic variations. Those geckos living outside the buildings, under tree bark, etc., are greenish gray, with 5 distinct dark wavy crossbars; tail is similarly barred. Dorsal pattern fades out in older geckos or in those living inside the buildings and rarely exposed to sunlight, usually they are almost uniformly grayish white; however, the ventrum is always light yellowish.

Natural history notes: It is the most common and most familiar house gecko throughout the plains of Punjab, Sind, and sub-Himalayan areas. It is a strict commensal with man and strongly edificial species. A very agile and excellent climber, it spends most of its time clinging to the ceilings and walls in dark

Plate 75. *Hemidactylus flaviviridis.*

Chapter 6. Lizards

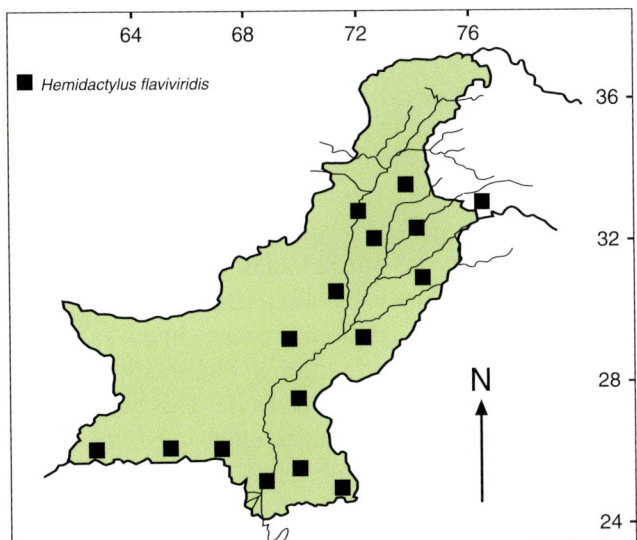

corners; it walks with ease on vertical walls and plain glazed glass surfaces. It spends the day hiding under electrical casings, loose cement plastering, wall hangings, curtains, and other household objects. It is often seen silhouetted against windows and doors. It climbs into large trees, like *banyan*, *Acacia*, and *Dalbergia*, where it lives and lays its eggs in holes or under bark. The arboreal invasion is a usual phenomenon during rains; however, old trees in inhabited areas always have a resident population of this gecko.

At night, it comes out of its hideout and feeds on photophilic insects. It runs after and stalks photophilic insects on vertical walls, often descending to feed on those falling onto the ground. It is a common sight for an adult gecko to chase a juvenile one as it comes into its feeding range. Often the juvenile escapes by autotomizing its tail, which is readily devoured by the pursuer. It almost gorges itself with insects and grows quite fat during the rainy season. The gecko is quite selective in its choice of insects upon which it feeds. It generally prefers the soft-bodied insects, avoiding large beetles, bees, and wasps. A feeding gecko appears to recognise the type of insect by carefully observing its shape and movements, sometimes appearing to decide at the last moment whether to catch or to leave an insect. Larger prey, grasshoppers and katydids, etc., are caught and they are thrashed several times against the substratum, until they are subdued, and their hard body coverings broken. Then the prey is engulfed as a whole, headfirst. For this the gecko sometimes has to exert considerable pressure on its throat muscles to swallow it down.

Fights among adults are common at feeding sites. The fights are partially to secure a good feeding site, a chosen prey or to get a female, since several females and males feed together. Normally every individual in the feeding ring maintains a certain distance from each other; as one changes its position to follow an insect, etc., the whole setup is changed. There is always a dominant male that fights with others trying to come within its activity range.

The gecko breeds from May to June. In the breeding male, grunts are like weak coughing, "skhh, skhh, skhh, skhh," with half-open mouth and blown out neck. The female, who runs before the male, is followed and hotly pursued until she comes to rest in a corner. The approaching male holds her neck, entwines its tail around that of the female, and effects the intromission. A gravid female lays 2, rarely 3, white-shelled round eggs, measuring 9.8 by 10 mm, in some protected dark place, usually among stacks of books or under household articles in remote dark corners. Juveniles are usually seen by late May–July (Mahendra, 1936).

Distribution: This species has a wide range in the Palearctic Region, from the Red Sea to the coasts of Arabia and Iran, Pakistan, and India to Bangladesh. Human agency has played an important role in its wide distribution. In Pakistan it is reported from throughout the plains below 1000 m, always in association with man.

Hemidactylus frenatus **Schlegel, 1836 (Plate 76)**
(Waif gecko: Awara chapkali)
1836 *Hemidactylus frenatus* Schlegel, *in:* Duméril and Bibron., Erpet. Gen. 3:366.
Type locality: Java, Southeast Asia.
Diagnosis:
1. First toe less than half the length of second.
2. Dorsal tubercles few or absent.
3. Supralabials 10–12, infralabials 8–10.
4. Subdigital lamellae under first 4–5, and 9–10 under fourth toe.

Plate 76. *Hemidactylus frenatus.*

5. Tail feebly depressed, with a series of enlarged subcaudals.
6. Male with a continuous series of 26–36 preanofemoral pores.
   Snout-vent length 60–62 mm, tail 60–67 mm.

Color: Dorsum grayish or pinkish, sometimes much darker. With indistinct dark spots, sometimes arranged in stripes. Ventrum white, tail sometimes red.

Natural history notes: It is a coastal species that does not venture deep into the plains. Its natural habitat coincides with that of *Hemidactylus brookii*. It inhabits parks and gardens where coconut and date palm trees are common. They have been collected from among palm fronds or beneath rubbish. It rarely ventures into buildings, apparently avoiding competition with the common yellow-belly house gecko and *Hemidactylus turcicus*.

Its call is a series of 4–5 loud staccato notes similar to the sound of a coin tapped against a window pane (Minton, 1966); it is said to be one of the subcontinent's noisiest geckos (Daniel, 1983). A pair of eggs measuring 9.8 by 8.3 mm is laid in palm axils, from April to June. Young hatch after 42 days.

Distribution: It is a pantropic species. In the subcontinent it is coastal in distribution; in Pakistan it has been collected from the lower Indus Delta. This gecko has been reported from different localities in North America (Norman, 2003).

Hemidactylus leschenaultii Duméril and Bibron, 1836 (Plate 77)

(Tree-bark gecko: Chaal chapkali)
1836 *Hemidactylus leschenaultii* Duméril and Bibron, Erpet. Gen. 3:364.

Type locality: Sri Lanka.

Diagnosis:
1. Round feebly keeled tubercles are scattered among granular dorsal scales which are much smaller than the interspaces, few on anterior, numerous on posterior half of the body.
2. Supralabials 10–12, infralabials 8–10.
3. Lamellae under first 6–7, and under fourth toe 9–11.
4. Tail strongly depressed, segmented, with a median series of enlarged subcaudals; 6 rows of dorsal pointed subcaudal tubercles.

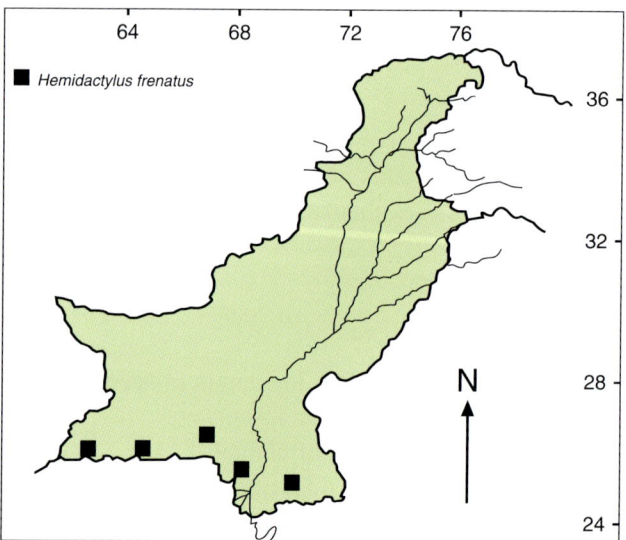

Plate 77. *Hemidactylus leschenaultii.*

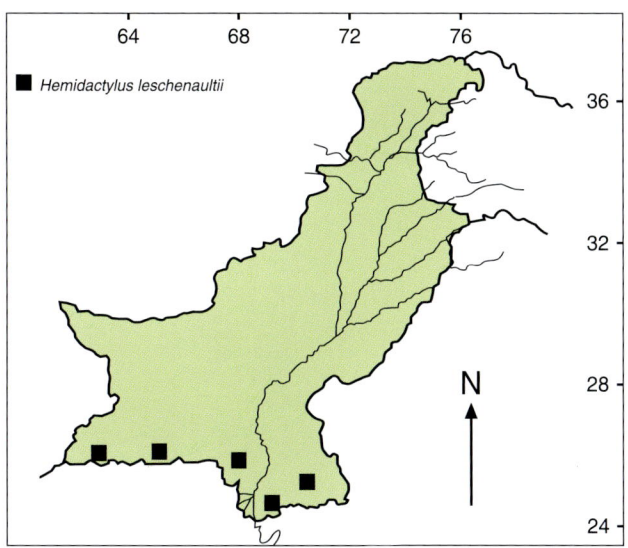
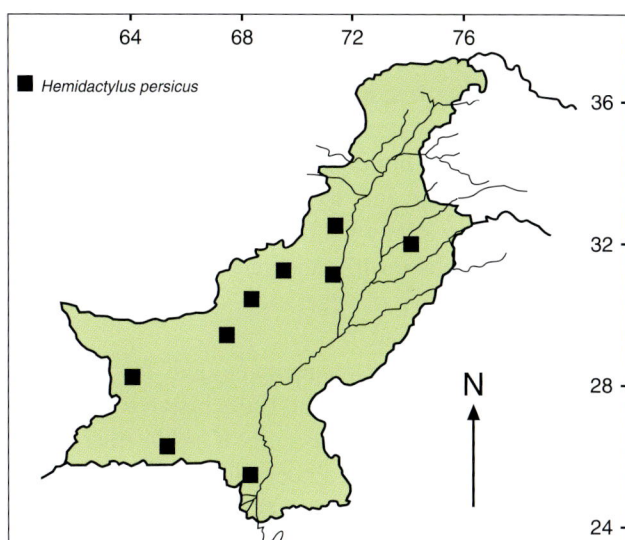

5. Male with 10–17 femoral pores, medially separated by several scales.
   Snout-vent length 85–86 mm; tail 85–87 mm.

Color: Dorsum gray, with dark brown crossbars or rhomboidal blotches. A dark streak from eye extends onto flanks. Dirty white belly.

Natural history notes: This is arboreal, preferring large mango and banyan trees, and rarely venturing inside buildings. It is a common house gecko in Bengal (Daniel, 1983). The gecko lives under loose tree bark or among branches. At nightfall it runs on the tree stems and branches, seldom descending to the ground.

It breeds from March to August. More than one clutch is laid, each of a pair of eggs measuring 7.9 by 9 mm.

Distribution: This gecko extends from Assam, Bangladesh, eastern and southern India, along the western coast, reaching the lower Sind in Pakistan, where it is recorded from various localities in the lower Indus Delta and Las Bela in southern Baluchistan.

***Hemidactylus persicus* Anderson, 1872 (Plate 78)**
(Persian house-gecko: Irani chapkali)
1872 *Hemidactylus persicus* J. Anderson, Proc. Zool. Soc.:378.

Type locality: Shiraz, Persia.

Diagnosis:
1. Supralabials 10–12, infralabials 8–10.
2. Lamellae under first 8–10, under fourth toe 12–14.
3. Preanal pores 9–13.
   Snout-vent length 68–70 mm, tail 84–86 mm.

Color: Dorsum light brownish gray, with irregularly scattered white, brown, or black keeled tubercles. Irregular light dark spots, a dark line on the side of the head, ventrum whitish.

Natural history notes: The Persian gecko inhabits trees and buildings in rocky deserts and xerophytic stony terrain. It is often found to be plentiful on trees

Plate 78. *Hemidactylus persicus.*

in oases. It invades buildings, especially old tombs, where it is the dominant gecko. Nocturnal, during the day it rests under tree bark, and in crevices in rocks and bricks; usually it remains clinging against walls in the darker parts of buildings.

It breeds from July to October. A pair of round eggs, measuring 8.2 by 9.5 mm, is laid under some protected dark place with moderate temperature and humidity like a heap of vegetation, debris, logs, leaf litter and tree bark, etc.

Distribution: The Persian gecko extends from eastern Arabia, through southern Iran, Baluchistan to Sind, northward to Waziristan. Recently it has been reported from the southern Potwar Plateau, in central Punjab, Pakistan (Khan and Tasnim, 1990b).

Hemidactylus triedrus **(Daudin, 1802) (Plate 79)**
(Blotched house gecko: Sahali chapkali)
1802 *Gecko triedrus* Daudin, Hist. Nat. Rept.:155.
Type locality: Unknown.
Diagnosis:
1. Dorsal large trihedral tubercles are arranged in 13–16 longitudinal rows. The tubercles are larger than the interspaces between them.
2. Supralabials 8–10, infralabials 7–8.
3. Subdigital lamellae slightly oblique, 6–7 under first, and 7–10 under fourth toe.
4. Tail slightly depressed, with a series of transversally enlarged subcaudal scales.
5. Preanofemoral pores 6–14, in male, interrupted medially by 1–3 scales.
6. Naris separated from first supralabial.
Snout-vent length 75–77 mm, tail 76–78 mm.

Color: Dorsum yellowish, with 3 large dark brown saddles which are narrowly edged with black; tail with dark rings. One or two yellow stripes behind eye, another on nape. Ventrum pale.

Natural history notes: This gecko has been collected from flat semiarid habitat with scrubby vegetation in the coastal region of Pakistan. It inhabits holes and crevices among rocks, bricks, holes in the ground, and under vegetation. It invades buildings and keeps to the holes and crevices in the compound walls of houses. It is said to be associated with termite hills in India (Daniel 1983). The gecko emerges at dusk and goes foraging in the surrounding vegetation.

Plate 79. *Hemidactylus triedrus*.

Its defense mechanism is typical gekkonid: lifting and arching of the body, swaying and twitching of the tail, snarling and lunging and striking with its tail, usually voiding excrement at the same time.

Females with eggs are collected from April to July and young are plentiful from July to October. Sympatric *Eublepharis macularius* preys on it (Minton, 1966).

Distribution: Its range extends from Sri Lanka and peninsular India to Karachi, in Pakistan, where it is known from the coastal localities in the lower Indus Delta.

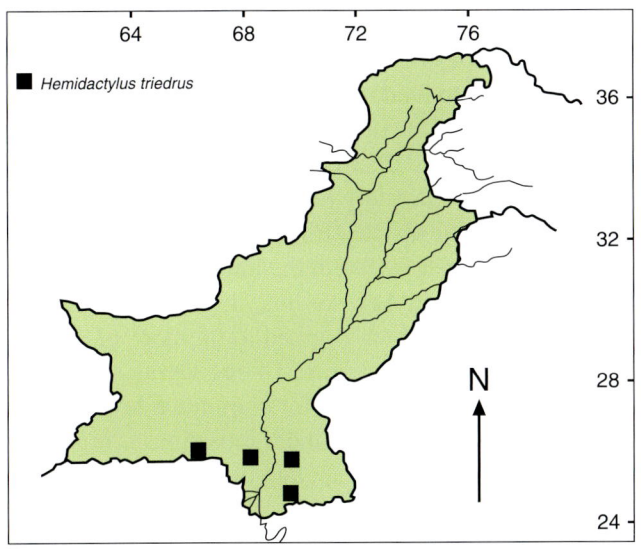

Hemidactylus turcicus (Linnaeus, 1758)
(Mediterranean house gecko: Turk chapkali)
1758 *Lacerta turcica* Linnaeus, Syst. Nat. Ed. (10)1:202.
Type locality: Asiatic Turkey.
Diagnosis:
1. Naris bordered by first supralabial.
2. Supralabials 7–10 and infralabials 6–9.
3. Dorsal tubercles large, subtrihedral arranged in 14–16 fairly regular longitudinal rows.
4. Subdigital lamellae under inner 7–8 and under fourth toe 8–11.
5. Tail subcylindrical, covered above with irregular, some what pointed scales and a series of 6–8 large pointed tubercles, median subcaudals are transversally enlarged.
6. Male has 4–10 (rarely 2) preanal pores. Snout-vent length 58–61 mm, tail 59–65 mm.

Color: Dorsum light brown or grayish, spotted with black, sometimes spots arranged in transverse series. A dark streak on sides of the head usually present. Ventrum dirty white.

Natural history notes: The gecko inhabits humid coastal environs. It invades cities where it is collected from suburban areas. It is picked especially from the base of palm fronds, from under rocks, piles of stones, and bricks and rubbish. Normally nocturnal, it is active during humid days.

No particular breeding period is recorded; gravid females are met with almost all year round. A clutch of a pair of eggs, measuring 9.2 by 8 mm, is laid in some protected place, also used for the same purpose by several geckos.

Distribution: Its range is quite extensive, due mainly to man, and includes the West Indies, eastern Mexico, southern United States, northern Africa, circum-Mediterranean countries and islands, the Middle East, Iran, and Afghanistan. In Pakistan it has been reported from various localities along coastal Sind. In Karachi city this gecko has been found to occur in houses side by side with *H. flaviviridis* (Minton, 1966).

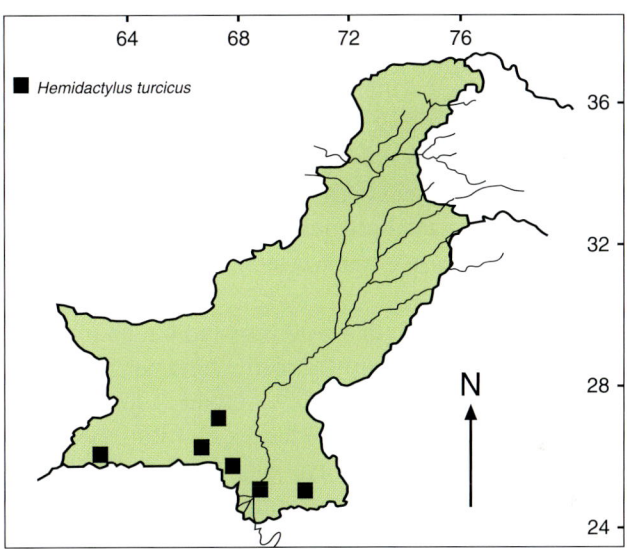

Genus *Indogekko* Khan, 2003
Sandstone geckos (Khan, 2001d). Large, thin geckos, with body and tail depressed, tail quadrangular, fragile at any point along its length, much longer than body, with very gradual taper; three rows of flat, keeled, laterally flared caudal tubercles, twice broadened subcaudals in a single series, two to a segment, laterally flanked by a row of elongated scales; supracilliary spines present on the posterior dorsal part of upper eyelids; limbs and digits long, tips of fingers reach snout tip or beyond. Dorsal granular scales flat, pentagonal, juxtaposed, laterally arranged roughly in transverse rows, interspersed with 2–3 times larger, flat, round to oval, smooth or slightly keeled tubercles. Interorbital scales 13–18; 2 or more often 3 pairs of

postmentals, first in contact by a long suture; scales across midabdomen 23–28; midventral scales 129–132; postfemoral tubercles absent; a continuous series of 4–9 preanal and 6–9 femoral pores in male (in some species female may have preanal and femoral pores).

Four species re represented in Pakistan.

Indogekko fortmunroi (Khan, 1993) (Plate 80)

(Fort Munro sandstone gecko: Munro reg-sang chapkali)
1993 *Tenuidactylus fortmunroi* Khan, Pakistan J. Zool. 25:217–221.
Type locality: Khar Gardens, Fort Munro, District Dera Ghazi Khan, western Punjab, Pakistan.
Diagnosis:
1. Dorsal granular scales tubercular, juxtaposed, interspersed with 12 rows of flat, keelless, round larger tubercles.
2. Body much depressed, habitus weak.
3. Tail longer than body, segmented, with 3 rows of trihedral caudal tubercles on sides.
4. Subcaudals in a transversely enlarged median series.
Snout-vent length 48–50 mm, tail 65–68 mm.

Color: Dorsum light brown, with a series of 14 median transverse dark bands, as broad as the interspaces. Head light brown with dark mottling. A dark stripe from snout through eye joins first band on nape. Dorsal tubercles of light and dark form a mosaic dorsal pattern. Limbs mottled with dark. The geckos found in buildings are light gray with darker spots, and no dark mottling on head and limbs.

Natural history notes: This gecko inhabits crevices and holes among sandstone slabs, around Fort Munro. It readily invades buildings where it gathers around lights at night, to feed on photophilic insects. Strictly nocturnal, during the day it is seen clinging to the roofs and walls in darker corners of the buildings. The gecko has a short activity period from late April to mid-August; it hibernates from late August to March. It is the only house gecko in the area.

It breeds from April to June, laying a pair of soft-shelled oval eggs which is glued to the sides of crevices which soon hardens. An egg laying site is usually used year after year for communal laying. In several such sites (usually under loose plastering on walls) several broken, fresh, and old eggshells have been discovered.

Distribution: This gecko is known from Fort Munro and Khar village, in the northwestern Dera Ghazi Khan District, Punjab, Pakistan.

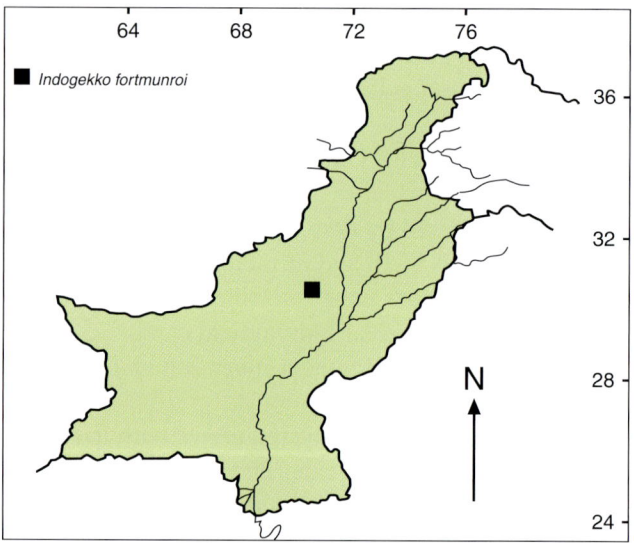

Plate 80. *Indogekko fortmunroi.*

Indogekko indusoani (Khan, 1980) (Plate 81)

(Soan gecko: Soan reg-sang chapkali)
1988 *Cyrtodactylus indusoani* Khan, J. Herpetol. 22:241–243.
Type locality: Pirpeahai, Iskinderabad, District Mianwali, northwestern Punjab, Pakistan.
Diagnosis:
1. Interorbitals 13–15.
2. Body depressed, tail longer than body.

Chapter 6. Lizards

Plate 81. *Indogekko indusoani.*

3. Dorsal tubercles flat, in 11 irregular longitudinal rows.
4. Scale rows across midabdomen 23–25.
5. Preanal pores 4–5.
   Snout-vent length 50–53 mm, tail 75–78 mm.

Color: Dorsum light to dark gray, with dark mottling on snout and labials. A dark streak behind eyes connected behind nape with that of other side. Seven transverse dark bands from neck to the level of vent, each composed of a series of 3 oval spots. Limbs mottled with dark. Ventrum white.

Natural history notes: The lesser sandstone gecko inhabits crevices among sandstone rocks along the bank of the River Indus, close to the site where Soan River opens in the Indus as it enters Punjab Province in District Mianwali. Though the gecko is nocturnal, it is often seen active during the day among sandstone slabs. It feeds on aquatic soft bodied-insects as they come to rest on rocks along the riverbanks.

A pair of soft-shelled eggs is glued to the underside of the sandstone rock slabs. The egg laying site is used by several females, and scattered debris of broken shells indicates the use of the site over several years.

Distribution: The gecko is known from the northwestern border of the Salt Range, Punjab, Pakistan.

Indogekko rhodocaudus (Baig, 1998)
(Red-tail sandstone gecko: Surakh-dum reg-sang)
1998 *Tenuidactylus rhodocaudus* Baig, Hamadryad 23:127–132.

Type locality: Tanishpa, District Kila Saifullah, Baluchistan.

Diagnosis:
1. Dorsum with 12–14 longitudinal rows of smooth or feebly keeled, enlarged tubercles.
2. Midabdominal scales 16–18.
3. Number of preanofemoral pores 23 in male; 5–9 preanal pores in female.
4. Dorsolateral rows of enlarged spinose scales on tail 2.
5. Midventral scales between mental and vent 92–106
   Snout-vent length 59.0 mm, tail length 74.0 mm.

Natural history notes: The gecko is found in crevices and under moist sandstone slabs near watercourses. It

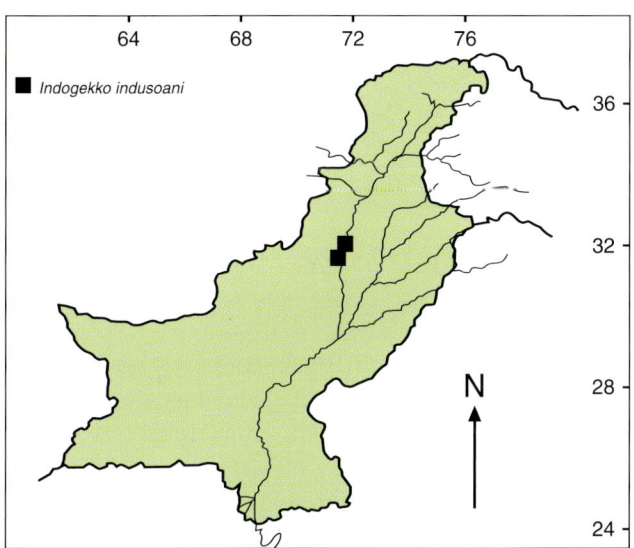

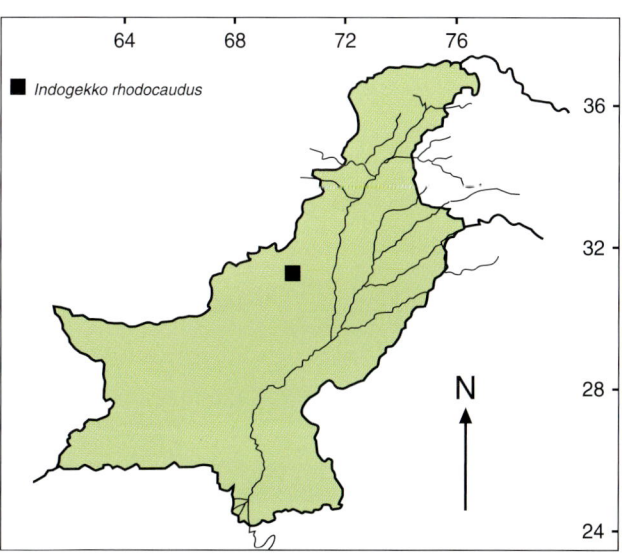

is crepuscular, rarely diurnal. Breeding season extends from April to July. The eggs are glued to the sides of protected stone slopes and under surfaces of sandstone slabs. The presence of debris of several broken shells at such sites suggests a communal habit of egg laying.

The type locality is Tanishpa, a small town, elevation 2400 m in the Toba Kakar Range. It is located in a game reserve established for the conservation of endangered urial goats (*Ovis orientalis*) and markhor (*Capra falconieri jerdoni*). The area is arid and dry, receiving little rain during summer; winter is severe and vegetation is sparse and desertic. In spring, the area is covered by short-lived seasonal flowers. Orchards are maintained along streams (Baig, 1998).
Range: The gecko is known from its type locality.

Indogekko rohtasfortai (Khan and Tasnim, 1990) (Plate 82)
(Rohtas gecko: Rohtas reg-sang chapkali)
1990b *Tenuidactylus rohtasfortai* Khan and Tasnim, Herpetologica 46: 142–148.
Type locality: Ahmadyyah Mosque, Goi Madan, Kotli, Azad Kashmir.
Diagnosis:
1. Middorsum with flat, round, slightly keeled tubercles; lateral sides with subtrihedral to conical tubercles, in 12–14 longitudinal rows.
2. Scale rows across midabdomen 24–33.
3. Males with a continuous series of 18–27 preanofemoral pores.
4. Midventral scales 103–135, from postmentals to the anterior lip of vent.

Snout-vent length 48–51 mm, tail 63–65 mm.
Color: Dorsum light gray, with 7 broken bands each formed of 3 rounded spots. Labials mottled. A light stripe from eye to the side of the neck. Tail barred.
Natural history notes: This gecko inhabits crevices among rocks and invades houses, where it rests clinging to walls in quieter dark corners. Nocturnal, those in inhabited houses gather around light at night to feed on photophilic insects, while those in the field, live in crevices and holes among rocks or under stones.

Breeding season extends from May to June; a pair of oblong eggs is glued to the sides of crevices, to the undersides of slabs, or to walls in dark parts of buildings. Juveniles are seen from late June to early July.
Distribution: *Indogekko rohtasfortai* is a widely distributed species in alpine Punjab and southeastern Azad Kashmir. It extends into the hilly terrain of the Potwar Plateau, especially from Jhelum to Islamabad.

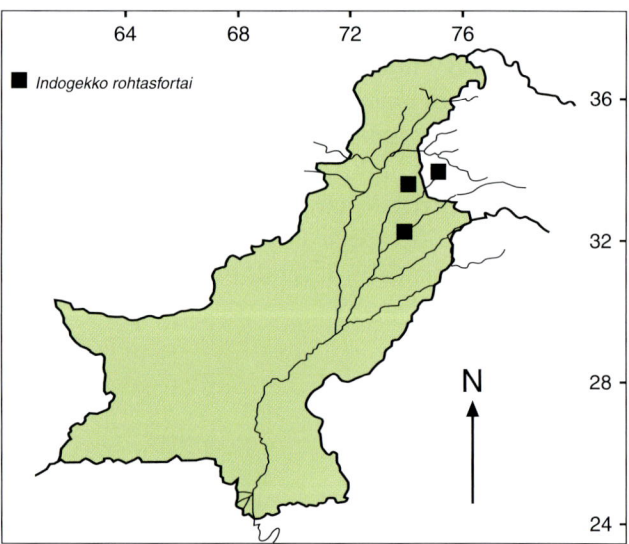

Plate 82. *Indogekko rohtasfortai.*

Genus *Mediodactylus* Szczerbak and Golubev, 1977
Diagnosis
Male with no more than 10 preanal pores, no femoral pores; first pair of postmentals in contact; dorsal tubercles oval or roundish, smooth; caudal tubercles spiny, do not contact each other; a row of enlarged subcaudals; supracilliary spines present.

A single species is represented in Pakistan.

Chapter 6. Lizards

Mediodactylus walli (Ingoldby, 1922) (Plate 83)
(Chitral gecko: Chatra chapkali)
1922 *Gymnodactylus walli* Ingoldby, J. Bombay Nat. Hist. Soc. 28:1051.
Type locality: Drosh Fort, Chitral, NWFP, Pakistan.
Diagnosis:
1. Body moderately depressed.
2. Dorsal tubercles are oval, keel-less scattered among granular scales.
3. Subcaudals are broader than long, in a single median series.
4. Caudal tubercles are given out from the middle of the caudal segments.
5. Preanal pores 4–5 in male.
   Snout-vent length 54–55 mm, 78–79 mm.

Color: Dorsum light gray to dark, with 9 wavy crossbars, slightly broader than the interspaces. A dark stripe from eye, joining the band on nape. Limbs barred, tail with alternating light and dark bars.
Natural history notes: The gecko has been picked from the walls of inhabited houses. Outside it is collected from under rocks, etc. It reproduces from May to June, and juveniles are seen by July.
Taxonomic notes: Smith (1935) described *Gymnodactylus chitralensis* and placed *Gymnodactylus walli* Ingoldby in its synonymy along with *Gymnodactylus stoliczkai*. He is followed by several subsequent workers (see Khan, 1992, for discussion). Khan (1992), on the basis of new material from the type locality, validated *Gymnodactylus walli* as senior synonym of *Gymnodactylus chitralensis* and has shown that it differs markedly in its morphology from *Gymnodactylus stoliczkai*.

The Mediterranean/mid Asian subgenus *Mediodactylus* (Szczerbak and Golubev, 1977) may represent an aberrant group of *Cyrtopodion* geckos with a mixture of *Altigekko* and *Indogekko* characters. Similarly *Gymnodactylus walli* Ingoldbyi, 1922, and *G. kirmanensis* Nikolsky, 1900, that Szczerbak and Golubev (1996) dumped in "Tibeto-Himalayan group of *Tenuidactylus*," may also belong to the Mediterranean/mid-Asian stock with *Altigekko* and *Indogekko* (Khan, 2003f).
Distribution: The gecko has been collected from Drosh Fort, Karakal village in Bumhoet Valley, and Ghariet village, Chitral, all in NWFP, Pakistan.

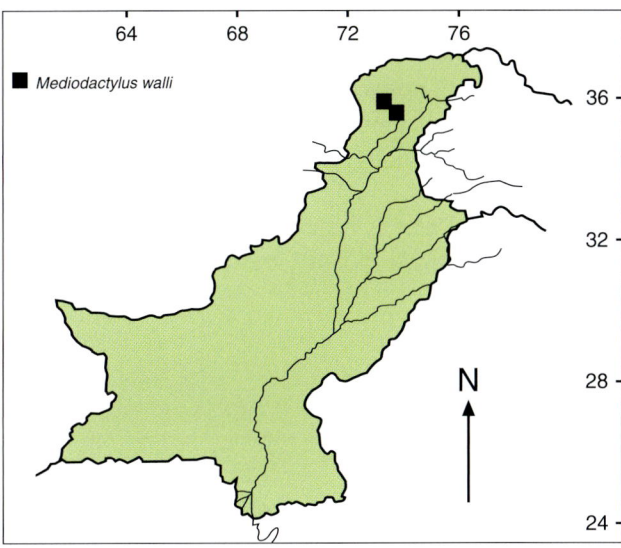

Plate 83. *Mediodactylus walli.*

Genus *Ptyodactylus* Goldfuss, 1820
Digits free, cylindrical at base, tips strongly dilated laterally; a series of subdigital lamellae under cylindrical basal part, while two diverging series on distal dilated part, a small fissure between two series; claws retractile between scales in the anterior notch in the apical expansion; dorsum with smooth juxtaposed, uniform granular scales, abdomen with subimbricate scales; tail subcylindrical, tapering.

A single species is known from Pakistan.

Ptyodactylus homolepis Blanford, 1876
(Fan-toed gecko: Pankh-ungusht chapkali)
1876 *Ptyodactylus homolepis* Blanford, J. Asiatic Soc. Bengal, Calcutta 45(2):19.

Type locality: Mahar Division, Shikarpur District, northwestern Sind, Pakistan.

Diagnosis:
1. Digits terminate in a subtriangular expansion, which is as broad as the diameter of eye; claws are centrally located (Figure 13B).
2. Supralabials 13–15, infralabials 12–15.
3. Mental and submental scales indistinct.
4. Rostral does not border naris.
5. Head and body with granular scales, no tubercles.
6. Tail subcylindrical, with smaller dorsal and larger ventral scales.
   Snout-vent length 100–108 mm, tail 85–89 mm.

Color: Dorsum light brown to gray, with broad transverse lighter wavy bands. Ventrum white.

Natural history notes: This gecko primarily inhabits crevices and holes among rocks in northeastern Kirthar Range. Nothing is known about its habits and natural history.

Distribution: There is no subsequent report of this gecko from Pakistan. It is yet known only from its type locality.

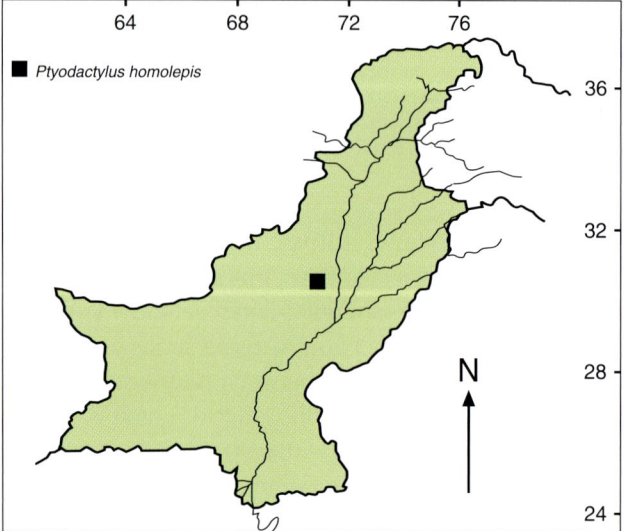

Genus *Rhinogekko* de Witte, 1973
Nostrils are carried at the tip of a prominent cylindrical caruncle on lateral side of snout, formed by three nasals and first supralabial, which are drawn into a prominent narial spout; digits straight, cylindrical, slightly compressed, with a row of smooth transverse lamellae; a row of enlarged subfemoral scales; tail thin, cylindrical, with a row of irregular squarish larger subcaudals; male with preanal pores.

The morphology of narial region of these geckos is enough to distinguish them from rest of the geckos.

One species occurs in Pakistan, a second in Iran, close to the Iran-Pakistan border.

Rhinogekko femoralis (Smith, 1933)
(Point-tail spider-gecko: Nook-dum makra-chapkali)
1933 *Agamura femoralis* Smith, Rec. Indian Mus. Calcutta 35:17.

Type locality: Kharan, Baluchistan.

Diagnosis:
1. Head about twice as long as broad.
2. Diameter of ear opening is half that of the eye.
3. Mental longer than adjacent labials, a pair of distinct postmentals, which are in contact with each other.
4. Gulars small, flat granules.
5. Naris on a caruncle formed by rostral and first supralabial.
6. Ventrals in 17–21 rows at midbody.
7. Males with 6 (rarely 5) preanal pores.
8. A series of 9–12 enlarged subfemoral scales.
9. Tail longer than body, sharp-tipped.
   Snout-vent length 60–62 mm, tail 60–64 mm.

Color: Light yellow gray, with 5 dark cross bars on body and 8–10 on tail.

Natural history notes: The gecko is nocturnal. It has been collected from sandy deserts with rocky outcrops and scarce vegetation of bushes. It is terrestrial and rarely climbs vertical surfaces. When encountered, it poises for defense by flattening its body, elevating its tail, and twitching it from side to side (Minton, 1966).

Breeding season extends from March to July; the female appears to lay more than two clutches of 2–8 eggs in a season. The eggs are laid under rock.

Distribution: It has been reported from Kharan and Chagai Deserts, close to the rocky outcrops.

Chapter 6. Lizards

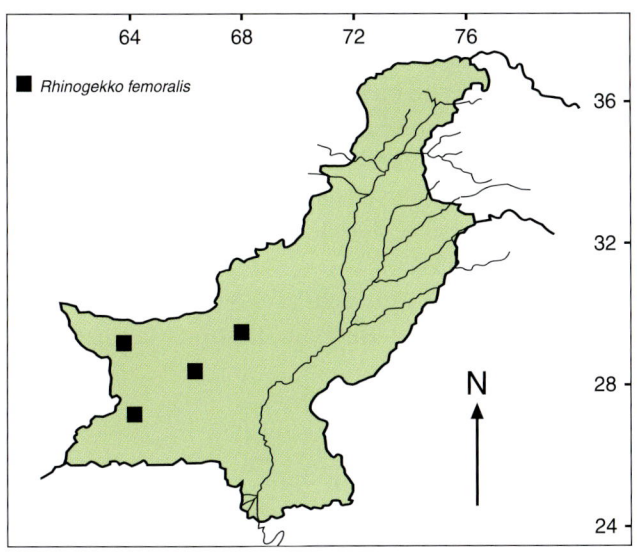

Rhinogekko misonnei de Witte, 1973
(Long-nose gecko: Nakali chapkali)
1973 *Rhinogekko misonnei* de Witte, Bull. Inst. R. Sci. Nat. Belg., Bruxelles 49:1.
Type locality: Dast-i-Lut Desert, Iran.
Diagnosis:
1. Nasal scales strongly elevated carrying naris at a tube on the snout.
2. scales across midbelly 26–28.
3. A row of 9–12 enlarged scales on thigh ventrum.

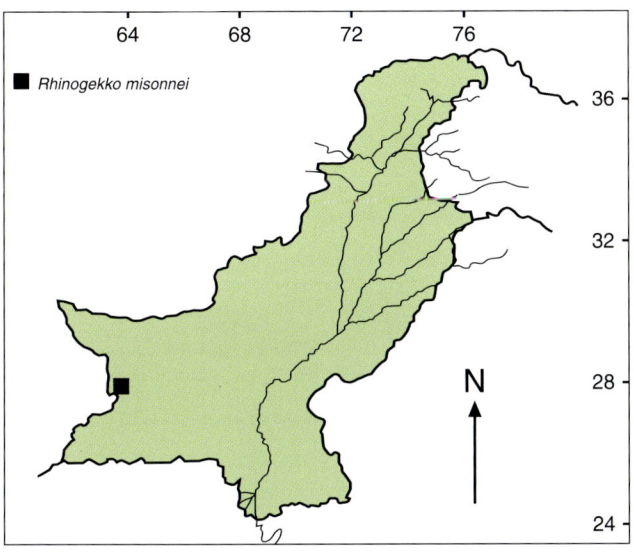

4. Four to eight very poorly developed preanal pores.
5. Tail slightly longer than the body.
Snout-vent length 56–61 mm, tail length 58–73 mm.
Color: Coffee gray dorsum, 5 wide dark transverse bands on body, 7 on tail, limbs similarly barred. Ventrum whitish.
Natural history notes: The lizard is reported from the gravel deserts with scanty vegetation.
Distribution: Known from Dasht-i-Lut along the Iran-Pakistan border.

Genus *Siwaligekko* Khan, 2003
Tibetanus subgroup (Khan, 2001d). Medium size geckos. Body and tail plump, nearly round, shorter or subequal of body, not whiplike, segmentation indistinct, indicated by 2–3 minute, blunt tubercles in anterior half of the tail; tail fragile at base, regenerated tail not swollen, subcaudals indistinct; no supracilliary spines on the posterior dorsal part of the upper eyelids; dorsal granular scales round to polygonal, juxtaposed, beaded (convex), interspersed with 3–4 large, beaded, smooth or slightly keeled tubercles, extending on neck and head, absent from limbs; postfemoral tubercles absent; interorbitals 21–35; midabdominals 36–56; midventrals 149–205; subdigital lamellae under basal part of digits somewhat broad, those under angular part narrower; subdigital lamellae under fourth toe 14–21; male with 8–10 preanal pores, no femoral pores, both absent in female.

Three species are represented in Pakistan.

Siwaligekko battalensis (Khan, 1993) (Plate 64)
(Reticulate plump-bodied gecko: Jal-dar goal-jasm)
1993 *Cyrtodactylus battalensis* Khan, Pakistan J. Zool. 25(1): 67–73.
Type locality: Batgram, Manshera, NWFP, Pakistan.
Diagnosis:
1. Dorsum with obtusely keeled flat tubercles scattered among granular scales.
2. Scales across midabdomen 50–52.
3. Scales along midventrum of body from postmentals to anterior of the vent 199–205.
4. Preanal pores 9–10 arranged in an arch, no femoral pores.

Plate 64. *Siwaligekko battalensis*.

5. Tail round, without marked segmentation.
6. Subcaudals small in several rows.
   Snout-vent length 60–64 mm, tail 63–64 mm.

Color: Dorsum light brown, with 7 dark brown transverse bands, much narrower than the interspaces, with irregular borders touching each other so to form a distinct dark reticulum on dorsum of the body. A vivid dark stripe from snout through loreal and eye joins first transverse band on the nape. Frontal with a dark, U-shaped mark, several small transverse stripes on head. Limbs with dark reticulation. Tail with 11 dark rings, extending onto the tail ventrum.

Natural history notes: The reticulate gecko is essentially an alpine gecko, living in crevices among rocks, close to the roots of vegetation. It invades buildings, living in holes and crevices among brick and stone walls. Its movements are deliberate, invades surrounding vegetation just after sunset, and returning to its retreats before dawn. Dipterous insects are predominant in its diet; it also feeds on insect larvae and worms.

Breeding season extends from April to early June; the juveniles are seen active by early May. A pair of oval eggs with calcareous shells is deposited between rocks or crevices in the brick walls of buildings (Khan, 1993a).

Distribution: The reticulate gecko is known only from its type locality, Batgram, District Manshera, NWFP, Pakistan.

***Siwaligekko dattanensis* (Khan, 1980) (Plate 66)**
(Plump banded gecko: Datta goal-jasm)
1980 *Gymnodactylus dattanensis* Khan, Pakistan J. Zool.12 (1):11–16.

Type locality: Datta, Manshera, NWFP, Pakistan.
Diagnosis:
1. A single pair of postmentals.
2. Dorsal tubercles round, keel-less, arranged in longitudinal rows.
3. Male with 8–9 preanal pores.
4. Tail unsegmented, round, with small flattish caudal tubercles, small subcaudals.
   Snout-vent length 60–62 mm, tail 56–57 mm.

Color: Darkish gray dorsum, with 10 dark brown wavy crossbars from nape to the vent, broader than the interspaces; bars tend to break into spots on flanks. Tail barred, bars extending onto tail ventrum. Belly dirty white.

Natural history notes: The banded rock gecko has been collected from rocky alpine terrain, where it lives

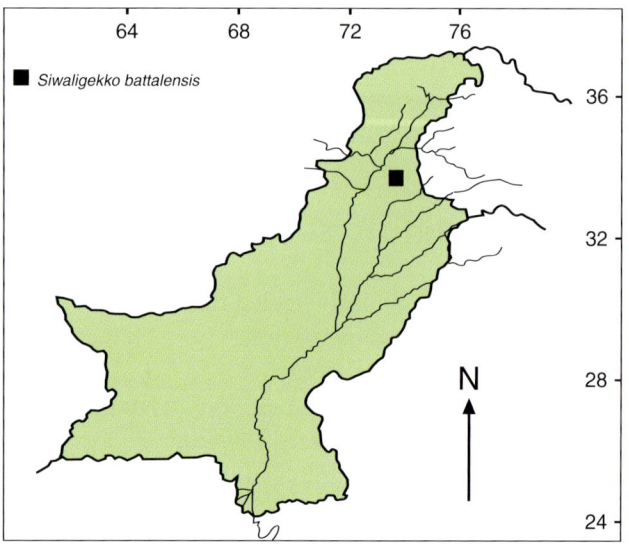

Plate 66. *Siwaligekko dattanensis*.

Chapter 6. Lizards

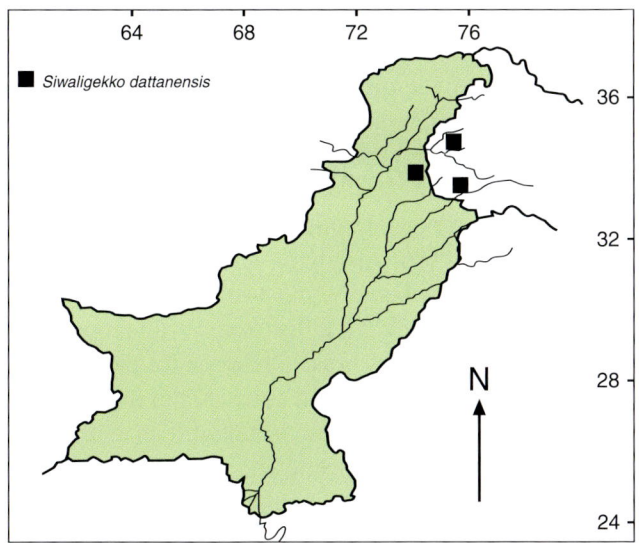

in crevices among rocks and among roots of pine trees. It feeds on soft bodied insects and their larvae.

When caught, it gives a low squeak and turns to bite, often voiding excrement. Breeding season extends from late March to May; juveniles are seen from late May to mid June. Clutch size is 1–2 eggs which are deposited in crevices among rocks (Khan, 1980a).

Distribution: The barred gecko is widely distributed in alpine Punjab and eastern Northwestern Frontier Province, Pakistan.

Siwaligekko mintoni (Golubev and Szczerbak, 1981) (Plate 67)

(Plump Swati gecko: Swati goal-jasm)
1981 *Gymnodactylus mintoni* Golubev and Szczerbak, Faun. Syst. 1981: 40–50.

Type locality: Udigram, Swat, NWFP, Pakistan.

Diagnosis:
1. Dorsal pattern of narrow wavy dark brown crossbars, much narrower than interspaces. Bands are broken into round spots on flanks.
2. Nasal scales 3.
   Snout-vent length 38–40 mm, tail 35–37 mm.

Color: Dorsum amber, shading to lemon on tail. A series of 8 irregular, wavy, broken, dark brown to black crossbars on body, mostly with posterior white edge. Sides of body, head, tail, and limbs with scattered dark brown dots. Ventrum pale yellow. Labials barred.

Plate 67. *Siwaligekko mintoni.* (Photo courtesy of S.A. Minton).

Natural history notes: The only known specimen of this species was collected curled up under a stone in the bed of a small ravine, in wooded hilly country at about 100 m of elevation (Minton, 1966). The gecko is sluggish with deliberate movements.

Distribution: This gecko is known by its type specimen, which was collected from Udigram, Swat, NWFP, Pakistan.

Minton (1966) tentatively identified this gecko as *Gymnodactylus stoliczkai*.

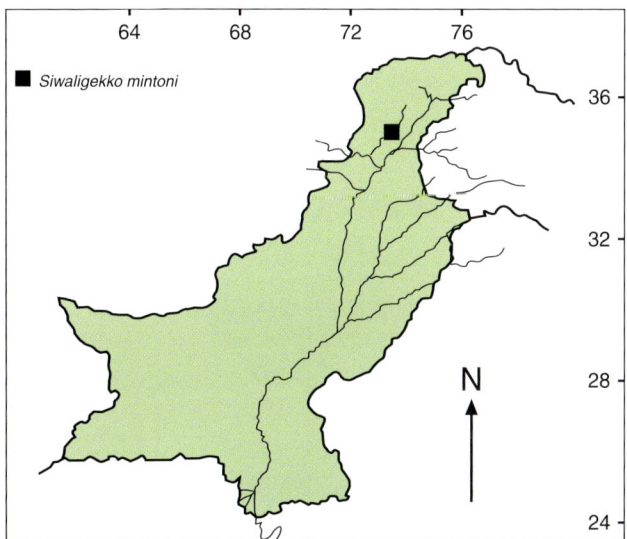

Genus *Teratolepis* Günther, 1870

Subdigital lamellae keeled, not divided; supranasal scales large, flat, not lining nostril; body with flat, rhomboid, imbricate, keeled or smooth scales; dorsal and abdominal scales subequal; no distinct subcaudals; tail cylindrical, slightly thicker and flat at the base, thinning at its last quarter, covered above and below by flat imbricate scales.

A single species occurs in Pakistan.

Teratolepis fasciata (Blyth, 1853)
(Flat-tail gecko: Chapti-dum chapkali)

1853 *Homonota fasciata* Blyth, *in*: Jerdon, J. Asiatic Soc. Bengal Calcutta, 22: 468.

Type locality: Jaulna, Hyderabad Province, southern India.

Diagnosis:
1. Digits are moderately dilated, with a terminal slender clawed phalanx arising angularly from the middle of the terminal expanded part of the digit (Figure 13E).
2. A series of transverse subdigital lamellae, anterior of which are slightly notched.
3. Body dorsum with large pointed imbricate scales, much larger than the ventrals.
4. Tail swollen, its posterior one-fourth tapers abruptly behind to a point, dorsally covered with large imbricate scales.
5. Subdigital lamellae 7 under first, 8–9 under fourth toe.
6. Male with 5–6 preanal pores.

Snout-vent length 45–46 mm, tail 32–33 mm.

Color: Light grayish brown dorsum, with 5 longitudinal dark brown stripes, crossed by 6 rows of large whitish spots. A pair of whitish transverse bands on the occiput. Tail brownish with whitish crossbars.

Natural history notes: The most striking feature of this gecko is its large thick flat tail, covered with large leaf-like scales. The tail serves as storehouse for the surplus fat, and is used as a shield and as an assault organ.

The gecko inhabits flat terrain with an average elevation of 10 m above sea level. The soil is generally loose gray silt with desert scrub vegetation, dominant plants are *Salsola* and grasses. The gecko hides in holes and crevices among bricks and under stones, rocks, and other scattered debris. It readily invades desolate old uninhabited edifices in deserticolous salty habitat, like old graveyard edifices. The gecko is nocturnal, and it is found curled up under stones and leaves during the day. It feeds on all types of arthropods including scorpions.

Terrarium studies have been made on this gecko (J.A. Anderson, 1964). When provoked the gecko immediately curls up and coils its tail around its body. As a defense it raises its tail and presents it as a shield toward its assailant. Then it arches its body, stands up, twitching its tail in jerks while waving its tip. The gecko strikes suddenly, with a barely audible rasping hiss. The tail is held high, always directed toward the assailant; it strikes repeatedly with it then scurries away swiftly, in spurts, covering a distance of a meter or so at a time. By its way of hissing and striking, the gecko appears to mimic the saw scale viper *Echis carinatus*, which is common in its habitat.

On land the gecko moves very slowly and deliberately; however, to escape capture, the male becomes furious, hisses, arches its back, strikes both with its open mouth with curled tail held high. It flees escape in 20–40 mm high leaps, covering a distance of 30–50 mm each time. When prevented from escape it flips its entire body alternately lengthwise and sideways, in quick jerks, hissing with each flip.

Females are very lethargic and they do not react to capture. The gecko climbs into the bushes or walls to hide among branches or in crevices. They are moderate swimmers. When in the terrarium particular

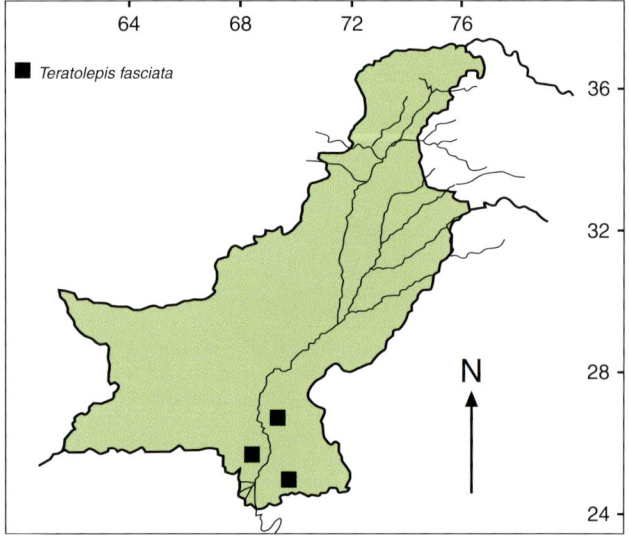

Chapter 6. Lizards

section of it is used for the deposition of excreta by all the geckos in the terrarium. The substratum material is shovelled backward with its forelimbs to form a cavity under stone, etc., where the gecko spends most of its time curled up. The gecko frequently laps water.

Courtship takes place mostly from February to June; however, courting pairs are met with almost throughout the summer. The male repeats clearly heard short clucks followed by a prolonged trill almost every 10 seconds; the responsive female calls back in a similar manner but in a lower tone. Breeding fights among rival males are frequent; males assault each other savagely, often snapping off each other's tails.

An adult female lays several clutches of a pair of white oval eggs, throughout the summer in deserted burrows, beneath piles of rocks, bricks etc. The eggs measure from 9.5 by 8 to 11.5 by 10.5 mm. Juveniles are met with throughout the summer. The female is known to guard the eggs and cares for juveniles for considerable time after they are hatched.

Distribution: The gecko is recorded from different localities in the lower Indus Delta in Pakistan.

Genus *Teratoscincus* Strauch, 1863

Digits straight, not bent or dilated, all clawed, with a lateral fringe of long pointed scales; subdigital surface with minute granules; body covered with uniform, cycloid, imbricate scales; tail covered above with large flat nail like transverse plates; male without preanal and femoral pores.

Two species occur in Pakistan.

Teratoscincus microlepis Nikolsky, 1899 (Plate 84)
(Baloch sand-gecko: Bloch reg-chapkali)
1899 *Teratoscincus microlepis* Nikolsky, Ann. Mus. Zool. Acad. Sci. St. Petersburg 4: 145.
Type locality: Duz Abad, Eastern Kirman, Iran.
Diagnosis:
1. Middorsal large cycloid scales do not extend beyond shoulder.
2. Scalesaround midbody 100.
3. Head with granular scales.
4. Postmentals absent.
5. Tail with a median dorsal series of large flat scales.
6. A prominent upper movable eyelid.
Snout-vent length 71–73 mm, tail 47–51 mm.

Plate 84. *Teratoscincus microlepis*. (Photo courtesy of S.C. Anderson).

Color: Dorsum cream, with dark brown, oblique or V-shaped bars on back, indistinct in older specimen. A dark U-shaped mark on the back of head. Tail barred with dark.

Natural history notes: This gecko lives in holes and crevices under xerophytic shrubs in loose sandy soil. It movements are slow and deliberate, however, to escape capture it sprints and takes to shelter. Captured individuals lunge and strike with the tail, which is kept twitching sideways. The geckos feed on soft-bodied arthropods and larvae of certain sand beetles. When moving on sand the tail leaves a streak between the tracks.

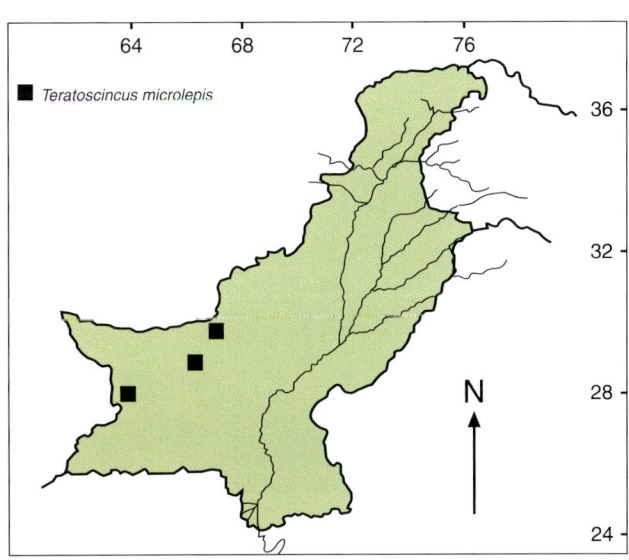

Distribution: The gecko has been collected from Nushki and Kharan in Baluchistan; it extends westward to Dasht-i-Lut, near Kirman, Iran.

Taxonomic notes: The name *Teratoscincus microlepis makranensis* has appeared on the trade lists of animal dealers, and at least one photograph bearing this name has been published in a popular magazine. However, it appears to be an unavailable name under the International Code of Zoological Nomenclature, never having been properly described, no holotype designated, etc.

Teratoscincus scincus keyserlingii Strauch, 1863
(Plate 85A, B)
(Turkish sand-gecko: Turkish reg-chapkali)
1863 *Teratoscincus keyserlingii* Strauch, Bull. Acad. Imp. Sci. St. Petersburg 6:480

Type locality: Ili River, eastern Turkestan.
Diagnosis:
1. A median dorsal series of large cycloid scales extends to the back of head.
2. No postmental scales.
3. Scalesaround the midbody 28–34.
4. Head with granular scales.
5. Tail with a median dorsal series of large cycloid scales (Figure 14E).
6. Upper eyelid moveable.
   Snout-vent length 116–120 mm, tail 89–92 mm.

Color: Dorsum cream, with faint dark brown transverse bars or with 4 reddish brown longitudinal stripes. Head with brown markings, lips barred. Sides and ventrum pinkish to white.

Plate 85B. *Teratoscincus scincus:* Juvenile.

Natural history notes: Restricted to fine windblown tracts of sand, it is nocturnal, becoming active just after sunset. It walks deliberately with the legs holding the body high, tail trailing behind, and leaving distinct spoor. When followed it sprints for a short distance, stopping occasionally to observe the pursuer. When cornered it turns around to attack with a snarl, twitching and swaying its tail, the scales of which produce faint rustling sound. It often lashes the intruder with its tail. The gecko, if cornered, bites powerfully. Its burrows are 20–28 cm deep and the entrances are plugged with sand (Minton, 1966).

Plate 85A. *Teratoscincus scincus;* Adult.

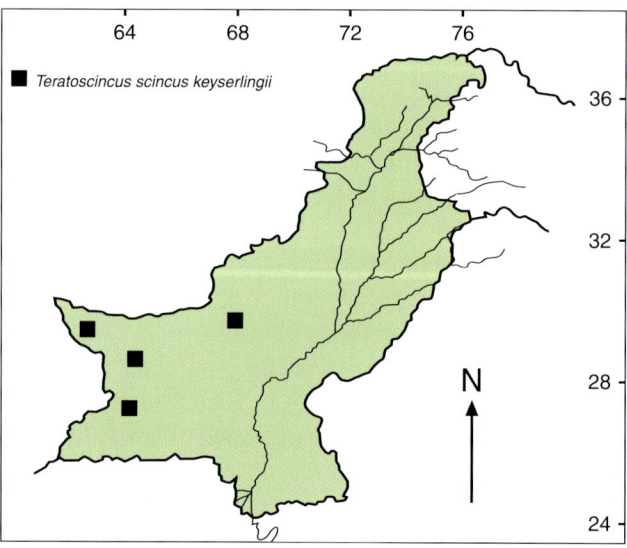

Chapter 6. Lizards

Feeds on soft-bodied insects and small lizards. The skin of the gecko is soft and extremely fragile, easily injured upon handling. It is repaired in due course of time. Bauer et al. (1993) believe that the fragility of the skin of this gecko is a self-mutilating anti-predator defense strategy.

The gecko breeds from April to July, with a pair of spherical eggs, of a diameter (14–15 mm in diameter) being laid in the moist layers of a burrow.

Distribution: This gecko has a wide range in the west from the Caspian Sea to Tajikistan, and extends into western Baluchistan, Pakistan.

Genus *Tropiocolotes* Peters, 1880

Small geckos, rarely exceeding 35 mm in snout-vent length; digits with slight angular bend, cylindrical, without a fringe, with a single series of transverse subdigital lamellae; dorsum with small, uniform, homogenous imbricate scales; postanal sacs present; in male preanal and femoral pores absent.

I have followed Szczerbak and Golubev (1996) for this genus.

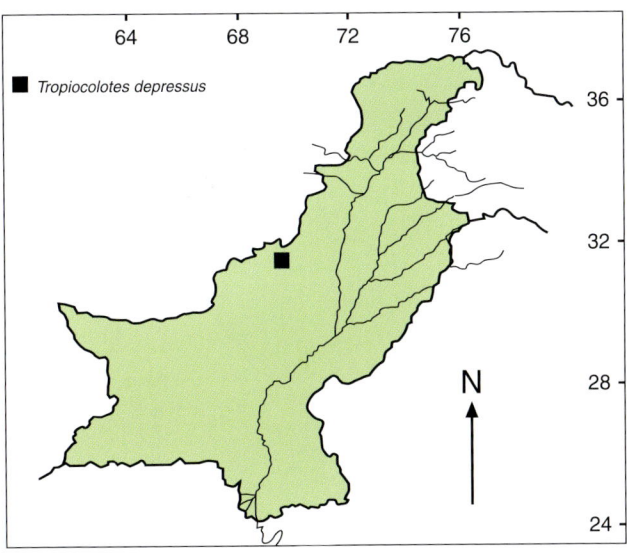

Tropiocolotes depressus Minton and Anderson, 1965

(Mountain dwarf gecko: Chattani chipolia)
1965 *Tropiocolotes depressus* Minton and Anderson, Herpetologica 21: 59.

Type locality: Kach, Quetta Division, Baluchistan, Pakistan.

Diagnosis:
1. Postmentals absent or a small pair may be present, not in contact with each other.
2. No enlarged internasal and postnasal scales, four scales border the naris.
3. Lamellae under fourth toe 17–18.
4. A pair of preanal pores in large preanal scales. Snout-vent length 31–34 mm, tail 30–31 mm.

Color: Dorsum saffron yellow, with a transverse dark band on neck, 3–5 on body, narrower than interspaces, 6 on tail. Ventrum pinkish white. A dark stripe from snout through eye onto neck.

Natural history notes: The mountain dwarf gecko has been collected from hillsides with sparse vegetation between 1800 and 2000 m. It inhabits crevices among rocks. The gecko is agile, running on rocks and stones with agility, feeding on soft-bodied arthropods.

Its eggs have been found under stones and rocky slabs in humid places (Minton and Anderson, 1965).

Distribution: The gecko is recorded from hilly tracts north of Quetta, Baluchistan.

Tropiocolotes persicus persicus (Nikolsky, 1903) (Plate 86)

(Persian banded gecko: Irani chipolia)
1903 *Alsophylax persicus* Nikolsky, Ann. Zool. Mus. Imp. Acad. Sci. 8: 95.

Type locality: Vikus Dehak, Iranian Baluchistan.

Plate 86. *Tropiocolotes persicus.*

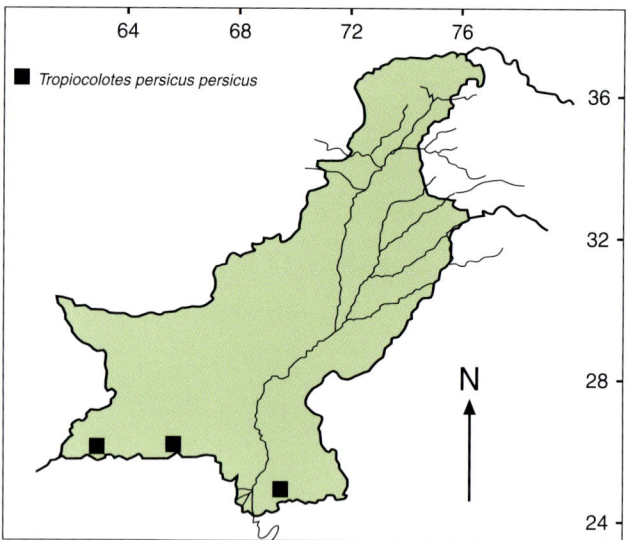

Diagnosis:
1. Supralabials 8–10.
2. Two pairs of postmental scales, usually first pair in contact with each other.
3. Interorbital scales 19.
   Total length 25.3 mm, tail 17–18 mm.

Color: In alcohol, dorsum is light yellowish. A wide dark brown stripe from nostril through eye extends along dorsolateral sides of the body; 5–7 similar crossbars between neck and base of the tail; 6–10 bands across tail dorsum, head and limbs without pattern, ventrum white.

Distribution: It extends from southwestern Iran to southeastern Sind, Pakistan.

Tropiocolotes persicus euphorbiacola **Minton et al., 1970 (Plate 86)**
(Sind dwarf-gecko: Sindi chipolia)
1970 *Tropiocolotes persicus euphorbiacola*, Minton, Anderson, and Anderson, Proc. California Acad. Sci. ser. 4, 37 (9):354.

Type locality: Las Bela, Pakistan.

Diagnosis:
1. Dorsal bands, on body, as wide as or narrower than the interspaces.
2. Postmentals 1–3, first in contact with each other.
3. Regenerated tail always yellow.
4. Subdigital lamellae under fourth toe 14.
   Snout-vent length 28–33 mm, tail 32–34 mm.

Color: Body dorsum pale yellow or amber. Five dark brown transverse bands on body, 8 on tail. A dark stripe from snout to the eye joins first dorsal bar. Ventrum pinkish.

Natural history notes: It is a widely distributed small gecko in Las Bela and found to be plentiful in the vicinity of drying dead plants of *Euphorbia caudicifolia* from mid-November to February. The gecko appears to have great affinity with the plant, perhaps it is well protected by its strong thorns. However, in rocky terrain below 200 m, the gecko invades rocks. In warm moist weather the geckos are very active on rocks, slipping from stone to stone with agility, hiding in crevices and holes. The tail of the gecko is very fragile. Food consists primarily of soft-bodied arthropods, larvae, etc. When disturbed the lizard quickly takes to cover, while on a *Euphorbia* bush, the lizard jumps from branch to branch.

Adults usually live in pairs. Breeding season is quite extended, from March to November. More than one clutch of a single egg measuring 8 by 5.5 mm is laid in holes and crevices under stones or in the roots of *E. caudicifolia*. The egg is elongated at the ends, glossy white with longitudinal stripes, has a soft pliable shell which becomes hard and brittle within a few hours of laying. Incubation period lasts 4–5 weeks, and hatchling size is 14–15 mm (Minton et al., 1970).

Taxonomic notes: Minton (1962, 1966) and Mertens (1969) treated this gecko as *T. helenae*. Later Minton et al. (1970) described it as an eastern race of *T. persicus*.

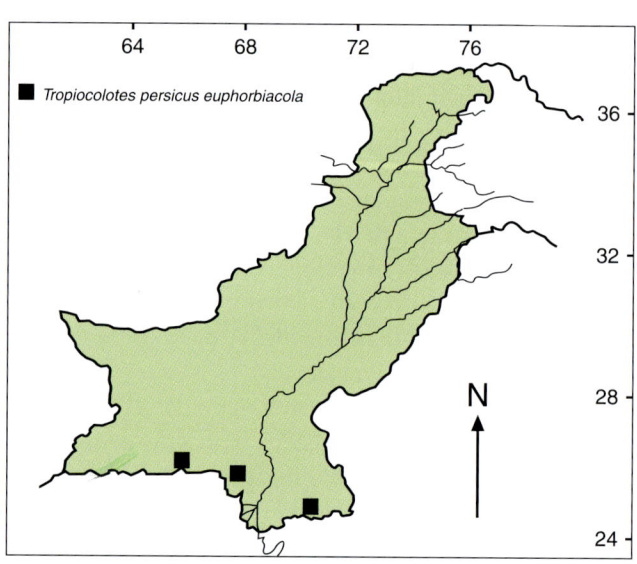

Chapter 6. Lizards

Distribution: This gecko has been reported from Las Bela, Baluchistan, and lower Sind, Pakistan.

Family Lacertidae

Despite recent several revisions of the family Lacertidae (Arnold, 1983, 1986; Mayer and Benyr, 1994), the increasing details and analysis of relationships within the family Lacertidae has failed to produce a widely accepted consensus; still more work is needed to establish an acceptable classification.

Four genera are reported from Pakistan.

Genus *Acanthodactylus* Wiegmann, 1834

Occipital scale absent; nostrils between two nasals and first supralabial; lower eyelid scaly; collar distinct; dorsal scales small, juxtaposed or large, imbricate; ventrals subquadrangular, smooth, imbricate; digits subcylindrical, with keeled subdigital lamellae, lateral denticulations, at least on the outer side of toes; femoral pores present.

Three acanthodactylid lizards are reported from Pakistan.

Acanthodactylus blanfordii Boulenger, 1918
(Red-tail sand lizard: Surakh-dum chalpaya)
1918 *Acanthodactylus cantoris* var. *blanfordi*, Bull. Soc. Zool. France 43:154.

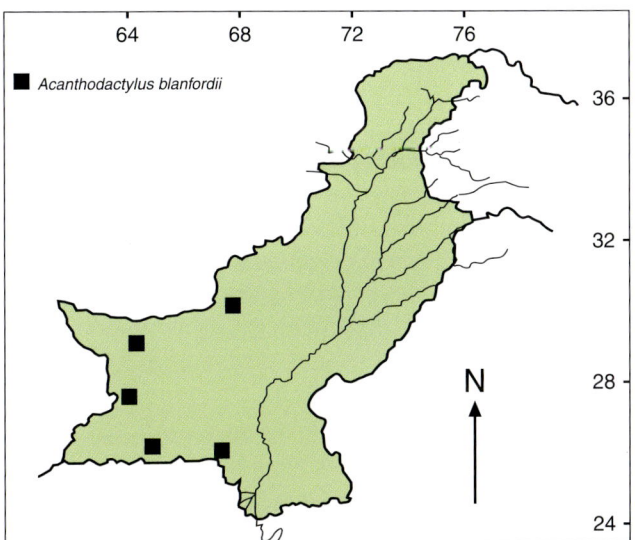

Type locality: Bam, Persia; Mand, Baluchistan, Pakistan.

Diagnosis:
1. Lateral body scales a little smaller than those of middorsum.
2. Scales at midbody 28–46, gulars 32–38.
3. Hindlimb reaches to ear or eye.
4. Femoral pores 21–29.
   Snout-vent length 70–73 mm, tail 172–175 mm.

Color: Dorsum reddish brown to gray, speckled with white, traces of longitudinal stripes present; tail reddish gray. Ventrum white. Juveniles with 6 longitudinal yellow stripes on dark dorsum, extending onto the tail; tail distally pinkish.

Natural history notes: The lizard frequents sand dunes, beaches, dry beds of water courses with sparse vegetation. It is diurnal, keeping to the shade of vegetation at midday. In the afternoon, the lizard has a second brief spell of daily activity; however, at sunset it retreats into its burrow at the root of vegetation. It feeds on arthropods and their larvae. To escape capture, it runs from bush to bush instead of retreating into a burrow or under cover. It feeds on various kinds of arthropods. The prey is quickly seized and, if of a larger size, is shaken and battered against the ground before being swallowed (Minton, 1966).

It breeds from March to May and fights among breeding males are frequent; 2–6 eggs are laid in burrows and young are active by late April to July.

Distribution: *Acanthodactylus blanfordii* is recorded from western Baluchistan and along the Makran coast, extending into southern Afghanistan and southeastern Iraq.

Acanthodactylus cantoris Günther, 1864
(Plates 87A, B)
(Blue-tail sand lizard: Neeli-dum chalpaya)
1864 *Acanthodactylus cantoris* Günther, Rep. Brit. India: 73.

Type locality: Ramnagar, India

Diagnosis:
1. Dorsals 26–36 across midbelly, gulars 26–38.
2. Hindlimb reaches between ear and collar in male, between collar and axilla in female.
3. Femoral pores 16–23.
4. Subocular does not border mouth, separated from it by fifth and sixth supralabial.

Plate 87A. *Acanthodactylus cantoris:* Adult.

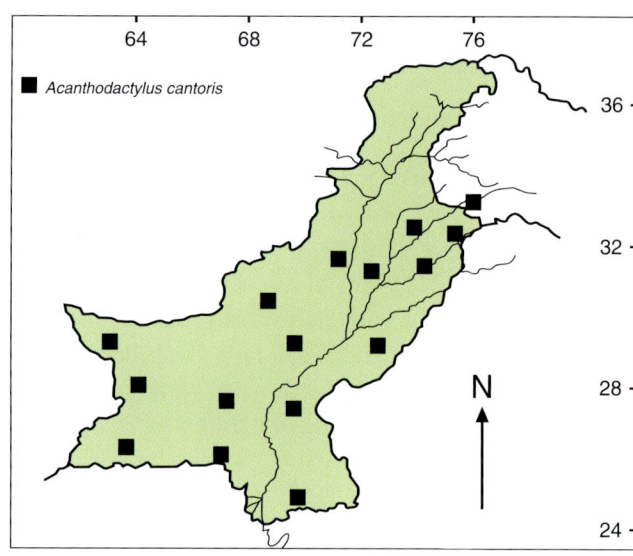

Plate 87B. *Acanthodactylus cantoris:* Juvenile.

Snout-vent length 75–78 mm, tail 180–188 mm.
Color: Dorsum reddish brown to gray, speckled with white, often with traces of longitudinal stripes; tail bluish gray. Ventrum white. Young with 6 longitudinal yellow stripes on dark dorsum, extending onto tail, distal part of the tail blue.
Natural history notes: The lizard frequents sand fields with sparse vegetation of bushes and grasses. It burrows in roots of bushes where the sand is compact. It is diurnal, usually active 2–3 hours after dawn, and retreats 2 hours before sunset. It escapes capture taking the nearest available shelter, then dashes toward its burrow. The fringe of flattish scales along digits enables it to move with considerable agility on loose sand. It is difficult to catch this lizard in the field. It is extracted from its 1–2 m deep sandy burrow by digging, often escaping when one is busy digging. It feeds on insects and their larvae. When followed in the field, it occasionally jerks its long tail in a characteristic way as if to distract the attention of its pursuer from itself (Minton, 1966).

Breeding season extends from March to May; 2–4 hard-shelled eggs are laid in a burrow at the roots of bushes. Young lizards with a vivid striped pattern are hatched within 3 weeks of egg laying and they soon make their own burrows.
Distribution: The blue-tail sand lizard ranges throughout the plains and deserts of Pakistan and India, from sea level to an elevation of 300 m. It is a common lizard along the beaches of Pakistani coastal areas.

Acanthodactylus micropholis **Blanford, 1874** (Plate 88)
(Yellow-tail sand lizard: Peeli-dum chalpaya)
1874 *Acanthodactylus micropholis* Blanford, Ann. Mag. Nat. Hist. London (4) 13: 33.
Type locality: Magas, Baluchistan.
Diagnosis:
1. Lower eyelid scaly.
2. Dorsals in 43–54 rows, less than twice the size of the laterals.
3. Ventrals in 9–10 straight rows.

Chapter 6. Lizards

Plate 88. *Acanthodactylus micropholis*. (Photo courtesy of S.C. Anderson).

3. Gular scales 30–32 in a median row.
4. Femoral pores 40–43.
    Snout-vent length 62–65 mm, tail 125–130 mm.

Color: Adult with seven bright yellow stripes separated by black or dark brown interstripe zones, lateral stripes are spotted with light; limbs grayish brown, spotted with yellow; tail bright yellow, ventrum white. Young similar to adult, but lack lateral spots on body.

Natural history notes: This sand lizard inhabits sandy places, stream beds with stones and sand. It is diurnal, becoming active 2–3 hours after dawn, resting in the hottest hours. It feeds on arthropods and their larvae. It breeds from March to May, when 2–4 hardshelled eggs are laid; juveniles are active by April to June (Minton, 1966).

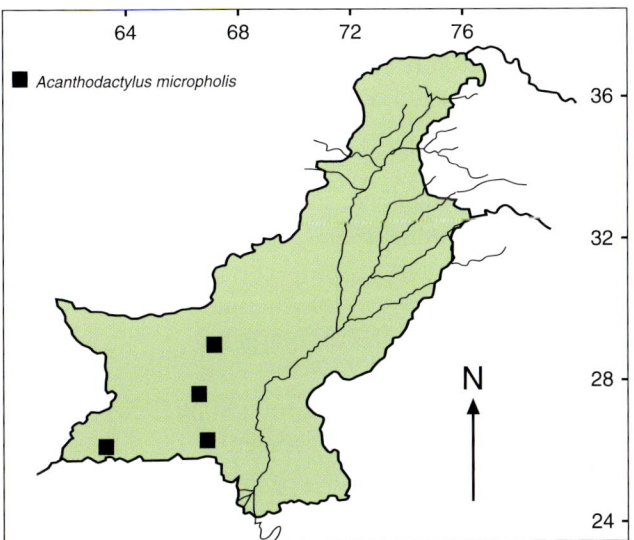

Distribution: *Acanthodactylus micropholis* extends from southeastern Iran along coastal Makran, Las Bela; northward it extends to central Baluchistan.

Genus *Eremias* Wiegmann, 1834

Occipital scale vestigial or absent; nostril between 3–4 nasals, not touching supralabials; collar complete or nearly so; dorsal scales granular, subimbricate or juxtaposed; abdominal scales subquadrangular, imbricate, smooth in oblique converging longitudinal rows; digits with or without fringe of lateral scales; usually femoral pores present.

Five species of eremiid lizards are reported from Pakistan.

Eremias acutirostris (Boulenger, 1887) (Plate 89)
(Lesser reticulate desert lizard: Lakeer-dar taiz-rao)
1887 *Scapteira acutirostris* Boulenger, Cat. Liz. Brit. Mus. 3:114.

Type locality: Between Nushki and Helmand, Baluchistan, Pakistan.

Diagnosis:
1. Dorsals granular, not enlarged on flanks.
2. Ventrals in 18–22, oblique longitudinal and 34–38 transverse rows.
3. Both fingers and toes with a well developed fringe of flat pointed scales.
3. An occipital scale absent.
4. Femoral pores 22–30.
5. The subocular does not border the mouth.
   Snout-vent length 68–70 mm, tail 148–150 mm.

Color: Dorsum light brown, with a fine reddish brown reticulum extending onto limbs, enclosing pale grayish spots. Ventrum pale yellow to white.

Natural history notes: This desert lacerta is very agile on loose fine sand dunes. When followed, it

Plate 89. *Eremias acutirostris*.

escapes by burrowing and tunnelling through the sand. It feeds primarily on deserticolous insects and their larvae (Minton, 1966).

Breeding males chase each other during breeding season which extends from March to July, 2–4 hard-shelled eggs are laid in burrows, at the roots of vegetation.

Distribution: It ranges from desert basins of north-western Baluchistan and adjoining Afghanistan and Iran.

Natural history notes: The reticulate lacerta has a robust body. It is very agile, frequenting sand fields with sparse marginal vegetation, also extending into sandy, rocky country. Its burrows are located in the roots of plants where the sand is a bit compact. It feeds on arthropods. It breeds from March to May; the breeding males chase each other and fights among them are common for the possession of a female. She lays 4–7 white-shelled eggs, and young are met from July to August.

Distribution: It has been recorded from around Koh Malik-do-Khand, at the Afghanistan-Pakistan border.

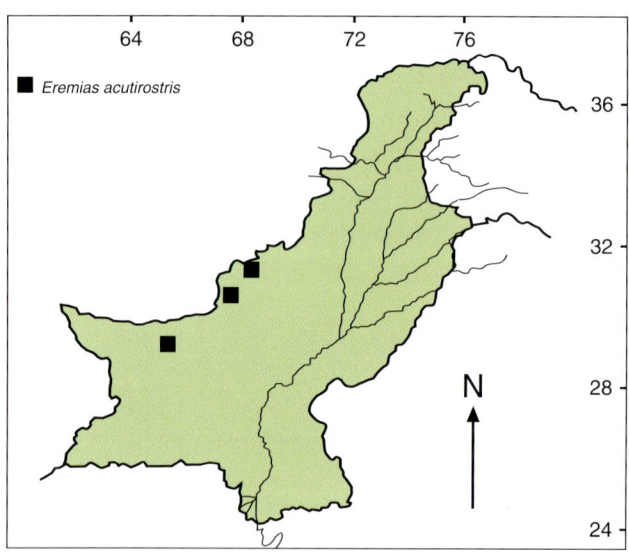

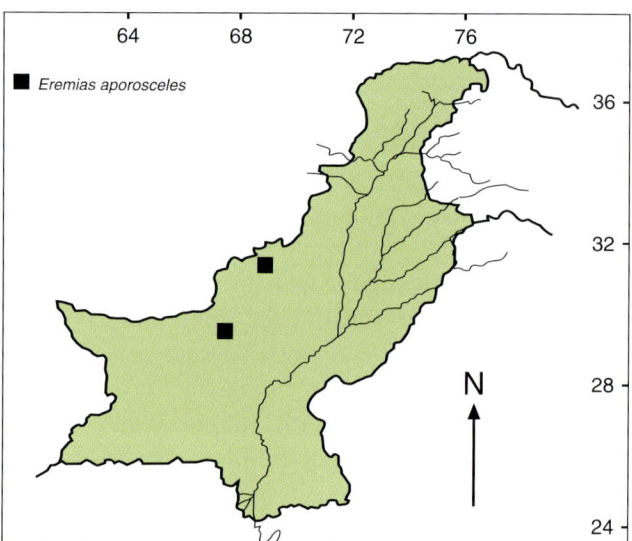

Eremias aporosceles (Alcock and Finn, 1896)
(Greater reticulate sand lacerta: Baloch taiz-rao)
1896 *Scapteira aporosceles* Alcock and Finn, J. Asiatic Soc. Bengal, Calcutta 65: 559.
Type locality: Near Nushki, northern Baluchistan, Pakistan.
Diagnosis:
1. Femoral pores absent.
2. Fringe of pointed scales on both sides of fourth toe.
3. Gular scales 28–33.
4. Dorsals granular, subequal, 68–82 across mid-dorsum.
5. Ventrals in 18–20 longitudinal, and 34–36 transverse series.
    Snout-vent length 78–80 mm, tail 135–140 mm.

Color: Dorsum sandy gray, with dark reticulation, head with symmetrical dark markings. Ventrum white.

Eremias fasciata Blanford, 1874
(Striped sand lacerta: Patti-dar taiz-rao)
1874 *Eremias fasciata* Blanford, Ann. Mag. Nat. Hist. London (4) 14: 32.
Type locality: Saidabad, southwest of Kirman, Iran (restricted by Smith 1935).
Diagnosis:
1. Ventrals in oblique longitudinal rows.
2. No fringe of scales on fourth toe.
3. No occipital scale.
4. Number of dorsal scale rows at midbody 44–50.
5. Caudal scales strongly keeled.
6. Femoral pores 15–18.
    Snout-vent length 63–65 mm, tail 120–124 mm.

Chapter 6. Lizards

Color: Dorsum pale gray to sandy, with 5–9 narrow dark, longitudinal streaks on neck, reduced to 5–7 on back. A lateral stripe from eye to supraocular region, another along outer margin of the ventrals. Ventrum white.

Natural history notes: The lizard inhabits sandy plains with bushes. It is very agile and runs from bush to bush, which makes its capture very difficult. Its linear dorsal pattern makes its detection difficult when it is in the shade of bushes. Its burrows are located at the roots of bushes where the substratum is a bit harder.

Diurnal, it feeds on various types of desert arthropods.

Distribution: It ranges from Iran, Afghanistan, and Baluchistan up to southern Waziristan. Recently it has been recorded from Dera Ismael Khan (W. Khan, 1997).

Plate 90. *Eremias persica.* (Photo courtesy of S.A. Minton).

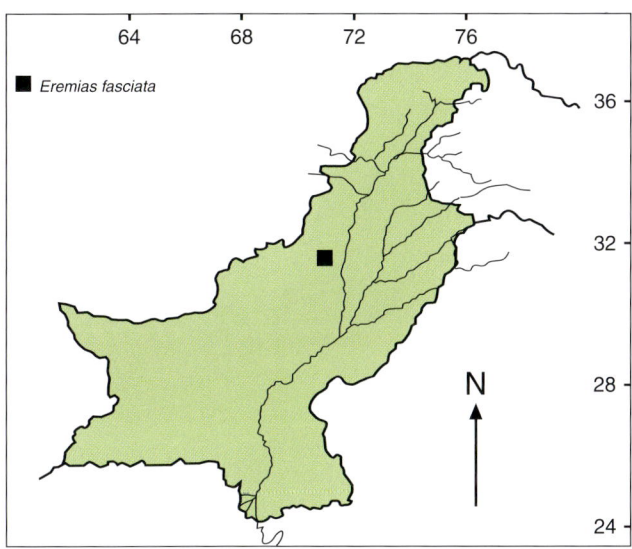

Eremias persica (Blanford, 1874) (Plate 90)

(Persian sand lacerta: Irani taiz-rao)

1874 *Eremias persica* Blanford, Ann. Mag. Nat. Hist. London (4)13: 31.

Type locality: Ispahan, Iran.

Diagnosis:

Largest and most robust of all Pakistani lacertids.
1. Interparietal is longer than the suture between parietals.
2. No fringe of scales along outer border of fourth toe.
3. Large supraoculars 2.
4. Caudals keeled, arranged in annuli.
   Snout-vent length 95–98 mm, tail 150–152 mm.

Color: Adult with grayish brown dorsum, a wide irregular black or dark brown stripe from shoulder to the base of tail. Additional pale gray, or white and dark markings and spots are arranged in longitudinal rows. Limbs with indistinct light spots. Ventrum white. Juveniles with 2–4 dark stripes alternating with yellow stripes. Black stripes enclosing small light spots. Distal half of tail bluish.

Natural history notes: The Persian lacerta inhabits scrubby and sparse grassland habitat with sandy substratum, clearly avoiding open sand dunes. Running at great speed from one clump of bushes to another is the characteristic behavior of this lizard. The lizard is difficult to detect when it is in the shade of bushes, where it usually skulks under leaf litter, searching for arthropods. While foraging in the leaf litter it continuously scratches, displacing dry leaves, looking for arthropods and their larvae which constitute the major part of its diet.

Breeding season extends from March to July, 6–8 elongated eggs are laid in burrows. Striped juveniles form the main population during August–September.

Distribution: This lacerta has been collected from elevations up to 1000 m. The range of this species is eastern and central Iran; in Pakistan it is reported from northern Waziristan, Quetta, and Nushki.

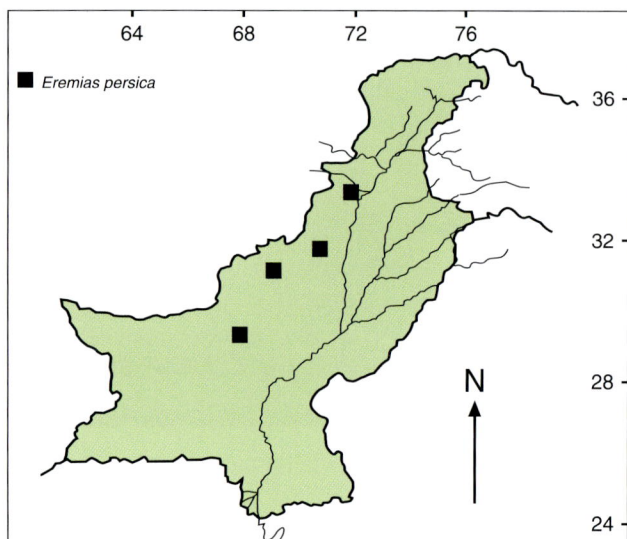

tation of small herbs and grasses. The lizard is very active during the early and late hours of the day. It burrows in sand to escape capture, readily climbing into branches of herbs or other vegetation to protect itself from the hot sand. At midday when on hot sand, the lizard lifts up its legs, arches its body, and rests on its belly for few seconds. The feet are then lifted alternately to protect them from burning, and the lizard sprints for shade.

Distribution: The vermiculate lacerta extends from Transcaspia to eastern Khazakistan, Iran, Afghanistan, and northern Baluchistan, to east of Nushki.

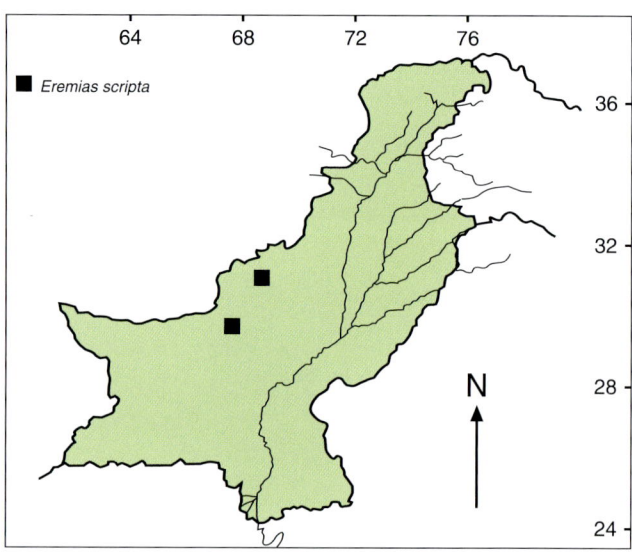

Eremias scripta (Strauch, 1867) (Plate 91)
(Vermiculate sand lacerta: Jal-dar taiz-rao)
1867 *Podarcis scripta* Strauch, Mel. Biol. Bull. Acad. St. Petersburg 4: 424.
Type locality: Aralo-Caspian desert.
Diagnosis:
1. Fringe of large scales only on outer border of fourth toe.
2. Ventrals in 14–16 oblique longitudinal and 32–34 transverse series (Figure 17A).
3. An occipital scale present.
   Snout-vent length 58–60 mm, tail 128–130 mm.

Color: Dorsum yellowish to golden brown, becomes pale bluish gray on tail, with 5–7 fine dark stripes, some of which are vermiculate. Sides and ventrum of the body cream to white.

Natural history notes: The vermiculated lacerta inhabits flat fields of loose fine sand, with sparse vege-

Genus *Mesalina* Gray, 1838
Occipital shield usually present; lower nasal scale in contact with first supralabial only; nostril between 3 nasals and is widely separated from supralabials; sometimes two or more transparent bodies in lower eyelid; abdominal scales in parallel longitudinal rows.

Two species are recorded from Pakistan.

Mesalina brevirostris Blanford, 1874 (Plate 92A)
(Short-snout desert lacerta: Chotta-sar taiz-rao)
1874 *Mesalina brevirostris* Blanford, Ann. Mag. Nat. Hist. London (4)13: 32.
Type locality: Kalabag, northwestern Punjab, Pakistan, and Tumb Island, Persian Gulf, Iran.

Plate 91. *Eremias scripta.*

Chapter 6. Lizards

Plate 92A. *Mesalina brevirostris.* (Photo courtesy of S.A. Minton).

Diagnosis:
1. Occipital separated from the interparietals.
2. Ventrals in 10–12 straight longitudinal series, median 4 rows distinctly broader than long.
3. Lower nasal scale rests on first supralabial only.
4. An occipital scale present.
5. No fringe of scales on digits.
   Snout-vent length 56–58 mm, tail 95–98 mm.

Color: Grayish brown dorsum with numerous white and black spots which are largest and more closely arranged on sides.

Natural history notes: The lizard has been recorded from the hard alluvial soil close to sand dunes, where vegetation is sparse in the form of clumps of thorny bushes and reeds. The lizard moves quickly from one clump to another and is caught with great difficulty.

Distribution: The short-snout lacerta is reported from District Mianwali, in northwestern Punjab, and coastal Las Bela. It has extensive distribution in the west to Syria.

Mesalina watsonana (Stoliczka, 1872) (Plate 92B)
(Spotted lacerta: Chittra taiz-rao)
1872 *Eremias watsonana* Stoliczka, Proc. Asiatic Soc. Bengal, Calcutta 1872:86.

Type locality: Sind, between Karachi and Sukkhur, Pakistan.

Diagnosis:
1. Ventrals distinctly broader than long, in 8–10 straight longitudinal series across midabdomen (Figure 17B).
2. Occipital scale in contact with the interparietal.
3. Lower nasal scale rests only on first supralabial.
4. No lateral fringe on fourth toe.
5. Dorsal and lateral scales of body granular.
   Snout-vent length 60 mm, tail 124 mm.

Color: Grayish to olivaceus dorsum, usually 2 pairs of indistinct light longitudinal stripes. Middorsum with longitudinal series of small white spots which are edged with black. Limbs with faint light spots. Ventrum white to yellowish.

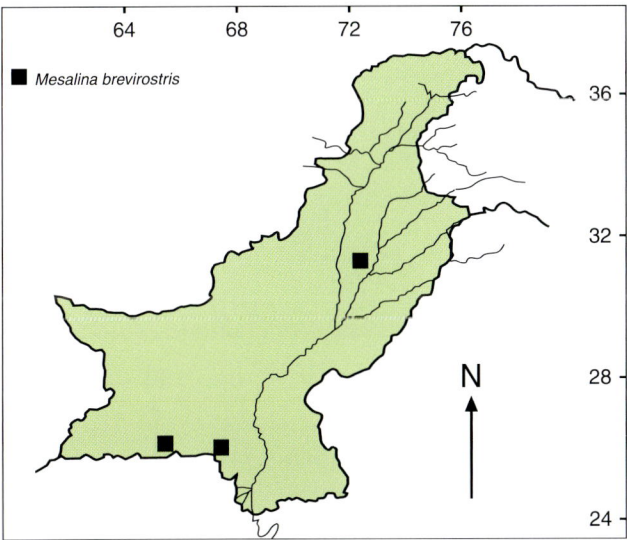

Plate 92B. *Mesalina watsonana.* (Photo courtesy of S.A. Minton).

Natural history notes: *Mesalina watsonana* prefers flat fields of hard soil, where the vegetation of herbs and grasses is sparse. It particularly avoids soft loose sandy soil. Its burrows are located among roots of low vegetation. It hides under stones and other available shelters at times of danger. It feeds on arthropods and their larvae, actively moving from one plant to another in search of food.

Two to four eggs are laid in burrows from April to July, and juveniles are generally seen during July and August.

Distribution: This species ranges from Rajputana to southern Afghanistan, and Iran. In Pakistan it occurs throughout the plains, excluding the deserts. It extends from sea level to 2000 m of elevation.

Plate 93. *Ophisops elegans.* (Photo courtesy of S.C. Anderson).

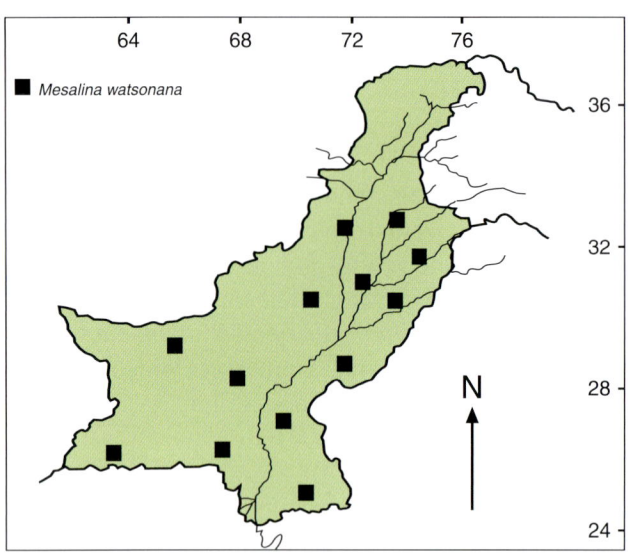

Genus *Ophisops* Ménétriés, 1832

Lower eyelid fused with upper, with a large transparent disc, covering the eye; collar weakly defined or absent in the middle; subdigital lamellae keeled; dorsal scales rhomboidal, imbricate, strongly keeled; femoral pores present.

Two species are reported from Pakistan.

Ophisops elegans Ménétriés, 1832 (Plate 93)

(Smooth spectacled lacerta: Naram chishma-chalpaya 1832 *Ophisops elegans* Ménétriés, Cat. Rais. Obj. Zool. Caucas.: 63.

Type locality: Near Baku, Caspian Sea.
Diagnosis:
1. Head scales smooth.
2. Scales around midbody 31–38 scales.
3. Femoral pores 10–12 on each side.
   Snout-vent length 55–58 mm, tail 102–107 mm.

Color: Olive greenish or brownish dorsum, with a pair of light dorsolateral stripes; first originating from the supraciliary margin; second from below eye, passing through ear, along body flanks to hindlimb. Upper stripe dorsally spotted with black. A series of vertebral spots. Ventrum greenish white.

Natural history notes: This small lizard frequents gullied scrubland, with sparse vegetation of low

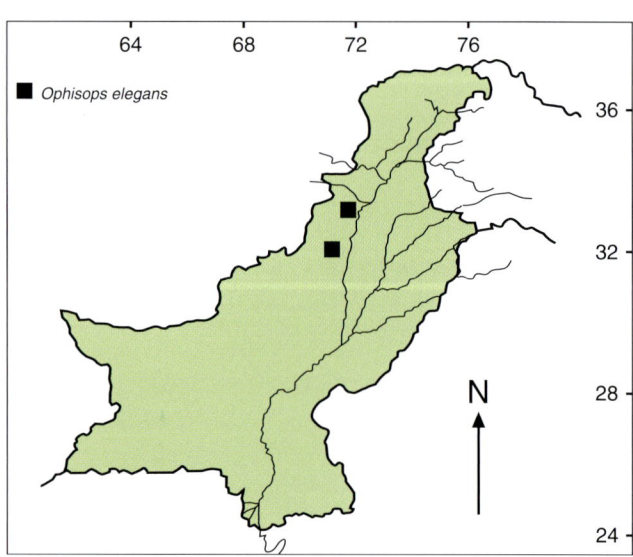

grasses, reeds, and thorny bushes. The lizard lives in crevices and holes in soil and moves very fast from one shelter to another.

The lizard is diurnal, feeding on insects and their larvae. Juveniles are generally seen during July and August.

Distribution: In Pakistan this lizard is reported from Mianwali District and Waziristan hilly tracts. However, it has wider distribution in the west, extending to the Caspian region and west to Turkey and the Levant.

Ophisops jerdonii **Blyth, 1853 (Plates 94A, B)**
(Rugose spectacled lacerta: Khurdra chisma-chalpaya)
1853 *Ophisops jerdonii* Blyth, J. Asiatic Soc. Bengal, Calcutta 22: 653.
Type locality: Alpine Punjab, Pakistan.
Diagnosis:
1. Head scales rugose, with strong striations.
2. Scales around mid body 28–35 scales.
3. Femoral pores 7–12, on each side.
   Snout-vent length 47–49 mm, tail 95–98 mm.

Color: Olive brown, golden or grayish above, with a pair of golden lateral streaks, upper one extending from supracilliaries to tail, lower bordering upper lip and extending along body flanks to the base of the hindlimb. The space between stripes, as well as the upper border of the upper stripe, usually spotted with black. Ventrum yellowish brown.

Natural history notes: It is common in the leaf litter, and inhabits moist environs of shady places in gardens, forests, grasslands, and the areas with much ground cover. Its body color blends with its habitat. It moves fast from one shelter to another and due to its small serpentine body and smaller diameter, it creeps with ease under stones and slabs, from under which it is not easily retrieved. Its diet consists mainly of ants and soft-bodied land crustaceans which are in abundance in the moist habitat around the leaf litter area. It is diurnal, mostly confined to the dense undergrowth.

It breeds from April to June, and gravid females with 2–7 eggs, measuring 6 by 4 mm are usual from April to May, after which adults become rare and are replaced by juveniles.

Distribution: It is a widely distributed lizard in the plains and semi-hilly regions throughout Pakistan and India. In Pakistan it has been recorded apart from various localities in the Punjab and Sind plains, from alpine Punjab, Salt Range (Khan, 1988), Waziristan, Quetta, and Khuzdar in Baluchistan (Khan and Ahmed, 1987).

Plate 94A. *Ophisops jerdonii.*

Plate 94B. *Ophisops jerdonii:* Side view.

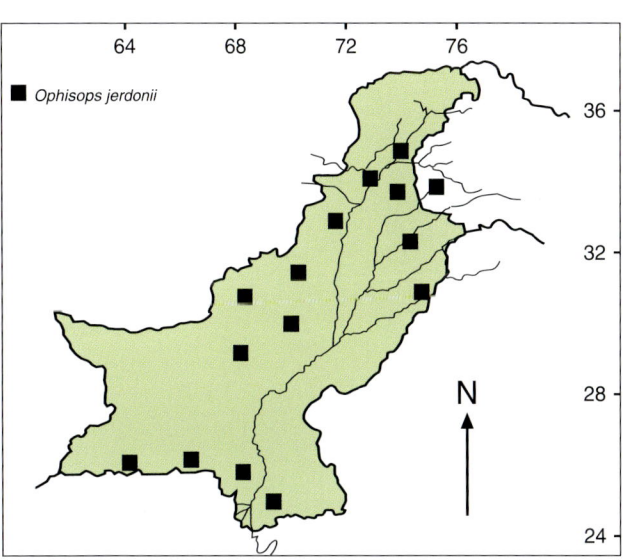

Family Scincidae

Limbs reduced in size and components, elongated cylindrical body and tail, with serpentine movements. Body scales polished, shiny, strongly imbricate, smooth or low-keeled. Head covered with paired large symmetrically arranged scales. Eyelids transparent, fused in genus *Ablepharus* forming transparent spectacle on the eye. In rest of the genera lower eyelid develops a transparent disc. The pupil is round. Preanal and femoral pores are absent. Tail is fragile; when broken, it regenerates. Scincid lizards frequent grasslands and sand deserts (Figure 15).

Seven genera of family Scincidae occur in Pakistan.

Genus *Ablepharus* Fitzinger, 1823

No supra-anal shields; eyelids fused with each other forming transparent spectacle covering the eye; ear opening reduced in size or absent. Limbs pentadactyl, weakly developed; body with smooth cycloid scales, in 18–26 rows around midbody (Fühn, 1969; Eremchenko and Szczerbak, 1986).

Two species known from Pakistan.

Ablepharus grayanus (Stoliczka, 1872) (Plate 95)
(Earless snake-eyed skink: Bahri saamp-chishm)
1872 *Blepharosteres grayanus* Stoliczka, Proc. Asiatic Soc. Bengal, Calcutta 1872: 74.

Type locality: Waggur District, northeast Kutch, India.
Diagnosis:
1. Ear opening obscure, position indicated by a shallow depression.
2. Scales around the midbody 18–20.
Snout-vent length 32–33 mm, tail 54–58 mm.

Color: Body olive greenish with distinct metallic luster, flanks speckled or streaked with dark brown. A light stripe from superciliary edge to the base of tail, edged above and below with black.

Natural history notes: This lizard frequents leaf litter and prostrate vegetation, inhabits the moist environs under fallen leaves in gardens and oases, grasslands, backyard gardens, and extend into the inhabited houses. It is diurnal, secretive, and very agile, moving from one shelter to another so quickly that the eye cannot follow it. It coils its long body to accommodate it in a limited space. Difficult to capture, its slippery serpentine body slips under stones and thorny bushes very quickly.

Food consists of ants, grasshoppers, nymphs, termites, and small dipterous larvae which abound in the leaf litter.

This lizard has been collected almost throughout the year. It basks in cold season in open fields at midday. Its burrows are shallow pits excavated under stones and other objects; mostly it hides under heavy objects. Usually, the lizard inhabits holes and crevices in walls and under stones covered with vegetation.

Plate 95. *Ablepharus grayanus.* (Photo courtesy of S.A. Minton).

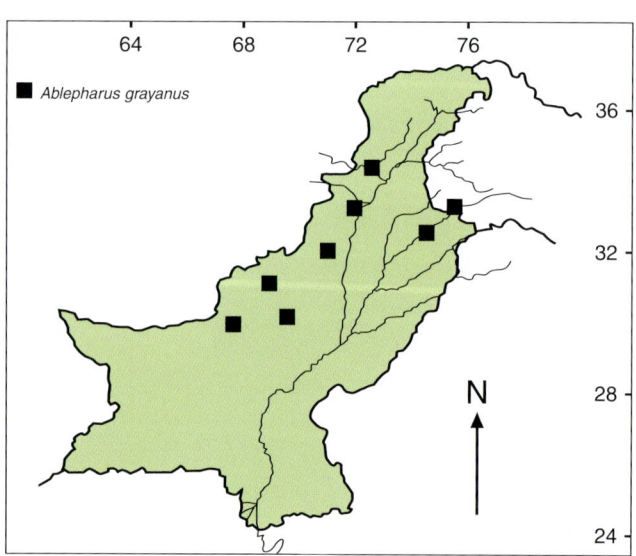

One to two eggs are laid during spring and young are active by April to August (Minton, 1966).
Distribution: This lizard is recorded from throughout Pakistan at low elevations, especially grass fields in the Indus riparian system. It has also been recorded from hills around Fort Munro, Dera Ghazi Khan at an elevation of 1800 m (Khan, 1993b).

Ablepharus pannonicus (Fitzinger, 1823)
(Red tail snake-eyed skink: Surakh-dum saamp-chishm)
1823 *Scincus pannonicus* Fitzinger *in* Liechtenstein *in* Eversmann, Reise von Orenburg nach Buchara Berlin: 103.
Type locality: Bukhara.
Diagnosis:
1. Small distinct ear opening, partially hidden by scales.
2. Number of scales around midbody 20, rarely 22. Snout-vent length 40–55 mm, tail 100–120 mm.

Color: Olive brown dorsum with metallic luster. A distinct dark brown dorsolateral stripe, edged above with white. Flanks with indistinct dark longitudinal lines. Supralabials cream, chin and throat whitish, flecked with blue, ventrum pale bluish gray; limbs and tail tinged with red.
Natural history notes: Generally this species inhabits stony hillsides, with thick growth of low grasses and herbs, under fallen needles of pine trees at an elevation of 2000 m. Its slippery body and small size make it difficult to capture.

It breeds from April to July.
Distribution: This lizard extends from the Arabian Peninsula and North Arabian Desert, through Iran to circum-Mediterranean region, Tajikistan and Afghanistan. In Pakistan it has been reported from around Quetta, Waziristan hills, Chitral and the Salt Range (Khan and Baig, 1988).

Genus *Chalcides* Laurenti, 1768
Nostril between nasal and rostral; supranasals present; prefrontals and frontoparietals absent; lower eyelid with a transparent disc; body elongate, cylindrical; limbs reduced or vestigial.

A single species known from Pakistan.

Chalcides ocellatus (Forskål, 1775) (Plate 96)
(Ocellate skink: Goal-jisam baamani)
1775 *Lacerta ocellatus* Forskål, Descript. Anim.: 13.
Type locality: Egypt.
Diagnosis:
1. Naris between nasal and rostral scales.
2. Supranasal scale present.
3. Prefrontal and frontoparietal scales absent.
4. Limbs short, pentadactyl.
   Snout-vent length 105–115 mm, tail 70–75 mm

Color: Dorsum bronze, sides pale yellow; dark scales are arranged in irregular transverse rows on body, in more regular annuli on tail. Each dark scale has a median light bar. Head is speckled with dark. Labials cream with dark bars.

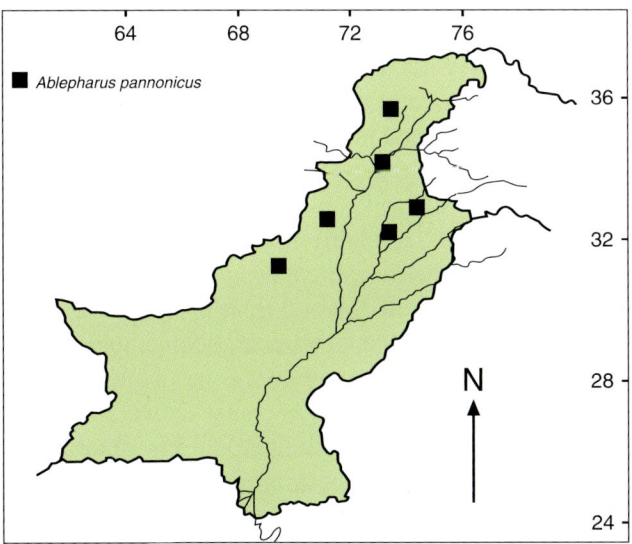

Plate 96. *Chalcides ocellatus.* (Photo courtesy of S.C. Anderson).

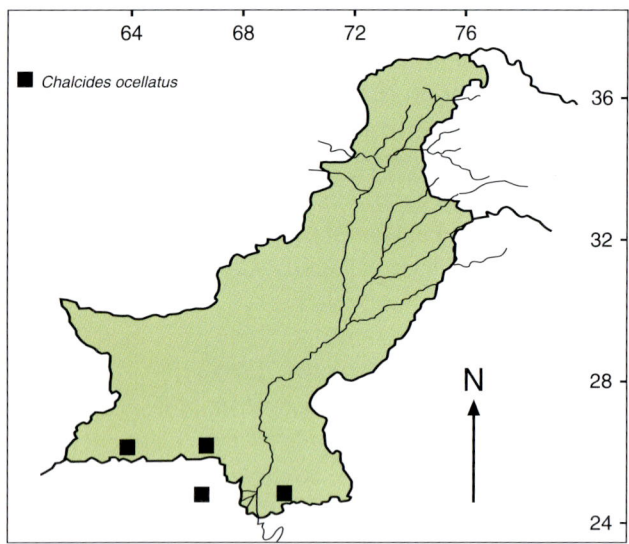

Natural history notes: It lives in warm humid areas with scrubby vegetation, in rocky and sandy terrain. It is secretive in habits. Viviparous, it gives birth to 4–6 young ones.

Distribution: Widely distributed in the west, from the Mediterranean through northern Saudi Arabia, the Persian Gulf to the Makran coast. In Pakistan it has been recorded from along the coastal strip, around Karachi and Astola Island (Minton, 1966; Mertens, 1969; Khan, 1999c).

Genus *Eutropis* (Fitzinger, 1826)

Eyelids moveable, lower with or without a transparent disc; ear opening distinct, tympanum deeply sunk; nostrils piercing single nasal scale; supranasals and prefrontals shields are present. Limbs well developed, pentadactyl, digits subcylindrical or compressed with transverse subdigital lamellae.

Two species are reported from Pakistan.

Eutropis dissimilis (Hallowell, 1860) (Plate 99)
(Striped grass skink: Lakeer-dar gaas-goodi)
1860 *Euprepes dissimilis* Hallowell, Transact. Amer. Phil. Soc. Philadelphia 11: 78.
Type locality: Bengal.
Diagnosis:
1. Supranasals in contact with each other.
2. Frontonasal almost as broad as long.
3. No postnasal scale.
4. An undivided transparent disc in lower eyelid.
5. Dorsal and lateral body scales subequal.
6. Dorsals with 2 or 3 laterals with 3 keels.
   Snout-vent length 53–55 mm, tail 69–70 mm.

Plate 99. *Eutropis dissimilis*. (Photo courtesy of S.A. Minton).

Color: Dorsum light brown, with 3 or more greenish white stripes, a median dorsal and 2 dorsolateral; stripes are edged with transversely arranged black spots. The sides of the body are spotted with white dark-edged spots. Ventrum yellowish white. Eyelids edged with bright yellow.

Natural history notes: The striped grass skink inhabits open moist grass fields and extends in tilled land where it is killed in large numbers, as people commonly believe it to be a "deadly poisonous legged snake" Its body color blends with the moist ground. Though terrestrial, it readily enters the water and may climb in low bushes, excellent adaptations in order to live in the periodically flooded riparian system of Punjab. It is plentiful in October and November, after monsoon rains, when temperature is moderate.

The lizard is diurnal: it becomes active 2–4 hours after dawn, moves under leaf litter and low grasses, feeding on a vast variety of arthropods, mostly crop pests, insect larvae, earthworms, and soil nematodes. Minton (1966) saw one with a small frog in its jaws.

Chapter 6. Lizards

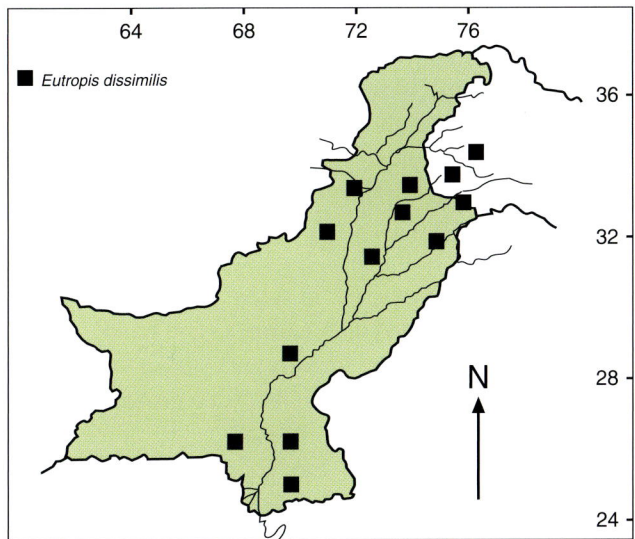

The skink usually drinks water. Burrows are located along the elevated margins of fields.

Three to seven oval white eggs with pliable soft shells are laid from March to July; juveniles are seen from May to August.

Distribution: This lizard has wide distribution from western Bengal, Bihar, across the Gangetic plain to Rawalpindi, upper Indus Valley to Waziristan, and extends into the Indus Delta.

Eutropis macularia (Blyth, 1853) (Plate 100)
(Bronze grass skink: Bhoori gaas-goodi)
1853 *Euprepes macularius* Blyth, J. Asiatic Soc. Bengal, Calcutta 22: 652.

Type locality: Rangpur, Bengal.
Diagnosis:
1. Supranasals are separated from each other.
2. Frontonasals about as long as broad.
3. A postnasal may or may not be present.
4. Scaly lower eyelid.
5. Dorsal and lateral body scales are subequal, with 5–9 keels.
   Snout-vent length 68–70 mm, tail 63–65 mm.

Color: Dorsum uniformly iridescent bronze, or specked with black. Sides of body darker, with rows of white dots. Female brown to gray, with sparse light dots, or dots may be absent. Ventrum white. Male with vermilion lips and flanks during breeding season.

Natural history notes: Diurnal and agile, it is so fast that one can have just a glimpse of it while looking for it in grass and leaf litter. It quickly moves in thick grass. The skink is commensal with man, it inhabiting grasslands, open fields, farmlands, and parks and gardens close to human habitations, seldom venturing into buildings. It is particularly active and abounds in rainy season. It sinuously slips through grass blades, often falling prey to the local snakes and varanids. A good swimmer, flood water drives it out of its burrow and carries it far and wide in the plains.

It feeds on arthropods; however, it becomes nocturnal in the hot dry season, often coming out to feed on photophilic arthropods under light posts. The breeding male develops a bright vermilion color on the lips and flanks of the body during May to September.

Plate 100. *Eutropis macularia.*

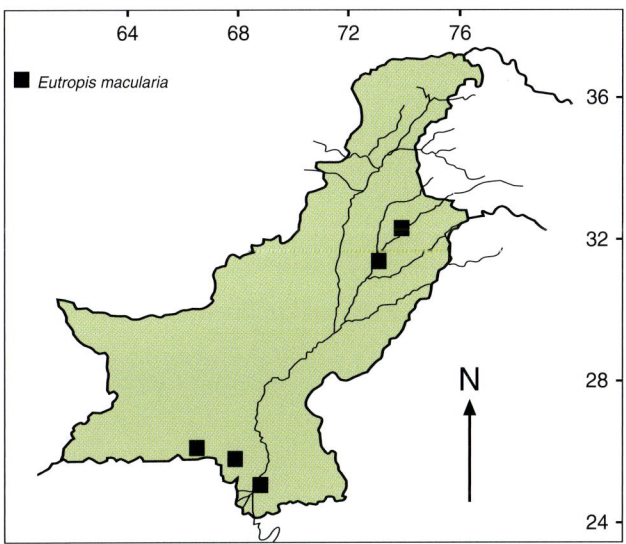

Four to six oval white-shelled eggs are laid under vegetation or under any other object, debris, or in holes. Juveniles are active by June to October (Minton, 1966).

Distribution: Range of this skink extends from Laos and Cambodia to northern Malaya, Myanmar, western Bengal, and Bihar, across the Indo-Gangetic Plain, to the Iranian Plateau. In Pakistan it has been reported from the plains and highland of the Salt Range, in Punjab, around Karachi and Las Bela.

Genus *Eurylepis* Blyth, 1854

"Elongate, 35 or more presacral vertebrae. Limbs relatively slender, lamellae not expanded. Head somewhat conical, dorsal surface convex in lateral view, pareietal bone with clear lateral indentations and suprafontanelle open. Sexual dimorphism in head proportions not distinct. Scales shiny, separated by shallow sutures. Two loreals, followed by two presuboculars. Postnasal scales present. Palbebral scales and superciliaries not separated by groove. Four or five pairs of nuchal scales, followed by several pairs of broadened mid-dorsal scales and broad row of fused mid-dorsal scales. Large median preanal scales overlie small lateral pair. Ear lobules conspicuous, but not covering ear opening. Color pattern consists of gray-brown background, with pale, broad dorsolateral stripes, more distinct anteriorly, brown rectangular spots dominating posteriorly" (Griffith et al., 2000). *Eurylepis* has been resurrected from the genus *Eumeces* in the recent revision by Griffith et al. (2000).

A single species is known from Pakistan.

Eurylepis taeniolatus taeniolatus Blyth, 1854 (Plates 97A, B)

(Common mole skink: Maidani reg-mahi)
1854 *Eurylepis taeniolatus* Blyth, J. Asiatic Soc. Bengal, Calcutta 23: 739.

Type locality: Salt Range, Punjab, Pakistan.
Diagnosis:
1. Number of scalesaround midbody 21–23.
2. A postnasal scale present.
3. Two median rows of dorsal scales fused into a broad median row.
   Snout-vent length 160–165 mm, tail 299–225 mm.

Color: Dorsum pale sandy to bronze, speckled with cream. Three dark brown stripes with scattered light flecks run on the body. Tail speckled and variegated black, brown, and pale gray. Ventrum vivid yellow. In juvenile each yellow scale is edged with black, and there is no stripe on the dorsum.

Natural history notes: The skink inhabits loose sandy soil or loamy with scrubby vegetation, in almost deserticolous habitat, mostly close to water courses. It is a diurnal secretive lizard, burrowing in the roots of thorny bushes and grasses. In stony sandy moist soil, its burrows are located under stones. The lizard burrows in loose soil by thrusting its strong snout into the loose soil particles. It has also been collected from silty moist soil along water courses.

On hot summer days the lizard moves about in the leaf litter and undergrowth; feeding on different kinds of arthropods and worms, rarely showing up in the open.

It breeds from March to June, 4–8 oval eggs with pliable shells are laid under stones or in a burrow in

Plate 97A. *Eurylepis taeniolatus*: Adult.

Plate 97B. *Eurylepis taeniolatus*: Juvenile.

Chapter 6. Lizards

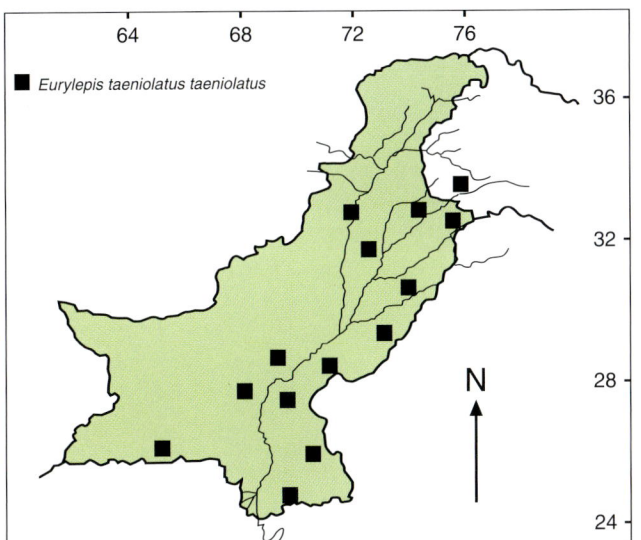

Plate 98. *Lygosoma punctata.*

moist soil. Juveniles are met with from April to August.

The lizard is in high demand by local physicians, the "hakims." The eviscerated dried skinks are used in preparation of some local medicines which are known as powerful sexual stimulants (M.S. Khan and M.R.Z. Khan, 1997).

Taxonomic notes: Szczerbak's (1990) breaking down of *Eurylepis taeniolatus* on color variation is not supported by morphological characters.

Distribution: *Eurylepis taeniolatus* is reported from Arabia to the Transcaspia, Kutch, and Sind to Kashmir. In Pakistan it has been collected from throughout the plains of Punjab and Sind to an elevation of 2000 m.

Genus *Lygosoma* Hardwick and Gray, 1827

Naris piercing nasal scale; supranasal scale present; dorsals uniform in size; body and tail elongated; limbs short vestigial, pentadactyl, clawed.

A single species is represented in Pakistan.

Lygosoma punctata (Linnaeus, 1766) (Plate 98)

(Spotted garden skink: Chitri baghban baamani)
1766 *Lacerta punctata* Linnaeus, Syst. Nat. (12)1:369.
Type locality: Asia.
Diagnosis:
1. Lower eyelid with undivided transparent disc.
2. Fingers and toes 5.
3. Number of scales around the midbody 24–28.
4. Tail longer than body.
   Snout-vent length 85–91 mm, tail 120–139 mm.

Color: Body brown, each body scale with a dark basal spot. In juvenile there are 4–6 longitudinal stripes down the back. A canthal yellowish streak strongly marked in young. Ventrum uniform yellowish white or each scale with a dark central mark. Tail reddish.

Natural history notes: This lizard inhabits grass fields with moderate moisture and shade, ideally available in gardens and orchards and around big cities. This skink is common in the lawns of large buildings around Lahore, where it extends into backyards of houses and often comes to live under household articles. It slips from one object to another and is difficult to catch. Diurnal, in winter it is mostly active by midday, basking in the sun, while in summer it rests around noon coming out in the cold hours of morning and afternoon.

The main foods taken are different arthropods and their larvae, earthworms, and millipedes.

It breeds from March to July; 2–6 elongated eggs are laid under stones, etc. Young are active by April to late August.

Distribution: The spotted garden skink has a wide range in the Indo-Gangetic plains, from Bangladesh through India and Sri Lanka. In Pakistan, it is restricted to the eastern strip of Punjab plain, northward it extends into Hazara Division.

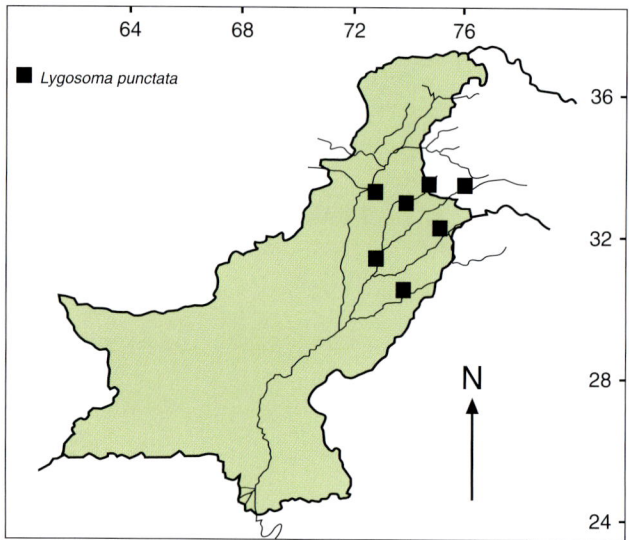

Genus *Novoeumeces* Griffith, Ngo, and Murphy, 2000

"26–28 presacral vertebrae. Relatively heavy limbs with robust pes, expanded lamellae on digits and phalanges. Head conical, dorsally convex in lateral view, parietal bone with clear lateral indentations and supratemporal fontanelle open. Sexual dimorphism in head proportions not distinct. Scales thick, separated by deep sutures. Two loreals, followed by two presuboculars. Postnasal absent. Palbebral scales separated from superciliary scales by deep groove containing small granular scales. Three or four pairs of nuchal scales. Two mid-dorsal scale rows broadened, not fused. Medial preanal scales overlie lateral preanal scales. Ear lobules conspicuous, but ear opening not fully covered. Color pattern usually consisting of small spots, broken longitudinal stripes or irregular narrow bars on brown or olive background" (Griffith et al., 2000). This genus has been separated recently from *Eumeces* in the revision by Griffith et al. (2000).

Three species are recorded from Pakistan.

Novoeumeces blythianus (Anderson, 1871) (Plate 101)

(Orange tail skink: Maltai-dum reg-mahi)
1871 *Mabouia blythiana* Anderson, Proc. Asiatic Soc. Bengal, Calcutta. 1871: 186.
Type locality: Purchased in Amritsar, Punjab, India.

Plate 101. *Novoeumeces* blythianus.

Diagnosis:
1. A single azygous postmental.
2. Nuchals 2–4 pairs.
3. Number of scalesaround the midbody 28–30. Snout-vent length 85–89 mm, tail 149–154 mm.

Color: Dorsum pale gray, with vermilion stripes from temporal to groin, tail vermilion to orange. Body and back with scattered orange scales, lips and chin chrome yellow, ventrum white.

Natural history notes: This skink inhabits rocky to hard soil terrain with low vegetation. It is diurnal, feeding on arthropods and worms.

It bites powerfully and holds on. Nothing is known about its reproduction.

Distribution: The orange tail skink has been collected from coastal areas of Pakistan. It has also been reported from near the Khyber Pass in north-

Chapter 6. Lizards

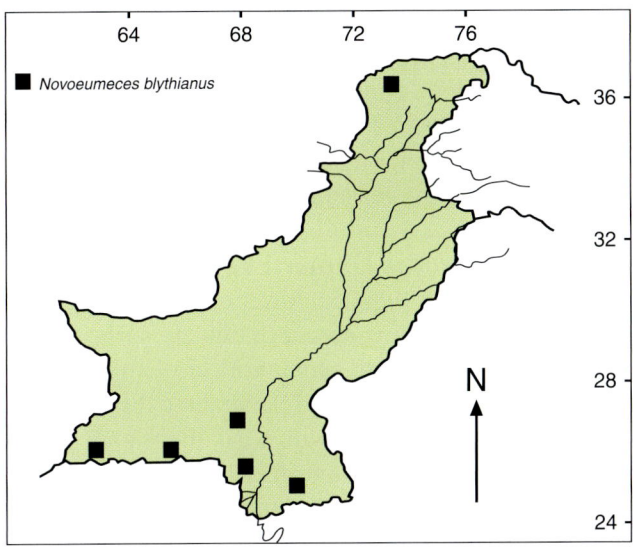
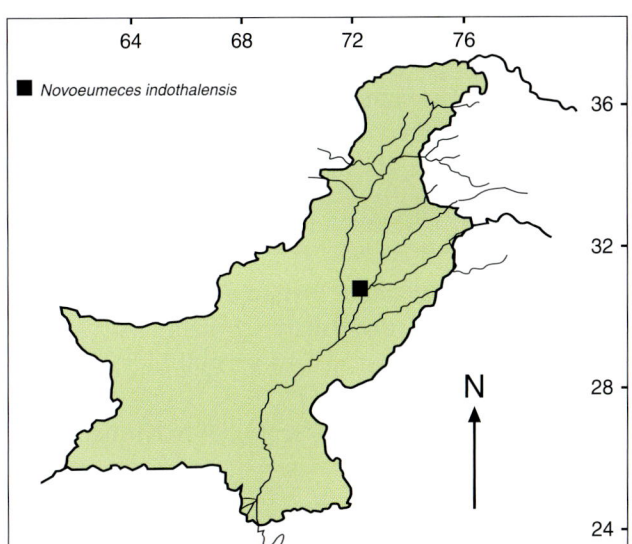

western Pakistan, close to the Afghan border (Minton, 1966).

Novoeumeces indothalensis (Khan and Khan, 1997) (Plate 102)

(Thal mole skink: Thal reg-mahi)

1997 *Eumeces indothalensis* Khan and Khan, Asiatic Herpetol. Res. 7:61–67.

Type locality: Bar Ginga Village, 9 km southwest of Bakkar western Punjab, Pakistan.

Diagnosis:
1. Dorsal pattern of 5–7 dark brown stripes, extending onto the tail.
2. Nasal scale is resting on first supralabial; nasal suture horizontal; no postnasal.
3. Interparietal about half the size of frontal and is of the same shape.
4. Azygous postmentals 2.
5. Middorsal row of 52–56 wide paired scales.
6. Posterior loreal and presuboculars are longer than deep.
7. No intercalary scales between subdigital lamellae.

Snout-vent length 54–57 mm, tail 40+ mm

Color: In formalin, 7 vivid dark brown longitudinal stripes on the body: median dorsal starts from behind interparietal and extends to the level of vent; second dorsal, paired, from behind parietals to the level of vent, join each other at the level of vent, do not extend onto the tail; third paired, from behind eye extends laterally to groin rather than along lateroventral side of tail; fourth paired from behind ear above shoulder to groin. The stripes are separated from each other by narrower lighter stripes, making the pattern distinct. Head is uniformly brownish, lips, chin, and ventrum light yellowish.

Natural history notes: The skink was collected from the roots of a common reed bush, *Sachharum moonja*, close to the tilled fields, from sandy soil. The reeds were uprooted to expose the lizards. *Eurylepis taeniolatus* abounds in the roots, while *Eumeces indothalensis* was rare: from a sack-full of *Eurylepis taeniolatus*, I found only two specimens of *Novoeumeces indothalensis*. Other reptiles collected from the area are *Crossobamon orientalis, Acanthodactylus cantoris, Ophiomorus tridactylus, Lytorhynchus paradoxus,* and *Trapelus megalonyx.*

Plate 102. *Novoeumeces* indothalensis.

Stomach contents of the skink revealed remains of beetles and crickets.
Distribution: The skink is known from the southwestern Thal Desert, close to the left bank of the River Indus, southwestern Punjab, Pakistan.

Novoeumeces schneiderii zarudnyi **(Nikolsky, 1900)**
(Red stripe skink: Surkh dahari reg-mahi)
1900 *Eumeces zarudnyi* Nikolsky, Ann. Mus. Zool. Acad. Sci. St. Petersburg 1899: 399.
Type locality: Bazman and Schur-Ab in Kirman; Labe-Ab in Seistan, Iran.
Diagnosis:
1. Postmentals paired, azygous.
2. Nuchals 3–5 pairs.
3. Number of scales around the midbody 26–30.
   Snout-vent length 110–115 mm, tail 220–240 mm.
Color: Body blue to steel gray, ventrum white. A broad bright orange red to vermilion stripe from temple to the groin. Tail orange-red.
Natural history notes: This skink has been collected from Island Astola, 10 km off the coast of Karachi, from a rocky terrain, with scrubby vegetation and in humid environments (Mertens,1969a). It has also been recorded from the Kharan Desert, around oases, where it burrows in the roots of thorny bushes.

The lizard is diurnal, feeds on different kinds of desert arthropods. The prey is crushed in powerful jaws. Breeding is quite extended from late February to September, 6–7 eggs with pliable shells are laid in burrows, young are quite common by May.
Distribution: Reported from central to southeastern Iran. In Pakistan, so far it is recorded from Astola Island, 10 km off the coast of Karachi (Mertens, 1969a).

Genus *Ophiomorus* Duméril and Bibron, 1839

Eyes small; eyelids reduced, lower eyelid with a transparent disc; ear opening absent or hidden; naris in a suture between nasal and supranasal, close to rostral; prefrontals, frontoparietal and parietal are distinct; body elongate, serpentine, with weak limbs and reduced digits, limbs may be absent.

Four species are reported from Pakistan.

Ophiomorus blanfordi **Boulenger, 1887**
(Makran sand-swimmer: Makran reg-tyair)
1887 *Ophiomorus blanfordi* Boulenger, Cat. Liz. Brit. Mus. 3: 395.
Type locality: Chah Bahar, Baluchistan, Pakistan.
Diagnosis:
1. Interparietals as long as broad.
2. Frontonasal not twice as long as broad.
3. Scales around midbody 20.
4. Fingers 4, toes 3.
   Snout-vent length 70–84 mm, tail 75–80 mm.
Color: In life, dorsum golden yellow to brownish, tail paler tending to gray, ventrum pearly white. Tail sky-blue in young. In preserved specimens, dorsum cream or pale brown, median 2 rows of dorsals with scattered dark spots, arranged in distinct longitudinal lines. Third and fourth rows from the vertebral line, on flanks, are with dark brown dots arranged in lines extending on to sides of the head. Head with or without a central dark streak.
Natural history notes: This sand lizard has been collected from coastal sand dunes, with sparse vegetation. It is nocturnal, glides sinuously under the sand, its movements visible from the surface by characteristically disturbed sand. It occasionally comes out to prey upon a passing arthropod. During the day it rests buried in shade at the root of vegetation.

It breeds from March to July; 4–8 young are recorded.
Distribution: It is known from coastal Makran, and extends into the adjoining coastal Iran.

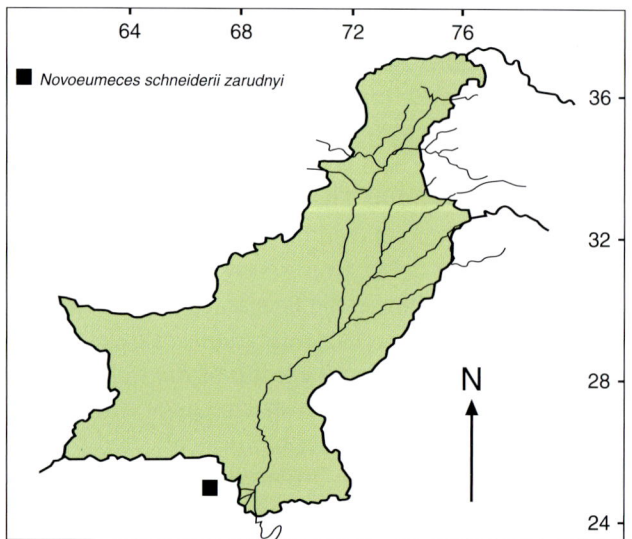

Chapter 6. Lizards

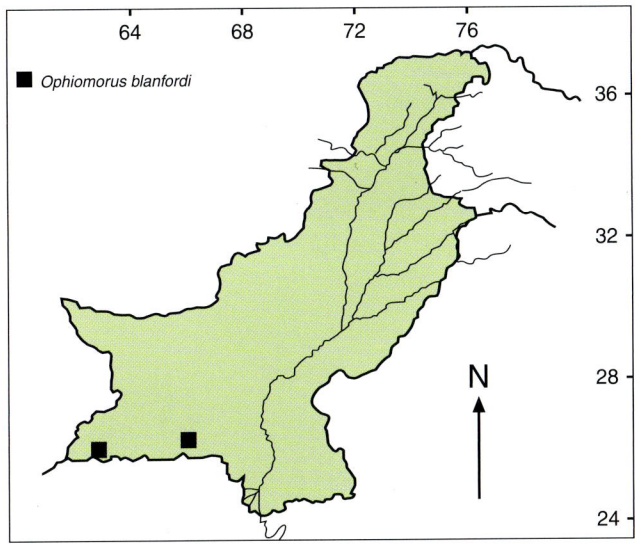

Ophiomorus brevipes (Blanford, 1874)
(Four-fingered sand-swimmer: 4-ungusht reg-tyair)
1874 *Zygnopsis brevipes* Blanford, Ann. Mag. Nat. Hist. London 14 (4): 33.
Type locality: Saadatabad, southwest of Kirman, Iran.
Diagnosis:
1. Interparietal considerably broader than long.
2. Frontonasal about twice as broad as long.
3. scales around midbody 22.
4. Fingers 4, toes 3.
   Snout-vent length 96–98 mm, tail 86–90 mm.

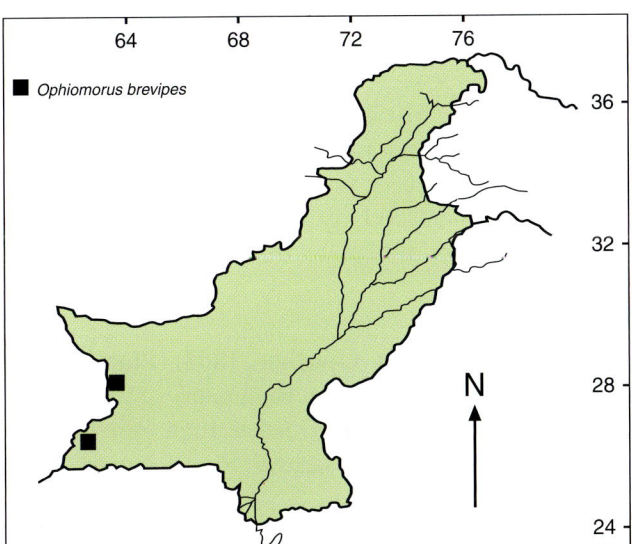

Color: Dorsum golden yellow, dorsal stripes occupy 10 rows of dorsal scales, arranged in 3 very distinct stripes, 1 vertebral and 2 dorsolateral.
Natural history notes: This sand lizard is found in similar habitat and as in the preceding species.
Distribution: Recorded from along the Iran and Pakistan border.

Ophiomorus raithmai Anderson and Leviton, 1966
(Three-fingered sand-fish: 3-ungusht reg-tyair)
1966 *Ophiomorus raithmai* Anderson and Leviton, Proc. Calif. Acad. Sci. San Francisco 33: 519.
Type locality: Ghizri, Karachi District, Pakistan.
Diagnosis:
1. Parietal not in contact with the anterior temporal.
2. Supranasal usually narrowly in contact and partially separated from each other by the tip of rostral scale.
3. Prefrontal in contact with upper labials.
4. Scales around midbody 22.
5. Both fingers and toes 3.
   Snout-vent length 82–90 mm, tail 74–79 mm.

Color: In life, dorsum pale brown, often suffused with yellow ocher. Upper center of each scale is intensely ocher with dark brown spots. In formalin, cream or pale brown, scales of 8–10 median rows with dark brown line formed by discrete dots, extending from nape to the level of hindlimbs, onto tail. Hindlimbs with dark brown dots. A dark brown line from nostril through eye, across temporals. Dark brown markings on median head scales. Ventrum cream.

Young have blue tails.

Natural history notes: The sand-fish is found in typical sand-dune habitat with sparse vegetation of grasses. It glides under the sand surface leaving a distinct sinuous track on the surface, occasionally emerging for a short time. It burrows to a depth of half meter, where the upper dry sand layer meets the deeper moist laycr. Tracks of skinks radiate out from in the roots of the bushes where the lizards have been resting for the day. Though the lizard is diurnal, its nocturnal activity has also been recorded (Minton, 1966).

The stomach contents, apart from copious amounts of sand grains, contain remains of termites and other soft-bodied arthropods (Minton, 1966).

Breeding season extends from March to May, and young are met with during June and August.

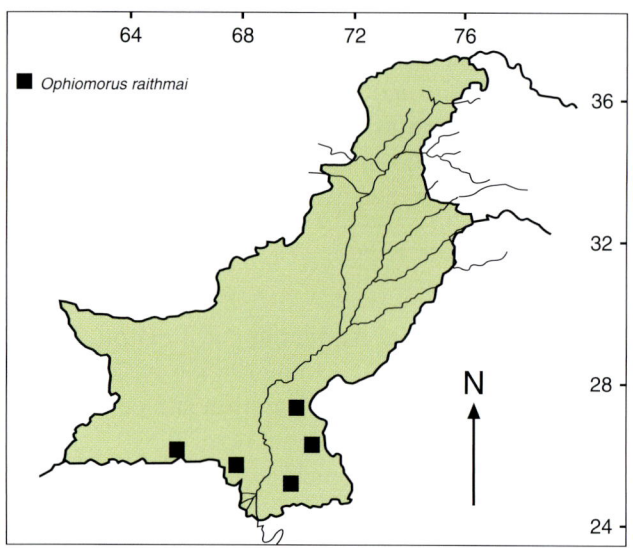
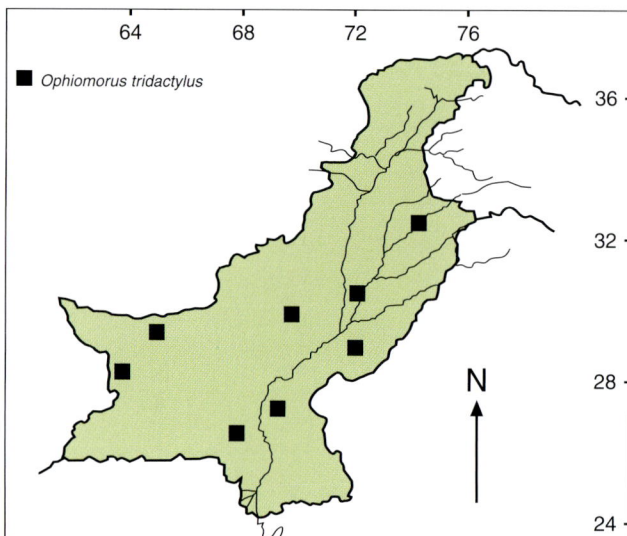

Distribution: Reported so far from Kathiwar, in India, various localities in Sind and Las Bela in Pakistan.

***Ophiomorus tridactylus* (Blyth, 1853) (Plate 103)**
(3-toed sand-swimmer: 3-ungusht reg-tyair)
1853 *Sphenocephalus tridactylus* Blyth, J. Asiatic Soc. Bengal, Calcutta 22: 654.
Type locality: Afghanistan.
Diagnosis:
1. Supranasals usually narrowly in contact, partially separated from one another by the posterior tip of the rostral scale.
2. Prefrontals in contact with upper labials.
3. Parietal in contact with anterior temporal.
4. Postocular about as large as posterior subocular.
5. Scales around midbody 22.
6. A distinct ventrolateral ridge from snout to groin.
7. Fingers and toes 3.
Snout-vent length 89–92 mm, tail 25–27 mm.

Color: Dorsum yellow to khaki, tail tends to be grayish, ventrum white. Tail in juvenile sky blue.
Natural history notes: The skink inhabits fine loose sand fields, usually along the moist soil of water courses where it is confined to the sandy banks, to the limit of the vegetation.
Distribution: Reported from along the borders between Afghanistan, Iran, and Pakistan. In Pakistan it is known from western Baluchistan, Dera Ghazi Khan, Thal and Cholistan Deserts, Punjab, Pakistan.

Genus *Scincella* Mittleman, 1950
Body rather short, not markedly serpentine; eyelids moveable; a supranasal scale absent; limbs more or less strong, pentadactyl, clawed.

Two species are known from Pakistan.

***Scincella himalayana* (Günther, 1864) (Plate 104)**
(Himalayan skink: Hamalayi baahmani)
1864 *Eumeces himalayanus* Günther, Rept. Brit. Ind.: 86.
Type locality: Kashmir; Garhval, Simla, India.
Diagnosis:
1. Scales around midbody 26–30.

Plate 103. *Ophiomorus tridactylus.* (Photo courtesy of S.C. Anderson).

Plate 104. *Scincella himalayana.*

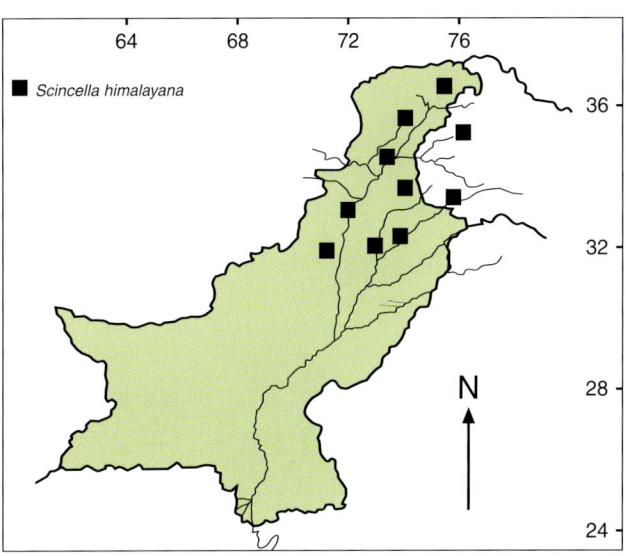

2. Prefrontals usually separated from each other.
3. Lamellae under fourth toe 14–20.
   Snout-vent length 67–70 mm, tail 64–70 mm.

Color: Dorsum iridescent bronze, with indistinct light or dark specks, a dark vertebral stripe present in some individuals. A narrow yellow lateral stripe with irregular edge, another narrow dark brown stripe extends from snout through eye, bordered below by a narrow irregular whitish stripe edged with black. All stripes continuing to the middle of the tail, distal part of it is bronze with light and dark flecks. Head and limbs bronze, with dark speckling, ventrum bluish white.

Natural history notes: This lizard is found under stones and other objects in moist grass fields along margins of lakes and other water bodies in the Kaghan Valley, at about 3000 m of elevation. It has also been collected along the sides of water courses in alpine Punjab and Azad Kashmir. In Salt Range, it is confined to grass fields in stony terrain, around Sakesar, in Soan Valley, Punjab, Pakistan.

The lizard slips from one stone to another and only a keen eye can follow its movement. In thorny bushy areas, as in Soan Valley, it takes to the roots of bushes from where it is difficult to retrieve. Its burrows are situated under stones; at places it descends into large holes and crevices in the ground. Food mainly consists of arthropods, predominantly butterflies and beetles, and worms have also been recorded from its stomach.

It breeds from April to July; 2–3 oval eggs with pliable shells are laid in moist environs under a stone, and juveniles are seen moving about from May to late August.

Distribution: A mountain lizard, it is recorded from Nepal to Turkmenistan. In Pakistan it abounds in marginal vegetation along water courses in alpine Punjab, Salt Range, Chitral, Waziristan, and Kalabag, District Mianwali.

***Scincella ladacensis* (Günther, 1864)**
(Ice field skink: Barfani baahmani)
1864 *Eumeces ladacensis* Günther, Rept. Brit. Ind.: 88.
Type locality: Ladak, Baltistan, Kashmir.
Diagnosis:
1. Scales at midbody 32–38.
2. Prefrontals in contact with each other, or just separated.
3. Lamellae under fourth toe 20–24.
   Snout-vent length 53–56 mm, tail 55–58 mm.

Color: Bronze or olive dorsum, profusely speckled with black, each speck with a white or golden central mark. A dark brown stripe along sides of the neck and body, spotted with light brown, bordered above with white. Sides with small dark dots. Ventrum bluish or gray, ventral scales bordered with gray.

Natural history notes: This lizard is alpine in habits, occupies almost the same habitat, and has the same habits as the preceding species.

Distribution: Reported from Kahajeng Khola, in Nepal, the highest altitude (6000 m) so far reported for any cold-blooded vertebrate. In Pakistan it has been reported from Baltistan.

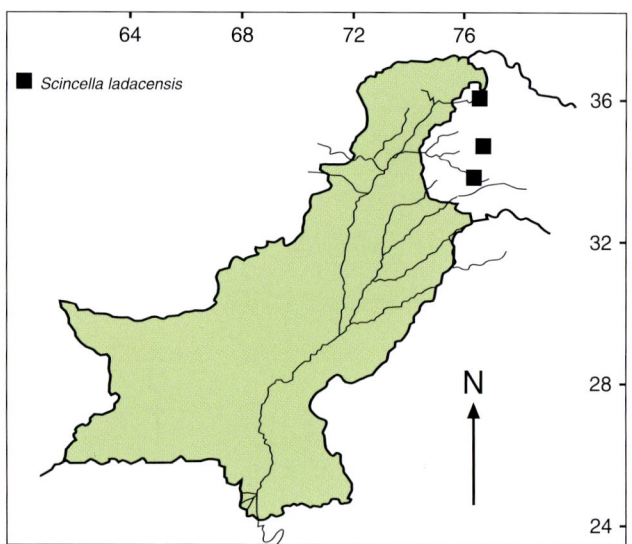

Plate 105. *Uromastyx asmussi.* (Photo courtesy of S.C. Anderson).

Family Uromastycidae

A family of large, strong, ground lizards, frequenting scrublands in plains. Body and tail depressed; head and body with granular scales, sometimes interspersed with tubercles; tail rather shorter than body, covered all over with transverse rows of round, spiny scales; eyelids moveable, round pupil, a broad supraocular shading the eyes; preanal and femoral pores in a series. Limbs strong, with large, strong, sharp claws.

A single genus is represented by two species in Pakistan.

Genus *Uromastyx* Merrem, 1820

Body depressed; dorsal scales granular, with or without larger tubercles; a transverse gular fold; tympanum distinct; tail flat with strong dorsal spinose scales arranged in transverse rows; preanal and femoral pores present.

Two species are recorded from Pakistan.

Uromastyx asmussi (Strauch, 1863) (Plate 105)
(Seistan spiny-tail ground lizard: Sestani sanda)
1863 *Centrotrachelus asmusse* Strauch, Bull. Acad. Sci. St. Petersburg 6:479.
Type locality: Sar-i-tschah, Persia.
Diagnosis:
1. Head with heterogeneous, obtusely keeled scales, nape with numerous spinose tubercles.
2. Dorsal scales granular, keeled, interspersed with numerous transverse rows of large, pointed tubercles.
3. Tail thick at base, strongly depressed, covered with large, spinose tubercles, 8–10 in a row, rounded at their bases. Rows of spines separated by 3–4 rows of keeled scales.

Snout-vent length 255–269 mm, tail 215–220 mm.
Color: Dorsum pale greenish brown; head, limbs and tail darker; belly dirty yellowish brown, heavily mottled, in female dirty brownish.
Natural history notes: The spiny-tail lizard has been collected from deserticolous habitat, at the fringe of

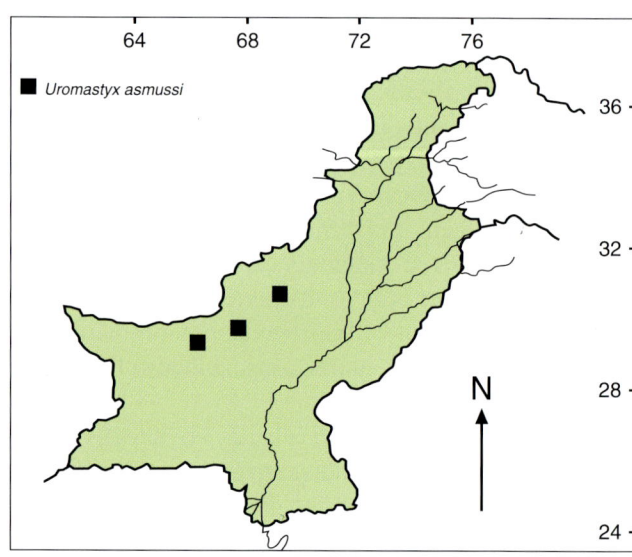

sand deserts along the Iran-Afghanistan border. The habitat is a gravel plain soil, with scattered patches of low thin bushes of barilla and tamarisk. The lizard is rather sluggish; however, it is a good runner. Its burrow has a wide opening. It feeds on leaves and flowers of grasses and other vegetation around.
The lizard uses its spinose tail as a weapon by lashing it out. It does not attempt to bite.

Distribution: The lizard has been collected from southern and central Iran and along the Afghan-Iran border. In Pakistan it has been reported from the Chagai area.

Uromastyx hardwickii Gray, 1827 (Plate 106)
(Indus Valley spiny-tail ground lizard: Maidani sanda)
1827 *Uromastix hardwickii* Gray, Zool. Journal London 3: 219.

Type locality: Kanauj District, United Provinces, India.

Diagnosis:
1. Head and body with uniform granular scales.
2. Caudal spines squarish at their bases, 20–24 in a row at the midtail.
   Snout-vent length 245–250 mm, tail 160–165 mm.

Color: Body dorsum uniformly light sandy gray or yellowish brown, with dark spots or dark vermiculations. Ventrum white, throat spotted with dark. A large distinct blue-black blotch on groin.

Natural history notes: The spiny-tail agama is a characteristic diurnal ground lizard of vast tracts of hard soil throughout the Indo-Gangetic plains. It avoids stony and sandy areas. It makes its burrows in desolate hard soil, areas with sparse vegetation of grasses and bushes. Colonies of lizards are located at slightly elevated areas, probably elevated by the excavated earth. The area is not likely to be flooded. The burrows are located in the open, avoiding the vicinity of bushes and trees. The burrow is a narrow sloping zigzag tunnel, about 2 m long, ending in a spacious chamber (Abdulali, 1960). The semicircular opening of the burrow almost equals the diameter of the lizard and is flush with the general surface of the soil. Burrows are 2–3 m apart from each other. The lizard lives singly in its burrow; however, hatchlings share the burrow with their mother for 1–3 months.

Homing instinct is very strong and the lizard does not venture far from its burrow. It runs straight to its own burrow and dashes into it. On the retiring for the night the opening of the burrow is plugged with a lump of earth.

The lizard is diurnal, emerging 2–3 hours after dawn and remaining active the whole day till just before sunset; however, at midday it retires to avoid the intensity of temperature. It hibernates during the coldest months (November-February) in the plains of Punjab. The lizard is vegetarian, feeding on leaves, twigs, flowers, and fruits. It feeds on nymphs and adult locusts in locust breeding areas. Juveniles are known to occasionally feed on insects.

It breeds from March to early May; 6–16 eggs, 25–30 mm diameter, are laid in a special side chamber in the burrow. Generally young ones are active by June (Parshad, 1916; Minton, 1966; Murthy and Arockiasamy, 1977; Daniel, 1983).

The spiny-tail agama falls prey to hawks, owls, jackals, and foxes. The lizard is used extensively in dissections for demonstration of vertebrate anatomy in colleges and schools throughout India and Pakistan. Moreover, its oil (fat) is said to have aphrodisiac and medicinal properties for which this lizard is extensively killed and exploited (Vyas, 1990; Khan, 1991d).

Falconers usually use *Uromastyx hardwickii* as bait to trap wild falcons in Punjab, Pakistan.

Distribution: The spiny-tail lizard is widely distributed in barren desolate tracts of hard soil, with moderate to sparse xerophytic vegetation, throughout the Indo-Gangetic plains. In Pakistan it has been recorded from throughout the Indus Valley and extends into Las Bela, southern Baluchistan.

Plate 106. *Uromastyx hardwickii.*

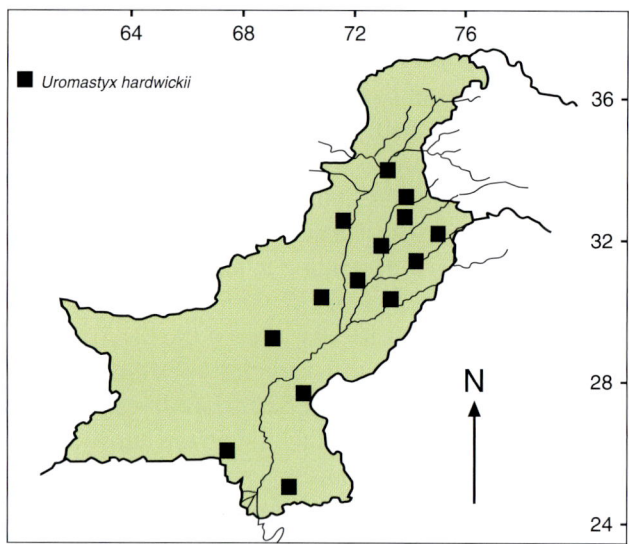

Taxonomic notes: The name *Uromastyx hardwickii baluchiana* has appeared on the lists of animal traders for several years. It does not appear to be an available name, having never been published within the rules of the International Code of Zoological Nomenclature.

Family Varanidae

Large, long tailed, terrestrial lizards, with elongated head, long neck, body, and tail; strong legs with strong claws. Mouth armed with long sharp teeth; tongue very long, smooth, cylindrical, bifid, retractile in a sheath at the base, it is flicked periodically like that of a snake. Eyes well developed, with moveable lids, pupil round. Head and body with small homogenous scales. Tail long, not fragile. Preanal pores may be present (Smith, 1935; Mertens, 1942).

Four species and subspecies of the genus *Varanus* are recorded from Pakistan.

Genus *Varanus* Merrem, 1820

Large lizards, with depressed body, strong limbs with strong claws. Tongue, very long, cylindrical, bifid into thin pointed branches, retractile in a basal sheath; large eyes with well developed eyelids, pupil round. Head scales polygonal, juxtaposed, back with roundish or oval scales, each surrounded with a ring of granular scales; abdominals quadrangular, in transverse series. Tail strong, long, nonfragile; preanal pores present.

Four species and subspecies of genus *Varanus* are recorded from Pakistan.

Varanus bengalensis (Daudin, 1802) (Plates 107A, B)
(Bengal monitor: Bengali goh)
1802 *Tupinambis bengalensis* Daudin, Hist. Nat. Rept. 3:67.
Type locality: Bengal.
Diagnosis:
1. Naris a little nearer to orbit than tip of the snout (Figure 21B).
2. Scales on head longer than nuchals, which are round, not keeled.
3. Supraoculars small subequal.
4. Abdominals smooth, in 90–110 transverse rows.
5. Digits elongate.
6. Tail strongly compressed with a double toothed dorsal crest.
7. Lateral caudals keeled, a little smaller than subcaudals.
 Snout-vent length 815–900 mm, tail 1230 mm.

Color: Dorsum olive to brown, with dark spots. Ventrum yellowish, with or without dark spots, especially under the neck.

Natural history notes: This large varanid frequents moderately dry forests, and extends into cultivated areas, where it inhabits tracts of barren badlands. It often invades inhabited houses, attracted by poultry and rodents. Essentially a burrower, it is also a good tree climber. During rainy season it lives in tree holes feeding on birds and eggs, otherwise it burrows in

Plate 107A. *Varanus bengalensis*: Adult.

Chapter 6. Lizards

Plate 107B. *Varanus bengalensis:* Juvenile.

hard soil. It often climbs into thatched houses to feed on nesting birds. In its burrow it wedges itself in by inflating its body and fixing its claws to the walls so that it is difficult to pull it out. It is a good runner and swimmer; it may remain submerged for a considerable time. When foraging at its leisure, it moves sinuously through the undergrowth, frequently flicking its tongue, looking for any moving object. It has a wide range of food items: arthropods, larvae, worms, frogs, lizards, snakes, birds, and mammals. It is known to munch on carrion and killed mammals (Kumar, 1992). When alarmed, it stands still and tries to slip away unnoticed, however, when cornered, it defends itself by hissing loudly, it elevates and arches its body toward the intruder, and lunges and lashes its tail from side to side. It inflicts a powerful bite with its long, strong, and sharp teeth (Auffenberg, 1983a, b, 1984).

Breeding activity is observed from April to June. Rival males wrestle to win over territories. Usually 6–12 leathery eggs averaging size of 29 by 15 mm in size and, weighing 20–23 g are laid in burrows. Minton (1966) records 30 eggs in this species.

Distribution: *Varanus bengalensis* has been recorded from Assam, Myanmar, Nepal, Sikkim, throughout India, and Sri Lanka. In Pakistan, it is reported from throughout the plains of Punjab and Sind, sub-Himalayan tracts, Waziristan and extends westward into southeastern Iran and eastern Afghanistan.

Varanus flavescens (Hardwicke and Gray, 1827) (Plate 108)
(Yellow monitor: Peeli goh)
1827 *Monitor flavescens* Hardwicke and Gray, Zool. Journal London 3:226.
Type locality: India.
Diagnosis:
1. Naris a little nearer to snout tip than orbit (Figure 21A).
2. Scales on head smaller than nuchals, which are strongly keeled.
3. Middle supraocular slightly elongated transversally.
4. Abdominals smooth, in 65–75 transverse series.
5. Digits short.
6. Tail strongly compressed, with a double-toothed crest.
7. Subcaudals not much longer than lateral caudal scales.
   Snout-vent length 500–515 mm, tail 575–600 mm.

Color: Dorsum reddish brown, body and tail barred with alternating dark-edged reddish brown and dirty yellow bars, ventrum light yellow.

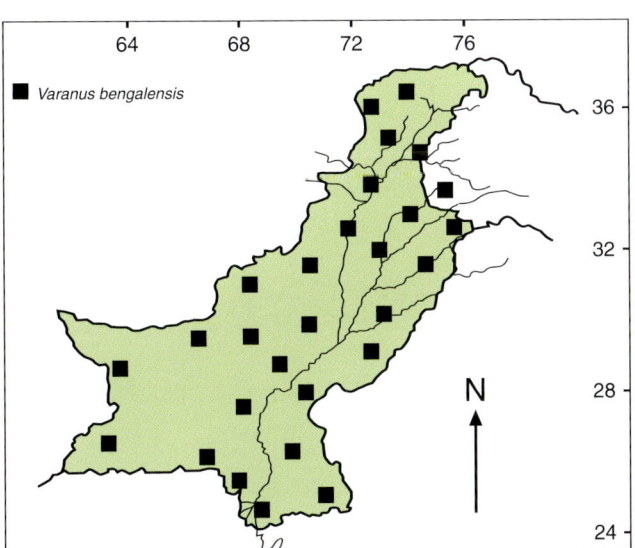

Plate 108. *Varanus flavescens.*

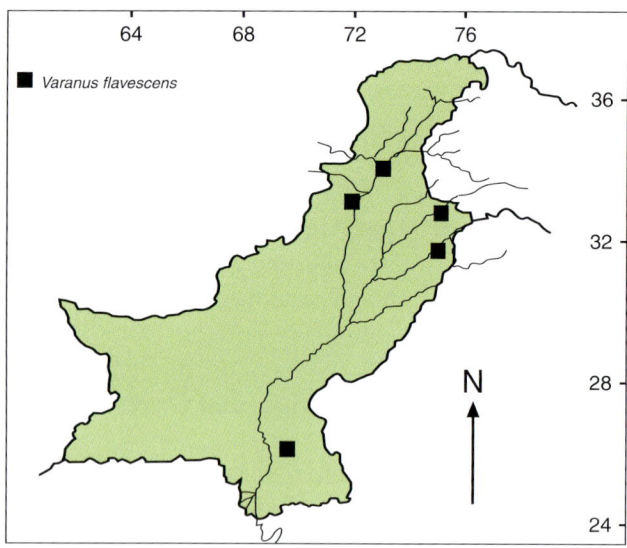

Natural history notes: It is a lizard of seasonally flooded forests, and marshy areas in flood plains of the Indus River and its tributaries. It extends into tilled fields along water courses with mesic habitat, and usually burrows in the roots of trees and other vegetation.

Breeding activity is observed from April to June; 15–30 eggs are laid in burrows.

Distribution: This varanid has a restricted distribution range. It has been reported from Salt Range and District Sialkot in northern Punjab, and the Sind Delta in Pakistan. It is known to extend to western Bengal (Khan, 1993c).

Varanus griseus **(Daudin, 1803)**
(Desert monitor: Patti-dar goh)
1803 *Tupinambis griseus* Daudin, Hist. Nat. Rept. 8:352.
Type locality: Egypt.
Diagnosis:
1. Naris much nearer to orbit than the end of snout.
2. Head scales longer than nuchals, those on neck conical.
3. Supraoculars small, subequal.
4. Dorsals obtusely keeled.
5. Abdominals smooth in 110–125 transverse rows.
6. Digits moderately elongated.
7. Tail round or slightly compressed at its posterior half.
8. Lateral caudal scales indistinctly keeled, slightly smaller than the subcaudals.
Snout-vent length 560–579 mm, tail 865–870 mm.

Color: Dorsum gray to yellowish brown; body and tail with brownish crossbars; 2–3 streaks on neck; ventrum yellowish. In adult vivacity of dorsal pattern is lost.
Natural history notes: This is a varanid of sandy fields with uneven surfaces and tracts of hard soil with sparse vegetation. It excavates burrows in the roots of trees and bushes. It is very secretive and is more active during the early hours of the day, spending most of its time in its burrow. It feeds on rodents, lizards, snakes, birds, eggs, frogs, and toads. Since it is a varanid of warm areas, it is not known to hibernate.

When encountered, it stiffens its legs to raise its body which is arched outward, and snarls and hisses vigorously by squeezing and inflating its body. The neck is puffed, and the tongue is extended fully, while the tail is vigorously lashed from side to side.

Breeding season extends from late March to late May; 10–25 eggs are laid in burrows or pits excavated for the purpose, which are then covered with vegetation and debris.

Distribution: The varanid has a wide range in the arid desert areas of India, Pakistan, and the Middle East. It ranges from Rajasthan to the Caspian Sea and North Africa. The Pakistani population of this varanid is distinguished in the following two races:

Varanus griseus caspius **(Eichwald, 1831) (Plate 109)**
(Caspian monitor: Koh-kaf goh)
1831 *Psammosaurus caspius* Eichwald, Zool. Spec. 3:190.
Type locality: Eastern coast of the Caspian Sea.

Plate 109. *Varanus griseus caspius.* (Photo courtesy of S.A. Minton).

Chapter 6. Lizards

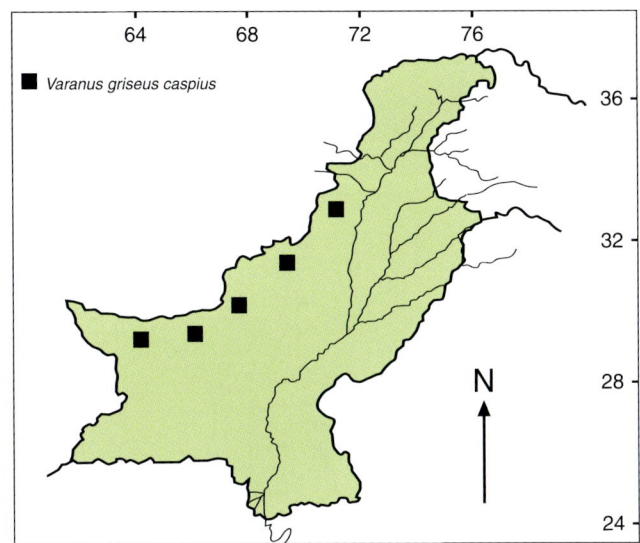
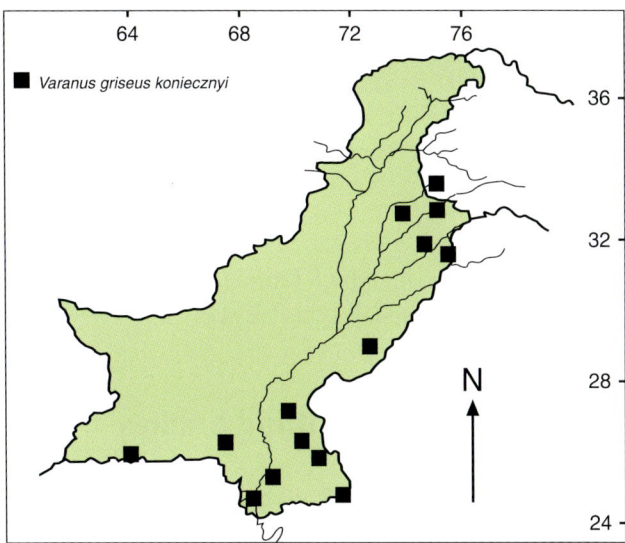

Diagnosis:
1. Only the posterior half of the tail is compressed in a dorsal crest.

Color: Body with 5–6, while tail has 16–19 transverse bands.

Distribution: The Caspian monitor extends from the Transcaspian Desert to southern Khazakistan and Afghanistan, extending into northern and western Baluchistan.

Varanus griseus koniecznyi **Mertens, 1954**
(Indo-Pak desert monitor: Hindo-Pak goh)
1954 *Varanus griseus koniecznyi* Mertens, Senckenb. Biol. Frankfurt a.M. 35: 355.
Type locality: Korangi, Karachi, Pakistan.
Diagnosis:
1. Tail round throughout, without a dorsal crest.

Color: Body with 4, tail with 10–15 transverse bands.

Distribution: From central India, to Cholistan, Sind, and the Kharan Desert in Pakistan.

Chapter 7
Ophidia: Snakes

Snakes are represented by 8 families, 34 genera, and 77 species and subspecies, in Pakistan.

Thread and Blind Snakes

Snakes of the families Leptotyphlopidae and Typhlopidae are small, with vermiform, delicate bodies with uniform diameter, covered with cycloid imbricate polished scales of uniform size. The eyes are reduced and vestigial, covered with scales, hardly visible. The tail is extremely short, often ending in a conical spine. These snakes readily burrow through loose moist soil, where they lead a subterranean life.

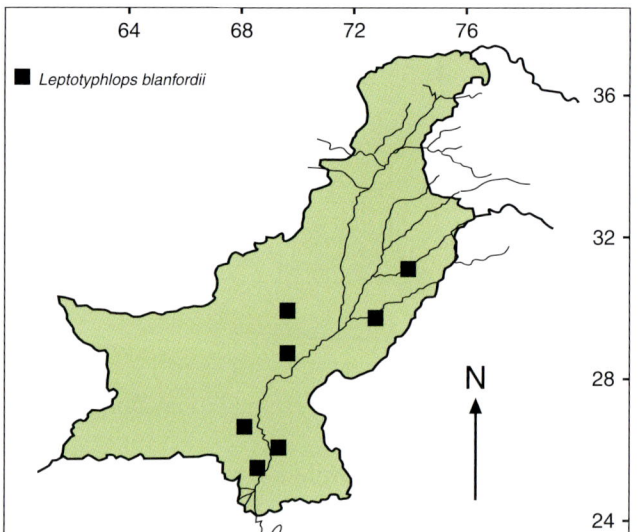

Family Leptotyphlopidae

There are 14 transverse rows of scales around the body. Two supralabials, ocular and nasal scales line the mouth, so that the first supralabial is separated from the second and rostral is separated from the first supralabial; nasal single or divided; preanal scale single and enlarged.

A single genus is represented in Pakistan.

Genus *Leptotyphlops* Fitzinger, 1843
Nasal and ocular scales border the mouth; an enlarged preanal scale present; 14 scales around the body.

Two species are reported from Pakistan.

Leptotyphlops blanfordii (Boulenger, 1890)
(Sindi thread snake: Sindi dahaga saamp)
1890 *Glauconia blanfordii* Boulenger, Fauna Brit. Ind., Rept. Batr.: 243.
Type locality: Sind, Pakistan.
Diagnosis:
1. Rostral, nasal, and ocular scales large, border the mouth.
2. Snout round, rostral not large, protruding not concave below.
3. Diameter of body contained 57–78 times in total length.
4. Scale rows around body 14.
Snout-vent length 220–225 mm, tail 21 mm.
Color: In life, pale pink to brown.
Natural history notes: This snake inhabits loose sandy soil with considerable moisture and feeds on termites and ant larvae. The snake spends its life underground and does not venture out in daylight. However, during rains, it is driven out of its subterranean burrows which lie along the underground sewer gutter lines etc. The snakes are seen moving around helplessly in broad daylight. Specimens with 2–8 elongated eggs have been collected; however, there is no data on egg laying and hatching.
Distribution: The snake has been frequently recorded from throughout the riparian system of Pakistan, at low elevations. Hahn and Wallach (1998), while commenting on the Old World leptotyphlopids, do not include Pakistan in the range of *Leptotyphlops blanfordii*, though the type

locality of this species is Sind, Pakistan (Gasperetti, 1988).

Leptotyphlops macrorhynchus (Jan, 1862)
(Plate 110)
(Beaked thread snake: Chonch-dar dahaga saamp)
1862 *Stenostoma macrorhynchus* Jan, Arch. Zool. Anat. Fis. Genova 1:190.
Type locality: Sennar, Egyptian Sudan.
Diagnosis:
1. Rostral scale prominent, hooked downward, and is concave from below (Figure 22C).
2. Diameter of body 80–110 times contained in the total length.
   Snout-vent length 243–245 mm, tail 21–22 mm.

Color: Light brown to pink or flesh color.
Natural history notes: It inhabits moist, sandy areas, especially along water courses: rivers, canals, lakes, and fast-flowing streams. It is collected from the roots of grasses and other low vegetation along banks of water courses. Food consists mainly of termites and ant larvae (Minton, 1966).

Snakes with 2–4 elongated eggs have been collected from November to February. Nothing more is known about their breeding habits.
Distribution: The beaked thread snake is reported from Sudan, Egypt, Saudi Arabia, Iraq, and Iran. In Pakistan it is recorded from Quetta to the Indus Valley. Hahn and Wallach (1998), while commenting on Old World leptotyphlopids, do not include Pakistan in the range of *Leptotyphlops macrorhynchus*.

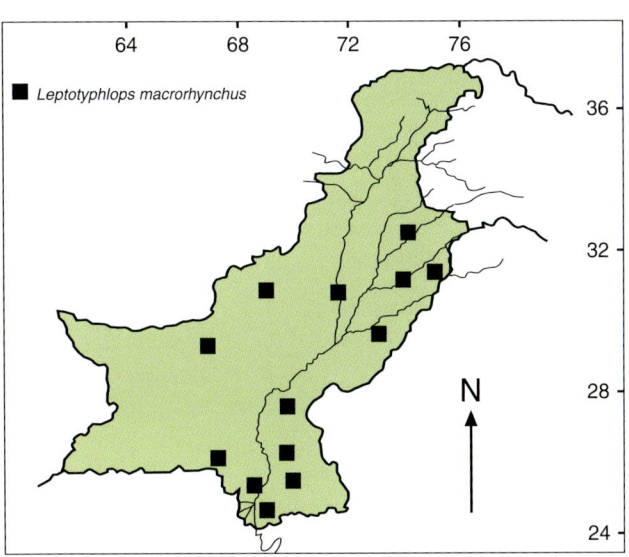

Plate 110. *Leptotyphlops macrorhynchus.*

Family Typhlopidae

More than 14 scales around the body; supralabials 4, ocular and nasal shields do not form a part of upper lip; no enlarged preanal scale.

Two genera are represented in Pakistan.

Genus *Ramphotyphlops* Fitzinger, 1843
Twenty scales around the midbody; nasal completely divided; penial retractor muscle is inserted at about the middle of the organ so that the distal part of the organ remains permanently drawn out as a solid awn. Usually hemipenis and the associated retractor muscle are helically coiled. A pair of retrocloacal sacs is present, opening in rectum posterior to the openings of vas deferens and ureter. A long rectal cecum is always present, opening insensibly in the rectum.

A single species is found in Pakistan.

Ramphotyphlops braminus (Daudin, 1803)
(Plate 111)
(Brahminy blind snake: Brahamni kainchwa saamp)
1803 *Eryx braminus* Daudin, Hist. Nat. Rept. 7:279.
Type locality: Vizagapatam, peninsular India.
Diagnosis:
1. Rostral scale large, round.
2. Supralabials 4, last 2 in contact with the ocular scale.

Plate 111. *Ramphotyphlops braminus.*

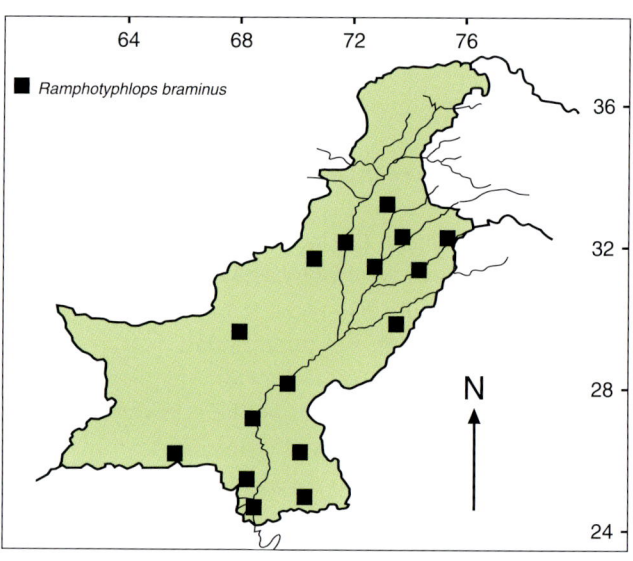

3. Nasal suture usually extends to the edge of the preocular, rarely between first and second labial.
4. Scales rows around the body 20.
Snout-vent length 138–145 mm, tail 11 mm.

Color: Dorsum reddish brown to dark brown, ventrum slightly pale.

Natural history notes: This vermiform little snake frequents humid shady fields with loose soil. Small plots in humid and shady environs of gardens and tilled fields, where the soil is periodically turned to keep it loose, are ideal situations for this snake. Its blunt round snout helps it to displace loose moist soil particles through which it disappears under the soil. Natural manure, which is periodically dispersed in the plots, attracts lots of soft-bodied scavenger arthropods. Stomach content studies of the blind snake have shown that these arthropods form their main dietary material. In arid areas the snake lives underground in moist soil along gutters and water supply lines.

This snake leads a strict subterranean life, seldom coming to the surface under normal conditions; however, during rains its subterranean burrows and passages are choked with water which drives them out, and they are seen wriggling helplessly on pavements, etc., in broad daylight.

Almost nothing is known about the natural life cycle of these secretive snakes. There is evidence to believe that these snakes reproduce parthenogenetically, laying 5–8 eggs during March–May.

Distribution: *Ramphotyphlops braminus* is almost cosmopolitan in its distribution, which is attributed to human activity. It is carried to remote areas in potted plants, etc. In Pakistan it is quite abundant in the upper and lower Indus Valley and except for the high mountains and deserts, it is reported from almost all over Pakistan.

Genus *Typhlops* Oppel, 1811

The midbody scales 18–28; long penial retractor muscle is inserted at the extreme tip of the relatively short thick hemipenis, so that the entire organ is pulled inside out when the muscle is contracted. The hemipenis and its retractor muscle both are straight. Retrocloacal sacs are absent. The supra nasal cleft complete or incomplete, extends horizontally so that little of the anterior nasal scale is exposed on dorsal side.

A rectal cecum may be present or absent; when present it is small, opening in the rectum through a narrow neck. In the genus *Typhlops*, imbrication of the head scales, i.e., which scales overlap which others, is important in species determination. This is expressed by means of a formula, wherein the symbol / indicates overlap of the following scale by the previous and the following abbreviations are used: N1 = anterior nasal; N2 = posterior nasal; Oc = ocular; Poc = preocular; Postoc = postocular; SL1–4 = 1–4 supralabials. Thus: N1/SL1/N2 indicates that the anterior nasal overlaps the first supralabial, which in turn overlaps the posterior nasal scale.

Four species are known from Pakistan.

Typhlops ahsanuli Khan, 1999
(Ahsan's blind snake: Ahsan's kainchwa saamp)
1999 *Typhlops ahsanai* Khan, Russian J. Herpetol 6(3): 238.
Type locality: Nadari village, 2 km east of Goi Madan, District Kotli, Azad Kashmir, 33° 30′N and 74° 00′E, elevation 1315 m.
Diagnosis:
1. The supranasal suture is 75% complete.
2. The anterior nasal scale overlaps the first supralabial and anterior ventral part of the posterior nasal scale.
3. The posterior nasal scale is deeply scalloped behind so that the preocular scale extends deep in the postnasal concavity.
4. The preocular scale is larger than the ocular.
5. The preocular nasal suture is studded with thick squamous glands.
6. Preocular scale is in contact with only third supralabial.
7. The posterior nasal scale is in contact with second and third supralabials.
8. The supralabial imbrication formula is: N1/SL1/N2, N2/SL2/Poc, Poc/SL3/Oc, Oc+postoc/SL4.
9. Body scales have micropits on their surface.
10. Tail with terminal spine.
    Total length 170 mm.

Color: Dorsum dark brown, ventrum slightly lighter.
Natural history notes: This blind snake was collected from an ant hole among stones. When the stones were removed, the ant nest was disturbed and the snake was moving among scattered ants; apparently the ants were not heeding its presence.
Distribution: Known only from its type locality, Nadari village, 2 km east of Goi Madan, District Kotli, Azad Kashmir.

Typhlops diardii Schlegel, 1839
(Thick body blind snake: Moota kainchwa saamp)
1839 *Typhlops diardii* Schlegel, Abbild. Amphib. p. 39.
Type locality: India, Orientales.
Diagnosis:
1. Stout-bodied *Typhlops*.
2. Scales around the midbody 22 or 26.
3. Rostral one-fourth as broad as head breadth.
4. Nasal incompletely divided, cleft passing to second supralabial.
5. A distinct subnarial pit.
   Total length 430 mm.

Color: Brown to blackish above, ventrum pale.
Distribution: Known from Nepal, Bengal, Assam, Myanmar, Singapore, Thailand, Malaysia, Cambodia, and Laos.

Typhlops diardii platyventris Khan, 1998 (Plate 112)
(Kashmir blind snake: Kashmiri moota kainchwa saamp)
1998 *Typhlops diardii platyventris* Khan, Pakistan J. Zool. 30(3):213–221.
Type locality: Goi Madan, Kotli, Azad Kashmir.
Diagnosis:
1. Body stout.
2. Number of scales around midbody 22–25.
3. Body ventrum distinctly flat.
4. A distinct subnarial glandular patch.
5. Supralabial imbrication formula: N1/SL1, N2/SL2/PrOc, PrOc/SL3/Oc, Oc/SL4.
6. Tail ends in a spine.
   Total length 214–295 mm.

Color: Light to dark brown dorsum, light yellowish below.
Natural history notes: The snake was collected from under stones along fast-flowing streams.

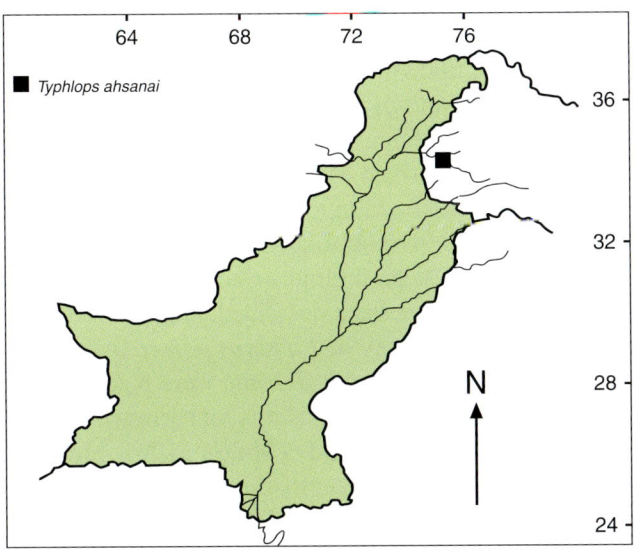

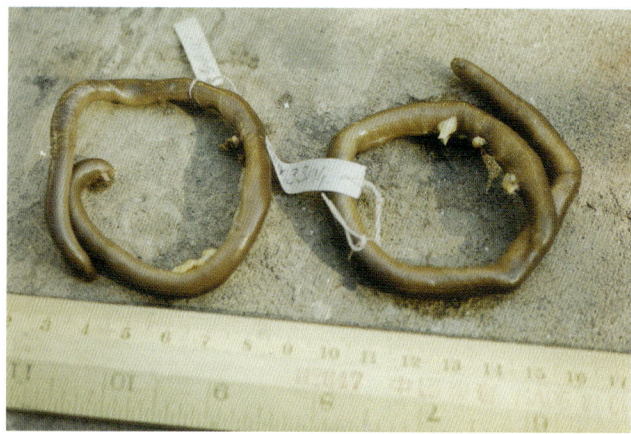

Plate 112. *Typhlops diardii platyventris.*

Plate 113. *Typhlops ductuliformes.*

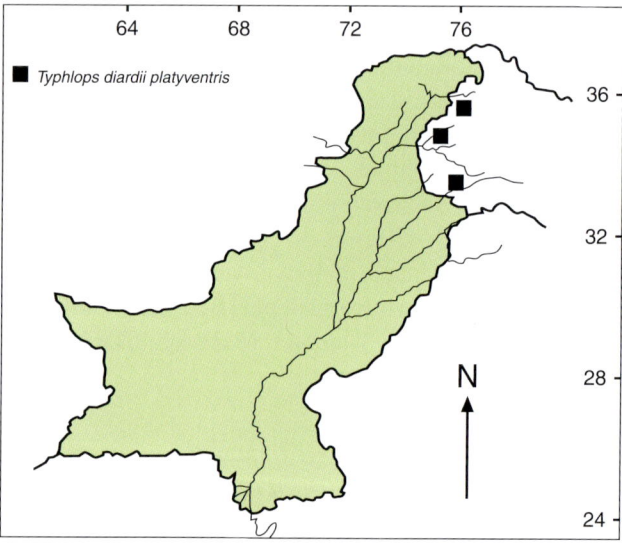

Distribution: *T. diardii platyventris* is known from its type locality.

***Typhlops ductuliformes* Khan, 1999 (Plate 113)**
(Slender blind snake: Patla kainchwa saamp)
1999 *Typhlops ductuliformes* Khan, Pakistan J. Zool. 31(4):385–390.
Type locality: Jhangir's Tomb, Lahore.
Diagnosis:
1. Nasal suture usually touches second labial, and does not divide nasal scale completely.
2. Eyes hardly visible.
3. Scales around midbody 18.
4. Diameter of body less than 2 mm.
5. Middorsals 412–461.
6. The supralabial imbrication formula is: N1/SL1, N2/SL2/Poc, Poc/SL3/Oc, Pto/SL4.
7. Tail ends in a spineless cone.
8. Microstriations on the surface of body scales. Total length 90–202 mm.

Color: Body dorsum light chocolate, ventrum dirty white; exposed part of body scales dark brown, distinctly marked from light chocolate flared part of rest of the scale. The surface of flared part has longitudinal microstriations. Rostral, labial, gular regions lighter. Circumanal and subcaudal regions are nonpigmented.

Natural history notes: Nothing is known about the natural history of this blind snake except that it is collected along well-watered grassy plots.

Taxonomic notes: *Typhlops porrectus* Stoliczka, 1871 was described from the western Himalayas. It is distinct from *Typhlops filiformis* Duméril and Bibron, 1844, due to its greater body diameter (2.5–3 mm) and from recently described *Typhlops madgemintonae* Khan and *Typhlops ahsanai* Khan by having more middorsal scales (416–440) along its body.

Since 1871 several blind snake taxa with 18 midbody scales have been described, and mostly placed in the synonymy of *Typhlops porrectus* (see Khan, 1999e, for review), and some were placed with *Ramphotyphlops braminus*, as it is long known to have 18–21 scales around the body (Minton, 1966; Mertens, 1969a; Khan, 1982a). Recently described *Typhlops* species from Azad Kashmir and Pakistan have uncovered the fallacy of these taxonomic decisions (Khan,

Chapter 7. Ophidia: Snakes

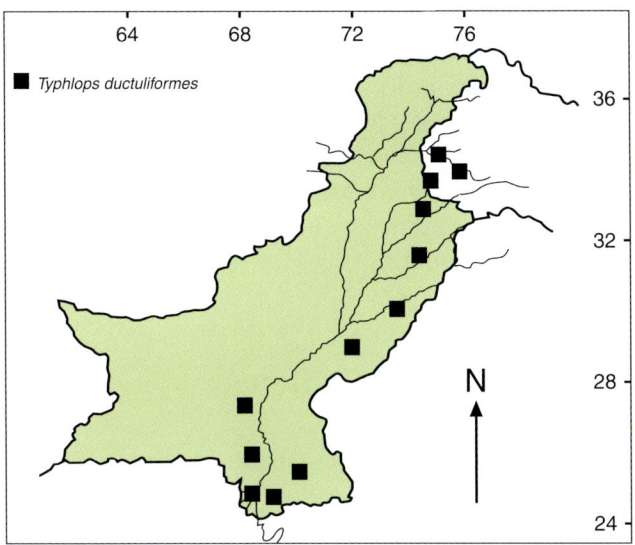

Plate 114. *Typhlops madgemintonae madgemintonae.*

1998a, 1999b, c). The new snakes with 18 midbody scales differ from *Typhlops porrectus* due to smaller body diameter (less than 2 mm), tail ends in a spineless cone, and N2 is strongly scalloped behind. So that recent studies on present and earlier collections of blind snakes from Pakistan (Minton, 1966; Mertens, 1969a; A.Q. Khan and M.S. Khan, 1996) strongly suggest that *Typhlops porrectus* is an extralimital species which is confined to the eastern Himalayas; it does not belong to the herpetofauna of Pakistan. Apparently, what several of the past authors have been regarded as *Typhlops porrectus* in Pakistan, actually involved more than two species: *Typhlops madgemintonae*, *Typhlops madgemintonae shermanai*, *Typhlops ahsanuli*, and *Typhlops ductuliformes* (Khan, 1999b, e, 2005c).

Distribution: The slender blind snake is known from Lahore, Hyderabad, and Karachi, Pakistan.

***Typhlops madgemintonae madgemintonae* Khan, 1999 (Plate 114)**
(Kashmir's slender blind snake: Kashmir patla kainchwa saamp)
1999 *Typhlops madgemintonai* Khan, Russian J. Herpetol. 6(3): 233–236.
Type locality: Goi Madan, District Kotli, Azad Kashmir.
Diagnosis:
1. Nasal almost completely divided.
2. Middorsals 336–364.
3. Caudal spine solid cuspidate, with thick embossed base.
4. Subcaudal colorless area extends dorsally to include base of the caudal spine.
5. Posterior one-fourth of the body strongly coiled ventrally.
6. Supralabial imbrication formula: N1/SL1, N2/SL2/Poc, Poc/SL3/Oc, Oc+Postoc/SL4.
7. Micropits on the surface of the body scales.
8. Tail ends in a spine.
   Total length 170–200 mm.

Color: Dorsum uniformly dark brown, lighter ventrum. Snout, labia, circum-anal and subcaudal

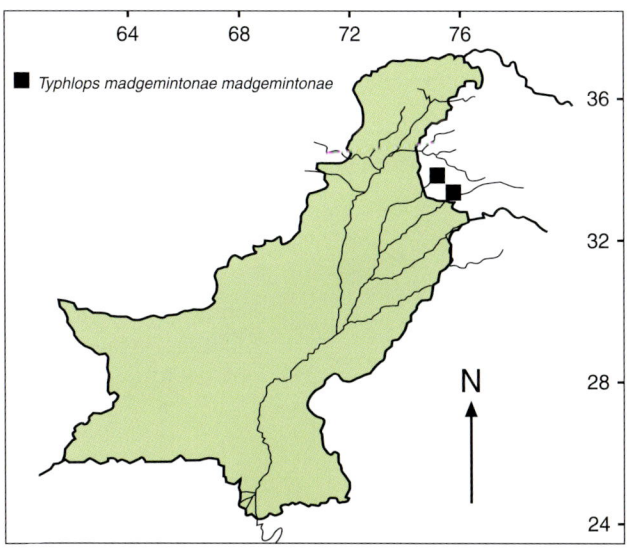

regions are colorless. The subcaudal colorless area extends dorsally to include tail spine and its base.
Natural history notes: Habitat of this blind snake is rocky, pine countryside with lush green vegetation of grasses. The collection of this snake coincides with the rainy season when rainwater floods subterranean animal life.
Distribution: Known from Goi Madan and Barmoach, District Kotli, Azad Kashmir.

Typhlops madgemintonae shermanai **Khan, 1999**
(Sherman's slender blind snake: Sherman's patla kainchwa saamp)
1999 *Typhlops madgemintonai shermanai* Khan, Russian J. Herpetol. 6(3): 236–238.
Type locality: Charnali village, 2 km west, Goi Madan, District Kotli, Azad Kashmir.
Diagnosis:
1. The supranasal suture is 50% complete.
2. Narial opening round, anterolateral.
3. A subnarial pit in the inferior nasal suture.
4. Preocular and ocular scales subequal.
5. A row of thick squamous glands, along preoculonasal suture.
6. Labial imbrication formula: N1/SL1/N2, N2/SL2/Poc, Poc/SL3/Oc, Oc+Postoc/SL4.
7. Surface of body scales with micropits.
8. Tail ends in a spine.
Total length 120–150 mm.

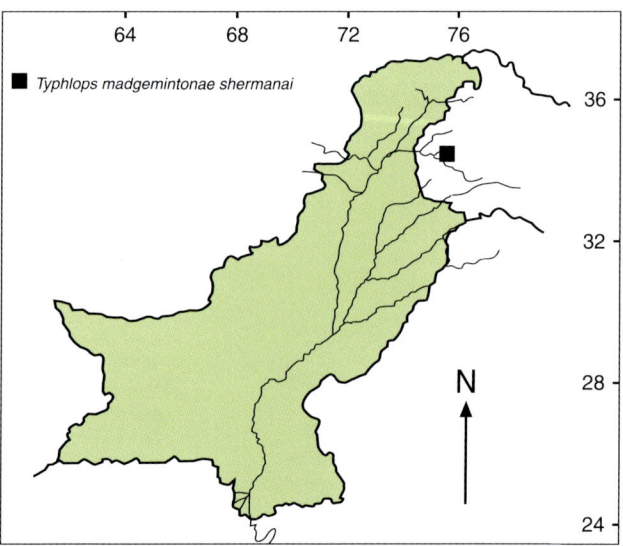

Color: Dorsum dark brown, ventrum lighter. Circumanal, subcaudal, and base of spine colorless.
Natural history notes: Collected from under leaf litter at the roots of a pine tree on a rocky slope, near an ant nest.
Distribution: Kashmir and Rabwah, Punjab, Pakistan. Probably washed down by flood water in the Chenab River from Kashmir.

Family Boidae

Includes largest and heaviest snakes of the world, as well as several smaller species. Dorsal scales smooth, imbricate, more than 50 rows around the midbody, ventrals not extending across the body ventrum. Vestiges of pelvis and hindlimb are present, consisting of ilium, pubis, ischium and femur, the femur usually terminates in a claw-like spur on each side of vent.

Two genera occur in Pakistan.

Genus *Eryx* Daudin, 1803
Head not distinct from neck, covered with uniform small scales; body cylindrical, stout; eyes small, with vertically elliptical pupil; ventrals broader than long, but do not extend across body; tail very short; subcaudals in a single row, anal tripartite.

Three species are known from Pakistan.

Eryx conicus **(Schneider, 1801) (Plate 115)**
(Chain sand-boa: Sindi du-muhi)
1801 *Boa conica* Schneider, Hist. Amphib. 2: 268.
Type locality: Madras.
Diagnosis:
1. Head slightly distinct from neck.
2. Rostral wide, not wedge-shaped, without free edge.
3. Nostril slit between nasals and internasal.
4. Supralabials 11–13, infralabials 14–17.
5. No mental groove.
6. Body tapering abruptly just posterior to vent, tail very short and blunt.
7. Scales at midbody 45–56, dorsals keeled, keels particularly heavy in posterior part of body.
8. Ventrals 161–178, subcaudals 18–22.
Snout-vent length 730–732 mm, tail 68–70 mm.

Chapter 7. Ophidia: Snakes

Plate 115. *Eryx conicus conicus.*

Color: Light gray to yellowish or rusty. A series of vertebral irregular dark brown to sooty oval blotches, narrowly edged with dull yellow, usually confluent with each other. A similar lateral irregular series. A dark stripe behind eye. Belly white with scattered gray spots.

Natural history notes: The chain sand boa frequents comparatively moist sandy and silty soil. The snake is timid and shy; it forages at night. Its primary reflex in the face of danger is to bury itself in the soil. When captive it remains buried for several days, keeping its head close to the surface. When provoked, it throws itself in a coiled ball; when touched it flinches, occasionally slashing to strike with its whole body. The snake allows moderate handling, is of uncertain temperament, and its bite is painful.

Lizards, rats, and mice are the main food items, occasionally nestlings, and eggs are included in the menu. Prey is constricted.

Breeding season extends from April to May; clutch size varies from 4–16 young, depending on the size and age of the female.

Distribution: The chain boa extends from Bihar and Orissa through India and Sri Lanka. In Pakistan it has been reported from southern Sind and the Cholistan Desert.

Eryx johnii (Russell, 1801) (Plate 116)

(Common sand boa: Du-muhi)
1801 *Boa johnii* Russell, Ind. Serp. 2: 18, 20.
Type locality: Tranquebar, India.
Diagnosis:
1. Rostral wide, thick with a ridge.
2. Nostril slit between large nasals.
3. Supralabials 9–12, infralabials 13–18.
4. Scales in circumorbital ring 9–12 (Figure 23A, B).
5. Scales at midbody 51–61; smooth or weakly keeled in anterior half, while heavily keeled in posterior half of the body.
6. Tail blunt, terminating in small scales (Figure 23D).
7. Ventrals 190–212, subcaudals 20–40.

Snout-vent length 940–945 mm, tail 89–92 mm.
Color: Juveniles characteristically light reddish in color, with 11–17 dark crossbars on body. Tail lighter

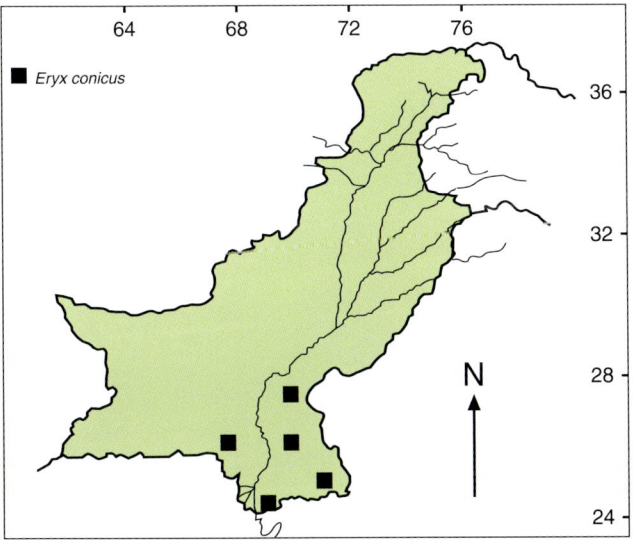

Plate 116. *Eryx johnii.*

with 2–5 crossbars. Ventrum cream, lightly speckled with pale and dark gray.

Body becomes darker with age, with slight traces of bands on a uniformly brown or dark brown dorsum; however, the caudal bars are retained.

Natural history notes: This boid snake is characteristic of moderate deserts of loose soil with sparse vegetation; it becomes rare in sandy stony and damp soil. It readily invades human habitations, attracted by mice and rats, where it resides in their burrows. It is a powerful constrictor, usually following its prey into its burrow where the prey is killed by pressing against the walls of the burrow.

The sand boa is predominantly nocturnal; however, it is active during day in rainy season. It is a slow-moving reptile, but when alarmed, it quickly escapes by burrowing in loose soil. When cornered it does not show an active defense, rather, it throws its body into a loosely coiled ball, hiding its head in the middle of the coils, while banded tail is exposed and kept elevated. When the balled body is touched, it flinches, and hisses at each contraction and expansion of the body (Minton, 1966; Khan and Tasnim, 1986a, 1987b).

The sand boa comes out of hibernation by March and is active until late September. Reproductive activity occurs from March to May; being viviparous, 2–8 young are born from mid-April to early June.

Distribution: The common sand boa occurs throughout central India. In Pakistan it has been recorded from the plains of Punjab, Sind, and Baluchistan. It is reported from eastern Afghanistan and Iran. It does not extend above 200 m.

Eryx tataricus speciosus Zarevsky, 1915 (Plate 117, 118 [*E. t. tataricus*])

(Tartary sand boa: Tatar du-muhi)

1915 *Eryx speciosus* Zarevsky, Ann. Mus. Zool. Acad. Sci. St. Petersburg, 20: 361.

Type locality: Bukhara, Uzbekistan.

Diagnosis:
1. Rostral much wider than high.
2. Supralabials 11–12, infralabials 16.
3. Circumorbital scales 10–11.
4. Tail blunt, terminating posteriorly in a large shield.
5. Dorsals 49 at midbody, smooth anteriorly, slightly keeled posteriorly.
   Snout-vent length 335–338 mm, tail 45–47 mm.

Color: Dorsum yellowish brown, lighter on sides. A median series of 45–50 brown irregular blotches. Body flanked with dark flecks and spots. Ventrum lighter.

Natural history notes: The Tartary sand boa has been collected from sandy terrain with sparse vegetation of clumps of grasses and scattered bushes. It is a timid species which allows considerable handling. However, when alarmed, it throws itself into a ball. It feeds on

Plate 117. *Eryx tataricus.*

Chapter 7. Ophidia: Snakes

Plate 118. *Eryx tataricus tataricus.* (Photo courtesy of S.C. Anderson).

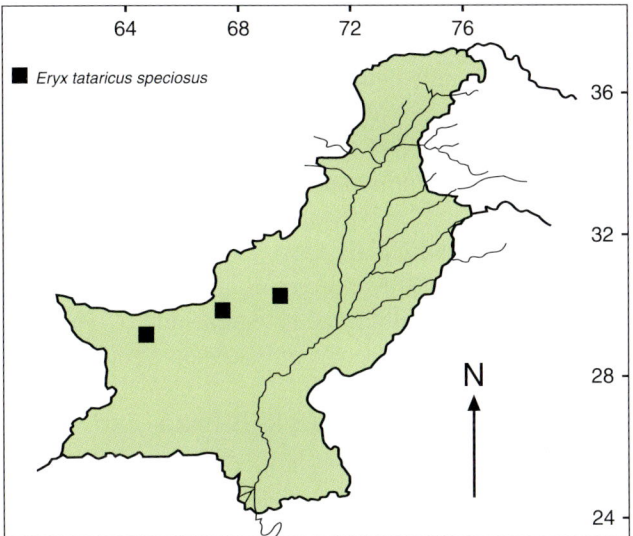

mice, lizards, and bird nestlings. It is nocturnal, seldom seen during the day.

Distribution: This boid extends from the Aral Sea to the Altai Mountains and Iran. In Pakistan it is reported from northwestern Baluchistan.

Genus *Python* Daudin, 1803

Large, heavy bodied snakes; head distinct from neck, covered with large shields; eye with vertical pupil; Rostral, anterior supralabials and infralabials pitted; body with smooth scales in 60–75 rows; ventrals narrower than body width; subcaudals generally paired; anal single. In both sexes vestigial hindlimbs present, which are longer in male.

A single species is represented in Pakistan.

Python molurus (Linnaeus, 1758) (Plate 119)

(Rock python: Sindi azh-daha)
1758 *Coluber molurus* Linnaeus, Syst. Nat. 1(10): 225.
Type locality: India.
Diagnosis:
1. Large, thick-bodied heavy snake.
2. Dorsals smooth, at midbody in 60–75 rows.
3. Prefrontals two pairs, posterior pair smaller often broken in small scales (Figure 23E).
4. Ventrals not extending to the full width of ventrum, 245–278, subcaudals 58–75.
Total maximum length reached about 6 m.

Color: Dorsum light yellowish to cream, grayish or brownish above. A median series of large elongated subquadrangular dark gray, brown, or reddish brown dark edged blotches becoming more irregular in posterior half of the body. Flanks with similarly colored irregular spots. A spear-shaped mark on head top, extending on to the nape. A postocular and another subocular streak. Ventrum marbled with yellow and dark spots.

Natural history notes: The rock python inhabits undisturbed thick forests, rocky edges, near marshy areas, close to lakes and streams. It lives close to the water in large dens, in dense clumps of vegetation. In forests it generally occupies hollows of tree trunks. Swims fairly well and appears to be semiaquatic in habits. It often takes refuge in large caves and unattended ruins of old buildings with clumps of vegetation around, and is reluctant to move away from its established territory.

Plate 119. *Python molurus.*

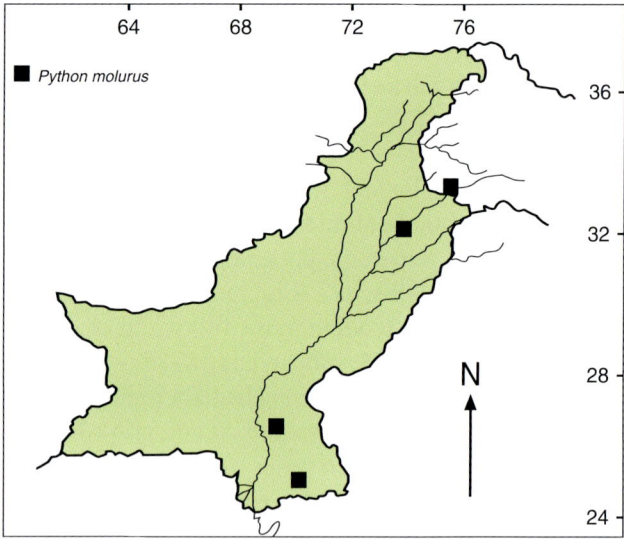

This big snake is nocturnal. In winter it basks and sleeps in sun close to its hideout. It is sluggish on land, however, it climbs well in trees, often suspending itself from branches, waiting for its prey, or lying in ambush of its prey near water edges or on a regular mammal pathway. It is a powerful constrictor, feeding mainly on warm-blooded animals from small rodents to young deer, birds, eggs, dogs, jackals, lizards, etc. After a meal, the snake rests for several weeks, as the food it has taken is in the digestive process.

A cornered python hisses loudly and does not take any apparent defensive posture. It strikes suddenly and powerfully, inflicting a large painful wound with its long curved teeth. The snake can knock down a fully grown adult man by coiling around his legs (Daniel, 1983).

It hibernates in winter, when mating takes place. From March to June, 60–100 eggs are laid in some secure place. Eggs are of the size of a duck's egg or larger. Females are known to lie coiled around the eggs for 60–80 days. It often contracts and expands its body to generate heat in order to keep the eggs at a constant temperature or it basks in the sun to warm its body, and then coils it around the eggs.

Distribution: About a century ago the rock python was common in India, Sri Lanka, and Pakistan, from the plains to an elevation of 2000 m. Its merciless persecution at the hand of man has reduced its population and now the snake is lingering only in some remote areas.

In Pakistan most of the records are from the Indus Delta. Occasionally, there are reports of pythons washed down by the flood water and stranded in the upper Indus Valley, which indicate that there is still a population existing in the upper valley of the Indus and its tributaries (Khan, 1993d). As human intervention in its natural habitat is increasing, the python is becoming more and more rare, almost on the verge of extinction.

Family Colubridae

Largest family of medium-sized snakes, includes majority of snakes found throughout the world. They exhibit a variety of morphological structures due to adaptations to aquatic, subaquatic, terrestrial, arboreal and fossorial habits. The interrelations between genera are vaguely defined, making taxonomic determinations difficult.

In general the head is distinct from neck, covered with large symmetrically arranged scales; eyes large with round pupil; body elongated, cylindrical, with smooth or keeled imbricate scales, mostly in 19 or 15–21 rows at midbody; ventrals extend across the body ventrum, tail mostly long, with divided or entire subcaudals, anal scale entire or divided.

The family Colubridae is represented by 16 genera and about 40 species in Pakistan.

Genus *Amphiesma* Duméril, Bibron, and Duméril, 1854

Medium-sized terrestrial and semiaquatic snakes, body more or less elongated, cylindrical; head distinct from the neck, eyes large with round pupil, naris non-valvular, lateral; body scales in longitudinal rows, pointed strongly imbricate, keeled; subcaudals and anal paired.

Three species are reported from Pakistan. *Amphiesma platyceps* and *Amphiesma sieboldii* are sibling species (Malnate, 1966).

Amphiesma platyceps (Blyth, 1854)

(Spotted keelback: Chitra khar-pusht)
1854 *Tropidonotus platyceps* Blyth, J. Asiatic Soc. Bengal, Calcutta 23:297.
Type locality: Assam and Darjeeling, Eastern Himalayas.
Diagnosis:
1. Supralabials 8, third to fifth touching eye.
2. Midbody scale rows 19, 5–7 middorsal scale rows weakly keeled.

Chapter 7. Ophidia: Snakes

3. Ventrals in male 205–234, in female 191–216, subcaudals divided male 88–98, female 78–96, anal divided.
Snout-vent length 880–895 mm, tail 230–235 mm.

Color: The spotted keelback is variable in coloration. Olive brown dorsum, with small dark spots; rarely a dorsolateral series of white spots. Usually a pair of white dark-edged parallel lines on dorsum, or a black line from eye to gape. Labials white or yellow. Ventrum yellowish, bordered with bright red; tail and throat mottled with black.

Natural history notes: The spotted keelback is recorded from Changla Gali, Pakistan. The known habitat of this snake is fast-flowing stream beds where it feeds on fishes, tadpoles, and frogs.

Distribution: It extends from the western to the eastern Himalayas. It is known to be common in Darjeeling District, between 1500 and 3000 m of elevation. In Pakistan it is known only from Changla Gali, western NWFP.

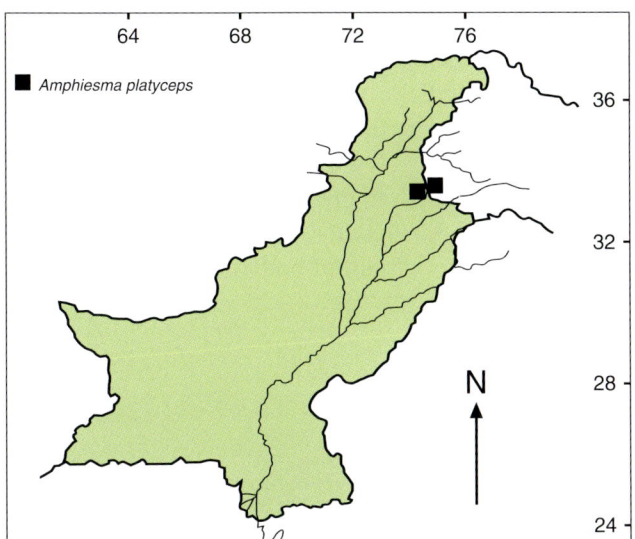

Amphiesma sieboldii **(Günther, 1860)**
(Sub-Himalayan keelback: Damni khar-pusht)
1860 *Herpetoreas sieboldii* Günther, Proc. Zool. Soc. London 1860:156.
Type locality: Sikkim, 2500 m.
Diagnosis:
1. Ventrals: male 191–207, female 168–190.
2. Subcaudals: male 97–111, female 81–110.

Distribution: Recorded from Ghora Gali in alpine Punjab, Pakistan. It has been recorded from Assam in the eastern Himalayas (Malnate, 1966).

Taxonomic notes: Morphologically close to *Amphiesma platyceps*. Boulenger (1890) and Smith (1943) regard these two species as synonymous. Malnate (1966) treats them as sibling species.

Amphiesma stolatum **(Linnaeus, 1758) (Plate 120)**
(Striped keelback: Lakeer-dar khar-push)
1758 *Coluber stolatus* Linnaeus, Syst. Nat. 1(10):219.
Type locality: Asia.
Diagnosis:
1. Head elongate, distinct from neck.
2. Supralabials 7, rarely 8; third and fourth touching eye; infralabials 10, rarely 9 or 11.
3. Posterior genial longer than anterior, not in contact with each other.
4. Scales at midbody 19, reduced to 17, at the level of vent; all scales keeled, except last row.
5. Ventrals 143–150, subcaudals 64–73, divided; anal divided.
Snout-vent length 490–497 mm, tail 85–90 mm.

Color: Body light olive brown, with a pair of dorsolateral light yellow stripes extending onto the tail. Anteriorly stripes tend to break into spots and dashes. Dark brown to black interstripe spots, more prominent ante-

Plate 120. *Amphiesma stolatum*.

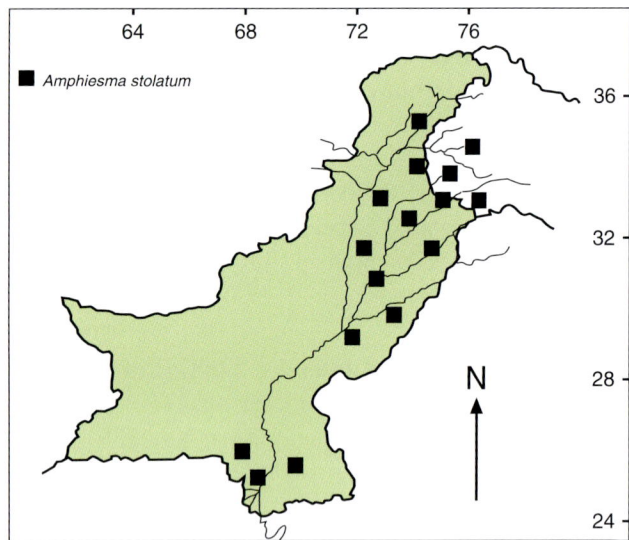

riorly. Head bluish gray to olive. Frontal and parietals are edged with black. A V-shaped dark mark on neck. Dark bars in front, below and posterior to the eye. Ventrum white.

Natural history notes: The striped keelback is characteristically a snake of open damp grassland. It frequents mesic habitat with moderate vegetation. It is diurnal and common in tilled fields, barns, under heaps of reaped crops and fallen leaves, logs, heaps of debris, etc. It is a gentle snake which seldom bites, and proves to be a good pet.

Breeding from March to May, a receptive female is often attended by several males; 10–16 elongated eggs are laid under stones, in crevices, and holes in the ground. Eggshells are soft and pliable; eggs adhere to each other to form clusters.

Amphiesma stolatum feeds on frogs, toads, fishes, lizards, mice, and nestlings. Juveniles feed on insects, froglets, and tadpoles.

When alarmed, the snake raises the anterior part of its body at an angle of 45°, and flattens it to form a narrow "hood"; it then sways the hood. Apparently it cannot keep its body in this posture for a long time; it is soon lowered and the snake crawls to safety (pers. obs.).

Distribution: The striped keelback is one of the most widely distributed snakes in Southeast Asia. It extends from the valley of the Mekong River in Laos, Thailand, north to southern China, then throughout India, Bangladesh, Myanmar, Sri Lanka, to the Indus Valley in Pakistan. It does not cross west of the Indus River into Baluchistan (Khan, 1987).

Genus *Argyrogena* Werner, 1924

Colubrid snakes with lower jaw countersunk; head a little distinct from neck; eyes large with round pupil; body scales smooth in 21 or 23 rows; subcaudals and anal divided.

A single species is known from Pakistan.

Argyrogena fasciolata (Shaw, 1802) (Plate 121)
(Banded racer: Patta taiz-rau)
1802 *Coluber fasciolatus* Shaw, Gen. Zool. 3:528.
Type locality: India.
Diagnosis:
1. All scales smooth.
2. Nostrils between nasals and internasals.
3. Supralabials 8, fourth and fifth touching eye; 9–10 lower labials (Figure 25C).
4. Both genials of equal size, posterior narrower.
5. Midbody scales 23, reduced to 17 just anterior to vent.
6. Ventrals 212–234; subcaudals 82–94, divided; anal divided.

Snout-vent length 1100–1120 mm, tail 268–275 mm.

Color: Body reddish to grayish brown. Lower sides of scales dark. Adults unicolor, indistinct light grayish crossbars on anterior part of body in juveniles, ventrum cream.

Plate 121. *Argyrogena fasciolata.* (Photo courtesy of S.A. Minton).

Chapter 7. Ophidia: Snakes

Natural history notes: The banded racer inhabits vast grasslands with moderately loose soil and bushes, infested with rodent burrows. It lives in rodent burrows or crevices among rocks and old walls, or in fissures and holes in ground. It is a very alert, agile, and diurnal snake, moving swiftly through grass and bushes. It is moderately bad tempered; when cornered it flattens its neck, hisses loudly, and attempts to bite furiously. It soon becomes a good pet, allowing moderate handling.

Food consists mainly of insects, field rodents, and lizards. Rodents are followed into their burrows and are killed by constriction by pressing them against the walls of the burrow.

It breeds twice a year, from April to May, then from August to September; 23–30 eggs are laid in burrows (Minton, 1966; Wilson, 1967).

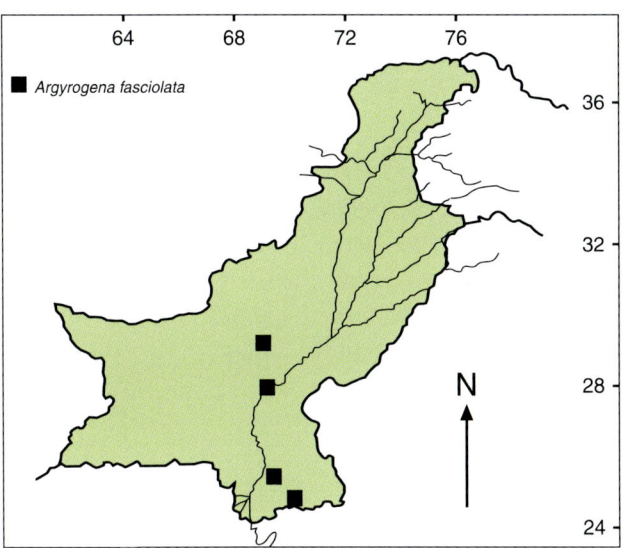

Distribution: It extends from Bangladesh throughout India, Sri Lanka. In Pakistan it is known from Sind and southern Punjab.

Genus *Boiga* Fitzinger, 1826

Body laterally compressed; head large, sharply distinct from the neck; eyes large with vertical pupil; scales smooth with apical pits, in oblique 19 to 23 rows at midbody; the median dorsal row of scales distinct in shape and size from rest of the dorsals; subcaudals paired, anal scale single.

Two geographically isolated, but morphologically very close, species are reported from Pakistan.

Boiga melanocephala (Annandale, 1904)
(Dark head cat-snake: Siah-sar billi-chisham)
1904 *Dipsadomorphus trigonata melanocephala* Annandale, J. Asiatic Soc. Bengal, Calcutta 73: 209.
Type locality: Border between Iran and Baluchistan.
Diagnosis:
Morphologically close to *Boiga trigonata* in pholidosis and habits, however, differs from it in body coloration, choice of habitat and in its range.
1. Supralabials 8–9, third to fifth in eye; 10–11 infralabials.
2. Rows of strong diagonally arranged dorsal scales at midbody 21.
3. Ventrals 222–238, 74–94 subcaudals; anal single. Snout-vent length 950–955 mm, tail 172–174 mm.

Color: Head top dark gray, with traces of Y-shaped light mark, labials and chin dark. Median dorsal crossbars without dark edges. Ventrum white without speckling.

Natural history notes: It inhabits desert fields with bushy vegetation and rocky areas with loose soil; it has occasionally been collected from low to moderately high bushes, trees in grassland, and gardens (Wall, 1914).

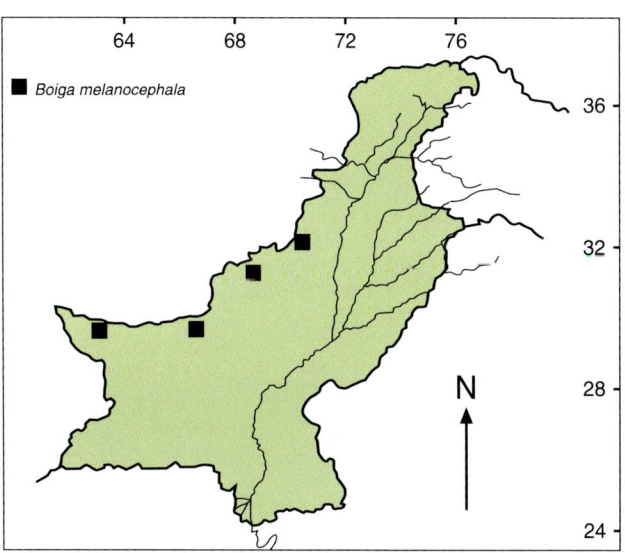

Oviparous, its 4–8 eggs are laid under stones or in burrows in the roots of vegetation during May and June; young are seen by July.

Food consists mainly of lizards, arthropods, eggs, and birds.

This snake is known to mimic *Echis carinatus* in its habits and behavior (Gans and Latifi, 1973).

Distribution: The dark head cat snake extends from western Baluchistan through eastern Iran, Saudi Arabia, north to southern Uzbekistan.

***Boiga trigonata* (Schneider, 1802) (Plate 122)**
(Common cat snake: Maidani Billi-chisham)
1802 *Coluber trigonatus* Schneider, Naturgesch. Amph. 4:156 (256).
Type locality: Vizagapatam, southern India.
Diagnosis:
1. Head triangular, flat, very distinct from neck (Figure 26E).
2. Eyes large, protruding, with vertically elliptical pupil.
3. Body slender, laterally compressed.
4. All scales smooth, with apical pits, 21, at midbody.
5. Supralabials 8, third and fifth in eye; infralabials 10–11.
6. Ventrals 212–239, subcaudals 76–92; anal not divided.

Snout-vent length 990–995 mm, caudal 187–190 mm.

Color: Dorsum light brown, a median series of 35–50 irregular transverse oblique white bars margined with black, more distinct anteriorly, fading on tail. Head brownish, with distinct Y-shaped light mark. A postorbital dark bar to angle of mouth. Ventrum dirty white speckled with gray.

Natural history notes: The natural habitats of the common cat snake are thick jungles, plantations, and grasslands with moderately thick bushes. It tolerates high temperatures; however it is common in mesic to humid habitat. It is typically arboreal and lives among branches of trees and bushes, hiding under bark or in holes in tree trunks. At the onset of winter, it descends to hibernate under heaps of stones or in crevices in stone walls, or in rodent burrows. It is mainly nocturnal; however, it is sometimes seen active during the day in rainy season. When cornered, it coils tightly and puts up a defensive stance by putting its neck in an S-shaped loop, posing like the common saw-scale viper *Echis carinatus*. It strikes, and twitches its tail rapidly (Wall, 1907c; Minton, 1966; Khan, pers. obs., 1987).

Food of the cat snake consists mainly of tree lizards, birds, and their eggs and nestlings; grasshoppers and mice are also preyed upon.

Plate 122. *Boiga trigonata*.

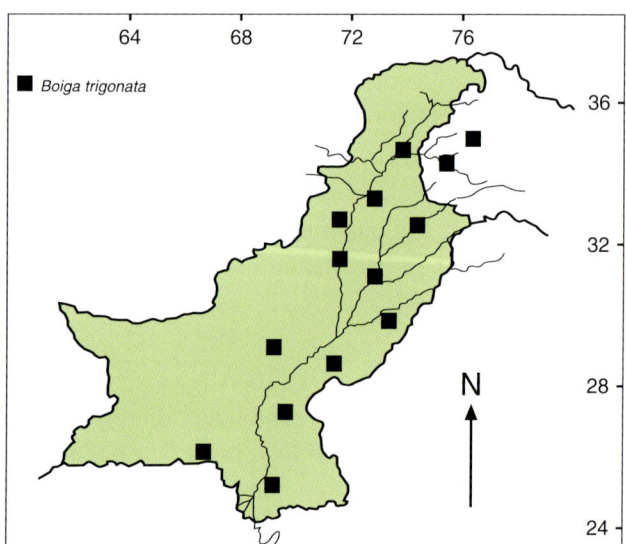

It breeds from March to May; 7–10 eggs are laid under tree bark, under stones, or in burrows in the ground.

Distribution: From Bangladesh throughout India, Sri Lanka, Kashmir, and Pakistan, exclusive of higher mountains and Baluchistan highland.

Genus *Coluber* Linnaeus, 1758

Midbody scale rows 19, reduction to 12–13 just anterior to vent. Almost always 5th (rarely 6th, or 5th,6th, or none) supralabial in contact with orbit. Ventrals 192–207, subcaudals 90–110. Vivid sooty pigment uniformly deposited on scales of the dorsal pattern. Vivid oculo-labial and oculo-temporal stripes always present.

Coluber karelini karelini Brandt, 1838 (Plate 123)

(Banded desert racer: Patti-dar koluber saamp)
1838 *Coluber (Tyria) karelini* Brandt, Bull. Acad. Sci. St. Petersburg 3: 243.

Type locality: Southwestern Asia.

Diagnosis:
1. Head long, distinctly wider than neck.
2. Nostrils bordered with nasals and internasals.
3. Two preoculars, upper more than twice the size of lower; 2–3 postoculars, rarely 1.
4. Supralabials 9, rarely 8, only fifth in contact with eye, a large postocular prevents sixth to coming into contact with eye; infralabials 9–10, rarely 11.
5. All scales smooth, 19 at midbody, reduced to 13 at vent.
6. Ventrals 196–209, subcaudals divided 93–97; anal divided.

Snout-vent length 510–512 mm, tail 140–143 mm.

Color: Dorsum pale gray to yellowish brown, with a vertebral series of 52–57 sooty crossbars (Figure 28Biii), with one or two series of alternating similarly colored smaller spots on sides, outermost row touches tips of ventrals. The pattern is very vivid on body, fades on tail. Head unicolor, a subocular and temporal vivid dark band, loreal and postocular region cream, ventrum milky white.

Natural history notes: The spotted racer frequents plain deserts between elevations of 1500–3000 m in northwestern Baluchistan. It lives in rodent and lizard burrows or under heaps of vegetation, logs, and stones. It is diurnal, moving about in sparse vegetation of grass and bushes. When disturbed it takes to nearby bushes, where it freezes, the barred pattern of its body blending with the light-shade pattern created under the bushes by the light.

It feeds on insects, lizards, field rodents, bird eggs, and birds.

Distribution: It has been recorded from Transcaspia to Kirghiz and southward to Iran and northern Baluchistan, where it is recorded from Quetta and Pishin area. Specimens have also been collected from northwestern Punjab, from Sulaiman Range.

Plate 123. *Coluber karelini.*

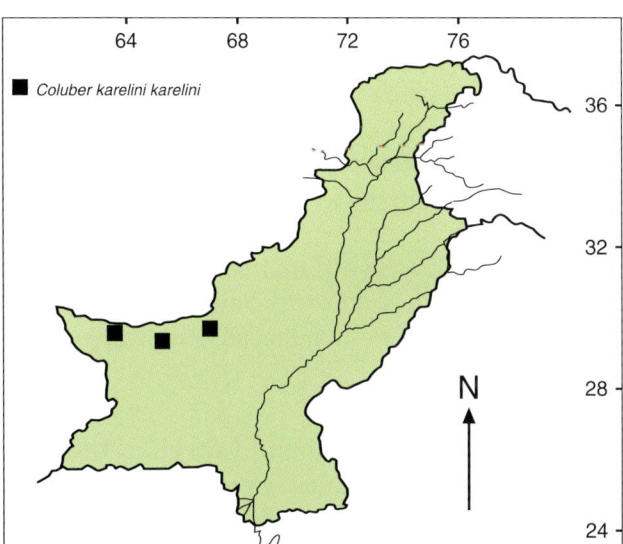

Coluber karelini mintonorum Mertens, 1969

1969 *Coluber karelini mintonorum* Mertens, Stuttg. Beitr. Naturk. 197:56–60.

Type locality: Zangi-Nawar, 27 km southwest of Nushki, District Chagai, Baluchistan.

Diagnosis:
1. Only fifth supralabial in eye.
2. Ventrals 192–220.
3. Subcaudals 114–123.
   Snout-vent length 1190 mm, tail length 360 mm.

Color: Body with closely set, diluted transverse bands, broader than interspaces. Scales in the bands with peripheral deposition of pigment.

Natural history notes: The snake is a true racer, very agile on sand.

Taxonomic notes: Mertens (1969a) described a new taxon, *Coluber karelini mintonorum,* from Zangi-Nawar, 27 km southwest of Nushki, District Chagai, Baluchistan, with only fifth supralabial in eye, ventrals 192–220, subcaudals 114–123, and body dorsum with close-set, diluted, transverse bands, broader than interspaces, and scales in the bands with peripheral deposition of pigment. Khan (1997e) regards *Coluber karelini mintonorum* conspecific with *Platyceps rhodorachis* and remarks, "Often *Coluber rhodorachis* is confused with *Coluber karelini* due to occasional specimens of *rhodorachis* with one supralabial (5th) in orbit and dorsum with alternating light and dark cross bars. Gasperetti gives photographs of similar specimens (1988, Figure 29) from Afghanistan and Nushki (Baluchistan). These specimens neither have dorsal vivid dark pattern nor orbitolabial and temporal stripes as in *Coluber karelini* from Quetta-Peshin, Baluchistan." Merten's (1969a) *Coluber karelini mintonorum* is a color variant of *Platyceps rhodorachis rhodorachis* and has also been collected from rocky areas in the Sulaiman Range, in the western Punjab, Pakistan" (Khan, 1997e).

Distribution: It is reported from sandy desert in the Chagai area. I collected two specimens from the Sulaiman Range, in western Punjab, Pakistan, in a rocky area.

Genus *Enhydris* Sonnini and Latreille, 1802

Body more or less cylindrical, with loose skin and smooth scales, in 19–29 at midbody; head depressed, hardly distinct from the neck; eyes small with vertically elliptical pupil; nasal scales in contact, semi-divided, the nasal cleft extends to the first and second supralabials; loreal present; parietals entire, not fragmented; subcaudals divided, anal usually divided.

A single species is reported in Pakistan.

Enhydris pakistanica Mertens, 1959 (Plate 124)

(Sindi ditch snake: Sindi jheel saamp)

1959 *Enhydris pakistanica* Mertens, Senckenb. Biol. Frankfurt a.M. 40:117.

Type locality: Jati, Sind, Pakistan.

Diagnosis:
1. Body stout, cylindrical, skin loose, scales at midbody 27–31, smooth and glossy.
2. Head small, slightly distinct from neck.
3. A pair of internasals, 1 pre- and 2 postoculars.
4. Supralabials 8, fourth in eye, 11 infralabials (Figures 25A, B).
5. Ventrals small, not extending to full breadth of body, 156–164; subcaudals 75–92; anal 2–3.
   Snout-vent length 500–515 mm, tail 210–215 mm.

Color: Dorsum light olive brown, with 3 black stripes, median broadest. Ventrum and sides dirty white to yellowish. Head dark, chin and throat white clouded with brown.

Natural history notes: This peculiar ditch snake frequents seepage pools along water channels. Ponds and ditches so formed are rich with emergent vegetation. This snake is almost entirely aquatic, not known to venture away from water. It is shy; during winter when the water dries up, it buries itself in moist mud and hibernates. It spends most of its time

Plate 124. *Enhydris pakistanica.* (Photo courtesy of S.A. Minton).

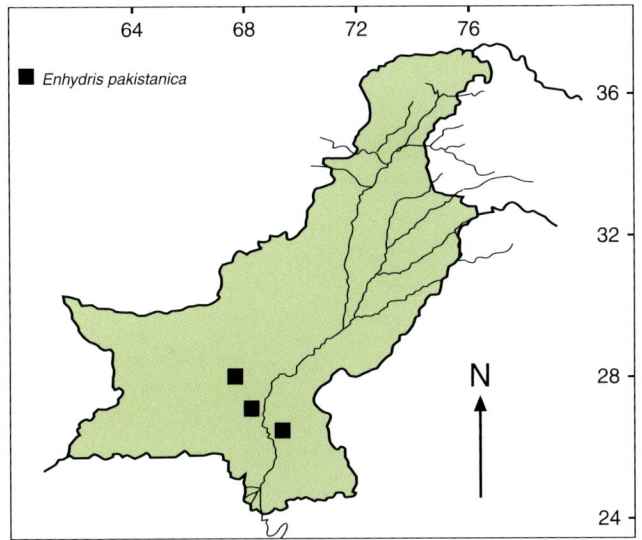

Plate 125. *Hemorrhois ravergieri.*

in crevices and pits made in the moist and soft margin of water bodies. It feeds on fishes, frogs, tadpoles, and aquatic arthropods. It is swift in water, but on land it moves in jerks for short distances. Its tongue is not protruded as frequently as in other snakes; however, during swimming it is protruded slowly. It is nonaggressive and does not attempt to bite (Mertens, 1969a).

Distribution: The Sindi ditch snake is so far reported only from the Indus Delta in lower Sind, Pakistan.

Genus *Hemorrhois* Boie, 1826

Supralabials 9, 5th, 6th in eye, midbody scales in 21 rows, 15 just anterior to vent. Temporals 3+3. Ventrals 205–212, subcaudals 82–89.

Hemorrhois ravergieri (Ménétriés, 1832) (Plate 125)
(Mountain racer: Pahari koluber saamp)
1832 *Coluber ravergieri* Ménétriés, Cat. Rais. Obj. Zool.: 69.
Type locality: Baku, Azerbaijan.
Diagnosis:
1. Supralabials 9, fifth and sixth touching eye.
2. Anterior and posterior temporals both 3.
3. Scales at midbody 21, reduced to 15 just anterior of vent.
4. Ventrals 205–212, subcaudals 82–89.
   Snout-vent length 990–995 mm, tail 195–198 mm.

Color: Dorsum light gray brown, with 61 olive vertebral transverse bars, alternating with lateral row of similarly colored spots, both tending to fuse at tail. Head with indistinct dark mottling, a dark bar from below eye to edge of mouth. Belly pale yellow to dirty white.

Natural history notes: The mountain racer is known to inhabit sparsely forested rocky hillsides at 2000–4000 m of elevation. Diurnal, it feeds on lizards, eggs, amphibians, and mice.

It breeds from April to May; 4–12 elongated eggs are laid in burrows or under stone, and juveniles are seen till late June.

Distribution: A western Asian snake, it extends from Transcaucasia to Israel to western Mongolia. In Pakistan it is recorded from mountainous areas in northern Baluchistan to Chitral.

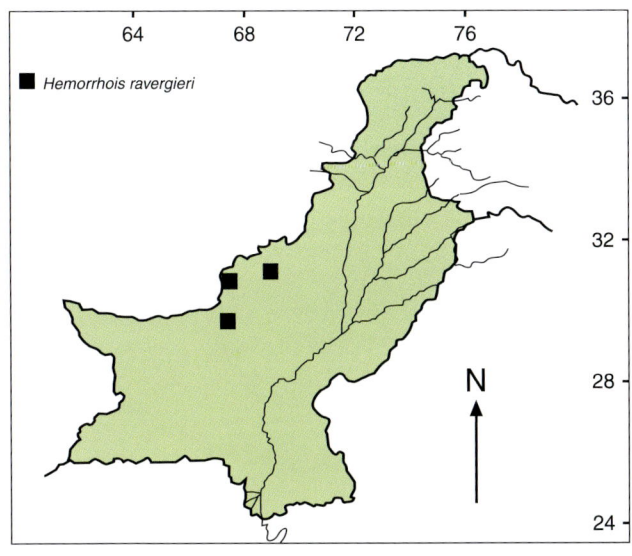

Genus *Lycodon* Boie, 1826

Head elongated, depressed, distinct from the neck, snout spatulate with obtuse tip; rostral wider than high, loreal (if present) twice longer than high; third, fourth, and fifth supralabials in eye; scales smooth, in 17 to 19 rows at midbody; subcaudals and anal divided.

Three species are known from Pakistan.

Lycodon aulicus aulicus (Linnaeus, 1758) (Plates 126A, B, C)

(White-spotted wolf snake: Chitra fraakh-dahan saamp)

1758 *Coluber aulicus* Linnaeus, Syst. Nat. Ed. 10, 1: 220.

Type locality: America.

Diagnosis:
1. Head depressed, snout spatulate, projecting beyond lower jaw.
2. Loreal elongated, in contact with internasals, not with eye.
3. Supralabials 10, third to fifth in contact with eye (Figure 26B).
4. Smooth shiny scales, at midbody in 17 rows.
5. Ventrals 171–214, laterally angulate, subcaudals 55–83, anal divided.
   Snout-vent length 780–785 mm, tail 143–148 mm.

Color: Dark brown to grayish brown dorsum, with slight gloss. Light crossbars 10–20, expanding laterally, and enclosing triangular patches of original body color. Jet black slightly protruding eye, pupil invisible. Upper lip and ventrum white.

Plate 126B. *Lycodon aulicus aulicus:* Balling behavior.

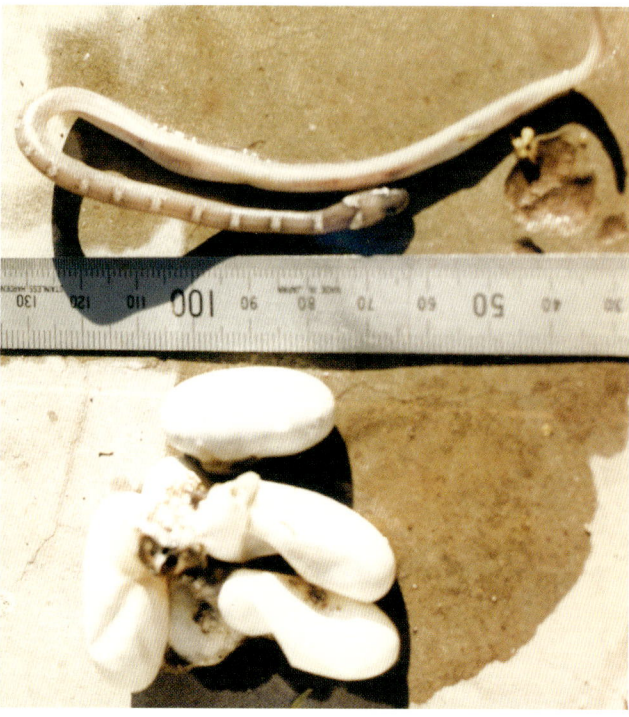

Plate 126C. *Lycodon aulicus aulicus:* Eggs and a juvenile.

Plate 126A. *Lycodon aulicus aulicus:* Adult.

Natural history notes: The common wolf snake has been collected from under piles of broken bricks partially covered with soil and debris. It frequents bush forests with mesic to dry environs and often enters human habitations, hiding under household articles.

It is a sluggish snake and does not attempt to bite. It feeds on small lizards, young mice, and arthropods. It is nocturnal, hibernates in winter, and breeds from April to July; 6 to 8 eggs are laid (Wall, 1909a).

Distribution: The nominal subspecies is widely distributed throughout India, Sri Lanka, Nepal, Assam, and Myanmar. In Pakistan it has rather spotty distribution, reported from different localities in Sind, Lahore and Rabwah, District Jhang, Pakistan (Khan, 1993d).

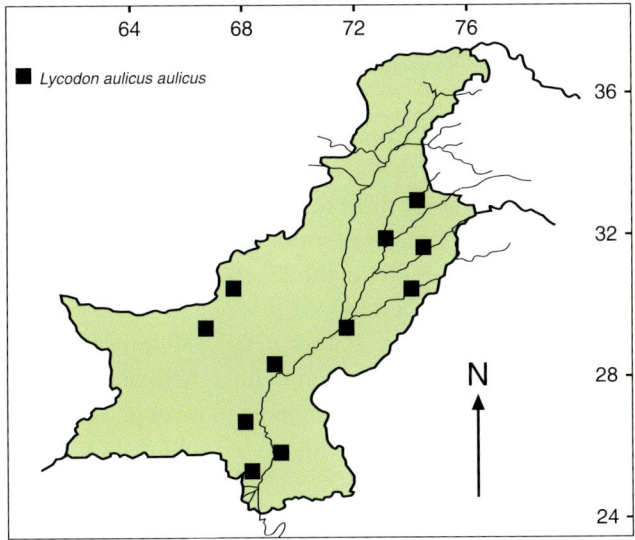

Lycodon striatus **(Shaw, 1802)**
(White-spotted wolf snake: Chittra fraakh-dahan saamp)
1802 *Coluber striatus* Shaw, Gen. Zool 3: 527.
Type locality: Vizagapatam and Hyderabad, southern India.
Diagnosis:
1. Loreal in contact with internasals.
2. One pre- and 2 postoculars.
3. Supralabials 8–9, fourth and fifth in eye; infralabials 9–10.
4. Scales smooth in 17 midbody rows.
5. Ventrals 173–205; subcaudals 40–60.
   Snout-vent length 400–405 mm, tail 90–95 mm.

Color: Dorsum light dull brown, with 15–23 white or yellow crossbars from nape to the base of tail, alternating with similarly colored spots or dashes. Nape dark, ventrum white.

Natural history notes: The white spotted wolf snake has been collected from both cultivated land and open fields with sparse bushes in mesic habitat. It is nocturnal, feeds on small lizards, young mice, and insects (Wall, 1909b).

It breeds from March to July; 4 to 8 eggs are laid, and the eggs are hatched within a month.

Distribution: It ranges from Chota Nagpur in the east throughout India and Sri Lanka. In Pakistan it is collected from different localities in Punjab, Sind, Baluchistan, and NWFP. It extends westward to Transcaspia and Central Asia. In Pakistan its two races are distinguished as follows:

Lycodon striatus striatus **(Shaw, 1802) (Plate 127)**
(Indus Valley wolf snake: Sind fraakh-dahan saamp)
1802 *Coluber striatus* Shaw, Gen. Zool. 3: 527.

Plate 127. *Lycodon striatus striatus.*

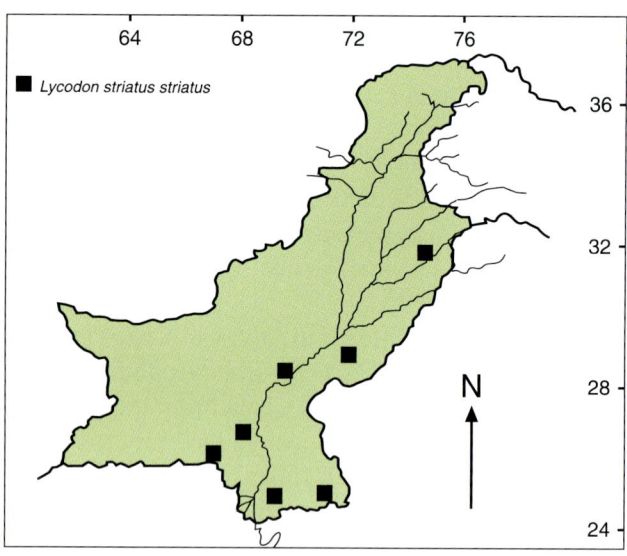

Type locality: Vizagapatam and Hyderabad, southern India.
Diagnosis:
1. Dorsum black to light chocolate brown with 16–21 light crossbars.
2. Subcaudals 45–60.

Distribution: Mainly the Indus Valley and along the coastal strip.

Lycodon striatus bicolor (Nikolsky, 1903)
(Golden wolf snake: Sunahra fraakh-dahan saamp)
1903 *Contia bicolor* Nikolsky, Ann. Mus. Zool. Acad. Sci. St. Petersburg, 8: 96.

Type locality: Transcaspia.
Diagnosis:
1. A distinct light band on nape.
2. Golden yellow (light in preserved specimens), transverse bars 35 or more, from nape to level of vent.
3. Subcaudals 50–70.

Distribution: it occupies the northwestern part of the range of the species, extending into Pakistan from upland Baluchistan and Waziristan, NWFP (Khan, 1987).

Lycodon travancoricus (Beddome, 1870)
(South Indian wolf snake: Dakhni fraakh-dahaan saamp)
1870 *Cercaspis travancoricus* Beddome, Madras Monthly J. Med. Sci. 2: 169.
Type locality: Travancore Hills, southern India.
Diagnosis:
1. Head distinct from neck.
2. Loreal not in contact with internasals and eye.
3. Supralabials 9, third to fifth in eye in eye.
4. Scales smooth, 17 at midbody.
5. Ventrals laterally angulate, 162–206; subcaudals 64–78, anal not divided.
   Snout-vent length 630 mm, tail 130 mm.

Color: Dark purplish dorsum, with a median row of 28–32 yellow crossbars from nape to level of vent. Bars bifurcate on sides enclosing triangular spots of body color. Ventrum white.

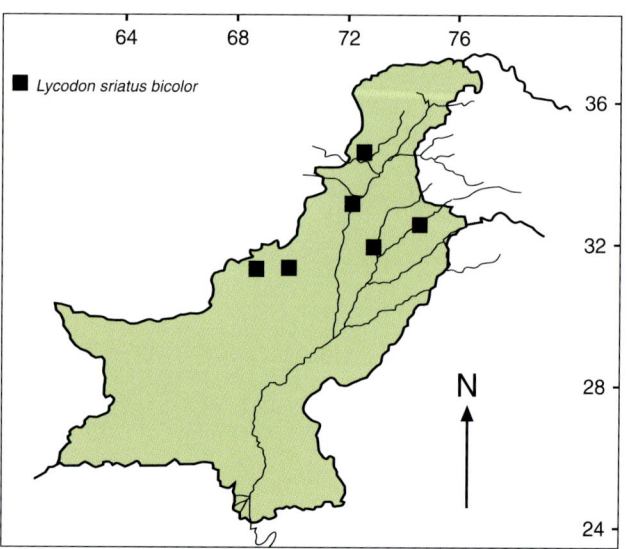

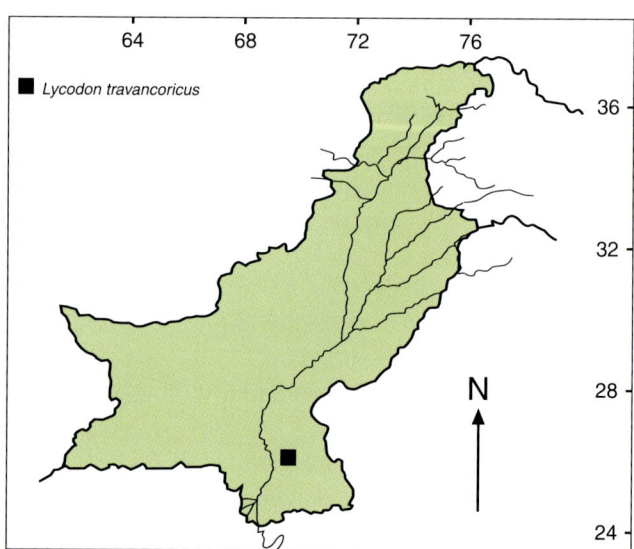

Natural history notes: The south Indian wolf snake has been reported from the garden of Hyderabad College, Sind. Nothing is known of its habitat and habit in Pakistan (Loveridge, 1959).
Taxonomic notes: Mahendra (1984) treats *L. travancoricus* as subspecies of *L. aulicus*.
Distribution: This snake is widely reported from southern India. In Pakistan it has been reported once from Hyderabad, Sind.

Genus *Lytorhynchus* Peters, 1862

Body rather short and cylindrical; head slightly distinct from the neck; eyes large with vertically elliptical pupil; rostral projecting, angularly bent in profile, concave below; nostrils oblique slits between a pair of large nasals; scales smooth or feebly keeled without apical pits, in 19 to 21 rows at midbody; subcaudals paired, anal single or paired.

Three species are known from Pakistan.

Lytorhynchus maynardi Alcock and Finn, 1896 (Plate 128)

(Balochi awl-head sand snake: Baloch crotia-sar saamp)
1896 *Lytorhynchus maynardi* Alcock and Finn, J. Asiatic Soc. Bengal, Calcutta 65:562.
Type locality: South of Koh Malik-do-Khand, Afghan-Baluchistan border, Pakistan.
Diagnosis:
1. Head long, slightly distinct from neck, snout sharply pointed.
2. Scales smooth, at midbody 19.
3. Rostral extends on top of head, separating internasals from each other almost completely (Figure 26A, Figure 26 I.i, ii).
4. Loreal present.
5. Eyes large, bordered by 7–8 circumocular scales, no supralabial in contact with eye, pupil round.
6. Supralabials normally 7, occasionally 6 or 8; infralabials 9–11.
7. Mental with a small anterior projection fitting into a notch at the base of rostral.
8. Ventrals 186–202, subcaudals 52–65.
Snout-vent length 200–205 mm, tail 56–60 mm.

Color: Dorsum pale orange to pinkish, with 35–47 black crossbars on body and 12–15 on tail. There is an alternating lateral series of dark spots. A characteristic elongated thick dark median stripe from frontal to nape. A dark postocular stripe. Ventrum white.

Natural history notes: This sand snake has been collected from fine, windblown sand dunes. It burrows in sand with the help of its pointed snout, usually close to roots of bushes and other vegetation. It is nocturnal, feeding on small lizards and arthropods; the prey is constricted. It is bad tempered and does not easily allow handling; it coils defensively, setting its neck in an S-shaped loop, attacks rapidly, all the while vibrating its tail (Minton, 1966).

It breeds from March to May; 2–4 elongated eggs are laid.

Plate 128. *Lytorhynchus maynardi.*

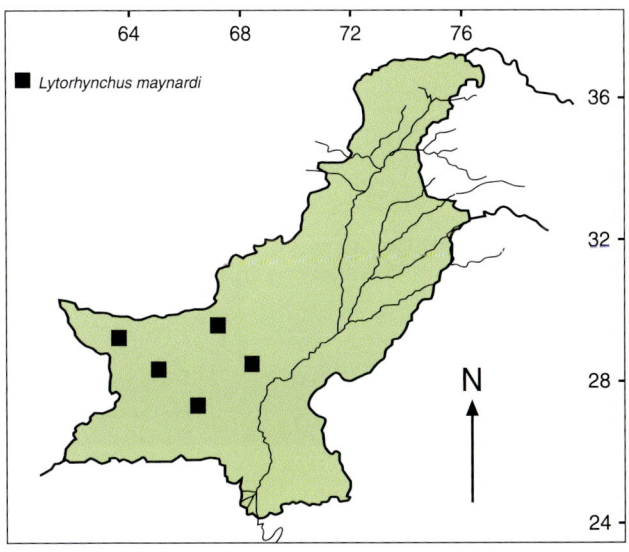

Distribution: This is a characteristic snake of wind-blown, sand dune habitat of western Baluchistan. It has been recorded from Nushki to eastern Iran in the west.

Lytorhynchus paradoxus (Günther, 1875) (Plates 129A, B)

(Sindi awl-head sand snake: Sind crotia-sar saamp)
1875 *Aconitophis paradoxus* Günther, Proc. Zool. Soc. London 1875: 232.

Type locality: Northern India restricted to Zangipur, northern Sind, Pakistan.

Diagnosis:
Morphologically close to the preceding species, differing from it in the following characters:
1. Snout less projecting (Figure 26A IIIi, ii).
2. Supralabials 8, fifth touching eye; 10–11 infralabials.
3. Ventrals 169–185, subcaudals 40–53; anal divided.
   Snout-vent length 360–375 mm, tail 63–67 mm.

Color: Dorsum brownish gray, with a median series of 40–52 brown to sooty blotches, as broad as or broader than interspaces, alternating with a lateral series of smaller spots. Head top with an elongated dark blotch and a dark postocular stripe, ventrum white.

Natural history notes: The nominal species frequents sand deserts with sparse vegetation of grass and bushes. It feeds on small lizards, young mice, and arthropods. When alarmed, it throws itself into a ball, hiding its head and vibrating its tail and does not attempt to bite. Sensing that it is out of danger, it quickly dives into sandy soil, making its way by its projecting snout.

Plate 129B. *Lytorhynchus paradoxus:* Head in profile, note the shape of snout.

Plate 129A. *Lytorhynchus paradoxus.* (Photo courtesy of S.A. Minton).

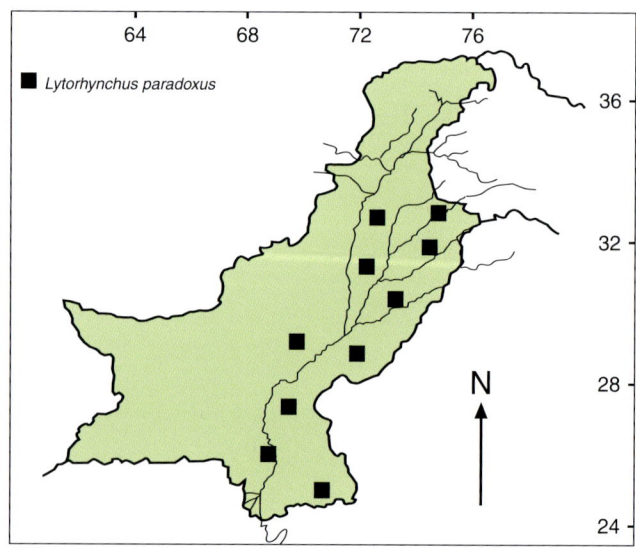

Chapter 7. Ophidia: Snakes

It breeds from March to May; 2–4 elongated eggs are laid in burrows at roots of bushes or under a stone.
Distribution: This sand snake extends from the Thar, Cholistan, and Thal Deserts in northwestern Punjab, Pakistan (Khan, 1999c).

Lytorhynchus ridgewayi Boulenger, 1887 (Plates 130, 131)

(Afghan awl-head sand snake: Afghan crotia-sar saamp)
1887 *Lytorhynchus ridgewayi* Boulenger, Ann. Mag. Nat. Hist. London 20(5): 413.
Type locality: Chinkalok, Afghanistan.
Diagnosis:
1. Snout long, sharply pointed.
2. Rostral extends onto head, almost completely separating internasals from each other (Figure 26A IIi, ii).
3. A circumocular ring of 7–8 scales.
4. Supralabials 7–8, none in contact with eye.
5. Ventrals 170–190, subcaudals 40–55.
   Snout-vent length 505–510 mm, tail 80–87 mm.

Color: Pale gray to light brown dorsum, with a median series of 40–49 blotches, which are anteriorly black to dark brown and posteriorly light brown with dark edges. A pair of lateral series of light brown spots. Head top with dark anchor-shaped mark, arms of which extend through eye to the angle of mouth. Ventrum of body white.

Plate 131. *Lytorhynchus ridgewayi.* (Photo courtesy of S.C. Anderson).

Natural history notes: This little snake is characteristic of gravel fields with scrubby vegetation. It avoids open fields of windblown sand dunes, and keeps close to roots of vegetation. It is nocturnal and feeds on small lizards and arthropods; the prey is constricted and violently shaken before engulfing. When disturbed, its first reaction is to dive into soft soil; if on hard soil, it throws itself in coils hiding its head between the coils (Minton, 1966).

Plate 130. *Lytorhynchus ridgewayi.* (Photo courtesy of S.A. Minton).

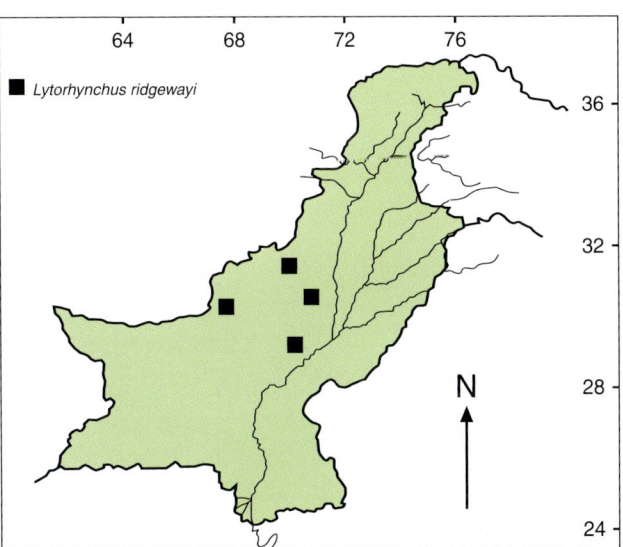

It breeds from March to July, 2–4 elongated eggs are laid in a recess in hard soil among roots of vegetation (Andrushko and Mikkau, 1964).

Distribution: This species extends to Transcaspia, central Iran, Afghanistan, and in Pakistan it is reported from western Baluchistan up to an elevation of 2000 m.

Genus *Oligodon* Boie, 1827

Body short, rather stout, cylindrical; head short, not distinct from the neck, with blunt tip; the snout short, tipped with much enlarged rostral; eye with round pupil; scales smooth, slightly imbricate, 13 to 17 rows at midbody; subcaudals paired, anal divided or entire.

The posterior maxillary teeth are very strongly enlarged and curved like a long curved knife carried by Gurkhas of Nepal, known as *kukri* (=*khanjar*, in Urdu), so the common name of the genus.

Two species are known from Pakistan.

Oligodon arnensis arnensis (Shaw, 1802) (Plate 132)

(Banded kukri snake: Patta kukri saamp)
1802 *Coluber arnensis* Shaw, Gen. Zool. 3: 526.

Type locality: Vizagapatam and Arni, southern India.

Diagnosis:
1. Head indistinct, snout blunt.
2. Rostral large, extends well on top of head, almost separating internasals (Figure 26D).
3. Loreal may be present or absent.
4. One preocular, 2–3 postoculars.
5. Supralabials 7, third and fourth touching eye; 7 or 8 lower labials.
6. Anterior genial about three times longer than posterior.
7. Body scales smooth glossy, 17 at midbody.
8. Ventrals 175–195, subcaudals 48–57; anal divided.

Snout-vent length 640–643 mm, tail 98–102 mm.

Color: Dorsum reddish to dark brown, with 32–42 black crossbars, narrowly edged with white. Nape with a pair of V-shaped marks, extending backward. Ventrum white.

Natural history notes: The brown kukri snake has been collected from mesic to semidesert habitat with bushy to dense forests. It lives in caves, crevices, and tree holes. It is common in hot weather, and is especially quite active at noon; otherwise, it is nocturnal. It has special liking for eggs, which it pierces with its specially enlarged posterior pair of upper teeth, the "kukri" (dagger). The contents of the broken egg are sucked in, and the eggshell is disgorged (Minton and J. A. Anderson, 1963; pers. obs.). The snake does not attempt to bite on handling.

It breeds from April to July; 3–8 hard-shelled eggs are laid which are glued to each other in a bunch. Juveniles are seen from June to September (Minton and J. A. Anderson, 1963).

Distribution: The barred kukri snake extends to East Bengal, central Nepal, central and peninsular India,

Plate 132. *Oligodon arnensis*.

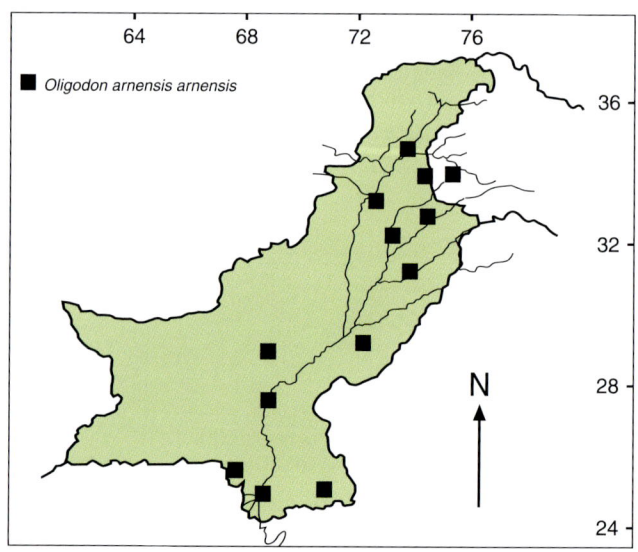

and Sri Lanka. In Pakistan it has been reported from Sind, throughout Punjab, and southern NWFP.

Oligodon taeniolatus taeniolatus (Jerdon, 1853) (Plate 133)

(Streaked kukri snake: Dahari-dar kukri saamp)
1853 *Coronella taeniolata* Jerdon, J. Asiatic Soc. Bengal, Calcutta 22:528.
Type locality: Vizagapatam, peninsular India.
Diagnosis:
1. Loreal present (Figure 26C).
2. Scales at midbody in 15 rows.
3. Ventrals 190–218, subcaudals 29–58.
   Snout-vent length 580–590 mm, tail 89–73 mm

Color: Dorsum khaki to brown, with 36–47 narrow irregular dark brown bands edged with white. A narrow vertebral and lateral dark brown stripe. Nape with W-shaped dark collar. A dark band across head at level of eyes. Ventrum white.

Natural history notes: This snake has been collected from flat clay deserts, oases, and suburban gardens to elevations of about 200 m. It lives in crevices and holes in the earth, under stones and rocks, or in brick walls. It hides under piles of wood and trash in gardens. It is nocturnal; in hot weather it moves about just after sunset until dawn. When cornered it flattens against the substratum and puts its tail in a spiral.

This snake feeds primarily on the eggs of lizards and snakes, even its own eggs. Larger eggs are punctured with the help of posterior dagger-shaped enlarged teeth which are brought into play by lateral kinking movements of the head and neck. The punctured egg is either collapsed so that its swallowing is facilitated, or the snake gains entry to consume its contents by thrusting its head into it. Smaller eggs are engulfed whole (Minton and J.A. Anderson, 1963). Except for small lizards and arthropods, no other animal is taken. It breeds from March to August; 4–8 hard-shelled eggs glued in a bunch are laid in some secluded place.

Distribution: The streaked kukri snake has been recorded from Bihar, throughout India and Sri Lanka. In Pakistan it has been recorded throughout the plains from Rawalpindi to Las Bela, at low altitudes (Khan, 1993d).

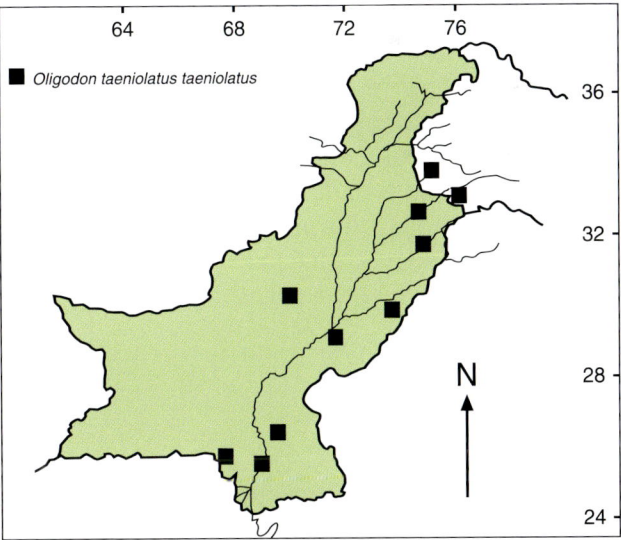

Plate 133. *Oligodon taeniolatus*. (Photo courtesy of S.A. Minton).

Genus *Platyceps* Blyth, 1860

Supralabials 9, rarely 8, usually two $5^{th} + 6^{th}$ (rarely $4^{th} + 5^{th}$ or 5th, or three $4^{th}+5^{th}+6^{th}$) supralabials in eye. Single posterior subocular may or may not be present. Single pair of parietals, single loreal, no subtemporal scale; midbody scale counts 19, reduced to 11–13 at the level of vent. Ventrals 195–277, subcaudals 82–144.

Platyceps rhodorachis rhodorachis (Jan, 1865) (Plate 134)

(Cliff racer: Chattani koluber saamp)
1865 *Zamenis rhodorachis* Jan, in: de Filippi, Note viaggio. Persia: 356.
Type locality: Persia (Iran).
Diagnosis:
1. Body slender, head distinct.
2. Supralabials 9, rarely 8, fifth and sixth in eye; infralabials 9 or 10, rarely 11.
3. Ventrals 206–220, subcaudals 125–140.
   Snout-vent length 947–1055 mm, tail 164–170 mm.

Color: Dorsum pale to dark gray, with brownish, short, median crossbars formed of 4 rows of alternating roundish spots, posteriorly unicolor (Figure 28Bi). Head brownish, with lighter pre- and postoculars. Ventrum white, with dark speckling on lateral sides.

Natural history notes: The cliff racer prefers stony highland, with sparse grass and bushes, and extends from sea level to 350–3000 m of elevation. It inhabits holes and narrow crevices among rocks, or burrows of rodents and lizards. It is diurnal, venturing into inhabited houses attracted by house geckos, birds, and mice. It is very agile and is a moderate climber. It climbs into crevices in walls or thatched roofs for eggs and nestlings. Its diet includes insects, small birds, eggs, mice, and small lizards. Moderately bad tempered; when cornered, it hisses and bites.

It breeds from April to May; 4–8 elongated eggs are laid in crevices among rocks or under stones, and juveniles are active by July–August.

Distribution: The cliff racer has a wide range in the west, from Turkmenistan, Syria, Iraq, Iran, Saudi Arabia, and extends to Somalia. In Pakistan it is recorded from Baluchistan, and is common around Karachi (Minton, 1966). However, it does not extend in plains of Punjab (Khan, 1997e).

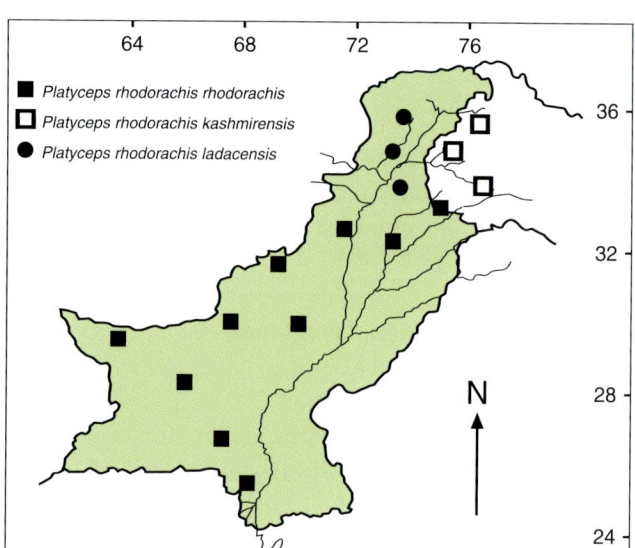

Plate 134. *Platyceps rhodorachis rhodorachis.* (Photo courtesy of S.A. Minton).

Platyceps rhodorachis kashmirensis (Khan and Khan, 2000) (Plate 135)

(Kashmir cliff racer: Kashmir koluber saamp)
2000 *Coluber rhodorachis kashmirensis* M.S. Khan and A.Q. Khan, Pakistan J. Zool.32 (1):49.
Type locality: Goi Madan, Kotli, Azad Kashmir.
Diagnosis:
1. Infralabials almost always 10 in number.
2. Dorsals almost always reduced to 13 at vent.
3. Supralabials 9, fifth, sixth in eye, rarely fourth and fifth.
4. Single preocular and presubocular, 2 postoculars.
5. Posterior genials are separated by a patch of 5–8 granular intergenial scales, which are sometimes arranged in a double row.
6. Ventrals 210–239 mm, subcaudals 119–135. Snout-vent length 385–660, tail 190–228 mm.

Plate 135. *Platyceps rhodorachis kashmirensis.*

Color: Head and anterior half of the body dark without light spotting, except a pre- and postorbital bar. On the side of neck a series of 2–4 roundish dark spots with white ocelli. Posterior half of the body lighter with irregular dark dashes formed by the black pigment deposited between scales. Ventrum dark, pigment especially deposited between ventrals.
Natural history notes: It lives in submountainous stony habitats with hard reddish soil, cut into ravines and gullies. Alpine vegetation consists of grasses, pine trees, and hedges.

The snake was collected from under stones and slabs. It forages around, feeding mostly on skinks and other lizards. It usually climbs in trees and thatched roofs to feed on eggs and nestlings.
Distribution: The cliff racer was collected from several localities in southern Azad Kashmir, Mirpur, Bhimbar, Dulliah Jattan, Kotli, Goi Madan, Aram Bari, Palandri, Punch, Bagh, and Muzaffarabad (A.Q. Khan and M.S. Khan, 1996).

Platyceps rhodorachis ladacensis **(Anderson, 1871)**
(Ladakh Coluber: Ladak koluber saamp)
1871 *Zamenis ladacensis* Anderson, J. Asiatic Soc. Bengal, Calcutta 40:16.
Type locality: Ladakh, Baltistan, northeastern Pakistan.
Diagnosis:
1. A typical *rhodorachis* with a red middorsal line.
Taxonomic notes: Unaware of Jan's (1865) description, Anderson (1871) described *Zamenis ladacensis* from Ladakh, Baltistan. Later, (1895) he compared *rhodorachis* with *ladacensis* and found them identical. Despite proven conspecificity, a *rhodorachis* with a pinkish red median dorsal (Nikolsky, 1916), drab, light orange to vermilion (Minton, 1966), red (Mertens, 1969a) vertebral line is regarded as *ladacensis*. The "colored vertebral line" is discernible only in a living snake; it is soon lost upon preservation and the identity of taxa is confused. It is why the validity of *ladacensis*, as a separate taxon, has always been questioned (Kramer and Schnurrenberger, 1963; Leviton and Anderson, 1961; Kral, 1969; Ataev, 1985; Mertens, 1969a). *Platyceps rhodorachis ladacensis* is an alpine race; it has never been collected from Punjab highland. It appears to be simply a color variant, however, I recognize it as a subspecies until more material is available from its type locality.
Distribution: Minton (pers. comm., 1998) records the uniform gray form of *rhodorachis*, with orange or red vertebral stripe, as common in Quetta (Baluchistan) and Peshawar (NWFP). Mertens (1969a) placed the Pakistani population of *Platyceps rhodorachis* in the *ladacensis* race.

Platyceps ventromaculatus ventromaculatus **(Gray and Hardwicke, 1834)**
(Plains racer: Maidani koluber saamp)
1834 *Coluber ventromaculatus* Gray and Hardwicke, Ill. Ind. Zool. 2, Plate 80, Figure 1.
Type locality: Not stated.
Diagnosis:
1. Body moderately robust, cylindrical with even taper.
2. Supralabials 9, rarely 8, fifth and sixth touching eye; infralabials 9 or 10, rarely 11.
3. Ventrals 195–213, subcaudals 97–115.
Snout-vent length 1090–1100 mm, tail 205–212 mm.
Color: Sandy white or light reddish brown, with a median series of 57–76 crossbars formed of gray scales edged with black (Figure 28Bii), alternating with one or two lateral series of spots, the outermost touching tips of ventrals. Dorsal pattern fades out in

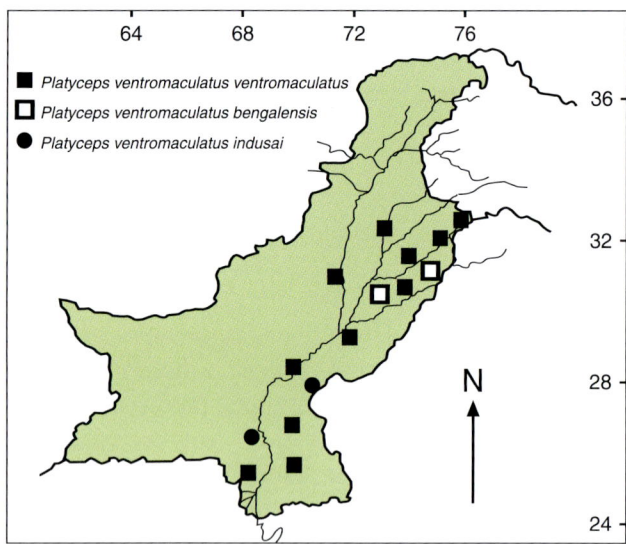

posterior one-third of body. Head with symmetrical dark mottling on top. Ventrum glistening white.
Natural history notes: The racer frequents grasslands, forests with alluvial soil and moderate scrubby vegetation. It does not extend into more humid areas along water courses. Primarily terrestrial, it climbs into low bushes and roofs of thatched houses for eggs and nestlings. It feeds on frogs, toads, lizards, eggs, nestling, birds, mice, musk shrew, and insects. It is nocturnal and is more active in the early hours of the night. It hibernates from November to late February in rodent burrows. When encountered in the open, it defends itself by coiling its body, throwing its neck in an S-shaped coil, and striking vigorously.

It breeds from April to June; 6–10 elongated eggs with soft shells are laid under stones and other objects. Young snakes are hatched by July–August.
Distribution: The plains racer is widely distributed in the Indo-Gangetic plain through India and Pakistan, below 200 m. It extends westward to Uzbekistan and Israel.

Platyceps ventromaculatus bengalensis (Khan and Khan, 2000)

(Bengal racer: Bangali koluber saamp)
2000 *Coluber ventromaculatus bengalensis* Khan and Khan, Pakistan J. Zool., 32(1):50–51.
Type locality: Bengal.

Diagnosis:
1. Supralabials 9, fifth to sixth in eye, 10–11 infralabials.
2. Ventrals 220–230, 69–70 subcaudals.
3. A median dorsal row of 60+, 1–2 scale thick dark cross bands, much narrower than interspaces, replaced on tail by irregular narrow transverse streaks formed by approximation of dark edges of adjacent scales.
4. Distinct orbitolabial and temporal stripes.

Distribution in Pakistan: A rare snake in the Punjab riparian system of Pakistan.

Platyceps ventromaculatus indusai (Khan and Khan, 2000) (Plates 136A, B)

(Sindi racer: Do-ab koluber saamp)
2000 *Coluber ventromaculatus indusai* Khan and Khan, Pakistan J. Zool., 32(1): 50–51.
Type locality: Specimens from upper and lower Indus Valleys were reported by Minton (1966), Mertens (1969a), and Khan (1997e).

Diagnosis:
1. Supralabials 8–9, fifth (rare) usually fifth to sixth in eye.
2. Preoculars 2, 2–3 postoculars (rarely 1).
3. Ventrals 195–220, subcaudals 82–119.
Total length 738–913 mm, tail 120–150 mm.

Color: Dorsum grayish white to pale sandy. Head with indistinct dark mottling. Body with a median row of 3–4 scale thick 50–70 rhombs or saddles,

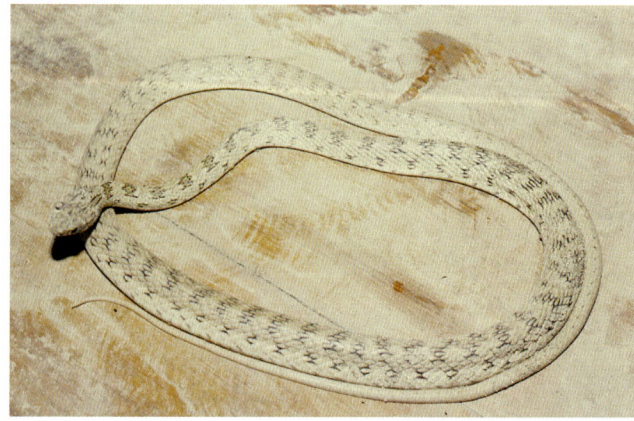

Plate 136A. *Platyceps ventromaculatus indusai:* Adult.

Chapter 7. Ophidia: Snakes

Plate 136B. *Platyceps ventromaculatus indusai*: Juvenile.

Plate 137. *Psammophis condanarus condanarus*.

broader than interspaces, alternating with two lateral rows of spots, outermost row of smaller spots usually touching ventrals. A 2–4 scale thick nuchal streak always flanked by large temporal blotch, which obscures oculotemporal streak. The sooty pigment is confined to the scale's periphery (Minton, 1966).

Genus *Psammophis* Fitzinger, 1826

Body and tail long, cylindrical; head narrow, elongated, distinct from neck, with angular canthus and concave loreal region; nostril between two nasals, completely or incompletely divided; eyes large with round pupil; scales smooth, in 17 more-or-less oblique rows at the midbody; tail long with divided subcaudals, anal single or divided.

Four species are reported in Pakistan.

Psammophis condanarus condanarus (Merrem, 1820) (Plate 137)

(Indo-Burmese sand snake: Burman teer-mar)
1820 *condanarus* Merrem, Tent. Syst. Amp.: 107.
Type locality: Ganjam District, Orissa, India.
Diagnosis:
1. Head long, narrow, distinct from neck, snout blunt, eyes large.
2. Loreal region concave, loreal scale elongate, twice as long as high.
3. One pre- and 2 postoculars, preocular not in contact with frontal.
4. Nasal incompletely divided.
5. Supralabials 8–9, fourth and fifth in eye; infralabials 11.
6. Midbody scales 17.
7. Ventrals 165–179, subcaudals 83–93; anal divided.

Snout-vent length 800–825 mm, tail 254–258 mm.

Color: Dorsum pale olive to yellowish. A pair of dark brown stripes on fifth and sixth scale rows does not extend to head. Reddish brown stripes on 1-3 scale rows, extends on head passing through eye and snout. Lateral interstripe space cream to straw yellow. Supralabials greenish, green color extends posteriorly

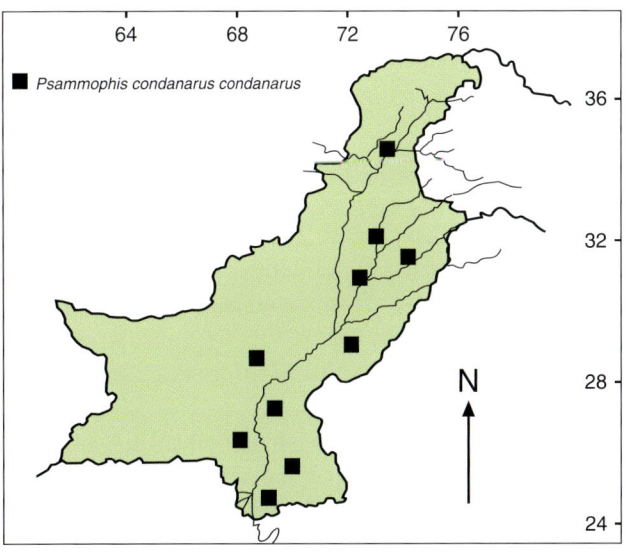

as a band on lateral tips of ventrals, ventrum medially cream, margined with a narrow reddish brown stripe.
Natural history notes: This snake frequents mesic habitat with long grass and sparse bushes. It is diurnal, very agile, and moves very fast. It feeds on lizards. The snake freezes and the prey is taken by surprise. When in danger, it freezes and its light color makes it inconspicuous in dry yellow grass in broad daylight. The snake climbs into bushes and smaller trees to feed on birds and their eggs. Wall (1911a) records *Echis carinatus*, frogs, and skinks of genus *Mabuya* as its food.

The snake rests for the night under stones or fallen trees. Usually it uses the burrows of sand lizards which are in abundance at the roots of bushes. The snake remains active most of the day in summer; however, its activity is curtailed in winter to the warmer hours of the day.

It breeds from March to July; 6–8 eggs are laid in burrows.
Distribution: This sand snake is known from western Bengal, Orissa to Bombay, Sind, and Punjab. Pakistan records are from Jacobabad, Lahore, and the Indus Delta. In the western Himalayas it is one of the commonest snakes at altitudes between 1000 and 2000 m (Wall, 1911b).

Plate 138. *Psammophis leithii*. (Photo courtesy of S.A. Minton).

Psammophis leithii leithii Günther, 1869 (Plate 138)
(Sindi ribbon snake: Sindi teer-mar)
1869 *Psammophis leithii* Günther, Proc. Zool. Soc. London 1869: 505.
Type locality: Sind, Pakistan.
Diagnosis:
1. Preorbital touching frontal.
2. Temporals 1 + 2.
3. Supralabials 8–9, fourth and fourth in eye; infralabials 9–10.
4. Nasal divided (Figure 27B).
5. Posterior genial a little longer than anterior.
6. Ventrals 164–187, subcaudals 104–109; anal not divided.

Snout-vent length 770–778 mm, tail 225–230 mm.
Color: Ground color yellowish, darker on sides. A pair of dark to brown stripes starting from internasals and continuing along body occupying fifth to seventh scale rows becomes wider at midbody and fusing on tail. The dark lateral color of body forms a stripe, confined on first and second scale row, lateral tips of ventrals white, central part lemon yellow. Head top with median dark, longitudinal stripe, another stripe through eyes. Labials cream with reddish brown mottling, reddish brown spots on neck.
Natural history notes: This snake frequents desert grassland with scrubby vegetation near marshy areas in lower Sind. It extends into clay deserts along the

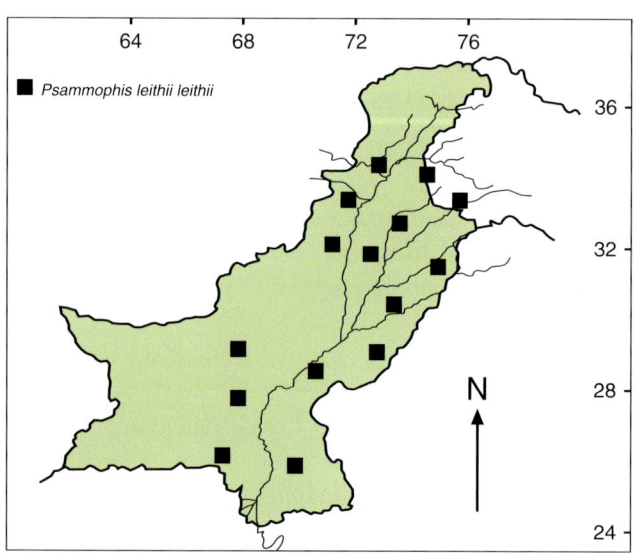

sea coast. Habitats of all the three ribbon snakes of Pakistan overlap in the deserts of Pakistan.

It is diurnal, feeding on lizards and mice. It climbs in bushes to feed on birds and eggs. It breeds from March to June; 4–10 eggs are laid in burrows of field mice and lizards.

Distribution: This is a northern Indian snake, ranging from the United Provinces of India, Poona to Waziristan and southeastern Baluchistan, in the west. In Pakistan it is recorded from Azad Kashmir, the Thar, Cholistan, and Thal Deserts to Kalat District in Baluchistan, below 800 m of elevation. It is fairly common along lower the Indus (Minton, 1966).

Psammophis lineolatus lineolatus (Brandt, 1838) (Plate. 139)

(Steppe ribbon snake: Patta teer-mar)
1838 *Coluber (Taphrometopon) lineolatus* Brandt, Bull. Acad. Imp. Sci. St. Petersburg 3: 243.
Type locality: Transcaspia.
Diagnosis:
1. Preocular touching frontal.
2. Temporals 2 + 2 or 3.
3. Supralabials 9, 4th to 6th touching eye; 9 infralabials.
4. Midbody scale rows 15.
5. Nasal divided.
6. Ventrals 174–190, subcaudals 70–90; anal divided.

Snout-vent length 875–900 mm, tail 190–194 mm.

Color: Dorsum light grayish buff, a pair of dark stripes from eye running along the length of body. At midbody a medium brown stripe, with several black-tipped scales. Head with a median dark stripe. Labials cream, with reddish brown mottling. Central part of belly lemon yellow, with median stippling and a dark line at lateral tips of ventrals.

Natural history notes: This snake inhabits scrublands in sandy deserts or close to foot hills. Near Quetta it is recorded at 1500–2100 m (Minton, 1966).

Food consists mainly of sand lizards and arthropods. It is diurnal, oviparous; 8–10 eggs are laid from March to July, under stones or burrows of rodents and lizards.

Distribution: It is a central Asian species extending from Transcaspia to northern Iran, east to Mongolia and northwest China. In Pakistan it has been recorded around Quetta, Baluchistan.

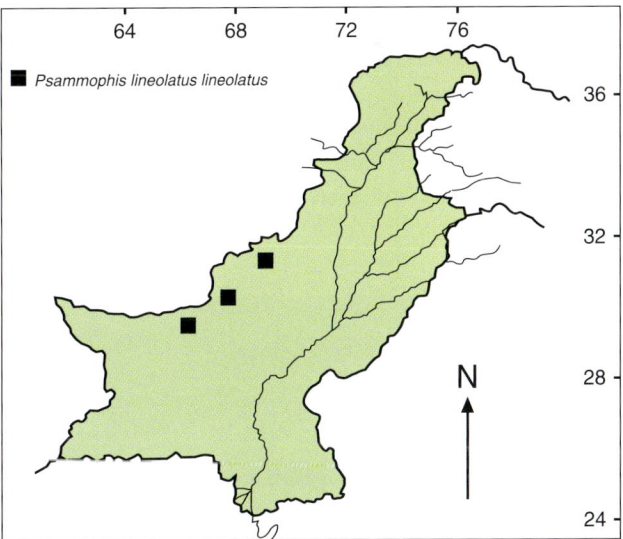

Plate 139. *Psammophis lineolatus.*

Psammophis schokari schokari (Forskål, 1775) (Plates 140, 141)

(Saharo-Sindian ribbon snake: Saharae teer-mar)
1775 *Coluber schokari* Forskål, Descr. Anim.:14.
Type locality: Yemen, South Arabia.

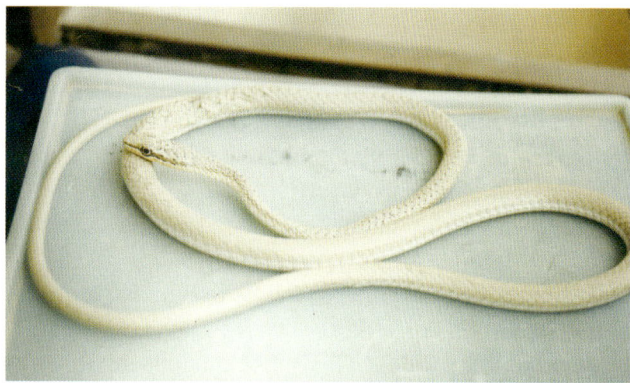

Plate 140. *Psammophis schokari.*

Diagnosis:
1. Anterior end of frontal just touching preorbital.
2. Temporals 2 + 2 or 3.
3. Supralabials 9, fifth and sixth in eye; 10 to 11 infralabials.
4. Both genials equal in size and in contact with each other.
5. Midbody scale rows 17.
6. Ventrals 173–186, subcaudals 118–130; anal divided.

   Snout-vent length 1285–1290 mm, tail 460–465 mm.

Color: Color of this snake is very variable: dorsum light olive, chestnut, dark brown, or gray. Head top with symmetrical dark markings; a dark stripe from snout through eye; labials white, flecked and spotted black; central part of belly usually bluish gray, occasionally reddish or almost black.

Natural history notes: This snake inhabits sand deserts with sparse grass and occasional bushes. It lives in lizard or rodent burrows at the roots of bushes or under rocks. It climbs in branches of bushes to bask in the sun, or to avoid heat during the hottest hours.

It is a diurnal, alert, speedy, and very agile snake, and to capture one uninjured is difficult. It feeds on various species of sand lizards and rodents. It breeds from March to July; 4–10 eggs are laid in burrows.

Distribution: The geographical range of *P. schokari* is continuous across northern Africa through Somalia, through southwestern Asia to Soviet Central Asia, Baluchistan, and Sind (Marx, 1988). It extends in Cholistan and Thal Deserts of Punjab, Pakistan (Khan, 1987).

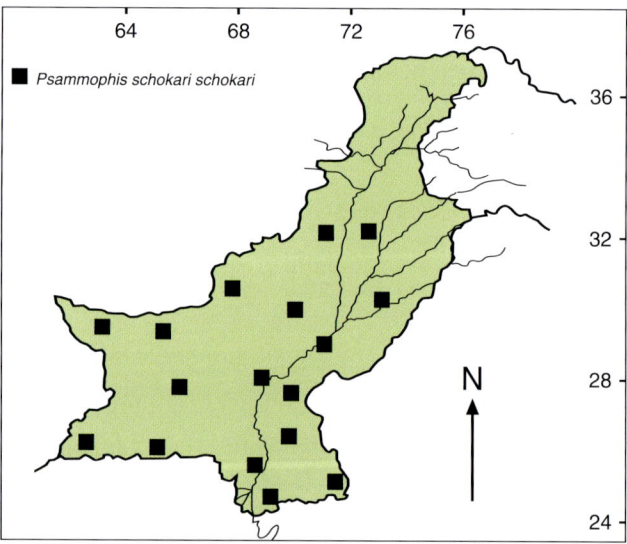

Plate 141. *Psammophis schokari.* (Photo courtesy of S.C. Anderson).

Genus *Pseudocyclophis* Boettger, 1888

Slender snakes with cylindrical body with little taper. Head depressed more or less distinct from neck; eyes with round pupil, nostrils in a single elongated nasal; loreal may be absent; scales smooth, with or without apical pits, in 15 rows at midbody; subcaudals and anal paired.

A single species is found in Pakistan.

Chapter 7. Ophidia: Snakes

Pseudocyclophis persicus (Anderson, 1872)
(Plate 142)
(Dark head dwarf-racer: Irani bauna taiz-rau saamp)
1872 *Cyclophis persicus* J. Anderson, Proc. Zool. Soc London 1872:371–404.
Type locality: Neu-Serachs, northeastern Persia.
Diagnosis:
1. Head long slightly distinct from neck, body slender.
2. Loreal scale generally absent, if present very small.
3. Eyes small, round pupil, not distinct in life.
4. Single pre- and postocular.
5. Supralabials, third and fourth in eye; 7–8 infralabials (Figure 27C).
6. Dorsals smooth, 13–17 at midbody, with apical pits.
7. Ventrals 206–238, subcaudals 75–99.

Snout-vent length 380–390 mm, tail 107–110 mm.
Color: Dorsum light brown, scales at base dark. Sometimes unicolor or with a neural series of 50–65 narrow transverse dark bars which become broader and fainter posteriorly. Head uniformly gray to dark from snout to

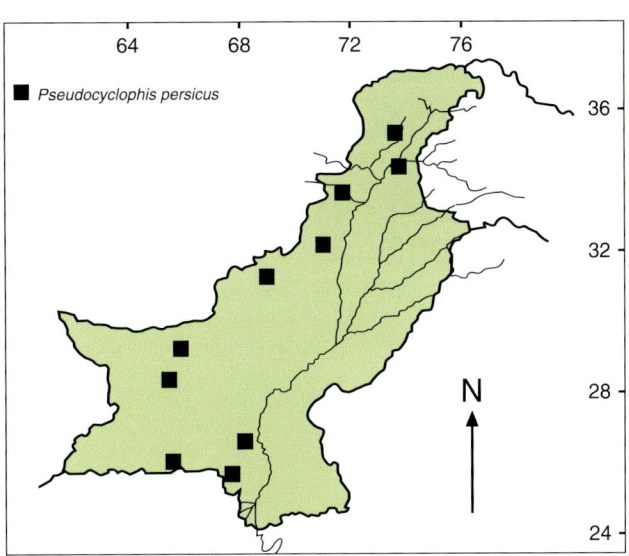

Plate 142. *Pseudocyclophis persicus.* (Photo courtesy of S.A. Minton).

nape, with a postorbital light transverse band, another crossing tips of parietals, ventrum cream.
Natural history notes: The dwarf racer frequents rocky and hilly terrain with xeric to mesic environments with scrubby to forest vegetation. It is very agile retreats into crevices among rocks, and is nocturnal. It is found usually hiding under stones and rocks during the day.

Food includes small lizards, and arthropods are frequently taken. It breeds from April to June; 4–8 oblong eggs are laid in crevices.
Distribution: The dwarf racer has been reported from Turkmenistan, Iraq, and Iran. In Pakistan it is reported from Baluchistan, western Sind to Swat in the north.

Genus *Ptyas* Fitzinger, 1843
Large snakes, with elongated, subcylindrical body. Head elongate, distinct from neck; eyes large, with round pupil; loreal 2–5, concave, single subocular, fourth or fifth supralabial in eye; scales smooth or keeled in 15 to 19 rows at midbody, with apical pits; tail long, subcaudals and anal divided.

A single species is found in Pakistan.

Ptyas mucosus mucosus (Linnaeus, 1758)
(Plates 143A, B)
(Rope-snake or Dhaman: Dahaman)
1758 *Coluber mucosus* Linnaeus, Syst. Nat. 1(10): 226.

Plate 143A. *Ptyas mucosus:* Adult.

Type locality: India.
Diagnosis:
1. Large snake with elongated head, distinct from neck, snout blunt.
2. Loreal concave, loreal 2–4, usually 3.
3. Preoculars 2 and 2–3 postoculars (Figure 27A).
4. Nostril between nasals and first supralabial; internasals shorter than prefrontals.
5. Supralabials 8, fourth and fifth touching eye, 9 or 10 infralabials.
6. Temporals 2 + 2.
7. Median rows of dorsals distinctly keeled, 17 rarely 19 at midbody.
8. Ventrals 190–197, subcaudals 110–117, anal divided.

Snout-vent length 2250–2260 mm, tail 550–560 mm.

Color: Dorsum gray to dark olive brown. Edges of most of the scales of body light, dark, or yellowish. Labials and scales of lateral sides of body dark-edged. Belly yellowish white, free edges of ventrals and subcaudals dark.

Natural history notes: The dhaman snake frequents damp and marshy situations along water courses, grasslands, cultivated land, gardens, mango groves, forests, mesic hilly environs, etc. It is attracted into human settlements by the rodent population. It often invades houses and lives in rodent holes. Its loud and resonant hiss is sufficient to chill the hearts of inhabitants. Due to its large size and dorsal body color, it is generally mistaken for *Naja naja*. It is a good swimmer, and climbs into branches of trees and roofs of thatched houses, attracted to bird's nests. Occasionally its presence is indicated by flocks of calling birds. It is a diurnal, alert, and active large snake; when on the prowl, it moves about in search of food. When cornered, its first priority is to slip away, however, if cornered it coils its body and hisses loudly, lunging furiously emitting a loud growling hiss (Abdulali 1935). Its bite is painful but nonvenomous.

Wall (1906) describes its feeding habits, "The dhaman is very catholic in its taste, devouring almost anything that chance brings within its reach, but it displays a very marked partiality to a batrachian diet, doubtless because toads, and more especially frogs, are extremely plentiful, easily captured, and too defenseless to offer much resistance. The possibility of taste influencing its selection may be dismissed, since flesh, however toothsome, must fail to impart its relish when clothed in feathers, fur, or integuments." Diet of the dhaman consists mainly of lizards, birds, eggs, rodents, and frogs. Wall (1906) includes a young soft-shell turtle (*Aspideretes*) in its diet. Its voracity is recorded by the same author by quoting the example of a specimen that had consumed 18–22 full-sized *Euphlyctis cyanophlyctis*. The snake is known to press the struggling prey against the ground with its body while devouring it (Wall, 1906). It is known to be cannibalistic and ophiophagus.

It is not killed by the local peoples because it is known to exterminate rats from village houses and barns. In Punjab and Sind it hibernates from December to February.

The dhaman breeds from March through August. During August and September 8–16 oblong eggs (50 by 40 mm) are laid in burrows, or under some protected shelter which gets sufficiently warmed by

Plate 143B. *Ptyas mucosus:* Juvenile.

Chapter 7. Ophidia: Snakes

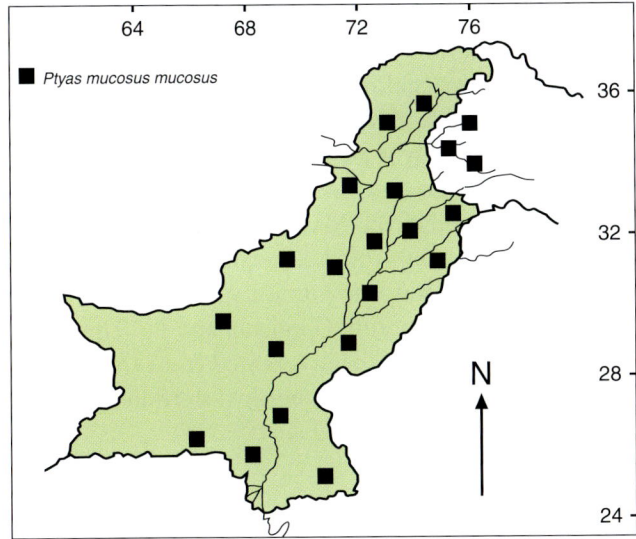

the sun to keep the eggs at an optimum temperature. Juveniles hatch between September and December, and are very secretive and alert (Daniel, 1983).

Distribution: The range of the dhaman is quite extensive in Southeast Asia; from Java, Sumatra, Vietnam, and China, it extends throughout India, the Andaman Islands, Sri Lanka, Pakistan, Iran, and Afghanistan. It particularly avoids deserts throughout its range. It is common at moderate elevations of 200 m; however, it has been recoded about 1600 m of elevation near Quetta (Minton, pers. comm., 1983).

Genus *Sibynophis* Fitzinger, 1843

Slender small snakes; head short, slightly distinct from the neck; eyes moderate with round pupil; body cylindrical, elongated; scales smooth, without apical pits, in 17 rows throughout the body length; ventrals rounded, subcaudals paired, anal divided.

A single species is found in Pakistan.

Sibynophis sagittarius (Cantor, 1839)

(Golden snake: Sunahra saamp)
1839 *Calamaria sagittaria* Cantor, Proc. Zool. Soc. London: 49

Type locality: Western Himalayas.
Diagnosis:
1. Supralabials 7–8, third and fourth or third to fifth in eye.
2. Single anterior temporal touching fifth, sixth, and seventh supralabial.
3. Ventrals 197–238; subcaudals 54–70.
   Total length 350 mm, tail 93–112 mm.

Color: Dorsum light brown. A vertebral series of black dots. Head and nape dark brown or black, with a large elongated oval yellow patch on each side of the back of the head. Snout variegated with yellow. Ventrum yellow, a dark dot on the outer edge of ventrals.

Natural history notes: The only specimen of this snake was recorded from Dulliah Jattan, southern Azad Kashmir, from under a rocky slab.

The terrain is stony with hard soil. Vegetation is sparse of grasses, with occasional low bushes.

Distribution: Western Himalayas; Ganges basin, lower Bengal (Mahendra, 1984). It has recently been recorded from southern Azad Kashmir (A.Q. Khan and M.S. Khan, 1996).

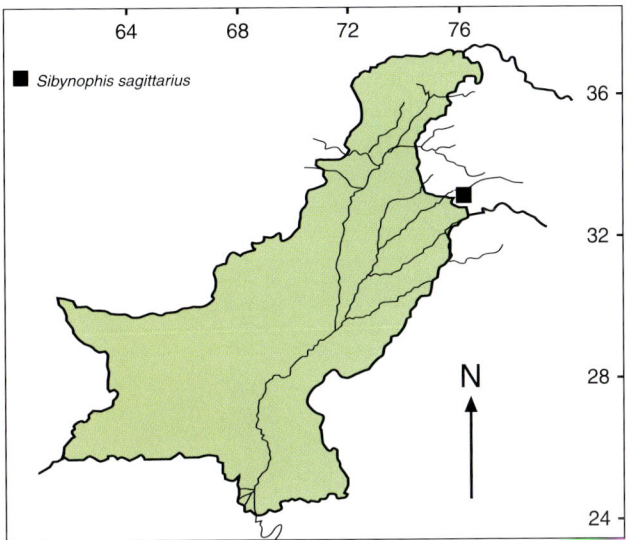

Genus *Spalerosophis* Jan, 1865

Large, heavy snakes; head distinct from the neck, rather flat; prefrontal scale broken in several pieces, loreal 2–5, temporal broken in several scales; eyes large, with round pupil, separated from supralabials by a series of suboculars; body scales feebly keeled in 25 to 43 rows at midbody, subcaudals divided, anal entire.

Three species are known from Pakistan.

Spalerosophis arenarius (Boulenger, 1890)
(Plate 144)

(Red-spotted diadem snake: Regasthani surakh saamp)
1890 *Zamenis arenarius* Boulenger, Fauna. Brit. Ind.: 329.

Type locality: Sind and Karachi, Pakistan. Restricted by Marx (1959) to Karachi.

Diagnosis:
1. Prefrontal broken into 3–6, usually 4 scales, arranged in a transverse series.
2. Snout pointed, rostral extends well on top of snout separating internasals from each other, more than length of suture between them.
3. Loreal 2, one behind other.
4. Scales in circumocular ring 6–8.
5. Supralabials 9–11, infralabials 10–12.
6. Scales keeled, 25 rows at midbody.
7. Ventrals 226–257, subcaudals 71–86, anal not divided.
   Snout-vent length 930–935 mm, tail 175–180 mm.

Color: Dorsum grayish to reddish brown, with seven series of alternating reddish spots of various sizes, those of median series largest, 54–72 from nape to the base of tail, where they tend to fuse in longitudinal stripes. Ventrum glossy white.

Natural history notes: This snake frequents sandy country with scrubby vegetation. It is nocturnal and feeds on various sand lizards, birds, field mice, and eggs. When cornered it hisses loudly and attacks. Wounds inflicted by it are painful.

It climbs into low bushes to feed on eggs and nestlings. It breeds from March to June, 8–10 elongated eggs are laid in some protected place having optimum temperature and humidity.

Distribution: Apart from Cholistan and the Thar Deserts, it has recently been reported from Dera Ghazi Khan and the Thal Desert in northwestern Punjab, Pakistan (Khan, 1999a).

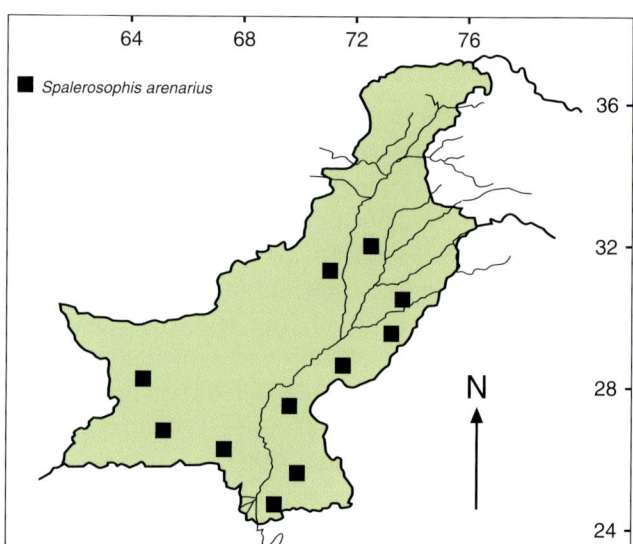

Plate 144. *Spalerosophis arenarius.* (Photo courtesy of S.A. Minton).

Spalerosophis diadema diadema (Schlegel, 1837)
(Plate 145A)

(Blotched diadem snake: Surakh chittra chua khana saamp)
1837 *Coluber diadema* Schlegel, Phys. Serp. 2: 148.

Type locality: Buchier "near Bombay," India.

Diagnosis:
1. Loreal two, one behind other.
2. Circumocular ring of 8–10 scales.
3. Prefrontal segmented in 5–8 scales, arranged in two transverse series (Figure 25D).
4. Supralabials 10–13, infralabials 11–14.
5. Dorsals keeled, in 27–31 rows at midbody.

Chapter 7. Ophidia: Snakes

Plate 145A. *Spalerosophis diadema*.

6. Ventrals 232–254, subcaudals 96–114, anal not divided.
   Snout-vent length 1220–1230 mm, tail 328–232 mm.

Color: Color varies with age: young snakes are thin-bodied with a median row of dark brown large rhomboidal blotches on body, and an alternating lateral series of similar smaller spots. There is a dark bar between eyes, and an oblique stripe from behind the eye to the angle of mouth. Older specimens are thick-bodied, yellow dorsum with scattered dark brown or sooty black spots, usually confined to individual scales. Head either blood red with tinge of black or entirely sooty black. Ventrum deep red, visible through ivory white ventrals. A transitory color phase is observed in subadults. Melanistic adults are not uncommon (Mertens, 1969a).

Natural history notes: This large snake inhabits forests along water courses, and open fields with scrubby vegetation and moderately hard soil. In rocky area it lives in crevices in loose rock. It is the usual snake frequenting old uninhabited buildings with untended natural vegetation. It is often seen in groves, barns, and suburban gardens where it is attracted by rats, mice, and nesting birds. While in inhabited houses, it lives in rat holes and often climbs in roofs to feed on birds and their nestlings. It is a bad tempered snake; when cornered it hisses loudly and strikes repeatedly, biting savagely and inflicting a painful nasty bite.

It is nocturnal; however it is it is occasionally active during the day also. Its natural diet consists of frogs, rats, birds, eggs, and lizards.

It breeds from March to September; 3–12 eggs are laid in more than one clutch, each of 2–6 eggs measuring 68–78 mm by 16–28 mm.

Distribution: The diadem snake has been recorded from sea level to 2000 m, in Gilgit, northern Pakistan; however, it is widely distributed throughout India, Bangladesh, and northern Sri Lanka. In Pakistan it has been recorded from throughout NWFP, Punjab, Sind, and Baluchistan.

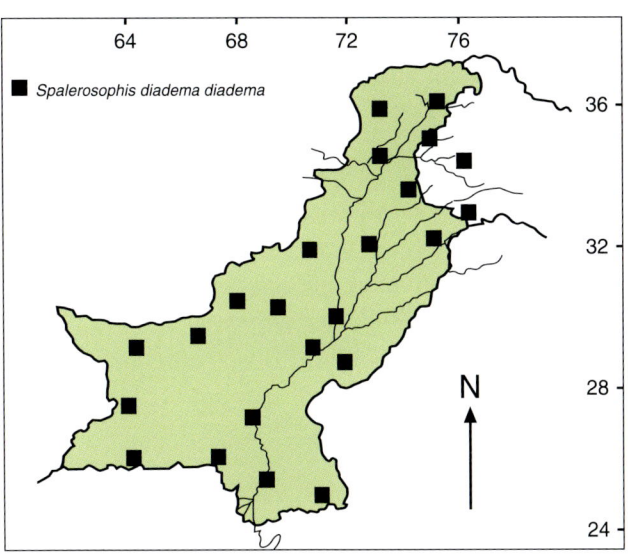

***Spalerosophis diadema* var. *atriceps* (Fisher, 1885)**
(Plate 145B)
(Variegated diadem snake: Dabba chua khana saamp)
1885 *Zamenis diadema* var. *atriceps* Fisher, Jahrb. Hamburg Wiss. Anst. 2: 82–119.
Type locality: Himalayas.
Taxonomic notes: Differs from *Spalerosophis diadema* in dorsal color and pattern which is highly variable. Dorsum is straw yellow to orange or dusky pink, with irregular black flecks and blotches, head black or red, mottled with red or black, shading to dark red on nape and temples. Belly peach to pink, immaculate or mottled with dark gray, chin and throat white (Minton, 1966). However, Mertens (1969a) has shown that *Spalerosophis diadema*

Plate 145B. *Spalerosophis diadema* "*atriceps*" phase.

changes its color and pattern to *atriceps* (Plate 145B) phase, as it grows old.

Spalerosophis schirazianus (Jan, 1865) (Plate 146)
(Persian diadem snake: Irani surakh saamp)
1865 *Periops parallelus* var. *schiraziana* Jan, in: de Filippi, Note viaggio Persia: 356.
Type locality: Shiraz, Iran.
Diagnosis:
1. Scales in ocular ring 8–10.
2. Prefrontals 6–9.
3. Dorsals smooth or with traces of keels, 25 or 27 at midbody.
4. Ventrals 231–244, subcaudals 78–79; anal not divided.

Snout-vent length 975–980 mm, tail 195 mm.
Color: Dorsum pale gray, yellowish khaki. A median series of 50–54 light olive to dark gray blotches. Head marking like young *S. diadema*. Ventrum white.
Natural history notes: The Persian diadem snake frequents flat clay deserts with sparse vegetation. However, it extends into fruit groves hiding under fallen trees, leaf litter, in crevices and holes in the soil. It usually occupies burrows of lizards and field rats. It is primarily a nocturnal species; however, it is occasionally seen active during the day.

Food includes birds, eggs, lizards, and field rats. The snake climbs in low bushes in search of prey.

It breeds from March to August; 4–12 elongated eggs are laid in some protected place with optimum temperature and humidity.
Distribution: It has wider range in the west, from Transcaspia to the Zagros Mountains in southern Iran and Tajikistan. In Pakistan it is at its easternmost distribution range and has been recorded from western Baluchistan and Las Bela (Khan, 1987).

Plate 146. *Spalerosophis schirazianus*.

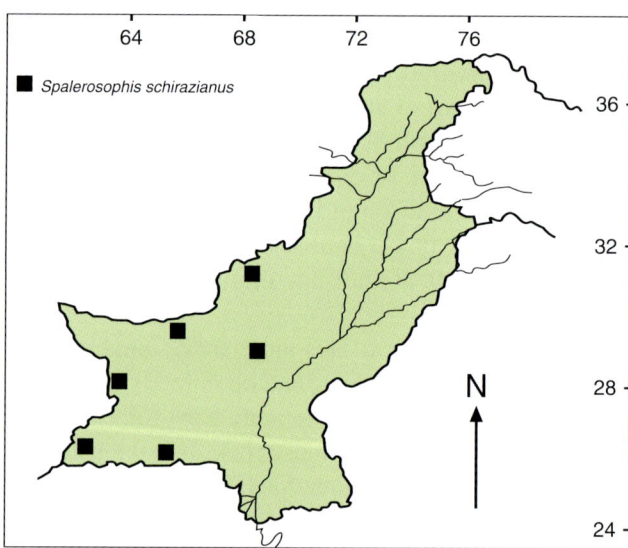

Genus *Telescopus* Wagner, 1830
Body subcylindrical, slightly compressed, with even taper; head subtriangular, flat, broad at the temporal region, very distinct from neck; eyes large with

Chapter 7. Ophidia: Snakes

vertical elliptical pupil; scales smooth in oblique rows, with apical pits, in 19–23 rows at midbody, vertebral scales not enlarged; subcaudals paired, anal entire.

A single species in Pakistan.

Telescopus rhinopoma (Blanford, 1874)
(Desert cat snake: Regasthani billi-chishm saamp)
1874 *Dipsas rhinopoma* Blanford, Ann. Mag. Nat. Hist. London 13:34.
Type locality: Kirman, southern Iran.
Diagnosis:
1. Triangular head, distinct from neck, snout flat, broad.
2. Body cylindrical, slender.
3. Eyes large with vertical pupil.
4. Nostril small, in partially divided nasal, loreal elongate touching eye, single preocular in contact with frontal.
5. Anterior temporals 3, posterior temporals 3–4.
6. Supralabials 8–9, fourth to sixth in eye; 12 infralabials.
7. Dorsals smooth, oblique with apical pits, 23 at midbody reduced to 17 at vent, middorsals broader than laterals.
8. Ventrals 247–280, subcaudals 77–99, anal entire. Snout-vent length 830–840 mm, caudal length 160 mm.

Color: Head sandy, with dark Y-shaped mark, stem of Y a dark nuchal color. Body dorsum pale grayish, with a middorsal series of large dark brown squarish blotches, broader than interspaces, posteriorly median row of blotches may divide in 2 rows of spots. A lateral alternating series of smaller poorly defined spots. Ventrum dark brown.

Natural history notes: A rare snake, it is nocturnal, taken from arid rocky hills at an elevation of 700 m. Related species of desert cat snake feed almost entirely on lizards (Minton, 1966).

Distribution: South-central Iran to Waziristan. Kacha Thana, Baluchistan (Wall, 1914); Miranshah, Tochi Valley, Waziristan, NWFP. A specimen BMNH 94.210.4.4. from Sind is in the British Museum, London (Minton, 1966).

Genus *Xenochrophis* Günther, 1864
Moderately large snakes; body cylindrical; head narrow, elongated, distinct from the neck; canthus rostralis angular; eyes moderate with round pupil; nostril in a single nasal, directed upward and outward; scales strongly keeled in 19 rows at midbody; subcaudals and anal paired.

Four species are recorded from Pakistan.

Xenochrophis cerasogaster cerasogaster (Cantor, 1839) (Plate 147)
(Red-belly marsh snake: Surakh dhoobi saamp)
1839 *Psammophis cerasogaster* Cantor, Proc. Zool. Soc. London 1839:52.

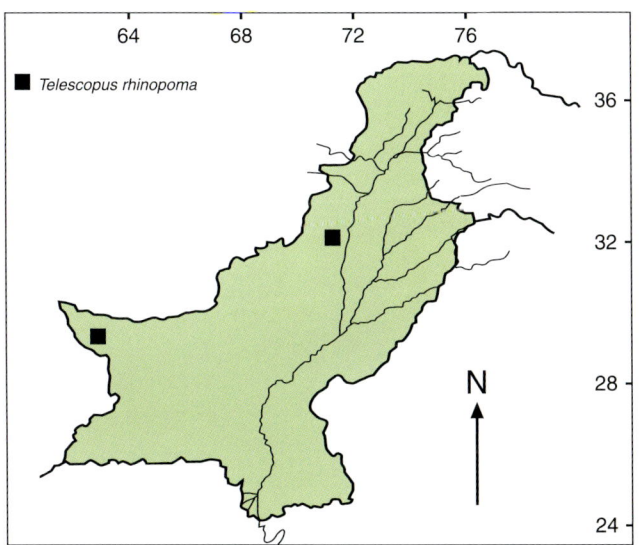

Plate 147. *Xenochrophis cerasogaster.*

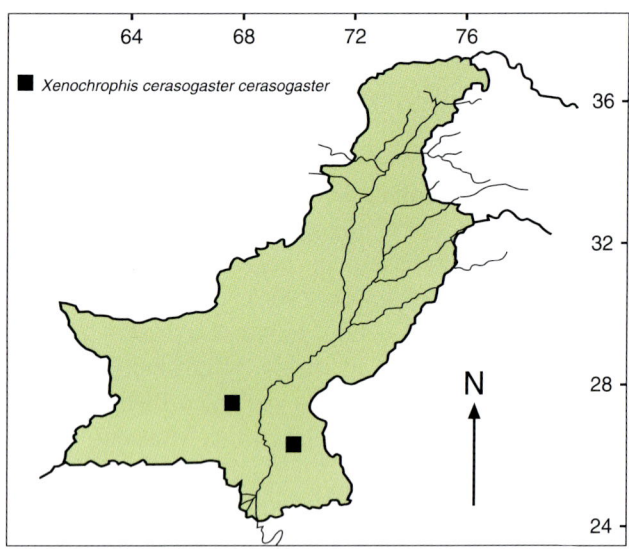

Type locality: Near Calcutta, India.
Diagnosis:
1. Head long, narrow, distinct from neck.
2. A pair of anterior and 3 posterior temporals (Figure 28A).
3. Supralabials 8–9, fourth touches eye; infralabials 9–11.
4. Scales at midbody 19, weakly keeled, last 3–4 rows smooth.
5. Ventrals of males 140–149, females 144–158; subcaudals paired, males 63–78, females 63–71, anal divided.

Total length of males 577–668 mm, females 924–973 mm.

Color: Dorsum dark brown, olive, or russet, with a dull faint dorsolateral stripe. A bright yellow stripe along outermost scale row and lateral tips of ventrals, extending on to the lips and temporal region. Ventrum posteriorly purplish black, anteriorly becoming gray, heavily marbled with red. Chin and throat white, with red mottling.

Natural history notes: This water snake avoids main water bodies, and confines itself to shallow quieter, side pools with much emergent vegetation. It often rests on the edge of the water body or on Lily pads. It is met with from late June to late November (Minton, 1966).

It is diurnal, shy, quite active in water, and moves about in emergent marginal vegetation in search of fish, which is its main dietary item. It also feeds on shrimps and other water creatures including tadpoles and frogs. When alarmed it slips into deep water. It has a moderate temperament and rarely bites upon handling.

It breeds from March to July; it is oviparous and 20–25 eggs, average size 22 by 12 mm, are laid in early April in shallow holes in moist banks of ponds. They hatch during June and July (Malnate and Minton, 1965).

Distribution: The red-belly water snake ranges widely in the Indo-Gangetic plains, from Assam through Bangladesh, to the Indus Delta. Northward it extends up to Nepal. In Pakistan it is known only from lower Sind (Khan, 1999c).

Xenochrophis piscator piscator (Schneider, 1799) (Plate. 148)
(Checkered keelback: Chittra dhoobi saamp)
1799 *Hydrus piscator* Schneider, Hist. Amphib.1:247.
Type locality: East Indies.
Diagnosis:
1. Head slightly flattened, distinct from neck.
2. Supralabials 8–10, fourth and fifth in eye; 9–10 infralabials (Figure 28C).
3. Anterior genial tuberculated in adult male.
4. Ventrals 135–152, subcaudals 62–78.

Snout-vent length 930–940 mm, tail 175–180 mm.

Color: Dorsum light green-gray or light reddish brown, with 5 rows of blackish blotches, smaller than interspaces, often fused with each other to form a reticulation, more marked in anterior half of body, the

Plate 148. *Xenochrophis piscator piscator.*

pattern fades posteriorly. A pair of dark post-oculo-supralabial bars. Ventrum white or cream.

Natural history notes: The checkered keelback is more common in large ponds with thick emergent vegetation; it confines itself mainly to side pools, avoiding the main stream. In winter when most of the water bodies are dry, the helpless snakes are killed in large numbers by people and other animals like mongoose and kites in large numbers. Water-visiting bird are said to take a high toll on young snakes. The snakes that have survived attack usually have broken tails, which are common in this species (Auffenberg, 1980b).

The water snake is strong and active, moving briskly on land and in water. It is reported to move in jumps on land (Wall, 1907d). It is known to be bad tempered; when cornered it rears up and flattens its body ready to bite. It strikes with great determination and rapidity. It bites viciously, holding on with such tenacity that it is difficult to dislodge. It leaves nasty wounds. During winter it is diurnal, while in summer it becomes crepuscular and nocturnal. It is often seen swimming close to the upper warmer layers of pond in winter and basks on dry ground.

It feeds on fishes, frogs, and tadpoles (Minton, 1966; Auffenberg, 1980b). The prey is ambushed, with the large teeth of the snake playing an important role in retaining a firm hold on slippery prey and subduing it.

The water snake breeds from February to May; 50–80 eggs are laid in adhering clusters in holes away from water. Eggs measure 27–31 by 15–18 mm in dimension.

Distribution: This snake has wide distribution in Southeast Asia, from Borneo, Taiwan, throughout India, westward to the Indus Delta. It is quite common in all major drainage systems in the upper and lower Indus Valley.

Xenochrophis sanctijohannis (Boulenger, 1890) (Plate. 149)

(Olive water snake: Zatooni dhoobi saamp)
1890 *Tropidonotus sancti-johannis* Boulenger, Fauna Brit. Ind.:350.

Type locality: Kashmir.
Diagnosis:
1. A single (fourth) supralabial touching eye.
2. Single pre- and 3 postoculars.
3. Dorsals feebly keeled, outer 4 rows smooth.
4. Ventrals 149–159, subcaudals 85–92.
5. Postoculosupralabial stripes absent, or very feebly indicated.

Snout-vent length 830–835 mm, tail 273–275 mm.

Color: Dorsum yellowish green without pattern or slight indication of spots on anterior half of body.

Natural history notes: This water snake was first recorded from a fast-flowing stream in Kashmir; recently it has been collected from irrigation channels from the Sutlej River, in Cholistan, southern Punjab.

Plate 149. *Xenochrophis sanctijohannis*.

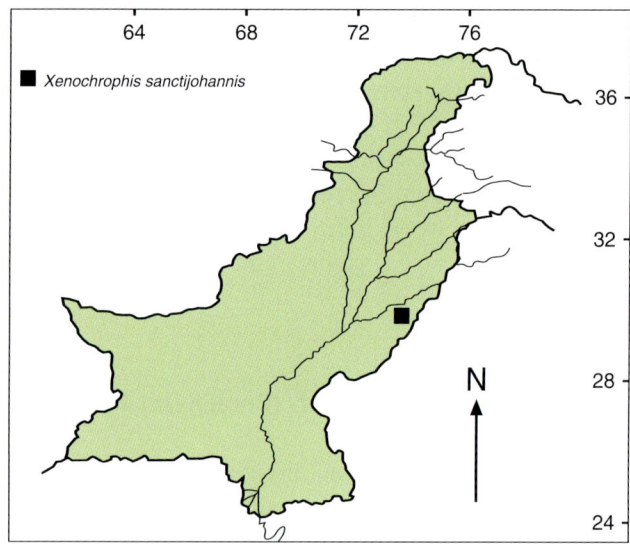
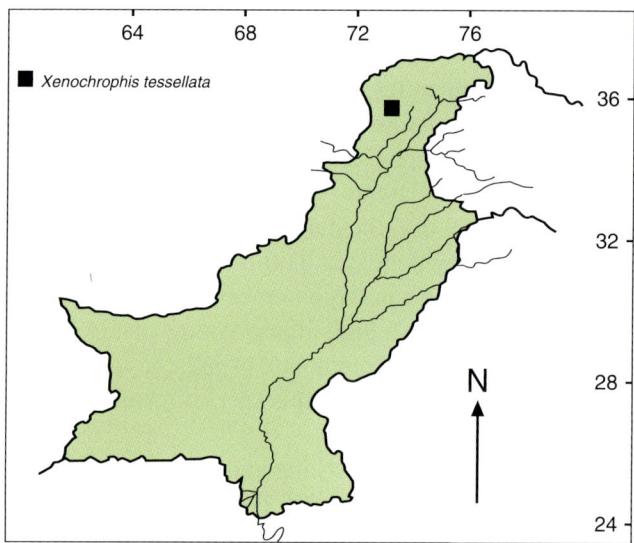

Distribution: Primarily a highland water snake, it is washed down the Sutlej River and finds its way into irrigation channels in areas surrounding Bahawalnagar and Bahawalpur, southern Punjab, Pakistan (Khan, 1984c, 1985a).

***Xenochrophis tessellata* (Laurenti, 1768)**
(Blotched water snake: Chittra nadi-wala saamp)
1768 *Coronella tessellata* Laurenti, Spec. Med., Syn. Rept.:87.
Type locality: "in Japidia (= Lapydia), vulgo Cars (= alpine meadow)" (probably Italy).
Diagnosis:
Morphologically close to *Xenochrophis piscator*; however it differs from it as follows:
1. Internasals as long as or longer than broad, triangular, truncate in front.
2. Preoculars 2, rarely 1 or 3, with or without a subocular.
3. Supralabials 8, rarely 7.
4. Dorsal strongly keeled, in 19 midbody rows.
5. Ventrals 160–198, subcaudals 45–87.
Snout-vent length 760–768 mm, tail 150–156 mm.
Color: Dorsum light green, olive gray, to almost black, with rows of small square alternating darker spots. Ventrum white, yellow, or orange-red. An inverted V-shaped dark mark on nape.
Natural history notes: It inhabits water bodies, feeding on fishes and amphibians.

Distribution: This snake is widely distributed in central and southeastern Europe to Western Asia, as far as Turkey, across southwestern Asia to Pakistan. It has been reported from Iraq, Syria, and Jordan to the Nile Delta in Egypt.

This species was reported from Mastuj in Chitral, NWFP, Pakistan by Wall (1911a), with no subsequent confirmation.

Family Elapidae

A family of deadly venomous terrestrial snakes. Characterized by a permanently fixed fang. The fang is an enlarged canaliculate tooth, grooved anteriorly, permanently fixed in erect position fitting in a pocket on the lower jaw between the mandible and the lip when mouth is closed; the fang is replaced like the other teeth; loreal scale absent from the head shields.

Two genera are found in Pakistan.

Genus *Bungarus* Daudin, 1803
Medium-sized venomous snakes; head indistinct from the neck; eye black with round pupil, hardly visible in life; no loreal scale; body scales smooth, polished, with bluish luster, median dorsal scale row distinctly enlarged, in 13–19 rows at midbody; subcaudals in a single row or paired in the terminal part of the tail, anal entire.

Two species are known in Pakistan.

Chapter 7. Ophidia: Snakes

Bungarus caeruleus caeruleus (Schneider, 1801)
(Plates 150A, B)

(Common krait: Sangchoor saamp)

1801 *Pseudoboa caerulea* Schneider, Hist. Amphib. 2: 284.

Type locality: Vizagapatam, southern India.

Diagnosis:
1. Medium-sized snakes, head barely distinguishable from neck.
2. Small eyes, round pupil, barely visible in life.
3. Loreal absent, one pre, two postoculars.
4. Temporals 1+2.
5. Supralabials 7, 3rd and 4th in eye; 8 infralabials (Figure 30Ai).
6. Body scales smooth, shiny, in 15 rows at midbody (Figure 30Aii).
7. First transverse light stripe appears between 42nd and 43rd ventral.
8. Ventral 205–217, subcaudals 43–54, undivided. Snout-vent length 1120–1125 mm, tail 150–157 mm.

Color: Dorsum jet black to deep blue, which upon preservation becomes dark brown. A series of 3–9 light vertebral spots on anterior part of body followed by 38–56 narrow transverse bands, usually in pairs. Supralabials and body ventrum white.

Natural history notes: The krait frequents open grasslands, semi-deserts with alluvial soil. It is common in the marginal vegetation along tilled fields and extends into barns, farms, groves and gardens. It is camouflaged efficiently by the light-dark effect formed under vegetation in daylight. The snake lives in holes and crevices in the ground, piles of cut vegetation, bricks, debris, etc.

Plate 150B. *Bungarus caeruleus caeruleus*: Balling behavior.

It is nocturnal, active just after sunset until dawn. Food consists of toads, frogs, snakes, lizards, and mice.

There is marked change between day and night behavior of this snake: during the day it does not attempt to escape; instead, it throws its body in a loosely coiled ball, keeping its head well concealed, the tail is kept upward and is periodically turned and twisted. When touched, the ball flinches and hisses, making jerky movements. The balled snake often attempts to bite the source of annoyance; however, if the "ball" is lifted gently, the snake allows considerable handling, but may inflict a fatal bite without provocation. Its treacherous behavior during the day has caused many casualties (Wall, 1908; Khan and Tasnim, 1986b). On the other hand, at night it is very active; it escapes or keeps still, occasionally biting the source of annoyance. It is one of the deadliest snakes of the area. Most snakebite cases at night are due to this snake (Khan and Tasnim, 1986b).

Breeding season extends from March to July; 6–8 eggs are laid at some protected place.

Distribution: The common krait is reported from Bangladesh, throughout India, and northern Sri Lanka. In Pakistan it has been reported from throughout Punjab

Plate 150A. *Bungarus caeruleus caeruleus*.

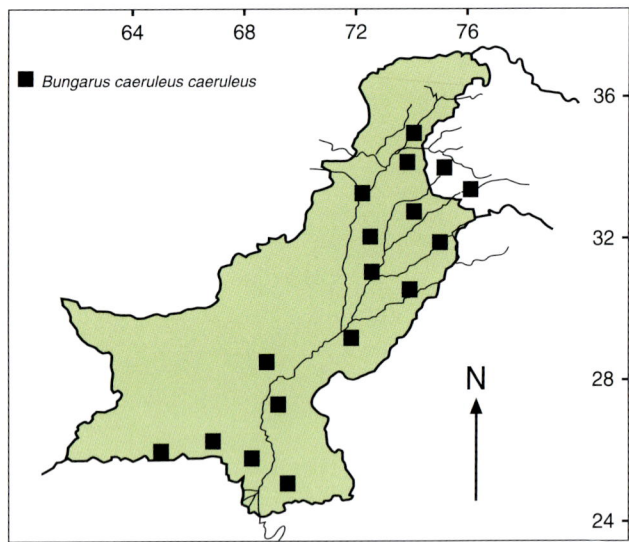
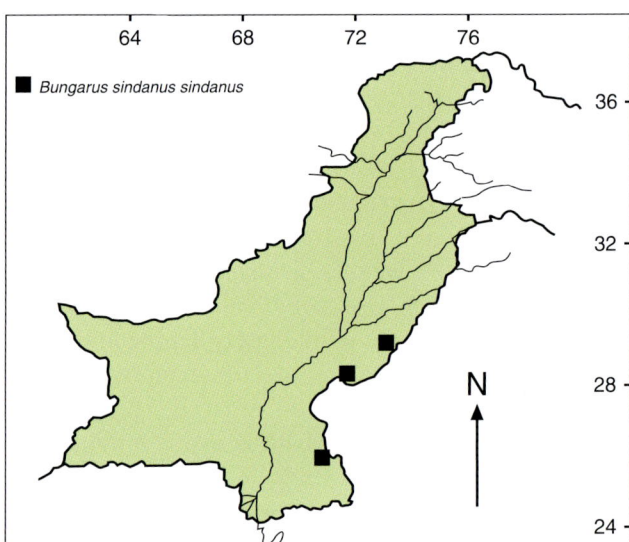

including the alpine area, NWFP, Azad Kashmir, Sind, and southern Baluchistan (Khan, 1993d).

***Bungarus sindanus sindanus* Boulenger, 1847
(Plate 151)**
(Sindi krait: Sindi sangchoor saamp)
1897 *Bungarus sindanus* Boulenger, J. Bombay Nat. Hist. Soc. 11: 73–74.
Type locality: Umarkot and Sukkhur, Sind, Pakistan.
Diagnosis: In general morphology, the nominal taxon is very close to the preceding species. However, the following characters are sufficient to diagnose it as follows:

Plate 151. *Bungarus sindanus sindanus.*

1. Midbody scale rows 17.
2. Ventrals range from 220–237, subcaudals 49–52.
3. First light transverse band appears at the level of eleventh to fifteenth ventrals.
   Snout-vent length 1034–1038 mm, tail 132–137 mm.

Color: Dorsum with similar color and pattern as in preceding species, except that first light stripe appears at the level of eleventh to fifteenth ventral.
Natural history notes: This snake has been collected from more deserticolous situations. In the Thar, Cholistan, and Thal Deserts it is collected from close to fine sandy alluvium with sparse vegetation. In the northwestern highland of Punjab, it has been collected from stony deserticolous situations at an elevation of 300 m, close to badlands cut into deep gullies on the slopes of low hills. This habitat has only sparse grass with occasional bushes (Khan, 1985a, b).

This snake is nocturnal.
Distribution: The Sindi krait has been reported from Umar Kot and Khanpur in Sind, Bahawalpur, Bahawalnagar, Rhimyar Khan, Mianwali, and Makerwal in Punjab, Pakistan (Wall, 1907a; Khan, 1990c).

***Bungarus sindanus razai* Khan, 1985**
(Northwestern Punjab krait: Punjab sangchoor saamp)
1985 *Bungarus sindanus razai* M.S. Khan, The Snake 17:71–78.

Chapter 7. Ophidia: Snakes

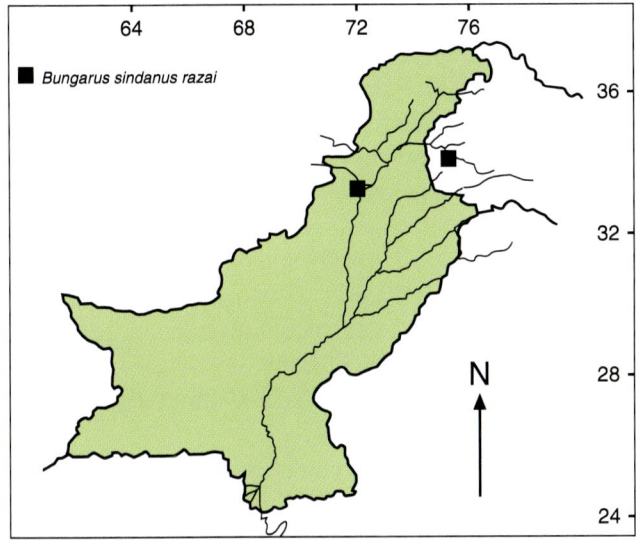

Type locality: Dandot, District Mianwali, Punjab, Pakistan.
Diagnosis:
1. Midbody scales 17.
2. Ventrals 214–221.
3. Subcaudals 44–47.

Range: Northwestern highland, Punjab, Pakistan. Recently reported from Azad Kashmir (Khan, 1997b).

Genus *Naja* Laurenti, 1768

Deadly, venomous, terrestrial snakes. Permanently fixed fangs with a distinct anterior groove, followed by 1–3 solid teeth.

Naja naja naja (Linnaeus, 1758) (Plate 152)
(Black cobra: Sheesh nag)
1758 *Coluber naja* Linnaeus, Syst. Nat. (10)1:221.
Type locality: India, Oriental region.
Diagnosis:
1. Medium-sized, heavy snake.
2. Head not distinct from body, neck dilatable into an expanded hood.
3. Loreal absent (Figure 30Bi)
4. No maxillary teeth (very rarely 1).
5. Body scales smooth, in 21–23 at midbody (Figure 30Bii).
6. Ventrals 182–196, subcaudals 53–67.
Snout-vent length 1658–1670 mm; tail 269–272 mm.

Plate 152. *Naja naja naja*.

Color: Adult jet black, dark olive or dark brown, or variegated. Ventrum pale gray to yellowish, with heavy slate gray or dark brown clouding. A spectacle mark on hood dorsum and a large, round, dark ocellus ringed with yellow on hood ventrum, medially interrupted by dark, subhood ventrals. Usually, variegated specimens with light-edged brown scales are met with in Punjab, while jet black snakes are rare.

Natural history notes: The cobra frequents various habitats, grassland, vegetation along tilled fields, along water courses, semidesert forests, barns, ruins with grassy growths, and growths around villages. It is plentiful in paddy growing areas, where it is attracted by mice and poultry into inhabited houses. It climbs into branches of trees in search of nesting birds. In inhabited houses it lives in rat holes.

The cobra, though diurnal, usually prefers coming out at the time of least disturbance. It is particularly shy of human beings. It is a restless creature, moving from place to place with agility in search of its prey, which are mainly mice, rats, poultry, frogs, and snakes. It avoids confrontation with man; upon seeing an enemy its first priority is to escape undetected. To have a good look at its surroundings, it lifts the anterior part of its body above ground and distends its neck into a hood. If the disturbance is not provocative, it levels its body and moves quietly away; however, if the disturbance is provocative, it hisses loudly and sways its hood, fixing its gaze on its adversary, following its movements keenly, all the while nervous and looking for a chance to avoid confrontation. Swaying the hood with its dorsal spectacle mark and yellow ventral ocellus, and the loud resonant hiss are measures just to look as fearsome as possible to impress its enemies. However, when cornered it attacks viciously, striking with full strength and biting and chewing at its adversary savagely (Daniel, 1983).

Breeding activity is observed from April to July; 12–30 eggs are laid in rat holes or in some protected place, and the female stays close until the eggs are hatched.

Distribution: The cobra has been reported from sea level to 4000 m in the Himalayas from Bangladesh, throughout India and Sri Lanka. In Pakistan it extends along the eastern border of the Indus from NWFP, Punjab, and Sind (Khan, 1993d, 1999c).

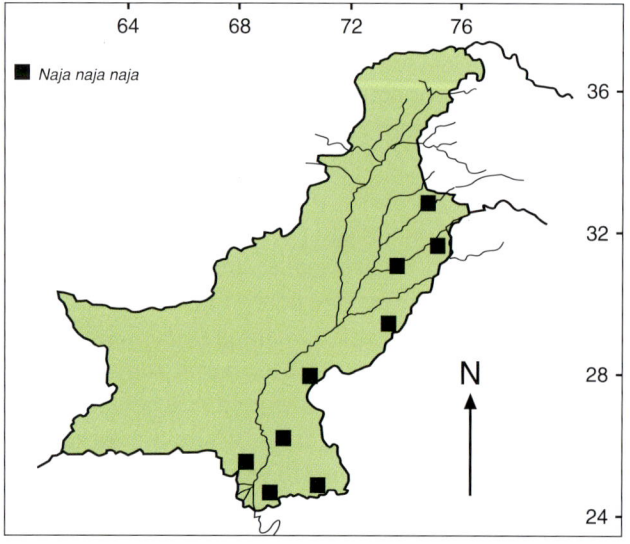

Naja oxiana (Eichwald, 1831) (Plate 153)
(Brown cobra: Bhoora nag)
1831 *Tomyris oxiana* Eichwald, Zool. Spec. 3: 171.
Type locality: Transcaspian Region.
Diagnosis: Both species of genus *Naja* are morphologically very close to each other. After careful study of the variables, *N. oxiana* is diagnosed as follows:
1. Ventrals 191–210, subcaudals 62–71.
2. One enlarged maxillary tooth.
3. Juvenile pattern conspicuously banded, bands extend on ventrum of body.

Snout-vent length 1512–1520 mm, tail 236–240 mm.

Color: Light yellowish to light brown, with or without a hood mark. Two or three dark ventrals under hood. Ventrum clouded with dark.

Plate 153. *Naja oxiana.*

Chapter 7. Ophidia: Snakes

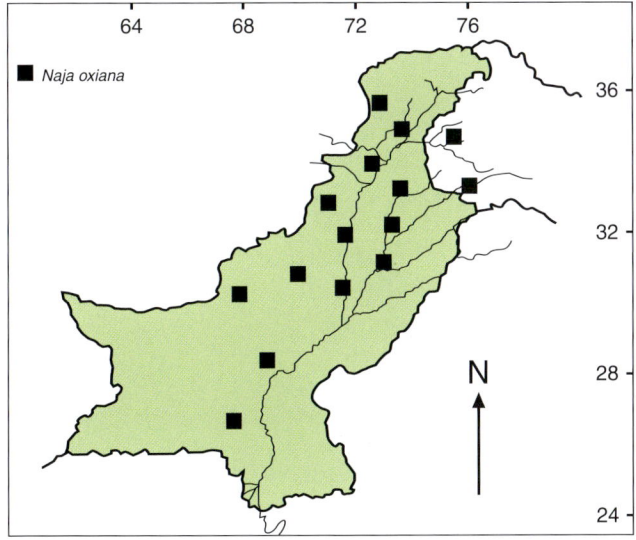

Natural history notes: The brown cobra inhabits dry wasteland where it lives in holes and crevices in uneven ground. In mountainous areas it lives in caverns, crevices, and holes in rocks.

It feeds on rodents, birds, snakes, lizards, and often enters inhabited houses attracted by rodents.

Distribution: It occurs from Transcaspia and southern Turkmenistan, eastern Iran, and Afghanistan. In Pakistan it is reported from throughout NWFP, northeastern Baluchistan to northwestern Punjab and Kashmir (Khan, 1993d; Wuster, 1998a).

Family Hydrophiidae

Marine to brackish water snakes, all venomous; recognized by their compressed, paddle-shaped tail, dorsal valvular nostrils, round pupil, tongue short, only cleft part protrusible; head scales large or small, no loreal scale; ventrals small, not extending across ventrum of body, sometimes indistinct.

Fourteen species of Pakistani sea snakes in seven genera have been collected widely in the coastal waters. It is not practical to plot precise localities for each species on distribution maps. However, they are more numerous in the mangrove swaps and at the mouth of Indus east of Karachi.

Genus *Astrotia* Fisher, 1856

Body short, stout, head large, neck more than half the greatest diameter of body. Head with large regular scales; naris dorsal, nasal scales in contact with each other with a suture reaching to the second supralabial; loreal absent, no distinct sublinguals; body scales strongly imbricate, pointed with a median bidentate keel, in 47 to 59 rows at midbody; a few anterior ventrals entire, rest longitudinally divided; anals two, enlarged.

A single species is found in Pakistan.

Astrotia stokesii (Gray, 1846) (Plate 154)
(Large-head sea snake: Bara-sar samundri saamp)
1846 *Hydrus stokesii* Gray, in: Stokes. Australia 1: 502.
Type locality: Australian Sea.
Diagnosis:
1. Head large, body short, stout with entire and regular scales.
2. Single preocular, 2 postoculars.
3. Nostril dorsal, nasal scales in contact with each other.
4. Supralabials 8–10, second sometimes third, in contact with prefrontal, fourth to sixth touching eye; 10–12 infralabials.
5. No distinct genials; body scales strongly imbricate, pointed, keeled, keels often broken in tubercles, 37–47 scale rows on neck, 47–59 on body.
6. Ventrals completely divided, except anteriormost, differentiated from rest of body scales with serrated margins; preanal strongly enlarged.

Plate 154. *Astrotia stokesii* (Photo courtesy of Harold K. Voris).

Snout-vent length 1028–1050 mm, tail 159–180 mm.

Color: Dorsum yellowish or pale brown, with 32–37 more or less complete broad dark brown rings, or the dorsal pattern may be of bars and ventrum with spots. Head dark olive or yellowish.

Natural history notes: This sea snake has been recorded from depth of 25–45 m. It is very aggressive in water and attacks repeatedly when captured.

Distribution: This snake has been reported from the Far East, Australia, Malay Peninsula, Bay of Patani, Singapore, Sri Lanka, along peninsular India, and extending along the Makran Coast in Pakistan (Mertens, 1969a).

Genus *Enhydrina* Gray, 1849

Body elongate, moderately stout, with posterior twice broad as the anterior; snout projecting, bent downward, beak-like, below the level of supralabials; head with large scales; nostrils dorsal, no internasal; nasals in contact with each other; scales imbricate or subimbricate with a short central keel, in 49 to 66 rows at the thickest part of the body; ventrals small, distinct throughout, a little broader than the adjacent scales.

A single species is reported from Pakistan

Enhydrina schistosa **(Daudin, 1803) (Plate 155)**
(Beaked sea snake: Chonchu samundri saamp)
1803 *Hydrophis schistosus* Daudin, Hist. Nat. Rept. 7: 386.

Type locality: Tranquebar, peninsular India.
Diagnosis:
1. Head of moderate size, slightly distinct from slender and much elongated neck, skin of which is loose, body moderately stout and laterally compressed.
2. Head scales densely tuberculated, rostral produced downward in a beak.
3. One anterior temporal and one postocular; 7–8 supralabials, third or fourth, fourth or fifth touching eye, 9 or 10 infralabials.
4. Genials not well differentiated.
5. Scales at midbody 51–65, subimbricate, with short central keel.
6. Ventrals small, 300–365; 2–6 preanals.
Snout-vent length 1020–1045 mm, tail 130–135 mm.

Plate 155. *Enhydrina schistosa.* (Photo courtesy of Harold K. Voris).

Color: Dirty white to pale greenish gray, with olive to black crossbars, distinct on posterior half of body. Ventrum white to light yellow, throat white. Sides of tail with few dark scales. Large adults often uniformly olive above. Newborn white with vivid black bands (Minton, pers. comm., 1994).

Natural history notes: It is the most important sea snake in terms of causing fatalities among fishermen. It frequents estuarine waters with muddy bottoms and does not extend into deeper waters. It extends deep into creeks well away from the open sea.

Distribution: It ranges from the Gulf of Oman eastward to the coast of southern Vietnam, along Australia to Rockhampton. It is a most common sea snake along the coastal waters of Pakistan.

Genus *Hydrophis* Latreille, 1802

Head with large scales; nostrils dorsal in undivided nasal scales which are in contact with each other. Body elongate, with imbricate, subimbricate or juxtaposed scales, in 20 to 57 rows at the thickest part; ventrals distinct, uniform, not broader than adjacent scales; anals 2 to 6 small or considerably enlarged scales.

Seven species are known from the shores of Pakistan.

Hydrophis caerulescens **(Shaw, 1802)**
(Blue-green sea snake: Neela samundri saamp)
1802 *Hydrus caerulescens* Shaw, Gen. Zool. 3:561.
Type locality: Indian Ocean.

Chapter 7. Ophidia: Snakes

Diagnosis:
1. Head small, not distinct from neck which is not markedly elongated.
2. Body strongly compressed laterally, about 2–3 times the diameter of neck.
3. Scales feebly imbricate or juxtaposed, distinctly keeled, 38–54 rows at midbody.
4. Supralabials 7, second in contact with the prefrontal, third and fourth are in eye, 9 infralabials.
5. Ventrals 301–310.
   Snout-vent length 770–785 mm, tail 97–104 mm.

Color: Dorsum bluish gray, ventrum yellowish, with 39–62 broad dark bands, about twice as broad as interspaces, becoming indistinct with age. Head dark in young, becoming dark gray in adult, usually with a light streak through eye.

Natural history notes: The blue-green sea snake has been recorded from mangrove swamps near Karachi, Sind, Pakistan.

Distribution: It has a wide range in Southeast Asia, from the northern coast of Australia, to Queensland, Borneo, Java, Malaya, Myanmar, and along the coasts of China, Siam, India. In Pakistan it has been reported from the Karachi coast (Minton, 1966).

Hydrophis cyanocinctus **Daudin, 1803 (Plate 156)**
(Annulated sea snake: Patta samundri saamp)
1803 *Hydrophis cyanocinctus* Daudin, Hist. Nat. Rept. 7:383.
Type locality: Sunderband, Bangladesh.

Plate 156. *Hydrophis cyanocinctus*. (Photo courtesy of S.A. Minton).

Diagnosis:
1. Head moderately small, slightly distinct from neck, rostral prolonged ventrally.
2. Body cylindrical anteriorly, laterally compressed posteriorly.
3. Scales imbricate with 2–3 keels, or row of tubercles, 38–48 at midbody.
4. Supralabials 7–8, second in contact with prefrontal, 2–3 touching eye, 9–10 infralabials.
5. Ventrals distinct 314–384, anteriorly about twice the size of adjacent scales, a little smaller posteriorly.
   Snout-vent length 1500–1650 mm, tail 130–140 mm.

Color: Dorsum dirty white, pale green or yellow, with 47–70 black cross bands, those at midbody widest, central part of the bands often include light scales. Markings on body usually disappear with age.

Natural history notes: It has been netted from shallow muddy mangrove swamps during monsoons, after which it migrates to the open sea.

Distribution: It is a wide ranging species which extends from the Persian Gulf along the coastal waters of Pakistan, the Indian coast, Sri Lanka, Bangladesh, Thailand, Islands of Indonesia to the Sea of Japan.

Hydrophis fasciatus **(Schneider, 1799) (Plate 157)**
(Small-head banded sea snake: Daula pattadar samundri saamp)
1799 *Hydrus fasciatus* Schneider, Hist. Amphib. 1:240.
Type locality: East Indies.
Diagnosis:
1. Head very small, body very slender anteriorly, deep, and greatly compressed posteriorly.
2. A series of 2–4 small scales behind parietals and between posterior temporals.
3. Supralabials 6–7 (rarely 5), third and fourth touching eye, last 1 or 2 very small.
4. A small "cuneate" scale between third and fourth infralabials.
5. Posterior body scales imbricate, slightly hexagonal, 49–53 at midbody, with a central tubercle or keel.
6. Ventrals with double keel, 323–514 at midbody.
   Snout-vent length 1080–1106 mm, tail 100–110 mm.

Plate 157. *Hydrophis fasciatus* (Photo courtesy of Harold K. Voris).

Plate 158. *Hydrophis mamillaris*.

Color: Dorsum black to olive, with pale or yellowish oval spots on sides. Ventrum whitish. In young 50–70 black crossbars which may or may not form complete rings.
Distribution: From coastal Pakistan it extends along the Indian coast, Sri Lanka, Myanmar to the Strait of Malacca.

Hydrophis lapemoides (Gray, 1849)
(Persian sea snake: Irani samundri saamp)
1849 *Auteria lapemoides* Gray, Cat. Sn. Brit. Mus.: 46.
Type locality: Madras, India, and Sri Lanka.
Diagnosis:
1. Temporals small, 2 + 3 or 3 + 3.
2. Supralabials 8, second in contact with prefrontal, third and fourth or third to fifth touching eye.
3. Scales at midbody 39–45, a feeble tubercle or keel on scales of posterior half of the body.
4. Ventrals 324–346, bicarinate.
Snout-vent length 930–950 mm, tail 94–100 mm.
Color: Young snakes with yellowish or whitish dorsum, with 32–43 dark crossbars. Head dark with a yellow curved mark. In older specimens markings on body become less pronounced.
Natural history notes: It invades estuarine waters, and is often thrown onto shore by sea waves.
Distribution: It ranges from the Persian Gulf, along the coasts of Pakistan, India, and Sri Lanka.

Hydrophis mamillaris (Daudin, 1803) (Plate 158)
(Broad-band sea snake: Patta dar samundri saamp)
1803 *Anguis mamillaris* Daudin, Hist. Nat. Rept.7:340.
Type locality: Vizagapatam, peninsular India.
Diagnosis:
1. Temporals variable, usually 2–3 imbricate scales.
2. Supralabials 7, second in contact with prefrontal, third and fourth touching eye; usually a small "cuneate" scale between third and fourth infralabial.
3. Scales at midbody 39–43, those of posterior part of body, hexagonal, with a central tubercle or short keel, juxtaposed, or feebly imbricate.
4. Ventral 305–384, bicarinate.
Snout-vent length 743–750 mm, tail 63–70 mm.
Color: Dorsum yellowish or grayish, with 43–57 broad crossbars, about twice as broad as their interspaces, usually connected along ventrals. Head entirely black with a yellow streak on temporal region.
Distribution: From the Karachi coast, along peninsular India.

Hydrophis ornatus (Gray, 1842) (Plate 159)
(Reef sea snake: Monga samundri saamp)
1842 *Auteria ornata* Gray, Zool. Mis.:61.
Type locality: Indian Ocean.

Chapter 7. Ophidia: Snakes

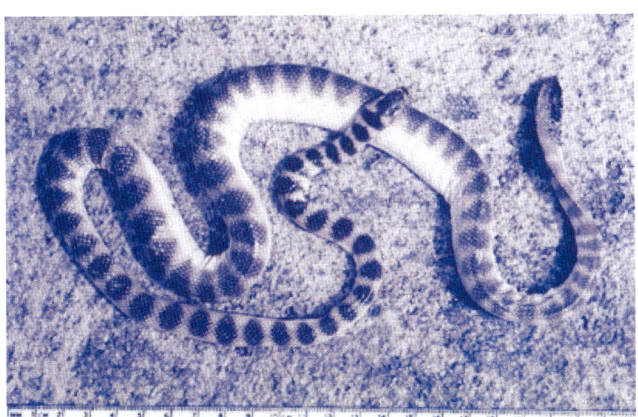

Plate 159. *Hydrophis ornatus.* (Photo courtesy of Harold K. Voris).

Plate 160. *Hydrophis spiralis.*

Diagnosis:
1. Head large, its breadth between eyes half or more than half its length; robust body, not markedly elongated, diameter of its posterior half about twice that of neck.
2. Supralabials 7–8, second in contact with prefrontal, third and fourth in eye.
3. A pair of overlapping anterior temporals.
4. Anterior genials well developed, in contact with each other, while posterior pair not well defined, separated from each other by small scales.
5. Body scales slightly imbricate or juxtaposed, with a central tubercle or short keel.
6. Rows at midbody 33–55.
7. Ventrals 209–312, distinct.
Snout-vent length 850–855 mm, tail 107–115 mm.

Color: Dorsum grayish, olivaceus, or white, with dark bars or rhomboidal spots. Ventrum yellowish or whitish. Head olivaceus.

Distribution: It ranges from the Persian Gulf, along coastal Karachi, peninsular India, Bangladesh, Sri Lanka, Siam, and New Guinea.

Hydrophis spiralis (Shaw, 1802) (Plate 160)
(Yellow sea snake: Peela samundri saamp)
1802 *Hydrus spiralis* Shaw, Gen. Zool, 3:564.
Type locality: Indian Ocean.
Diagnosis:
1. Head slightly distinct from neck, body slender and moderately compressed.
2. Supralabials 6–8, third and fourth or third to fifth in eye.
3. Anterior and posterior genials well developed and in contact with each other.
4. Scales feebly imbricate, smooth or with a small tubercle or short keel.
5. Ventrals 340–350, distinct, twice as broad as adjacent scales.
6. Considerably enlarged preanals 2–5.
Longest sea snake, snout-vent length 1896–1920 mm, tail 135–145 mm.

Color: Dorsum golden yellow to yellowish green, scales with dark borders, with 35–54 dark stripes which are narrower than interspaces, and are dark spotted. Flanks and ventrum of body pinkish white. Head in adult entirely yellow, in young blackish, with a yellow horseshoe-shaped mark.

Natural history notes: It is a deep-sea snake, not venturing into estuarine waters. A gravid female contained 7 embryos in uteri.

Distribution: It ranges from the Persian Gulf along coastal Pakistan and India, Sri Lanka, Bangladesh, Myanmar to Celebes and the Philippines.

Genus *Lapemis* Gray, 1835
Body short and stout; head large, with large scales, parietals divided; body scales squarish or hexagonal, juxtaposed, the lower most 3–4 rows on each side are larger than the others, with a central keel or tubercle, which in adult male becomes strongly spinose; ventrals small, usually distinct anteriorly, vestigial or absent posteriorly.

A single species is known from Pakistan coastal waters.

Lapemis curtus **(Shaw, 1802)**
(Pygmy sea snake: Bauna samundri saamp)
1802 *Hydrus curtus* Shaw, Gen. Zool. 3: 562.
Type locality: Unknown.
Diagnosis:
1. Body short, laterally strongly compressed, head short, slightly distinct from neck.
2. Rostral wider than high, trifid.
3. Single pre- and postocular, parietals broken into 5–6 small scales.
4. Nasal suture touching second supralabial. 6–7 supralabials, fourth touching eye; 9 infralabials.
5. Anterior genials small, not in contact with each other, posterior not differentiated.
6. Body scales smooth or with a central tubercle, male with short spine on lowest row, 35–39 at midbody.
7. Ventrals 165–180, anterior distinct, wider, posterior narrower, a pair of enlarged preanals.
Snout-vent length 625–630 mm, tail 50–55 mm.

Color: Dorsum pale olive, becoming white or pale yellow on sides. Black or dark olive crossbars 44–55 forming a median zigzag pattern, fading on sides. Head dark gray or olive; distal half of tail black.

Distribution: This snake has been recorded from the Persian Gulf, along the Makran and Karachi coasts, peninsular India, Sri Lanka to Madras.

Genus *Microcephalophis* Lesson, 1834

Head very small, elongate, snout projecting beyond the lower jaw; anteriorly body is long and very slender, much compressed 3–5 times that of the neck posteriorly. Head with large scales; scalearound the thickest part of body hexagonal, juxtaposed, in 29 to 48 rows. Ventrals entire anteriorly, posteriorly more or less divided, two halves of the ventrals alternating with each other; anals 2, not much enlarged.

I follow Gasperetti (1988) in recognizing the genus *Microcephalophis* rather than McDowell (1972), who considered it to be congeneric with *Hydrophis*. The two species are so distinctive as to be instantly separable from species assigned here to *Hydrophis*.

Two species are known from Pakistan.

Microcephalophis cantoris **(Günther, 1864)**
(Spotted small-head sea snake: Chittra daola)
1864 *Microcephalophis cantoris* Günther, Rept. Brit. Ind.:376.
Type locality: Penang, China.
Diagnosis:
1. Head very small, narrow, not distinct from long cylindrical neck.
2. Body stout, laterally compressed.
3. Single preocular, postocular, and temporals.
4. Supralabials 5–6, second and third or only the third in contact with the prefrontal, 7–8 infralabials.
5. Genials distinct, subequal
6. Ventrals 404–468.
Total length 1450–1800 mm, tail 83–90 mm.

Color: Anterior half of the body light olive to yellow above, pale below, with gray to dark dorsal bars, and a black midventral stripe. While posterior half dark olive above, laterally yellowish, with faint lateral bars.

Distribution: It ranges from Karachi to Cannore, and on the east from Orissa to Chittagong. Reports from Penang, China, have not been subsequently confirmed.

Microcephalophis gracilis **(Shaw, 1802) (Plate 161)**
(Banded small-head sea snake: Patta daola samundri saamp)
1802 *Hydrus gracilis* Shaw, Gen. Zool., 3: 560.
Type locality: Unknown.
Diagnosis:
1. Second supralabial in contact with prefrontal.
2. Third supralabial not in contact with the prefrontal.
3. Ventrals 220–350.
Total length 950–1020 mm, tail 80–96 mm.

Color: Head, chin, and throat region black to dark olive, with white to pale yellow spots or crossbars. Posteriorly pale yellow to greenish white with gray crossbars, or more or less uniformly gray above and light laterally and ventrally. Ventrum pale. Juveniles dark, with 40–60 light bars or paired spots.

Distribution: It ranges from Persian Gulf, around the coast of Pakistan, India to China, Hong Kong, Borneo, and northern Australia.

Chapter 7. Ophidia: Snakes

Plate 161. *Microcephalophis gracilis.* (Photo courtesy of S.A. Minton).

Plate 162. *Pelamis platurus.* (Photo courtesy of S.A. Minton).

Genus *Pelamis* Daudin, 1803

Long sea snakes with elongated head, indistinct from neck; head scales large, normally arranged; nostrils dorsal, nasal scales in contact with each other; temporals small; body scales hexagonal or squarish, juxtaposed, 49 to 67 around the thickest part of the body; ventrals very small, indistinguishable; anals 3–5, moderately enlarged.

Pelamis platurus (Linnaeus, 1766) (Plate 162)
(Pelagic sea snake: Azad samundri saamp)
1766 *Anguis platura* Linnaeus, Syst. Nat. 1(12): 391.
Type locality: Unknown.
Diagnosis:
1. Head long, narrow, distinct from neck which is not slender, body markedly compressed laterally.
2. One pre and 2–3 postoculars; anterior temporals 2–4.
3. Supralabials 6–8, with 1–3 small intercalated scales, fourth in contact with eye or separated from it by a subocular; 9–13 infralabials.
4. Three subequal pairs of genials widely separated from each other.
5. Midbody scales 45–55, juxtaposed, quadrangular, smooth, in male lowest row with minute tubercles, anals 3–5, moderately enlarged.
6. Ventrals 47–63, indistinguishable from adjacent scales.

Snout-vent length 880–900 mm, tail 83–100 mm.
Color: Dorsum light yellow to cream, ventrum pale, dorsum with 10–19-scale thick brown stripes. Head brownish. Tail with black and white bars or network.
Natural history notes: It is a very pelagic snake, rapid, graceful and an agile swimmer. It is often cast onto sandy beaches, where it dies helplessly.
Distribution: It ranges from the Persian Gulf westward to the Cape of Good Hope; eastward from coastal Pakistan to New Zealand, then northward to Possiet Bay, Kamchatka. In American waters it has been recorded from the Gulf of California to Ecuador.

Genus *Praescutata* Wall, 1921

Head depressed, distinct from the neck, snout broadly rounded; head shields normal and regularly arranged; naris dorsal, nasal scales in contact with each other, nasal cleft extends to the first supralabial and dividing it; prefrontals not in contact with supralabials; body scales more or less hexagonal, juxtaposed, in 37 to 50 rows on the thickest part of the body; ventrals distinct; anals enlarged.

Single species known from Pakistan.

Praescutata viperina (Schmidt, 1852)
(Spotted viperine sea snake: Samundri afi)
1852 *Thalassophis viperina* Ph. Schmidt, Abh. Naturwiss. Ver. Hamburg 2:79.
Type locality: Java.

Diagnosis:
1. Head short, wide, distinct from neck, body moderately thick, posteriorly laterally compressed, rostral with three short projections fitting into chin grooves.
2. One, rarely 2 preoculars, 2, occasionally 1, postocular.
3. Two anterior, 3–4 posterior temporals.
4. Supralabials 6–7, third or fourth, or both touching eye.
5. Infralabials 7–8.
6. Genials distinct, subequal, not touching each other.
7. Midbody scales 36–50, smooth, or with short keel, juxtaposed.
8. Ventrals 233–280, anterior-most about half the width of neck, decreasing in size, until at vent about the size of adjacent scales, 4–5 enlarged preanals.

Snout-vent length 642–650 mm, tail 74–80 mm.

Color: Dorsum greenish white, with a median series of 24–34 rhomboidal blotches, fused at midline. Head dark, labials and neck whitish, tail black.

Natural history notes: This species has been recorded from tidal creeks, several miles from the open sea.

Distribution: It ranges from Persian Gulf along Makran coast, around coastal India to southern China, thence to Borneo and Java.

Family Viperidae

Medium-sized snakes; head broad, distinct from neck, with small keeled scales, eyes are set forward nearer the tip of snout rather than corners of the mouth; body more or less stocky; tail short, tapering abruptly; pupil vertically elliptical. A pair of curved, foldable long fangs on each side of the upper jaw is present, with a completely closed channel inside for the transmission of venom. The fangs are enclosed in a membrane. When the snake opens its mouth to strike, the fangs come out of the membrane and are erected into vertical position; the body scales are strongly keeled, keels knobbed or serrated; no loreal pit; supralabials separated from orbit by more than two rows of scales.

Five genera of viperid snakes are found in Pakistan.

Genus *Daboia* Gray, 1842

Large, heavy-bodied snakes, with high, elongated head, long, stocky body, and a short tail; the large, crescent-shaped nostril is lined dorsally by a similar shaped elongated supranasal scale. Obst (1983) revived the genus *Daboia* Gray, 1842 assigning *Daboia russelii* to it (Wüster, 1998b).

A single species is found in Pakistan.

Daboia russelii russelii (Shaw and Nodder, 1797) (Plates 163A, B)

(Russell's chain viper: Koriala afi)

1797 *Coluber russelii* Shaw and Nodder, Vivar. Nat. or Nat. Mis. 8:291.

Type locality: Coromandel Coast, southern India.

Diagnosis:
1. Head longer than broad, distinct from neck, body stout, dorsoventrally flattened.
2. Nostril large, crescentic in shape, in a large nasal scale.
3. Supraocular scale undivided (Figure 31D).
4. Supralabials 11 to 12, separated from eye by 3–4 rows of small scales; 13–15 infralabials.
5. Anterior genial short and wide, posterior not differentiated.
6. Dorsals strongly keeled in 27–33 rows at midbody.

Plate 163A. *Daboia russelii russelii.*

Chapter 7. Ophidia: Snakes

Plate 163B. *Daboia russelii russelii*: Defense posture.

7. Ventrals 164–178, subcaudals 46–58, anal not divided.
Snout-vent length 1080–1090 mm, tail 220–225 mm.

Color: Dorsum light yellowish brown to sandy, with a median series of 22–32, large oblong chestnut blotches, bordered with black or dark brown, narrowly edged with cream, arranged lengthwise, mostly confluent with each other to form a middorsal chain. A lateral series of similar but smaller spots which on lower side are with scattered dark flecks with lighter edges. A pair of dark spots at base of head. A light, V-shaped canthal mark, its tip directed toward the snout. Labials mottled with brown and cream, ventrum pinkish white with curved dark spots. Throat white (Wüster, 1998b).

Natural history notes: The chain viper inhabits mesic habitat of grass fields with marginal trees, salty scrubby areas, along marginal growths of water bodies, hedges and fences along gardens and gallery forests.

It is nocturnal, becoming active just after sunset. It usually keeps to the shade of bushes and grass growing along paths where, due to its characteristic body color and pattern, it is difficult to locate. Sensing danger, the snake hisses loudly and throws itself into a defensive posture (Plate 163B). Generally the snake is heavy and sluggish with a placid temperament; it indicates its resentment by a loud hiss, occasionally striking.

It feeds on warm-blooded mammals like rats, mice, and birds. The prey is stalked, bitten, and released; within no time it is helpless, then devoured.

It breeds from April to July; gravid females are collected up to September, and 20–25 juveniles are born (Khan, 1995).

Distribution: Recorded from throughout the Indus Valley, from Karachi to Rawalpindi, at low altitudes. It is reported from throughout India to Bangladesh, extends into peninsular India and Sri Lanka. It is distinguished in several races throughout Southeast Asia to Taiwan (Wüster, 1998b).

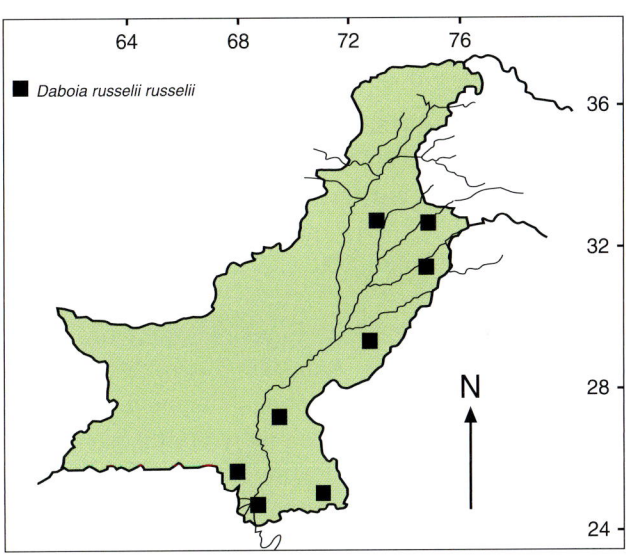

Genus *Echis* Merrem, 1820

Squarish short, flat head, very distinct from neck; short stocky body; short, abruptly narrowing tail; anteriorly placed small, dorsolateral nostrils; body scales strongly keeled, in 24–27 rows at midbody, lateral body scales smaller, oblique, with serrated keels; subcaudals and anal single.

Echis carinatus is very widely distributed species: from northern Africa to India and Seri Lanka. *Echis* has been split into 12 species (20 taxa) in 3 subgenera (Cherlin, 1984, 1990; Cherlin and Borkin, 1990), 3 subspecies of which occur in Pakistan (Khan, 1993d, 2002b).

Echis carinatus (Schneider, 1820)
(Saw scale viper: Khappra saamp)
1801 *Pseudoboa carinata* Schneider, Hist. Amphib. ii.: 285.

Type locality: Arni, near Madras, India.
Diagnosis:
1. Head short, distinctly wider than neck, covered with small strongly keeled imbricate scales, 8–12 between supraoculars.
2. Nostrils small, dorsolateral.
3. Supralabials 8–12, separated from ocular by 1 or 2 rows of small scales; 10–13 infralabials (Figure 31, B ii).
4. Anterior genial broad, followed by 3 pairs of slightly enlarged scales.
5. Dorsals at midbody in 24–37 rows, strongly keeled, median dorsals large, arranged in straight lines, while 3–7 rows of laterals smaller, oblique, pointing downward, with serrated keels (Figure 31Bi).
6. Ventrals 132–188, subcaudals 23–39, entire; anal single.
7. Body stocky, tail very short and abruptly tapering from vent

Snout-vent length 600–625 mm, tail length 51–56 mm.

Color: Dorsum light brown to yellowish brown, sometimes olive brown. A median row of 28–37 dark-edged, whitish blotches. Flank with undulating white line, dorsal loops of which are more prominent, but are diluted ventrally. Head with light arrowhead mark, posterior three prongs of which extend to considerable distance. A pale stripe from eye to angle of mouth. Labials light, dotted. Ventrum white to light pinkish, with fine dark gray spots.

Natural history notes: The saw-scale viper has been recorded from sandy and rocky alluvial habitat, with sparse xerophytic to moderately dense grass and scrub vegetation. In mountainous habitat it lives under rock blocks, while in submountainous regions it inhabits hedges and other scrubby vegetation, noticeably avoiding marshy areas and very dense vegetation.

The saw-scale viper is nocturnal; comes out just after sunset and rests close to the roots of a shrub, waiting for its prey. As it nears, the snake lunges at it furiously and bites it. Its diet includes common amphibians, *Bufo stomaticus*, young *Hoplobatrachus tigerinus*, *Sphaerotheca breviceps*, sand lizards belonging to genera *Crossobamon*, *Acanthodactylus,* and *Ophiomorus*, small snakes, birds, eggs, nestlings, and arthropods. Vyas (1998) reports to have fed captive snakes on young *Rattus rattus, Mus musculus, Hemidactylus flaviviridis, Hemidactylus brookii, Mabuya carinatus,* and *Calotes versicolor*.

The saw-scale viper is known to be bad tempered; it hisses loudly and goes on savagely striking at an intruder, often pursuing it for some distance. It causes fear in pedestrians as it follows them for some distance, while making characteristic rustling noise, and often taking a fatal bite. It is locally believed to jump 10–20 cm in the air, in order to attack its adversary! It climbs in branches of low bushes to avoid heat at the ground, and invade nests. When on the defensive, it throws itself into characteristic figure-8-shaped loops which work against each other and when mixed with loud hissing produce the characteristic rustling noise. A snake in this position is said to be dissolving its venom and is thought to be very venomous at that time.

Breeds from mid-February to late April. All races known to occur in Pakistan are viviparous, giving birth to 6–28 young, while oviparous races lay almost as many hard-shelled white eggs. Vyas (1998) records clutch sizes of 10–19 in captive snakes. Juveniles are met with from March to July. In northern areas of Pakistan this snake hibernates during winter. It basks on bright clear days during winter, while in warmer southern areas it is not known to hibernate.

Distribution: Saw-scale vipers are widely distributed from northern Africa, through the Middle East, southern former Russia, descending to Iran, Afghanistan, most of Pakistan excluding the high northern mountains, India, and Sri Lanka.

Echis carinatus has been distinguished in more than 10 supspecies, of which the African is now recognized as species. At least the following three supspecies are represented in Pakistan (Constable, 1949; Cherlin, 1983; Khan, 1993d), which Auffenberg and Rehman (1991) and Khan (2002c) do not validate.

Echis carinatus astolae Mertens, 1969
(Dark-blotched saw-scale viper: Astola khappra saamp)
1969 *Echis carinatus astolae* Mertens, Stuttg. Beitr. Naturk. 216: 3–4.

Chapter 7. Ophidia: Snakes

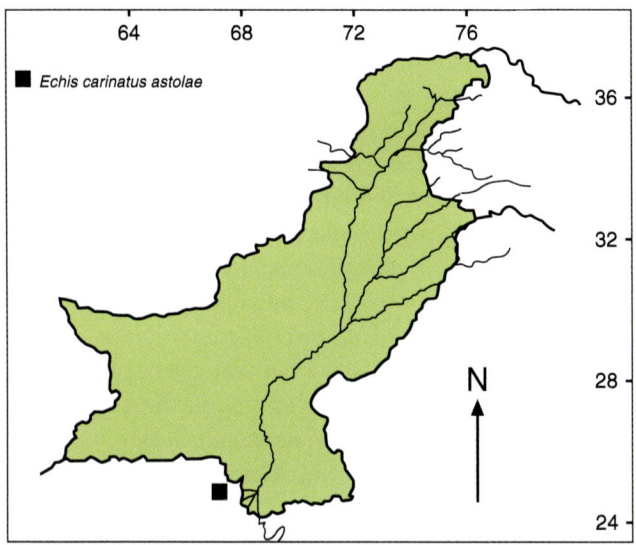

Type locality: Astola Island, 25 km SE off the Pasni coast, Baluchistan.

Diagnosis:
1. Dorsum whitish.
2. Dorsal pattern of dark brown blotches, with 1–3 lateral longitudinal rows of dark brown spots.
3. No lateral light arcs.
4. Head with 3 pronged light mark directed toward snout.
5. A light stripe from temporals, on each side, meeting each other at frontal region giving a branch to snout.

Distribution: This race has been described from Astola Island, 25 km southeast off the Pasni coast, Baluchistan, Pakistan.

Echis carinatus multisquamatus Cherlin, 1981 (Plate 164)
(Waziristan saw-scale viper: Waziristan khappra saamp)
1981 *Echis carinatus multisquamatus* Cherlin, Proc. Zool. Inst. Acad. Sci. 101:92–95.
Type locality: Bayram-Ali, Marysk area, Turkmenia.
Diagnosis:
1. Head mark is always cross-shaped.
2. Lateral continuous undulating white line present.
3. Narrow transverse white bands on middorsum.
4. Midbody scale counts 34–40 (highest of all subspecies).
5. Number of ventrals 169–199 (highest of all subspecies).

Plate 164. *Echis carinatus multisquamatus:* Strike posture.

Distribution: From the Caspian Sea, through Turkmenistan, into Bukhara, Samarkand and Tashkent to Tajikistan along the Afghan border, up to Hindu Kush, northeastern Baluchistan and eastern Iran.

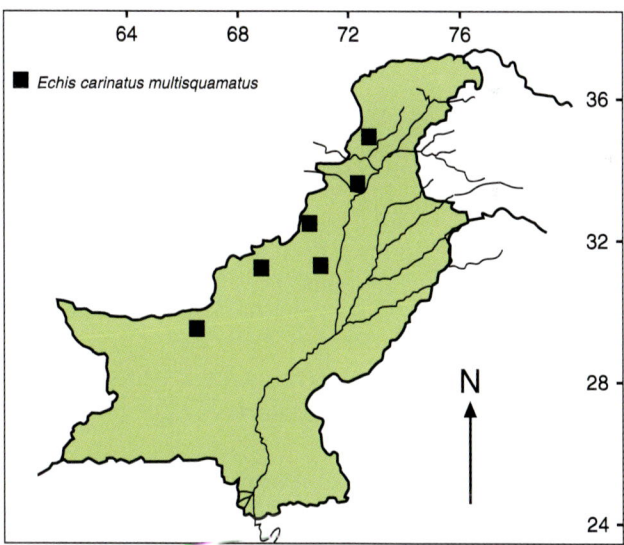

Echis carinatus sochureki Stemmler, 1964 (Plate 165)
(Sind Valley saw-scale viper: Sindi khappra saamp)
1969 *Echis carinatus sochureki* Stemmler, Aquaterra 6 (10):118–125.
Type locality: Band Kushdil Khan, Peshin, Baluchistan.
Diagnosis: It is the widely distributed race of saw-scale viper in Pakistan, distinguished from the rest of the races by the following characteristics:

Plate 165. *Echis carinatus sochureki:* Defense posture.

1. Head scales small, except 3–4 larger supraoculars.
2. Midbody scale rows 29–33.
3. Middorsals flat.
4. Oviparous.

Color: Dorsum tan, grayish, or brown, with a median row of 30 whitish (never yellowish) blotches with dark brown edges. Body sides with wide V-shaped ventrally open marks and with distinct dark spots. Ventrum whitish with dark gray spots. Head with light arrowhead, directed toward snout. A light loreal stripe extending to angle of mouth.

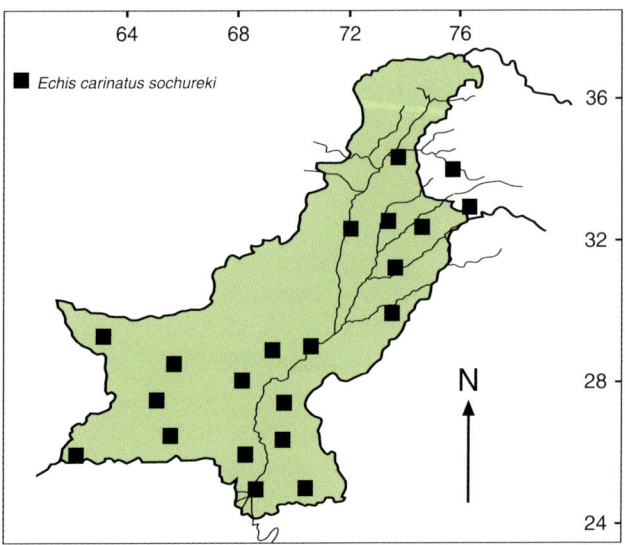

Distribution: It ranges to northern India, the whole of Pakistan except high mountains in north, southern Afghanistan, central Iran to the Iranian Gulf coast and Khuzestan.

Genus *Eristicophis* Alcock and Finn, 1896

Medium-sized viperid snakes; head elongate, more or less triangular, flat; stocky body; rostral scale much broader than high, crescentic and deeply concave, surmounted on either side by a large wing-like (butterfly) scale; nasal scales separated from the rostral by an enlarged scale; body scales arranged in straight rows, laterals slightly oblique, strongly keeled; ventrals with lateral keel; tail short prehensile, with paired subcaudals, anal entire.

A monotypic genus of snakes which are characteristic of Baluchistan desert.

Eristicophis macmahonii Alcock and Finn, 1897 (Plates 166A, B)

(Leaf-nose viper: Titli afi)

1897 *Eristicophis macmahonii* Alcock and Finn, J. Asiatic Soc. Bengal, Calcutta n.s. 65:564.

Type locality: Desert south of Helmand, Baluchistan, Pakistan.

Diagnosis:

1. Rostral much broader than high, crescentic and deeply concave, surmounted on each side by a wing-like, free edged, broad "butterfly" scale (Figure 31A).
2. Small, circumocular scales 16–25.
3. Supralabials 15–16, forming serrated dorsal lip, separated from eye by 3 rows of scales, about twice as large as those of the ocular ring.
4. Scales at midbody 23–29, arranged in straight regular rings.
6. Ventrals 140–145, ventrals with a keel on each side; anal undivided.

Snout-vent length 655–660 mm, tail length 72–78 mm.

Color: Dorsum light reddish brown to khaki, with a series of small dark brown lateral spots, each surrounded in its upper half by light dots. A thin light stripe from eye to angle of mouth. Base of tail with brown crossbars. Ventrum white.

Natural history notes: This snake is morphologically adapted to live in the fine loose sand of shifting dunes.

Chapter 7. Ophidia: Snakes

Plate 166A. *Eristicophis macmahonii.*

Its habitat is without any mentionable vegetation, except for very sparse growth of stunted bushes and grasses. The snake rapidly buries itself in sand by peculiar rocking and peristaltic movements of its body. As the snake disappears under the sand, it shakes its head peculiarly to free it from sand particles. Burying in sand is an escape as well as a defensive behavior; it also ambushes its prey in this fashion (Mertens, 1965).

It is a nocturnal, alert, and bad tempered snake. When alarmed, it throws itself into coils, which lie on top of each other, thus elevating its head considerably above ground, the neck is thrown into an S-shaped coil, the eyes are keenly focused on victim, and the snake is ready to attack. An attacking snake hisses loudly and strikes vigorously (Minton, 1966).

Its natural food consists of sand lizards and arthropods. The snake stays buried in sand, with only its eyes and nares exposed. As soon as prey approaches, it strikes, retaining its hold until the prey is almost dead (Tubex, pers. comm., 1993).

It breeds from March to May; 8–10 elongated, soft shelled eggs, 24–34 by 14–18 mm in measurement, are laid in pits excavated in sand (Tubex, pers. comm., 1993).

Distribution: So far, this snake has been recorded from Seistan in the extreme east of Iran into Afghanistan south of the Helmand River and southwestern Baluchistan, between the Chagai Hills and the Siahn Range, east to Nushki.

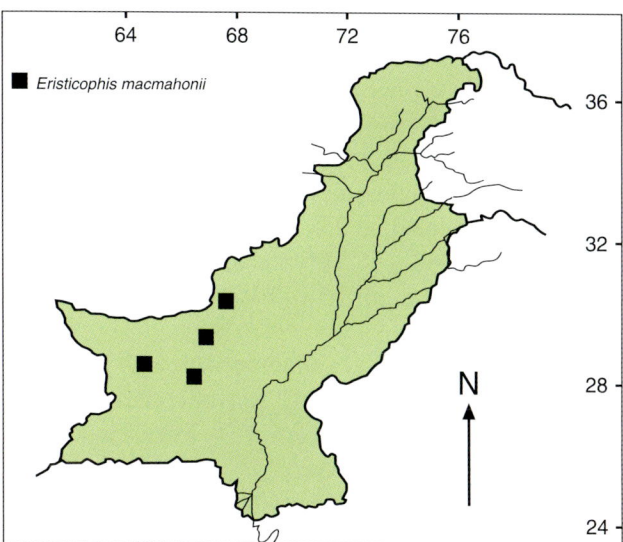

Genus *Macrovipera* Reuss, 1927

The head is short and wide; the supranasal scale not crescentic, nostrils lateral, in a large scale; the supraocular is divided; ventrals rounded, subcaudals divided, anal entire.

A single species is found in Pakistan.

Macrovipera lebetina obtusa (Dwigubsky, 1832) (Plate 167)

(Levantine viper: Roomi afi)

1832 *Vipera obtusa* Dwigubsky, Opyt estetv. Istorii 3:30.

Type locality: Jelisawetpol, Transcaucasia.
Diagnosis:
1. Head short and wide.
2. Supraoculars usually divided.

Plate 166B. *Eristicophis macmahonii:* Snout region.

Plate 167. *Macrovipera lebetina*. (Photo courtesy of S.A. Minton).

3. Supralabials 10 to 11, separated from eye by 2 to 3 rows of small scales, 12–14 infralabials.
4. Dorsals keeled, in 23–25 rows at midbody.
5. Ventrals 168–176, subcaudals 42–49.
   Snout-vent length 1040–1045 mm, tail length 143–148 mm.

Color: Dorsum khaki to yellow brown, with minute spots. A median row of 39–42 indistinct dark gray crossbars. Ventrum buff, anteriorly whiter, clouded with gray. Tail pinkish brown.

Natural history notes: This snake inhabits stony semiarid country with sparse scrubby vegetation, between 1000 and 2500 m of elevation.

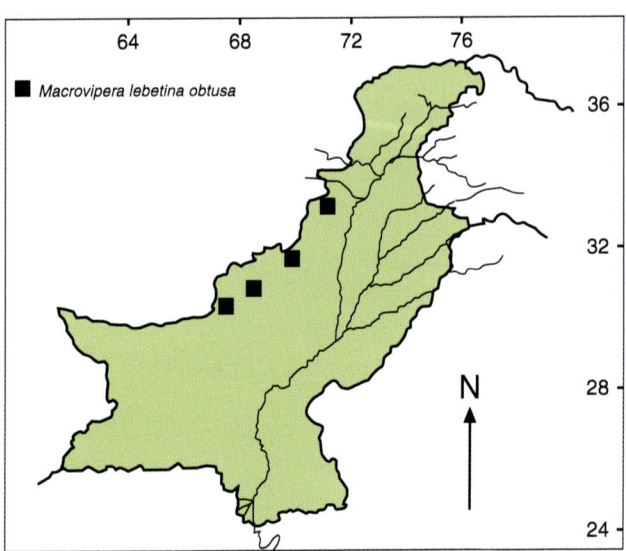

Distribution: Its range extends from Caucasus to Lebanon, through southern Turkmenistan and Uzbekistan. In Pakistan its records are from the Waziristan Hills, Quetta highlands, and Chitral.

Genus *Pseudocerastes* Boulenger, 1896

Typical viperid stocky body, short tail; dorsals keeled, keels with lateral fine striations ending in a knob in the posterior part of the scale; several scales in the supraorbital region pile up to form a horn-like structure; supralabials with serrated margin; nostrils directed outward and upward, are in a large undivided nasal scale.

Two species are known in Pakistan.

Pseudocerastes bicornis **Wall, 1913**
(Two horned viper: Du singha afee)
1913 *Pseudocerastes bicornis* Wall, Pois. Sn. Ind.: 64–65.
Type locality: Khajuri Kach, above Gwaleri kolal, Gomal Pass, Waziristan, Pakistan.
General characters: All that remains of the type specimen of this viper is preserved in the British Museum of Natural History in London, under registry number 1946.1.20.82, a viperid head with 2–3 head lengths of the anterior part of the body, up to 31 ventral. The rest of the body apparently is lost. The two elongated, free supraocular scales of *P. bicornis*, were erected in life as a pair of supraocular horns on each side.
Distribution: Reported only from its type locality.

Pseudocerastes persicus **(Duméril, Bibron, and Duméril, 1854) (Plate 168)**
(Persian horned viper: Irani seengh-wala afi)
1854 *Cerastes persicus* Duméril, Bibron and Duméril, Erpet. Gen. 7:1443.
Type locality: Persia.
Diagnosis:
1. A supraorbital horn on each side formed by several scales.
2. Body moderately stout, dorsoventrally flattened.
3. Circumocular scales 16 to 18 (Figure 31C).
4. Supralabials 12 to 14, separated from eye by 3 rows of small scales; 14–17 infralabials.
5. Anterior genial large, posterior not differentiated from surrounding scales.
6. Dorsals thin, weakly keeled, 23–25 rows at

Chapter 7. Ophidia: Snakes

Plate 168. *Pseudocerastes persicus*. (Photo courtesy of S.A. Minton).

midbody, laterals with a nodular prominence at posterior end.
7. Ventrals without lateral keels, 146–151, subcaudals 42–46.

Snout-vent length 688–690 mm, tail 80–85 mm.

Color: Dorsum pale gray or bluish gray to khaki. A median series of 28–33 rectangular gray or brownish blotches or cross bands, much narrower than interspaces. An alternating series of faint spots on sides. A dark band from nostril to the angle of jaw, lower edges of labials white. Belly white; in larger snakes the tip of tail dark, medially pinkish.

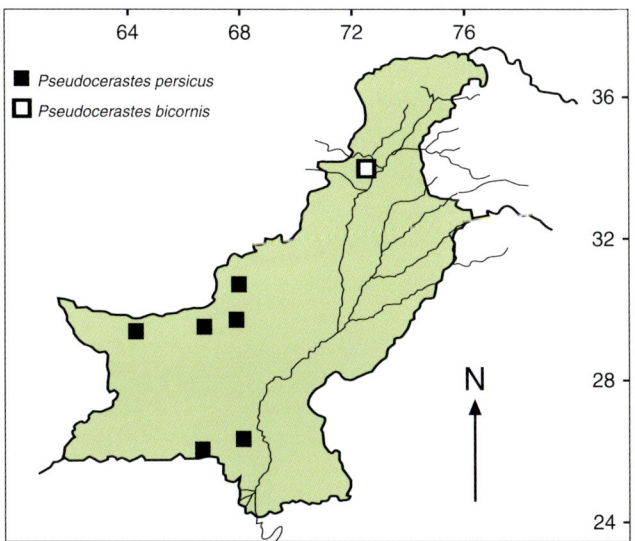

Natural history notes: This viper inhabits sandy and rocky terrain with moderate vegetation of bushes and grasses. It is nocturnal in habits, retreats during the day into crevices among rocks or burrows in comparatively hard soil near roots of bushes. Its diet consists of desericolous lizards, mice, and arthropods.

It breeds from March to July.

Distribution: This snake has been collected up to 2200 m of elevation. It has been recorded from Azerbaijan and northern Iraq through Iran south to Persian Gulf, eastward to central Afghanistan and western Las Bela.

Family Crotalidae

Head long, depressed, with large, symmetrically arranged scales, sometimes snout scale broken in small scales; naris lateral; eye with vertically elliptical pupil; a loreal pit between eye and nostril; subcaudals divided, anal scale entire.

A single genus is represented in alpine Pakistan.

Genus *Gloydius* Hoge and Romano-Hoge, 1981

Head with large symmetrically arranged shields; a "pit" between eye and nostril; eye separated from supralabials; body scales strongly keeled.

A single species is known from alpine Pakistan.

Gloydius himalayanus (Günther, 1864) (Plate 169)
(Himalayan pit-viper: Hamaliai hafra afi)
1864 *Halys himalayanus* Günther, Rept. Brit. Ind.: 393.

Type locality: Garhval, western Himalayas, India.

Diagnosis:
1. Distinct head with large symmetrical scales.
2. A distinct pit between eye and nostril (Figures 32A and B).
3. A pair of pre- and postoculars, lower postocular long, separating supralabials from eye.
4. Supralabials 7, posterior 2 united with temporals to form large post-temporal scales; 9–10 infralabials.
5. Only a single short broad genial.
6. Body scales strongly keeled, 21 rows at midbody.
7. Ventrals 147–175, subcaudals 32–52 divided.

Plate 169. *Gloydius himalayanus.* (Photo courtesy of S. Dattari).

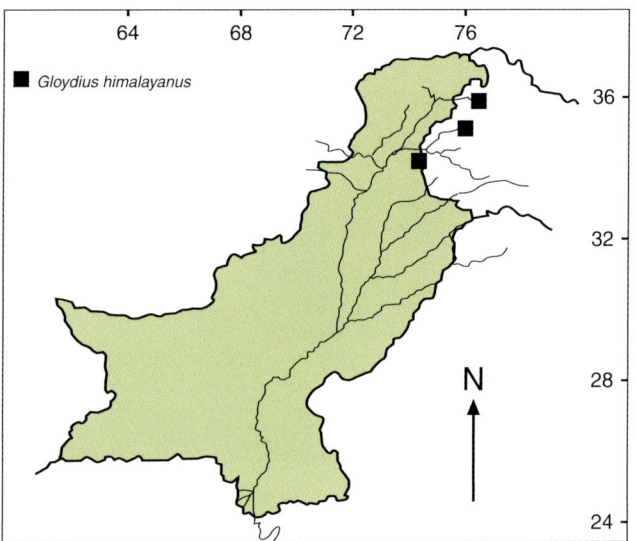

Snout-vent length 565–570 mm, tail 98 mm.
Color: Dorsum light brown-gray to dark brown. A median series of dark brown blotches, alternating with lateral row of spots. A broad dark band from eye to angle of mouth. Supralabials light with dark mottling. Ventrum light gray with dark clouding and fine spotting (Stoliczka 1870; Wall 1899; Khan and Tasnim 1986b).
Natural history notes: This mountain snake is restricted to an elevation of 1500 m; however, it has been reported from Dharmsala Glacier at 5000 m of elevation, the highest altitude for a snake (Wall, 1910; for discussion see Gloyd and Conant, 1990). It frequents rocky wooded hillsides where it lives in caverns and crevices in rocks. It takes refuge under leaf litter and fallen timber. It may be found in open fields, orchards, and hiding under marginal grass. It hibernates from October to April. Though nocturnal, it is known to bask on bright sunny days on hilly slopes. It usually selects a spot within easy reach of cover, to which it withdraws in a leisurely manner when disturbed since it is a sluggish slow-moving snake. When provoked the snake flattens its body; sometimes it coils up and vibrate its tail tip (Khan and Tasnim, 1986b).

It feeds mainly on skinks, other lizards, mice, millipedes, and centipedes. Adults often accumulate much fat, during summer body organs are packed with fat. Viviparous, it gives birth to 5–7 young from August to September (Khan and Tasnim, 1986b).

A bite by *Gloydius himalayanus* results in local intense pain and swelling subsiding within 2–3 days, having no dangerous effect, and needing no real treatment (Acton, 1921). In Kashmir local people do not fear this snake; rather they revere it (Dattatri, 1985).
Distribution: The Himalayan pit-viper is recorded from Nathia Gali in the western Himalayas and eastern NWFP, Pakistan (Khan and Tasnim, 1986b). I have seen a specimen in the Government College Natural History Museum, Lahore, Pakistan, from the Dosai Plains, Baltistan, northeastern Pakistan. The viper is known to be very common in the western Himalayas, less so beyond Nepal (Wall, 1910). However, its range extends from Sikkim to Chitral in the north, extending into eastern NWFP, Pakistan.

Chapter 8
Distribution and Affinities of Herpetofauna

Herpetological Survey

The complex diversity of habitat types of Pakistan (Chapter 9) is reflected in its varied herpetofauna. The number of species recorded from Pakistan has risen steadily since Minton's (1966) record of 144 species to the present 223 species (Table 8.1). The rapid increase in number of species is due to various collection reports which have accumulated over the years from different parts of the country, (Mertens, 1969a, 1970, 1971, 1974; Khan, 1974, 1977, 1979, 1980a, 1985a, 1986, 1988, 1989, 1991c, e, 1993a, d, 1997a, b, 1998, 1999b, e; Khan and Ahmed, 1987; Khan and Baig, 1988; Khan and Tasnim, 1989, 1990b; A.Q. Khan and M.S. Khan, 1996; Gvozdik and Radek, 1997).

There has been an increase in numbers in all major taxonomic groups (Table 8.1). In amphibians, the genus *Uperodon* has been added to the family Microhylidae (Baig and Gvozdik, 1998), and the snake genus *Sibynophis* has been added to the family Colubridae (A.Q. Khan and M.S. Khan, 1996). Similarly, several new taxa have been described in the family Bufonidae (Mertens, 1971; Eiselt and Schmidtler, 1973; Khan, 1997a, c, 1999c; Stock et al., 1999), and family Ranidae (Dubois and Khan, 1979, 1997; Khan and Tasnim, 1989). Several amphibians have been recorded for the first time from Pakistan (Mertens, 1969a; Khan, 1972a, 1974; Baig and Gvozdik, 1998). Of the reptiles, several new species, mostly geckos, have been described (Khan, 1980a, 1988, 1989, 1991d, f, 1992, 1993b, d, 1997a, b, c, 1998d; 2001b; Khan and Baig, 1992; Khan and Tasnim, 1990b; M.S. Khan and M.R.Z. Khan, 1997), and several doubtful taxa have been validated (Khan, 1989, 1992, 1994c, 1997f). New species and subspecies of snakes have been described (Mertens, 1969a, b; Khan, 1985a, 1999b, c; M.S. Khan and A.Q. Khan, 2000), and several doubtful species validated (Mertens, 1969a, b; Khan, 1984a, b, 1989), and added to the fauna of Pakistan (A.Q. Khan and M.S. Khan, 1996).

There are still several unconfirmed records from the territory now included in Pakistan (Table 8.2), the majority of which were reported by J.A. Murray (curator of Karachi Museum 1884–1892) from around Karachi, which is now the best herpetologically explored area within Pakistan (Minton, 1962, 1966; Mertens, 1969a). Murray's records are probably based on wrong identifications or on confused locality data.

Affinities of Herpetofauna

Sclater's (1858) Zoogeographical Regions are characterized by relatively homogeneous bird and mammalian faunae, resulting from long evolutionary interactions with changing ecological factors. Present worldwide distribution of amphibians and reptiles reflects the geohistorical process of breaking and fusion of land masses, changing climatic zones, as well as the rise and fall in sea levels (Wallace, 1876, 1881; Darlington, 1957; Voris, 2000).

The origin and history of the distribution of amphibians and reptiles in Pakistan and adjacent regions dates back to the Cretaceous, about 200 million years ago, when a huge landmass, Gondwana, fragmented. After breaking from Gondwana, the Indian Plate drifted northward across the Indian Ocean, sweeping along with it several small islands that fell in its way. Its collision with Eurasia resulted in the uplifting of trans-Himalayas: Hindu Kush, Karakorum, Himalaya, and the Pamir-Tien Shan ranges (Holmes, 1965; King, 1967; Courtillot and Vink, 1983). The sub-Himalayan mountain complex of Kohistan represents the crushed Indo-Pacific islands in eastern NWFP (Khan, 1980b; Buffetaut and Ingavat, 1985; Jaeger et al., 1989). The paleo-sutures were reactivated in the Eurasian landmass by the collision, resulting in the appearance of series of complex mountain systems, subdividing and genetically isolating populations of resident animals and plants (Axelrod, 1960; Sengör et al., 1988; Macy et al., 1999). The continued mountain building has resulted

Table 8.1. Increase in the Number of Species of Amphibians and Reptiles Reported From Pakistan, Since Minton (1966)

Source	Family	Genera	Species/Subspecies
Minton (1966):			
Salientia	2	2	7
Testudines	6	11	12
Crocodylia	2	2	2
Sauria	6	24	65
Serpentes	7	25	58
Mertens (1969a):			
Salientia	2	2	14
Testudines	5	11	14
Crocodylia	2	2	2
Sauria	6	26	82
Serpentes	8	31	66
Khan (1980b):			
Salientia	3	3	18
Testudines	5	12	15
Crocodylia	2	2	2
Sauria	6	28	88
Serpentes	8	34	72
Present work:			
Salientia	3	9	24
Testudines	5		5
Crocodylia	2	2	2
Sauria	8	35	103
Serpentes	8	36	79

in present day complex geomorphology and biodiversity in these regions (Anderson, 1999; Khan, 1997f, 1999b). Recent molecular studies on lizards: lacertids (Mayer and Benyr, 1994), geckos (Macey, et al., 1999a) and agamids (Macey et al., 2000) have supported the hypothesis of vicariance origin of the herpetofauna and its distribution in Eurasia and Southeast Asia.

The Himalayas have a profound effect on the geology, hydrology, and seasonal variations of the subcontinent: the wash-down debris settled to give rise to the vast Salt Range, Indo-Gangetic Plains, and regularly create a barometric vacuum over the subcontinent which brings annual spells of torrential monsoon rains. However, by the Pleistocene, the changes, brought about by the Himalayan barrier in climate, rise and fall in the sea level due to glaciation and deglaciation, created new pathways for the dispersal of eastern, western and southeast Asian fauna and flora to cross and colonise new areas across the subcontinent, and vice versa (Inger and Chin, 1962; Prater, 1965; Wadia, 1966; Khan, 1980b; Heaney, 1985, 1991; Voris, 2000).

By the Oligocene, extreme regions of temperature and precipitation developed in the south of the Indo-Gangetic plains. The dryness prevailing in

Table 8.2. List of Unconfirmed Species From Pakistan Reported in Literature

Species	Locality	Reference
AMPHIBIANS		
Bufo andersoni	All over Pakistan, except mountains	Minton, 1966
Rana ridibunda	Baluchistan	Boulenger, 1920
Sphaerotheca strachanii	Karachi	Murray, 1884
CHELONIANS		
Batagur fusca	Indus River	Murray, 1884
Batagur dhongoka	Indus River	Murray, 1884
Cabrita leschaultii	Indus River	Murray, 1884
Melanochelys trijuga	Indus River	Murray, 1884
Testudo leithii	Indus River	Murray, 1884
LIZARDS		
Alsophylax pipiens	Sind	Murray, 1884
Gymnodactylus brevipes	Southwestern Baluchistan	Murray, 1892
Gymnodactylus fedtschenkoi	Baluchistan	Smith, 1935
Mabuya aurata	Karachi	Murray, 1884
Pristuris rupestris	Karachi	Murray, 1884
Scincus arenarius	Karachi	Murray, 1884
Sitana ponticeriana	Thar Parker	Murray, 1884
SNAKES		
Cerberus rhynchops	Sind and Makran	Murray, 1886
Chersydrus granulatus	Sind	Murray, 1884
Coluber gracilis	Sind	Murray, 1884
C. florulentus	Sind	Wall, 1908
Dendrelaphis tristis	Sind	Smith, 1943
Elaphe helenae	Sind	Murray, 1884
Xenochrophis (Natrix) tessellata	Chitral	Wall, 1911
Ophiophagus hannah	Sind	Smith, 1943

western Eurasia extended onto the subcontinent. The subtropical temperate climate and low rainfall caused vast tracts of grasslands and the establishment of Cholistan and Thar deserts (Ahmed, 1951). Soon the North African, Central and Southeast Asian amphibians and reptiles became established in the present scenario of habitat types of Pakistan (Chapter 8).

In Pakistan, amphibians are represented by 23 species (10% of the total) comprising 11 ranids, ten bufonids, and two species of microhylid frogs. Frogs of the genus *Paa* and several high-altitude toads dominate in the Himalayas, while 8 wide-ranging plains amphibian species and 13 chelonians are mostly concentrated in the Indus Valley, along the riverine tracts in Punjab and Sind.

Lizards are dominant reptiles with 102 species (43%), while the number of snakes is 80 (36%). Of the lizards, geckos dominate with 38 species (39%), 28 species of agamids (30%), scincids 17%, and lacertids 12%, with single species of the families Eublepharidae and Chamaeleonidae. The 80 species of snakes (33% of the total) are dominated by colubrids (49%),

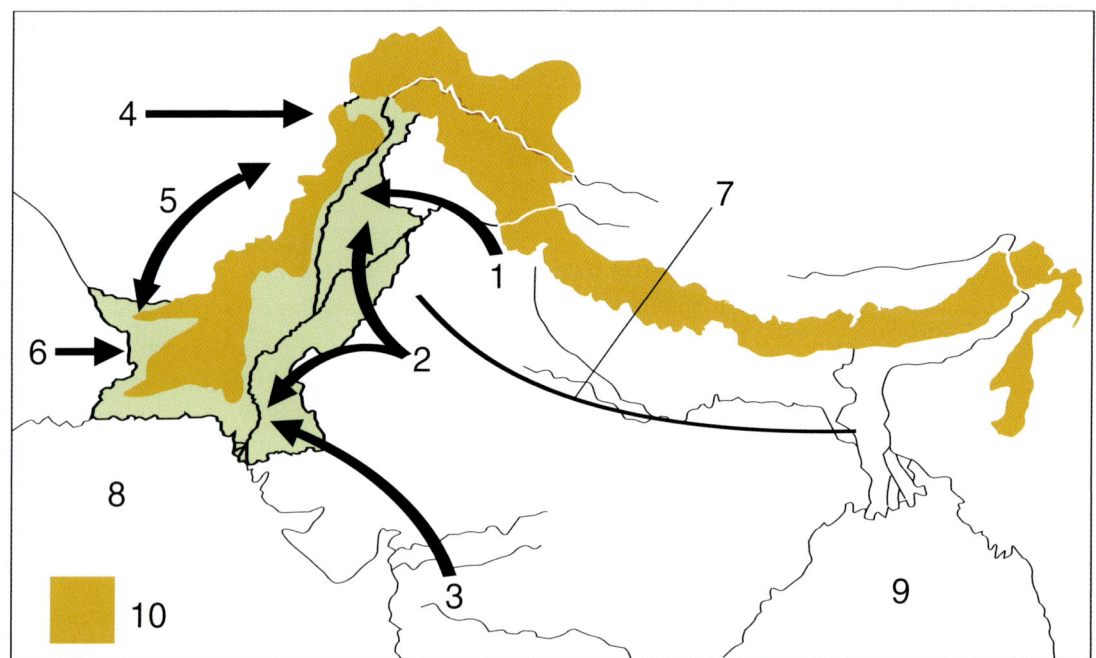

Figure 35. Pakistan, geographical position and faunal invasion routes (arrows):
1. Indo-Himalayan species; 2. Indo-Oriental species; 3. South Indian species; 4. Irano-Turanian species; 5. Seistanian species; 6. Saharo-Sindian species; 7. Indo-Gangetic Plain; 8. Arabian Sea; 9. Bay of Bengal; 10. land over 1000 m.

followed by 10 viperid species, 8 species of blind snakes and 4 of boids. From along the coastal strip, 14 sea snakes are reported.

Migration Routes and Distribution

Animal populations expand, contract, unite, break, and are isolated due to continuously occurring geo-ecological changes in ecosystems. The species flourish in suitable conditions for food and safety and expand, while those with maladaptive ecological variations perish or disperse as disjunct populations where suitable habitats are available. This phenomenon has repeatedly played a key role in the evolution and establishment of isolation in the phylogenetic history of species. New selective factors set up adaptive radiations in new situations, thus facilitating genetic isolations.

The Oriental region extends from the Indus Valley; it includes the Himalayas in the north and spreads across India to the southeast to include Indonesia and the Philippines (Sclater, 1858). After the formation of the Himalayas, as the ecobiological conditions became conducive in the subcontinent, the migration routes were established between neighboring regions. The northwestern part of the subcontinent (Pakistan) was invaded by several species from the west. Several species evolved here, some of them penetrated farther eastward into the Oriental region. Similarly, Oriental species invaded westward into the Palearctic region.

The issue of determination of boundaries between zoogeographical regions is complex. The range of certain species is taken into account to establish the boundaries. The boundary between the Oriental and the Palearctic regions is the longest; it is plotted through Pakistan, India, Nepal, and China (see Bobrov, 1997, for references), while Japan, Korea, Ryukyu, and Taiwan islands are included in the Palearctic Region (Voronov et al., 1985).

The Oriental species invaded westward through three corridors which have also been used by the Palearctic species to invade eastward (Figure 35). Roberts (1991) suggests similar invasion routes for birds across Pakistan. These routes are as follows.

The Siwalik Corridor

This invasion route ran westward along the humid foothills of the Himalayas. The Indo-Himalayan elements taking this westward route are *Euphlyctis cyanophlyctis*, *Hoplobatrachus tigerinus*, *Calotes versicolor*, *Laudakia agrorensis*, *Cyrtopodion watsoni*, *Eryx johnii*, *Xenochrophis piscator*, *Ptyas mucosus*, *Spalerosophis diadema*, *Bungarus caeruleus*, *Naja oxiana*, and *Gloydius himalayanus* (S. Anderson and Leviton, 1969; Clark et al., 1969; Kral, 1969; Clark, 1990). Palearctic forms taking this route eastward are *Bufo viridis*, *Pseudocyclophis persicus*, *Xenochrophis tessellata*, *Macrovipara lebetina*, and *Pseudocerastes persicus* (Khan, 1980b). The mesic-cold-tolerant Oriento-Gangetic species, *Bufo melanostictus*, and cyrtodactylid geckos differentiated into several forms are *Bufo himalayanus*, *Siwaligekko mintoni*, *Siwaligekko dattanensis*, and *Siwaligekko battalensis*. The wide-ranging Southeast Asian blind snake *Typhlops diardii*, has differentiated in its western race, *Typhlops diardii platyventris* (Khan, 1999a), while the eastern Himalayan genus *Typhlops* differentiated in several western forms as *Typhlops loveridgi*, *Typhlops madgemintonae*, *Typhlops madgemintonae shermanai*, *Typhlops ahsanuli*, and *Typhlops ductuliformes* (Khan, 1999d, e). The Himalayan amphibian genus *Paa* is represented in the "karez" channels in the western Himalayas by *Paa sternosignata*.

The Indus Valley Corridor

Recent laying of intricate canal systems in the Indus Valley has contributed to habitat suitability over the years so that several Southeast Asian frogs have extended westwards. The wide-ranging *Euphlyctis cyanophlyctis* differentiated in three races west of the Indus: *ehrenbergii*, *seistanica*, and *microspinulata* (Khan, 1997c). Other amphibian species are *Fejervarya limnocharis*, *Fejervarya syhadrensis*, and *Hoplobatrachus tigerinus*; geckos, *Hemidactylus brookii*, and *Hemidactylus flaviviridis*; snakes, *Ptyas mucosus*, *Amphiesma stolatum*, *Xenochrophis piscator*, *Boiga trigonata*, and *Daboia russelii*; Indian species of freshwater mud-turtles and frogs, *Bufo stomaticus* and *Sphaerotheca breviceps*; grass lizards of genera *Ophisops*, and *Novoeumeces* sand-lizards differentiated in *Novoeumeces indothalensis* and several races of *N. schneiderii*, and *Eutropis dissimilis*; garden lizard *Calotes versicolor*; yellow monitor lizard, *Varanus flavescens*, snakes, *Ramphotyphlops braminus*, *Eryx johnii*, several species of genera *Lycodon*, *Oligodon*, *Spalerosophis*, and *Bungarus*, while *Naja* has its wide-ranging representatve, *Naja oxiana*, in the west. The southern Indian *Sibynophis sagittarius* has recently been recorded from the Siwaliks (A.Q. Khan and M.S. Khan, 1996).

The Kach Coastal Corridor

The third penetration route is through the southeastern corner of the Thar Desert along the humid cooler coastal strip. It has been taken mostly by the following southern Indian species to penetrate westward: *Fejervarya syhadrensis*, *Crocodylus palustris*, *Gavialis gangeticus*, *Geochelone elegans*, *Chameleon zeylanicus*, *Hemidactylus leschenaultii*, *H. frenatus*, *H. triedrus*, *H. turcicus*, *Brachysaura minor*, *Chamaeleo zeylanicus*, *Eryx conicus*, *Python molurus*, *Lycodon travancoricus*, *Bungarus caeruleus*, *Naja naja*, and *Daboia russelii*.

The Palearctic region has significantly contributed to the fauna of Pakistan (Khan, 1980b; Roberts, 1977, 1991, 1992). The region embraces most of Europe, circum-Mediterranean countries, northern Africa, and the Near East, former Soviet Asia, Japan, the Koreas, and most of northern and central China. The Palearctic elements invaded eastward through western Baluchistan and along the Makran coastal strip. They spread out mostly along the western bank of the Indus River, from the Indus Delta all along the Kirthar, Sulaiman, and Hindu Kush Ranges, reaching the western Himalayas.

The Western Baluchistan Corridor

This invasion route was taken by the wide-ranging Palearctic forms: *Bufo viridis*, *Argrionemys horsfieldii*, *Cyrtopodion scabrum*, *Hemidactylus persicus*, lizards of genus *Eremias* and *Novoeumeces*, while *Pseudocyclophis persicus*, *Natrix tessellata*, *Pseudocerastes persicus*, and *Macrovipara lebetina* extended northward into the Safed Koh, Hindu Kush in the western Himalayas. Moreover, several Irano-Turanian and Seistanian genera, *Agamura*, *Bunopus*, *Teratoscincus*, *Ablepharus*, *Phrynocephalus*, and *Ophiomorus* invaded Baluchistan through this route.

The Makran Coast Corridor

The Ethiopian and Saharo-Sindian lizards of genus *Acanthodactylus* and snakes of genera *Echis*,

Lytorhynchus, and *Psammophis* have taken up this route to invade eastward.

The Hindu Kush Corridor

The Mediterranean and central Asian elements, *Platyceps rhodorachis, Agrionemys horsfieldii,* and *Telescopus rhinopoma,* took mainly the Irano-Hindu Kush route for eastward penetration.

Zoogeographical Analysis

Most of the Pakistani shared species, with neighboring faunal zones, are at the borders of their eastern or westernmost ranges. The Indus River marks the natural line across which most of them do not penetrate, so that the Palearctic elements are concentrated along its western bank in Baluchistan, Waziristan, NWFP, while the Indo-Oriental elements are dominant in Punjab and Sind, along the river's eastern bank.

The Oriental and Palearctic elements are derived from all major groups of amphibians and reptiles (Table 8.3). The 80 Oriental shared species are dominated by snakes (28%) while 60 Palearctic species are lizards (37%), and 55 (26%) endemic species are predominantly lizards. The Indo-Oriental elements are primarily mesic in habits and are widely distributed all over the Indo-Gangetic plains. The species are further distinguished, according to their origin, into Indian (59%) and Oriental (Southeast Asian) (41%) species. They are

Table 8.3 Affinities of Herpetofauna of Pakistan

Family	Genera	Species	Oriental	Palearctic	Endemic
Bufonidae	1	10	3	3	3
Microhylidae	2	2	2	—	—
Ranidae	6	11	6	1	4
Emydidae	3	4	4	—	—
Cheloniidae	3	4	3	—	—
Dermochelyidae	1	1	1	—	—
Testudinidae	2	2	1	1	—
Trionychidae	4	4	4	—	—
Crocodylidae	1	1	1	—	—
Gavialidae	1	1	1	—	—
Agamidae	6	28	9	11	8
Chamaeleonidae	1	1	1	—	—
Eublepharidae	1	1	—	1	—
Gekkonidae	11	38	7	12	19
Lacertidae	4	13	1	6	5
Scincidae	7	16	3	5	8
Varanidae	1	4	1	2	1
Typhlopidae	2	6	3	—	3
Leptotyphlopidae	1	2	1	—	1
Boidae	2	4	3	1	—
Colubridae	16	40	16	11	9
Hydrophiidae	7	14	14	—	—
Elapidae	2	5	2	1	2
Viperidae	4	9	2	5	2
Crotalidae	1	1	1	—	—

dominated by 24 species of lizards, the bulk of which is formed by geckos (35%), agamids, and scincids (25% each). Snakes are primarily colubrids (58%) and boids (12%). The 60 Palearctic elements (28% of the total) are primarily deserticolous. They are dominated by 44 lizards in which geckos and agamids are 26% each, lacertids 24%, and scincids 18%. The 16 species of Palearctic snakes are dominated by colubrids (62%) while viprids form 25% of the bulk. The 69 Palearctic elements are further distinguished into Mediterranean (4%), Irano-Turanian (51%), Central Asian (7%), Saharo-Sindian (16%), and Seistanian (22%).

The high number of Pakistan endemics (55 species, 26% of the total) indicates effective isolating mechanism provided by highly diversified habitat types available in Pakistan. Endemics are primarily gekkonids (19 species), which are followed by agamids (8), colubrids (6) and blind snakes (4).

Chapter 9
Herpetology of Habitat Types

Pakistan is not a geoecological entity; it is rather a man-made political division of the subcontinent. It is surprising that most of the world's major bioclimates are represented in this relatively small territory which is known as "the land of many lands" (Ahmad, 1951; Pfeffer, 1968; Khan, 1980b; F.K. Khan, 1996). However, its varied topography and numerous bioclimates are reflected in its rich species biodiversity (Khan, 1980b, 1997e, 1998a, c, 2002a, 2003a; Roberts, 1977, 1991, 1992; M.S. Khan and A.Q. Khan, 2000). The amalgamation of physical geography with its climates has played a key role in its habitat creation for its much varied fauna and flora (Mufti et al., 1997). In the past, several attempts have been made to sketch the overall biogeophysical features of Pakistan, to understand the nature of its physical, floral, and faunal characteristics: forest zones (Champion et al., 1965); climatic regions (Ahmad, 1951); Himalayan vegetation zones (Schweinfurth, 1957); habitat zones (Beg, 1975); mammalian ecological zones (Roberts, 1977); bird distribution (Roberts, 1991, 1992); fishes (Mirza, 1975); amphibians and reptiles (Khan, 1980b, 1987, 2002a, 2003a); termites (Akhtar, 1972; Akhtar and Ahmed, 1997), butterflies (Smith and Hasan, 1997), land snails (K. Auffenberg, 1997), etc.

Fifteen ecobiological habitat types with their distinctive herpetological assemblages have been distinguished in Pakistan (Khan, 1999e). In the following section of the book these ecobiological habitats are described in detail and their geographical position is marked (Figure 34). The associated herpetofauna of each habitat is described and given in a tabular form (Table 9.1). The conclusions are based mainly on data gleaned from different herpetological reports which have accumulated over several years.

Plants and Their Herpetological Associations

Pakistan is a forest-poor country; only about 3.8% of its total area is forested. The continued heavy logging and felling is still fast diminishing the existing forest cover. Recently several efforts have been made, at government as well as private levels, to increase the forested area by aforestation and regeneration (Kureshy, 1986; F.K. Khan, 1996). The major ecological factors affecting forest types are the arid and semiarid conditions prevailing over the Indus plains and Baluchistan Plateau; prevailing humidity in northern hills and mountains; and the diversity of topography from plains to the lofty mountains in the north, which rise above the snow line (Figure 34).

The present setup of the plant communities and associated herpetological assemblages is a commutative effect of the cohesion of physical geography, climatic conditions, and past transmigration of species. These factors have great bearing on the composition and distribution of the local herpetofauna. Except for a few species of chelonians and agamids, none of the Pakistani herps is known to forage on vegetation. However, every vegetation zone, described below, has its own characteristic resident herpetofauna. The vegetation not only provides animals with shelter against their enemies and against violent environmental changes, it also supports the dietary-required arthropods. Plants make the soil compact for burrowing and egg laying for several species. Moreover, the forest-floor covered by leaf-litter and other plant-debris, provides the ideal foraging site for several species of amphibians and reptiles. Types of vegetation differ from biotope to biotope; so do the leaf litter animals, amphibians and reptiles (Khan, 1999c).

An ecological zone is a biological entity which

Chapter 9. Herpetology of Habitat Types

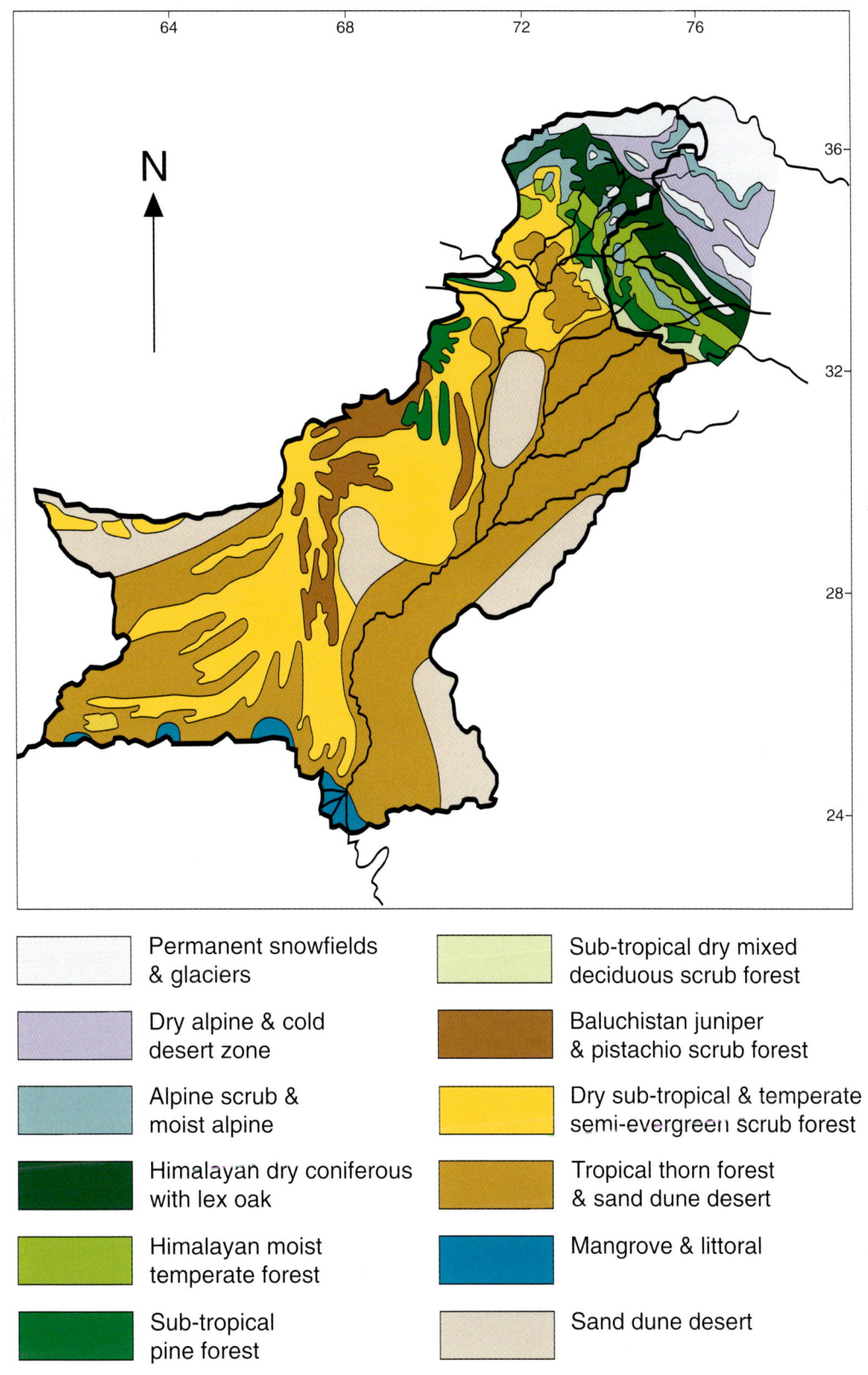

Figure 34. Pakistan, major habitat types.

is spread over a limited geographical area; its prevailing special ecological factors control the composition of its flora and fauna, while the interzone areas between different ecological zones represent the dilution areas where major zonal factors of adjacent habitats are not severe, but rather at their minimum. Here eurytopic satellite species aggregate, while relatively fewer stenotopic ones remain confined to the main habitat, where habitat-characterizing ecobiological factors are at their maximum. The eurytopic elements are widely distributed and they constitute the bulk of the herpetofauna. Though plant communities are fairly stable, since they are mainly altitude-dependent, natural disasters and human activity are violently altering factors. They devastatingly affect sensitive stenotopic elements, which either migrate in search of minimum suitable survival conditions or perish (Khan, 1980b).

The herpetological scenario in Pakistan is distinguished in three major geoecological regions, mountain, foothills, and plains (Figure 34). These three regions are further distinguished in the following 15 habitat types, which have distinct vegetation and herpetofauna.

Mountain Region

A mountainous rampart extends from the north in the western Himalayas to the northwest, all along the western border of Pakistan with Afghanistan and Iran to the Arabian seacoast in the south. It comprises mainly the western wing of the Himalayas and several ranges: Hindu Kush, Sulaiman, Kirthar, and several smaller associated ranges, covering about one 3rd of the area of the country (Figure 34).

Wide-ranging Herpetofauna
The following species of amphibians and reptiles are widely distributed in the northwestern highlands: *Bufo stomaticus, Euphlyctis cyanophlyctis microspinulata, Paa sternosignata, Argrionemys horsfieldii, Eublepharis macularius, Cyrtopodion scabrum, Laudakia caucasia, Laudakia melanura, Calotes versicolor, Acanthodactylus blanfordii, Mesalina watsonana, Eremias persica, Ophisops jerdonii, Eutropis dissimilis, Varanus griseus, Boiga melanocephala, Coluber karelini, Platyceps rhodorachis, Pseudocyclophis persicus, Lycodon striatus, Xenochrophis piscator, Ptyas mucosus, Spalerosophis schirazianus, Naja oxiana, Echis carinatus multisquamatus, Pseudocerastes persicus* and *Macrovipera lebetina* (Alcock and Finn, 1896; Hora and Chopra, 1923; Khan, 1987; Khan and Ahmed, 1987). In the southwest *Bufo viridis zugmayeri, Hemidactylus persicus, Laudakia nupta,* and *Trapelus ruderatus* are the wide-ranging species (Minton, 1962, 1966; Mertens, 1969a; Baig, 1977, 1988).

Eco-biological Zones
The following eco-biological zones are distinguished in the Mountain Region.

1. Cold, Dry Alpine Deserts
These include the northernmost regions of Pakistan, comprising parts of Chitral, Gilgit, Hunza, and Baltistan. These perpetual snow deserts lie at about 4500–6000 m above sea level, while bottom valleys lie at 2000 m. Here the climatic changes occur vertically. The mountains have a permanent ice cap and glaciers rise several hundred meters above the snow line. Annual rainfall varies from 50 to 100 mm, with heavy snow during winter. The diurnal range of temperature is very pronounced, fluctuating between −6° C and 25° C, throughout the year.

During the short-lived summer (June–July), the peripheral edges, alpine slopes, valley bottom, and stream beds are bared and they constitute a special cold arid habitat. They are sparsely overgrown during summer by meager herbaceous xerophytic vegetation. However, the slopes and stream beds have several associated plant species including *Hippophae rhamnoides, Salix* willow, *Mertensia tibetica,* and *Potentilla desertorum;* shrubs and forbs *Tribulus terrestris, Peganum harmala, Capparis spinosa, Sophora alopecuroides* and *Lycium ruthenicum;* grasses include *Festuca altaica* and *Paa attenuata.*

Associated herpetofauna: *Bufo siacheninsis, Bufo latastii, Bufo himalayanus, Altigekko baturensis, Altigekko boehmei, Altigekko yarkandensis, Laudakia badakhshana, Laudakia tuberculata, Laudakia himalayana, Scincella ladacensis,* and *Gloydius himalayanus* (Wall, 1911b; McMahon, 1901a, b; Khan, 1999c).

Chapter 9. Herpetology of Habitat Types

2. Himalayan Dry Coniferous Forests

They occur at 1500–3350 m of elevation, covering the inner or more northern ranges of the Himalayas, which are less subject to monsoons. This habitat gradually changes westward with moist temperate forests but is generally characterized by much fewer deciduous tree species and single-species stands of conifers. It is typified by Gilgit, Astor, and Chitral in the north, down to the Safed Koh Range and Takht-i-Suleiman in the south; it is distinguished further in the following four canopy associations:

a. Dry, temperate, evergreen oak deodar forests: Typified by lower Indus Kohistan, Swat Kohistan, northern Dir and parts of Chitral, and the inner valley of Hazara.

Vegetation: *Cedrus deodara, Pinus wallichiana, Gesrcus ilex, Juglans regia;* scattered bushes of *Artemisia maritima, Ephedra intermedia, Corylus corlurna, Parrotia jacquemontiana,* and *Sophora mollis.*

Associated herpetofauna: *Bufo pseudoraddei, Bufo himalayanus, Bufo melanostictus, Euphlyctis cyanophlyctis, Paa hazarensis, Limnonectes limnocharis, Microhyla ornata, Siwaligekko mintoni, Mediodactylus walli, Calotes versicolor, Scincella himalayana, Scincella ladacensis, Eutropis dissimilis, Hemorrhois ravergieri, Xenochrophis piscator, Spalerosophis diadema, Naja oxiana, Macrovipera lebetina,* and *Gloydius himalayanus* (McMahon, 1901a, b; Anderson, 1872; Wall, 1911b).

b. Dry zone blue pine and spruce forests: Typical of Naltar Valley and Astor in Gilgit and Takht-i-Suleiman.

Vegetation: *Picea smithiana, Pinus wallichiana, Rosa webbiana, Ribes grossularia, Prunus jacquemontii, Artemisia maritina, Berberis gambleana,* and *Colutea armata.*

Associated herpetofauna: *Japalura kumaonensis, Laudakia badakhshana,* and *Laudakia himalayana* (McMahon, 1901a, b; Baig, 1990).

c. Dry zone chilgoza and holly-oak: Chilas Dir, Gilgit Agency, Safed Koh, and higher ranges of Malakand Agency.

Vegetation: *Pinus wallichiana, P. gerardiana,* and *Cedrus deodara;* scattered shrubs of *Daphneo leoides, Sophora griffithii, Berbaris baluchistanica,* and *B. lycium.*

Associated herpetofauna: As recorded for the previous subdivision.

d. Higher or inner range Himalayan dry coniferous forests: Typified by Neelum Valley, Azad Kashmir, Salkhalla, and Machiara.

Vegetation: *Abies pindrow, Picea smithiana, Pinus wallichiana,* and *Quercus semicarpifolia.*

Associated herpetofauna: *Scutiger nyingchiensis, Paa hazarensis, Siwaligekko dattanensis, Indogekko rohtasfortai, Laudakia agrorensis, Typhlops ductuliformes, Typhlops diardii platyventris, Typhlops madgemintonae, Typhlops madgemintonae shermanai, Typhlops ahsanuli, Amphiesma stolatum, Oligodon arnensis, Oligodon taeniolatus, Boiga trigonata, Platyceps rhodorachis kashmirensis, Lycodon striatus bicolor, Ptyas mucosus, Sibynophis sagittarius, Spalerosophis diadema, Xenochrophis piscator piscator,* and *Echis carinatus multisquamatus* (Wall, 1910; A.Q. Khan and M.S. Khan, 1996).

3. Himalayan Moist Temperate Forests

This habitat occupies 3000–4000 m of elevation, in the inner ranges of the Himalayas and is predominantly coniferous forests with mixed deciduous broad-leaved species. It is being constantly modified by felling and overgrazing. It typifies the Murree Hills Range including Galis, lower Kaghan, and Neelam Valleys, extending westward to eastern Swat, along Indus Kohistan. It is the richest in plant and animal species.

Vegetation: *Abies pindrow, Pinus wallichiana, Taxus baccata, Ulmus wallichiana, Juglans regiaquercus, Prunus cornuta;* shrubs include *viburnum, Berberis lycium, Rosa moschata, Skimmia laureola,* and *Lonicera alpigena;* forbs include many species of genus *Impatiens* and *Euphorbia, Viola,* and *Gentiana;* creepers include *Hedera nepalensis* and *Clematis montana.*

Associated herpetofauna: *Laudakia tuberculata, Amphiesma sieboldii* and *A. platyceps, Macrovipara lebetina,* and *Gloydius himalayanus* (Khan and Tasnim, 1986).

4. Himalayan Moist Alpine Zone

This zone is richer in animal species and vegetation. It is typified by permanent grass cover and scattered junipers among tumbled boulders. This area has recently been overgrazed. It is spread throughout the higher slopes of the Kaghan Valley, Azad Kashmir, Swat, Dir, and Indus Kohistan (Plate 172).

Vegetation: *Betula utilis, Juniperus squamata, Salix himalayensis, Saxifragra sibirica, Draba trinervi;* many grasses of *Poa* genus and *Iris hookeriana*

Associated herpetofauna: *Eublepharis macularius, Mediodactylus walli, Siwaligekko battalensis, Varanus flavescens, Varanus bengalensis, Typhlops madgemintonae, Typhlops ahsanuli, Eryx johnii, Xenochrophis tessellata,* and *Bungarus caeruleus* (McMahon, 1901; Fenton, 1910; Wall, 1911b; Murthy and Sharma, 1976; Murthy et al., 1979; Khan, 1979; A.Q. Khan and M.S. Khan, 1996; Khan and Tasnim, 1986b).

5. Baluchistan Hill Ranges in Southern Latitudes, and Lower Slopes of Some Southern Ranges

Most of the original tree cover of steppic mountain scrub forests was lost due to felling and goat browse. A few relict patches are found only in wildlife reserves. It is typified by Chaman, Mashlak reserve, Hazar Ganji, the Harboi Hills in Kalat, and the Surkhab Valley near Peshin.

Vegetation: The original tree cover was lost due to heavy lopping for goat browse and felling for fuel. A few relict patches of wet steppic mountain scrub forest now found only in forest reserves or wildlife sanctuaries as mentioned above. The common plant species are *Juniperus macropoda* on higher ridges, *Olea cuspidata, Pistacia mutica, Pistacia khinjuk, Fraxinus xanthoxyloides,* with widely scattered bushes of *Sophora molle, Artemisia maritima, Ephedra major, Prunus eburnea, Stocksia brahuica; and* many bulbus perennials including Tulipa, *Ferula, Iris,* and *Allium* sp.; with grasses *Eleusine flagellifera* in the southern region, *Cymbopogon parkeri,* and many ephemerals of *Poa* sp. and *Bromus* sp.

Associated herpetofauna: *Laudakia caucasia, Laudakia nupta,* and *Laudakia ruderatus* (Khan, 1987; Khan and Ahmed, 1987).

6. Baluchistan Higher Ranges

A very arid mountain area with both diurnal and seasonal extremes of temperature, winter temperature dropping to $-14°C$, rising in summer above $35°$ C. Higher slopes experience light snowfall from January to March. Typified by the Takatu, Zarghun, Wam-Pilghar, Ziarat, and Toba Kakar Ranges, closely associated with northern Malakand and southwestern Chitral.

Vegetation: *Juniperus macropoda,* occasional *Fraxinus xanthoxyloide,* and *Pistacia khinjuk;* scattered bushes of *Prunus eburnea, Berbaris babaluchistanica, Rosa moschata, Salvia cabulica,* and *Sophora griffithi;* grasses include *Stipa pennata* and *Pennisetum orientalis.*

Signature herpetofauna: *Laudakia caucasia, Laudakia microlepis, Laudakia nupta, Acanthodactylus blanfordi, Eremias persica, Boiga melanocephala, Hemorrhois ravergieri, Coluber karelini, Spalerosophis schirazianus, Naja oxiana, Echis carinatus multisquamatus, Pseudocerastes persicus,* and *Macrovipera lebetina* (Alcock and Finn, 1896; Hora and Chopra, 1923; Khan, 1987; Khan and Ahmed, 1987).

Foothill Region

A strip of arid land extends along the southern face of the sub-Himalayas and western Baluchistan hilly tracts. It comprises ridges of sandstone and limestone escarpments. In the north it is interspersed with loess soil deposits, and is heavily overgrazed and heavily eroded into deep gullies. At places it is distinguished in several ecobiological habitats. Here in the submontane areas, plain and mountain species intergrade, giving rise to interesting taxonomic cases. In the north it is typified by the Salt Range and Kala Chitta Hills, and in the northwest by Waziristan Hills and Dera Ismael Khan (Plate 173); in the west are Dera Ghazi Khan, Rajanpur, and the "Pat"—the area between Kirthar Ridge and the Indus River (Plates 171, 172).

Wide-ranging Herpetofauna

In the foothills, wide-ranging herpetological species are as follows: *Bufo stomaticus, Bufo viridis zugmayeri, Microhyla ornata, Euphlyctis cyanophlyctis microspinulata, Hoplobatrachus tigerinus, Sphaeroteca breviceps, Fejervarya limnocharis, Lissemys punctata, Eublepharis macularius, Hemidactylus brookii, Hemidactylus flaviviridis, Trapelus agilis, Laudakia melanura, Acanthodactylus cantoris, Mesalina watsonana, Eremias persica, Ophisops jerdonii, Ablepharus pannonicus, Eurylepis taeniolatus, Scincella himalayana, Eutropis dissimilis, Varanus bengalensis, Varanus griseus, Ramphotyphlops braminus, Typhlops ductuliformes, Eryx johnii, Boiga trigonata, Platyceps rhodorachis, Platyceps ventromaculatus, Lycodon striatus, Amphiesma stolatum, Xenochrophis piscator, Oligodon arnensis, Oligodon taeniolatus, Psammophis leithii, Psammophis schokari, Ptyas mucosus, Spalerosophis*

Chapter 9. Herpetology of Habitat Types

Plate 171. Desolate, gullied mud flats, a common habitat between submontane areas and plains. In the Salt Range, Punjab, common species are *Eublepharis macularius, Cyrtopodion potoharensis, Cyrtopodion scabrum, Cyrtopodion watsoni, Cyrtopodion montiumsalsorum, Acanthodactylus cantoris, Eremias acutirostris,* and *Varanus bengalensis*; snakes include *Boiga trigonata, Ptyas mucosus, Echis carinatus,* and *Naja oxiana*.

Plate 172. High mud flat platforms, a habitat scattered throughout Punjab, Sind, North Western Frontier Province, and Baluchistan. Sides of deep gullies are eroded, with scattered low bushes and scrubby grasses. In Punjab a habitat for several lizards including *Cyrtopodion potoharensis, Cyrtopodion scabrum, Cyrtopodion watsoni, Acanthodactylus cantoris,* and *Mesalina watsonana*; snakes, *Echis carinatus, Naja naja, Oligodon arnensis, Boiga trigonata*; in Sind: *Cyrtopodion scabrum, Spalerosophis diadema, Novoeumeces blythianus, Eurylepis taeniolatus, Platyceps ventromaculatus indusai* and *Ptyas mucosus*; in North Western Frontier Province: *Cyrtopodion scabrum, Cyrtopodion watsoni, Acanthodactylus cantoris, Eremias fasciata, Echis carinatus, Eryx johnii, Boiga trigonata, Platyceps rhodorachis kashmirensis, Ptyas mucosus, Sibynophis sagittarius, Spalerosophis diadema, Bungarus sindanus razai, Naja naja, Naja oxiana, Echis carinatus multisquamatus,* and *Macrovipara lebetina*; in Baluchistan: *Trapelus agilis, Trapelus megalonyx, Novoeumeces zarudnyi, Bunopus tuberculatus, Cyrtopodion kachhense kachhense, Acanthodactylus blanfordii, Acanthodactylus cantoris, Eremias aporosceles, Eremias fasciata, Eremias persica, Mesalina brevirostris, Platyceps rhodorachis rhodorachis, Pseudocyclophis persicus,* and *Spalerosophis schirazianus*.

diadema, Bungarus caeruleus, Bungarus sindanus razai, Naja naja, Naja oxiana, and *Daboia russelii* (Minton, 1962, 1966; Mertens, 1969a, 1970, 1971, 1974; Khan, 1979, 1980a, 1986, 1988, 1997c; Khan and Baig, 1988).

The Foothill Region is distinguished in the following habitat types:

7. Dry, Subtropical, Semi-evergreen, Deciduous Scrub Forests

They are distinguished in two subzones:

a. Tropical, dry, deciduous forests: This habitat is characterized by subhumid climatic conditions in the submountainous Siwalik strip with recent alluvial deposits and is restricted to the narrow Jhelum Valley and outer Margala Hills, Salt Range, Kala Chitta Hills, and eastern Waziristan. Annual rainfall varies at 500–1000 mm. It is the most varied habitat, with moderate vegetation and climatic conditions.

Vegetation: Dominant vegetation comprises scattered stunted *Acacia jacquemontii, Acacia senegal, Commiphora mukul,* and *Ziziphus nummularia* trees. The tall bushes are *Rhazya stricta, Euphorbia caducifolia, Gewia tenax,* and *Bepharis sindica*.

Associated herpetofauna: *Bufo melanostictus, Indogekko rohtasfortai, Cyrtopodion montiumsalsorum, Hemidactylus persicus, Cyrtopodion watsoni, Cyrtopodion scabrum, Cyrtopodion potoharensis, Novoeumeces blythianus, Eurylepis taeniolatus, Argyrogena horsfieldii, Pseudocyclophis persicus,* and *Echis carinatus multisquamatus* (Ingoldby and Proctor, 1923; Khan, 1979, 1988; Khan and Baig, 1988).

Plate 173. Stony terrain at the foot of rocky outcrops in Dera Ismael Khan, southern North Western Frontier Province, and Dera Ghazi Khan, western Pakistan. It is a habitat for *Eublepharis macularius, Cyrtopodion kachhense ingoldbyi, Cyrtopodion scabrum, Cyrtopodion watsoni, Indogekko fortmunroi, Indogekko indusoani, Cyrtopodion kohsulaimanai, Cyrtopodion montiumsalsorum, Indogekko rohtasfortai, Ophisops jerdonii, Acanthodactylus cantoris, Eremias acutirostris, Eremias aporosceles, Eremias fasciata, Mesalina brevirostris, Mesalina watsonana, Ophisops jerdonii, Eryx johnii, Platyceps rhodorachis rhodorachis, Platyceps ventromaculatus ventromaculatus, Pseudocyclophis persicus, Lycodon striatus striatus, Oligodon arnensis, Oligodon taeniolatus, Psammophis schokari, Spalerosophis diadema, Bungarus sindanus razai, Naja oxiana,* and *Echis carinatus multisquamatus.*

b. Sind Kohistan and southern Baluchistan:
Moderate temperature, low rainfall desert, with pronounced humid monsoon winds during summer, but hot, dry, and relatively frost-free for the rest of the year; with mild but dry winters, dominated by a steady inflow of sea breeze throughout summer. Humidity is high, both annual and diurnal temperatures are low, and annual rainfall is 170–200 mm. Typified by Karachi, Malir, Sind Kohistan, Las Bela, Makran Range, Lakkhi and Pabb Hills, and Kirthar Range.
Vegetation: Clumps of cactus-like *Euphorbia* dominate the landscape of areas around Makran Range, Lakkhi and Pab Hills, and Kirthar Range. *Acacia jacquemontii, Acacia senegal, Commiphora mukul,* and *Ziziphus nummularia* form scattered stunted tree vegetation; tall bushes present are *Rhazya stricta, Euphorbia caducifolia, Grewia tenax;* and *Blepharis sindica.*
Associated herpetofauna: One of the most herpetologically rich and better known regions in Pakistan, in addition to the wide-ranging species listed above, several southern Indian species have been recorded from this zone: *Hemidactylus frenatus, Hemidactylus leschenaultii, Hemidactylus persicus, Hemidactylus turcicus, Geochelone elegans, Chameleon zeylanicus, Crocodylus palustris, Python molurus,* and *Eryx conicus.* While signature species are: *Bufo olivaceus, Bufo surdus, Ptyodactylus homolepis, Cyrtopodion kachhense, Tropiocolotes persica euphorbiacola, Acanthodactylus micropholis, Ophiomorus blanfordii, Uromastyx asmussi, Leptotyphlops macrorhynchus, Leptotyphlops blanfordii, Spalerosophis arenarius, Psammophis schokari, Psammophis leithii, Psammophis condanarus,* and *Pseudocerastes persicus* (Shockley, 1949; Minton, 1962, 1966; Mertens, 1962, 1969a).

8. Dry Temperate Semi-evergreen Scrub Forests
Characterized by long cold winters with some seasonal winter rain. They cover a wide latitudinal geographical range. Typified by the southern Chitral, Dir, Malakand Agency, Indus Kohistan, Amb, Bunir, and eastern fringes of northern Waziristan, Khyber, Mohmand Agency, Bannu, and Kohat (Plate 175). With hot dry summers, limited monsoon influence, and some spring and winter rains.
Vegetation: *Olea cuspidata, Acacia modesta, Artemisia maritima, Monotheca buxifolia, Dodonaea viscosa, Mallotus philippinensis, Lannea coromandelica, Rhazya stricta,* and *Wirhania coagulans;* with occasional trees, *Celtis eriocarpa,* and in ravines *Nannorrhops ritchieana,* dwarf palm; grasses include *Eleusine compressa, Chrysopogon aucheri, Cymbopogon jawrancusa,* and *Saccharum spontaneu;* shrubs include *Convolvulus spinosus* and *Adhatoda vasica.*
Signature herpetofauna: *Mediodactylus walli, Cyrtopodion scabrum, Scincella ladacensis, Eutropis dissimilis, Lygosoma punctata, Varanus flavescens, Typhlops ductuliformes, Hemorrhois ravergieri, Lycodon bicolor, Pseudocyclophis persicus, Bungarus caeruleus, Pseudocerastes persicus,* and *Naja oxiana* (Shockley, 1949; Minton, 1962, 1966; Mertens, 1962, 1969a)

Chapter 9. Herpetology of Habitat Types

9. Subtropical Pine Forests

Unique tall tree forests occurring at 910–1820 m of elevation. Typified by Kahuta, parts of Mangla Dam, lower Kaghan Valley around Kuwai, Batrasi Pass, lower Swat and lower Murree Hills around Tret. These forests are subject to periodic fires.

Vegetation: *Pinus roxburghii, Quercus incana, Ficus palmata, F. roxburghii, Punica granatum;* understory vegetation, *Ziziphus oxyphylla, Carissa opaca, Woodfordia fruticosa, Spiraea canescens, Buddleia paniculata, Beberis lycium, Indigofera pulchella;* grasses *Heteropogon contortus, Aristida cynantha, Apluda aristata,* and *Themeda anathera.*

Signature herpetofauna: *Uperodon systoma, Japalura kumaonensis, Python molurus, Lycodon striatus, Platyceps rhodorachis, Sibynophis sagittarius, Bungarus caeruleus razai, Naja naja, Echis carinatus sochureki,* and *Daboia russelii* (Khan and Tasnim, 1989, 1990b; Khan and Baig, 1988).

Plate 175. An aerial view of Goi Madan, Azad Kashmir, a mountain area with mainly alpine vegetation interspersed with clear areas for cultivation around fast-flowing streams. Rich in snakes, some of the species recorded are *Typhlops ahsanuli, Typhlops ductuliformes, Typhlops madgemintonae, Typhlops diardii platyventris, Eryx johnii, Amphiesma stolatum, Boiga trigonata, Platyceps rhodorachis kashmirensis, Lycodon aulicus, Lycodon striatus bicolor, Oligodon arnensis, Ptyas mucosus, Sibynophis sagittarius, Spalerosophis diadema, Xenochrophis piscator, Bungarus caeruleus, Naja oxiana, Gloydius himalayanus, Echis carinatus,* and *Daboia russelii.* The lizards recorded are *Eublepharis macularius, Eutropis dissimilis, Indogekko rohtasfortai, Scincella himalayana, Ophisops jerdonii, Siwaligekko battalensis,* and *Siwaligekko dattanensis.*

Plate 174. Rocky outcrops are scattered in Punjab as a part of the Karana Hills system. The rocks tower high above the flat Indus plains. Mostly these stony places are desolate, with sparse scattered low bushes and a few grasses arising among loose rocks. Recently these stony areas have been modified due to the stone-crushing industry. They were originally the natural habitat of several lizards, which are becoming scarcer day by day, such as *Eublepharis macularius, Cyrtopodion scabrum, Varanus bengalensis, Platyceps ventromaculatus, Platyceps rhodorachis, Oligodon taeniolatus, Spalerosophis diadema,* and *Ptyas mucosus.*

10. Tropical, Dry, Mixed Deciduous Forests

This habitat type is disjunct, confined to more sheltered ravines and northern-facing slopes; however, it represents an extension of the Siwalik zone farther east. It is typified by the Karot Valley draining into the Jhelum River, Kahuta, lower Lehtrar Valley, and ravine sides in the Margala Hills (Plate 174).

Vegetation: Forty species of plants recorded, of which the important spp. are *Acacia modesta, Bauthinia variegata, Cassia fistula, Celtis eriocarpa, Mallotus philippinensis, Pyrus pashia, Salmalia malabaricum,* and *Sterculia villosa;* understory of *Zizyphus mauritiana, Porana paniculata,* and *Woodfordia floribunda;* and grasses *Heteropogon contortus, Aristida cyanantha, Apluda aristata,* and *Themeda anthera.*

Associated herpetofauna: As recorded for the two previous zones (Khan and Tasnim, 1989, 1990b; Khan and Baig, 1988).

Indus Plains

The Indus plains are a characteristically flat, level stretch of land; their deep stone-free layer of alluvial silt has been washed down from the Himalayas through the centuries. It stretches from the sub-Himalayas in the north to the coast of the Arabian Sea in the south. It gradually sinks from sub-Himalayan heights to sea level; Rawalpindi is situated at 500 m of elevation, further south, Multan at 200 m, Sukkhur 100 m, and Hyderabad at 30 m. Its climate is of the subtropical continental lowland type, with high summer temperatures and late monsoon rains. Characteristic features of the Indus Plains are aridity and continentality. The range of annual rainfall is 350–400 mm, coming mostly in July–August, temperature ranges from 5°C to 48°C.

Wide-ranging Herpetofauna

In the plains the wide ranging herp species are *Bufo stomaticus, Microhyla ornata, Euphlyctis cyanophlyctis cyanophlyctis, Hoplobatrachus tigerinus, Sphaeroteca breviceps* (spotty), *Fejervarya limnocharis, Lissemys punctata, Eublepharis macularius, Hemidactylus brookii, Hemidactylus flaviviridis, Trapelus agilis, Laudakia melanura, Acanthodactylus cantoris, Mesalina watsonana, Ophisops jerdonii, Ablepharus pannonicus, Eurylepis taeniolatus, Mabuya dissimilis, Varanus bengalensis, Varanus griseus, Ramphotyphlops braminus, Typhlops ductuliformes, Eryx johnii, Boiga trigonata, Platyceps rhodorachis, Platyceps ventromaculatus, Lycodon striatus, Amphiesma stolatum, Xenochrophis piscator, Oligodon arnensis, Oligodon taeniolatus, Psammophis leithii, Psammophis schokari, Ptyas mucosus, Spalerosophis diadema, Bungarus caeruleus, Naja naja, Naja oxiana,* and *Daboia russelii* (Minton, 1966; Mertens, 1969a; Khan, 1980b).

The Indus plains are distinguished by the Punjab plain and Indus Delta, in which the following five life zones are recognized.

11. Riverine Tracts

The Punjab plain is traversed by five major tributaries of the Indus River, and is subdivided into four interfluves known as the "Duabs," which are belts of fertile alluvium. The area has extensively been modified and reclaimed by extensive ramification of the canal system. Due to vast flooding in all five rivers during monsoon rains, inundation of these interfluvial flats is almost a yearly feature. The scanty rainfall which rarely exceeds 300 mm and the high summer temperatures rising above 42°C make the summers hot. The monsoon rains bring relief.

Vegetation: Climax vegetation consists of *Acacia arabica;* less stable areas contain *Tamarix dioica* and *Populus euphratica;* grasses are *Erianthus munja, Saccharum munja,* and *Saccharum spontaneum.*

Associated herpetofauna: The regular annual flooding of the area extensively disturbs animal and plant life, so that herp species are common to all interfluves; no unique species has been recorded from this zone. All freshwater and mud turtles reported so far from Pakistan are represented. Apart from the wide-ranging amphibian species recorded for the previous zones, a south Indian frog, *Fejervarya syhadrensis,* prevails here. River turtles, *Hardella thurjii, Kachuga smithii, Kachuga tecta, Chitra indica, Lissemys punctata, Aspideretes gangeticus,* and *Aspideretes hurum,* frequent the waters of the lower Indus. Plain agamids, *Trapelus agilis, Trapelus megalonyx, Trapelus rubrigularis,* and arboreal agama, *Calotes versicolor,* abound. The rare *Brachysaura minor* was reported for the first time from Rabwah, Pakistan (Mertens, 1974; Khan, 2002a; Khan and Mirza, 1977). The psammophilous lizards consist of the wide-ranging species *Acanthodactylus cantoris, Mesalina watsonana,* and *Ophiomorus tridactylus.* Common grass skinks are *Ophisops jerdonii, Ablepharus pannonicus, Eurylepis taeniolatus, Novoeumeces indothalensis, Mabuya dissimilis,* and *Mabuya macularia.* Three species of varanids occur in the plains: *Varanus bengalensis, Varanus flavescens,* and *Varanus griseus koniecznyi.* Snakes in addition to three recorded for previous zones are *Eryx conicus, Python molurus, Platyceps ventromaculatus, Spalerosophis arenarius, Lytorhynchus paradoxus, Lycodon bicolor, Lycodon travancoricus, Xenochrophis sanctijohannis, Xenochrophis cerasogaster, Naja naja, Naja oxiana, Psammophis schokari, Psammophis leithii, Bungarus caeruleus sindanus, Naja naja,* and *Echis carinatus sochureki* (Loveridge, 1959; Khan, 1984a, b, 1986; J.A. Anderson and Minton, 1963).

Chapter 9. Herpetology of Habitat Types

12. Seasonal Inundation Zones, Seepage Areas, Lakes and Swamps

Subject to summer flooding, becoming dry by April–May. Typified by disjunct inter-riverine areas around Trimmu and Balloki Headworks, and Lal Suhanran near Bahawalpur in Punjab. Swampy areas around east Nara and Sanghar, Ghauspur in Jacobabad District, Manchar in Dadu District in Sind. It includes major lakes lying in the Indus Delta and lower Sind.

Vegetation: *Tamarix dioica, Phragmites karka, Typha angustata, Paspalum distichum, Imperata cylindrica,* and *Nelumbium nuciferum;* and in water pools, *Vallisneria spiralis* and *Hydrilla verticillata.*

Associated herpetofauna: In addition to the wide-ranging amphibian and reptilian species recorded for the previous zones, in the plains there is a preponderance of river turtles including *Geoclemys hamiltonii, Hardella thurjii, Kachuga smithii, Kachuga tecta, Chitra indica, Lissemys punctata, Aspideretes gangeticus,* and *Aspideretes hurum,* and snakes, *Leptotyphlops blanfordii, Leptotyphlops macrorhynchus, Eryx conicus, Python molurus, Argyrogena fasciolata, Xenochrophis cerasogaster,* and *Echis carinatus sochureki* which are the signature species of this region (Minton, 1962, 1966; Mertens, 1969a; Khan, 1980b).

13. Tropical Thorn Forests, Lower Indus Plain

It is the major habitat which some time back occupied the whole of the Indus plains from the foothills to the seacoast. However, due to human activity over more than 1000 years, most of the original forest cover has disappeared. The original tropical thorn vegetation survives in small pockets where human interference is restricted, around airfields, graveyards, and uncultivated lands such as saline flats or "pats."

Vegetation: In Punjab, *Prosopis spicigera, Capparis aphylla, Salvadora oleoides, Tamarix aphylla,* and *Ziziphus mauritiana,* and in lower Sind, *Euphorbia caducifolia, Culatropis procera,* and *Suaeda fruticosa;* grasses, *Aristida depressa* and *Eleusine compressa.* The first five plants form very scattered shrubby trees and are usually most affected by lopping.

Associated herpetofauna: As recorded for the previous region (Minton, 1962, 1966; Mertens, 1969a; Khan, 1980b; Akram, 1982).

14. Sand Dune Deserts

In Pakistan there are five sand dune deserts: in Punjab, the Thal Desert in the northwest, the Cholistan Desert in the southeast, The Thar Desert in the eastern Sind, the Sibi Desert in the northeast, and the Chagai-Kharan Desert in southwestern Baluchistan (Plate 176). Due to these arid lands, aridity is the hallmark of the Pakistani climate. The fauna and flora of deserts is largely drawn from neighboring faunal and floral associations.

Plate 176. Common desert habitat: the Thal and Cholistan Deserts in Punjab, and the Thar Desert in Sind, Pakistan. Fine, loose windblown aoline soil is compacted in places by growths of grasses and low bushes, making it a habitat for several psammophilous species including *Trapelus agilis, Trapelus megalonyx, Trapelus ruderatus, Novoeumeces blythianus, Novoeumeces indothalensis, Eurylepis taeniolatus, Novoeumeces zarudnyi, Ophiomorus tridactylus, Acanthodactylus cantoris, Eremias acutirostris, Eremias fasciata, Mesalina brevirostris, Mesalina watsonana, Argyrogena fasciolata, Platyceps ventromaculatus indusai, Lycodon striatus striatus, Lycodon striatus bicolor, Lytorhynchus maynardi, Lytorhynchus paradoxus, Oligodon taeniolatus, Psammophis schokari, Ptyas mucosus, Spalerosophis arenarius, Spalerosophis diadema, Spalerosophis schirazianus,* and *Telescopus rhinopoma* in Baluchistan.

Common desert plants: Plants shared among Pakistani deserts are: *Prosopis spicigera, Ziziphus mauritiana, Salvadora oleoides, Calligonum polygonoides, Calotropis procera, Aristida mutabilis,* and *Saccharum bengalense.* The first three plants grow to fair-sized trees. *Calligonum,* particularly, grows and colonizes sand dunes.

Associated herpetofauna: Desert habitat is made suitable for most of the wide-ranging plain amphibian species by recent reclamation efforts in the peripheral parts of the deserts. However, *Bufo olivaceus* and *Bufo surdus* are particularly known from the oases in the Baluchistan deserts. There are several reptilian genera which, in Pakistan, are exclusively represented in these deserts: *Agamura, Bunopus, Crossobamon, Teratolepis, Teratoscincus, Tropiocolotes, Phrynocephalus, Eremias, Acanthodactylus, Ophiomorus, Lytorhynchus, Psammophis,* and *Eristicophis.*

a. Thal Desert: It lies between the Jhelum-Chenab and Indus Rivers in northwestern Punjab. Occasional scanty rainfall, treeless loose sandy soil, and precarious and scattered pasturage are its characteristics. It lies 100–155 m above sea level. Dust storms and the hot wind or "Lu" are important summer features. Its peripheral parts are largely reclaimed.

Vegetation: *Prosopis spicigera, Ziziphus mauritiana,* and *Salvadora oleoides* grow to fair-sized trees, while *Calligonum polygonoides,* and *Calotropis procera* remain prostrated or bushy in formation. *Calligonum* is particularly adapted to grow and colonize sand dunes. Grasses are *Aristida mutabilis* and *Saccharum bengalense.*

Associated herpetofauna: Almost all the plains amphibians are represented in the well watered parts of the desert: Agamids belonging to genus *Trapelus, Brachysaura minor, Crossobamon orientalis, Acanthodactylus cantoris, Novoeumeces zarudnyi, Ophiomorus tridactylus, Varanus griseus, Lytorhynchus paradoxus, Psammophis leithii, Psammophis schokari, Spalerosophis arenarius, Bungarus sindanus razai, Naja oxiana,* and *Echis carinatus multisquamatus. Novoeumeces indothalensis* has recently been described from this desert.

b. Cholistan Desert: The arid strip of sand dunes lying to the east of the Sutlej River in southeastern Punjab, continuing with the Rajasthan Desert of India, is known as Cholistan. It lies 80–90 m above sea level. In the south it continues with the Thar Desert of Sind. Annual rainfall remains between 50–100 mm, and some years are without rain. Its part lying along the Sutlej River is reclaimed, and extensively canalized.

Vegetation: *Tamarix aphylla, Prosopis spicigera, Capparis decidua, Calligonum polygonoides, Leptadenia spartium,* and *Haloxylon griffithii,* grasses, *Aristida depressa, Saccharum spontaneum, Cymbopogon shoenanthus,* and *Pennisetum* spp.

Associated herpetofauna: Almost all the plains amphibian species are represented in the reclaimed parts of the desert. Similarly wide-ranging reptilian species are represented. Himalayan forms *Xenochrophis sanctijohannis* and *Sphaeroteca breviceps* are recorded from the irrigation water channels from the Sutlej River (Khan, 1984a, b; 1985a).

c. Thar Desert: It is the westernmost extension into southeastern Sind of the great Indian Rajasthan Desert. It consists largely of sand hills which vary from small dunes to 30–50 m high sand hills, overlying the Indus alluvium. At places they form sandstone rocks and ridges, separated by valleys of varying breadths (Plate 176). Both annual and diurnal ranges of temperatures are high. Summer dust storms are usual features. Wherever there is water available, thick forests occur.

Vegetation: In addition to common species; *Tamarix aphylla, Euphorbia caducifolia, Commiphora mukul, Ziziphus nummularia, Grewia tenax, Cassia angustifolia, Calligonum polygonoides,* and *Blepharis sindica* occur.

Associated herpetofauna: All wide-ranging plains amphibian and reptilian species are represented. Signature species recorded are *Cyrtopodion kachhense, Hemidactylus leschenaulti, Teratolepis fasciata, Tropiocolotes helenae, Mesalina brevirostris, Novoeumeces zarudnyi, Ophiomorus raithmai, Ophiomorus tridactylus, Eryx conicus, Python molurus, Lycodon travancoricus, Psammophis condanarus, Psammophis leithii, Psammophis schokari, Bungarus sindanus,* and *Echis carinatus sochureki* (Blanford, 1879; Murray, 1874, 1884, 1886; Mertens, 1969a; J.A. Anderson and Minton, 1963; Minton, 1966).

d. Sibi Desert: The Sibi Desert lies between the Sulaiman and Kirthar Ranges on the west bank of the Indus, as a westward arid strip. It is the hottest part of Pakistan, since it is sandwiched between two arid mountain ranges. The soil is hard, stony, and gullied.

Vegetation: The wide-ranging desert species, *Capparis decidua, Suaeda fruticosa, Tamarisk troupii,* and the grass, *Panicum antidotale,* are recorded.

Associated herpetofauna: All wide-ranging amphibian species except *Hoplobatrachus tigerinus, Sphaeroteca breviceps, Fejervarya limnocharis,* and *Fejervarya syhadrensis* are represented. Signature forms recorded are *Ptyodactylus homolepis, Cyrtopodion kohsulaimanai,* and *Cyrtopodion kachhense ingoldbyi* (Khan, 1991a, 1993b, 1997c).

e. Chagai-Kharan Desert: The Chagai-Kharan Desert lies at 610–1060 m elevation, and is one of the hottest and driest deserts of Pakistan. It is subjected to dust storms which constantly blow sand dunes here and there throughout the year. Annual rainfall, which is highly uncertain, is 0–60 mm. vast fields of sand dunes are interspersed with hard alluvial soil supporting stunted grasses and bushes which provide shade and hideouts for the local herps. The desert, in places, sinks down in great depressions which are filled by rainwater, resulting in the salt lakes—"Hamuns." These lakes store water and are fringed with thick vegetation. When dry, the salt-encrusted clay in the lake basins is sun-cracked, with a few persisting marshy patches. The fringe of reed and tamarisk patches provides humid and cooler environs for the local herp species. Moreover, they support several arthropods, gastropods, and seed-eating mammals and birds, thus a highly specialized desert community results.

Vegetation: *Haloxylon ammodendron, Rhazya stricta, Astragalus sericostachys, Peganum harmala, Salsola arbuscula;* grasses *Eleusine compressa, Pennisetum dichotomum, Andropogon halapensis,* and *Nepeta glomerulosa.*

Associated herpetofauna: The Chagai-Kharan is herpetologically the richest part of Pakistan, the species recorded are: *Bufo olivaceus, Bufo surdus, Bufo viridis zugmayeri, Euphlyctis cyanophlyctis microspinulata, Rhinogekko femoralis, Agamura persica, Bunopus tuberculatus, Crossobamon lumsdeni, Crossobamon maynardi, Teratoscincus microlepis, Teratoscincus scincus, Tropiocolotes depressus, Tropiocolotes helenae, Tropiocolotes persicus, Phrynocephalus clarkorum, Phrynocephalus euptilopus, Phrynocephalus luteoguttatus, Phrynocephalus maculatus, Phrynocephalus ornatus, Uromastyx asmussi, Acanthodactylus blanfordii, Acanthodactylus micropholis, Eremias acutirostris, Eremias aporosceles, Eremias fasciata, Eremias scripta, Eremias persica, Ophiomorus breviceps, Ophiomorus tridactylus, Eryx tataricus, Boiga melanocephala, Coluber karelini, Lytorhynchus maynardi, Lytorhynchus ridgewayi, Psammophis lineolatus, Psammophis schokari, Spalerosophis schirazianus,* and *Eristicophis macmahonii* (Minton, 1962, 1966; Mertens, 1969a).

15. Littoral and Intertidal Zones and Offshore Islands:
Typified by the mouth of the Indus, Somiani, and other bays along the Makran coast characterized by mangroves, and Astola Island.

Vegetation: *Avicenna officinalis, Ceriops tagal/candolleana, Halopyrum mucronatum, Bruguiera conjugata,* and scrub *Salsola fietida* and *Sueda fruticosa;* grasses are *Heleochloa dura* and *Halopyrum mucronatum.*

Associated herpetofauna: In this coastal biome, marine turtles and sea snakes are represented. The marine turtle species that regularly visit coastal beaches of Pakistan to lay eggs are *Chelonia mydas, Lepidochelys olivacea,* and *Dermochelys coriacea* (Groombridge, 1987).

Several freshwater turtles from the Indus River system are known to extend into the coastal mangroves.

Several marine snakes are recorded, some of which are *Hydrophis cyanocinctus, Hydrophis spiralis, Hydrophis lapemoides, Hydrophis caerulescens, Hydrophis mamillaris, Enhydrina schistosa, Praescutata* and *Lepemis curtus* (Myers, 1947; Minton, 1962, 1966; Mertens, 1969a).

There is strong indication of occurrence of some isolated populations of *Gavialis gangeticus* along the coastal swamps. However, a disjunct population of *Crocodylus palustris* has been reported from the Hab River along the Makran coast (Minton, 1966) and still farther west in southeastern Iran.

The offshore island herpetofauna consists of disjunct populations of reptiles, with morphology which varies from the mainland forms. The following species have been recorded from Astola Island, 25 km southeast off the Pasni coast, Baluchistan: *Acanthodactylus cantoris, Mesalina watsonana, Novoeumeces schneiderii zarudnyi Laudakia lirata, Trapelus agilis, Platyceps rhodorachis, Spalerosophis diadema.* A subspecies, *Echis carinatus astolae,* also has been described from the area (Mertens, 1969a).

Table 9.1 Distribution of Amphibians and Reptiles by Habitat

Legend:
1 = Cold, dry, alpine deserts
2 = Himalayan dry coniferous forests
3 = Himalayan moist temperate forests
4 = Himalayan moist alpine zone
5 = Baluchistan hill ranges in southern latitudes, and lower slopes of some southern ranges
6 = Baluchistan higher ranges
7 = Dry, subtropical, semi-evergreen deciduous scrub forests:
    a. Tropical, dry, deciduous forests
    b. Sind Kohistan and southern Baluchistan
8 = Dry temperate semi-evergreen scrub forest
9 = Subtropical pine forests
10 = Tropical, dry, mixed deciduous forests
11 = Riverain tracts
12 = Seasonal inundation zones, seepage areas, lakes, and swamps
13 = Tropical thorn forests, lower Indus plain
14 = Sand dunes in semidesert habitat:
    a. Thal Desert
    b. Thar Desert
    c. Sibi Desert
    d. Chagai-Kharan Desert
15. Littoral and intertidal zones, and offshore islands:
i = Astola Island
m = mainland
s = sea (Khan, 1999)

Taxa	Habitat Types														
	1	2	3	4	5	6	7	8	9	10	11	12	13	14	15
BUFONIDAE		+	+	+											
Bufo himalayanus	+	+													
Bufo latastii															
Bufo melanostictus hazarensis			+	+			+	+	+						
Bufo olivaceus							b							d	
Bufo pseudoraddei pseudoraddei		+		+			a	+		+					
Bufo pseudoraddei baturae		+													
Bufo siacheninsis		+													
Bufo stomaticus			+	+	+	+	ab	+	+	+	+	+	+	abcd	m
Bufo surdus															
Bufo viridis zugmayeri					+	+	b							cd	m
MEGOPHRYIDAE															
Scutiger nyingchiensis				+											
MICROHYLIDAE															
Microhyla ornata			+				a	+	+	+		+			

Chapter 9. Herpetology of Habitat Types

Taxa	Habitat Types														
	1	2	3	4	5	6	7	8	9	10	11	12	13	14	15
Uperodon systoma															
RANIDAE															
Euphlyctis cyanophlyctis cyanophlyctis			+	+	+	+	ab	+	+	+			+	+	ab
Euphlyctis cyanophlyctis microspinulata					+	+	a	+							m
Euphlyctis cyanophlyctis seistanica							a								
Fejervarya limnocharis			+	+			ab	+	+	+	+	+			
Fejervarya syhadrensis							ab			+	+	+	+		
Hoplobatrachus tigerinus						+	ab	+	+	+	+	+	+		
Paa barmoachensis				+	+										
Paa hazarensis					+				+						
Paa sternosignata					+	+									
Paa vicina			+	+											
Sphaeroteca breviceps					+		ab	+	+	+	+	+	+		a	m
CHELONIIDAE															
Caretta caretta															s
Chelonia mydas															s
Eretmochelys imbricata															s
Lepidochelys olivacea															s
DERMOCHELYIDAE															
Dermochelys coriacea															s
EMYDIDAE															
Geoclemys hamiltonii											+	+	+		
Hardella thurjii											+	+	+		
Kachuga smithii											+	+	+		
Kachuga tecta											+	+	+		
TESTUDINIDAE															
Agrionemys horsfieldii					+	+								d	m
Geochelone elegans								b					+	c	m
TRIONYCHIDAE															
Aspideretes gangeticus											+	+	+		
Aspideretes hurum							ab				+	+	+		
Chitra indica															
Lissemys punctata andersoni							ab	+		+	+	+	+		
CROCODYLIDAE															
Crocodylus palustris								b				+?			m
GAVIALIDAE															
Gavialis gangeticus								b				+?			m?
AGAMIDAE															
Brachysaura minor								b			+				
Calotes versicolor versicolor		+		+	+	+	ab	+		+	+	+	+	abcd	
Calotes versicolor farooqi					+										

Taxa	Habitat Types														
	1	2	3	4	5	6	7	8	9	10	11	12	13	14	15
Japalura kumaonensis		+					a	+							
Laudakia agrorensis		+	+				a	+			+				
Laudakia badakhshana	+	+	+					+							
Laudakia caucasia		+	+				+	a	+						
Laudakia fusca		+	+		+										
Laudakia himalayana	+	+	+												
Laudakia lirata							ab								mi
Laudakia melanura		+				+	a								
Laudakia microlepis														d	
Laudakia nupta			+		+	+	+								
Laudakia nuristanica			+	+						+					
Laudakia pakistanica pakistanica		+													
Laudakia pakistanica auffenbergi			+												
Laudakia pakistanica khani	+														
Laudakia tuberculata			+	+	+					+					
Phrynocephalus clarkorum														d	
Phrynocephalus euptilopus														d	
Phrynocephalus luteoguttatus														d	
Phrynocephalus maculatus														d	
Phrynocephalus ornatus														d	
Phrynocephalus scutellatus														d	
Trapelus agilis agilis					+	+	b							c	
Trapelus agilis pakistanensis						+	ab				+	+	+	+	a
Trapelus megalonyx							b					+		b	
Trapelus rubrigularis		+					a						+		
Trapelus ruderata					+	+									
CHAMAELEONIDAE															
Chamaeleo zeylanicus							b					+	+		
EUBLEPHARIDAE															
Eublepharis macularius		+			+	+	ab	+		+	+		+	abc	
GEKKONIDAE															
Agamura persica							ab							d	
Altigekko baturensis		+													
Altigekko boehmei		+													
Altigekko stoliczkai															
Bunopus tuberculatus		+					b						+	d	m
Crossobamon lumsdeni											+	+	+	d	
Crossobamon maynardi												+		cd	
Crossobamon orientalis							b				+	+	+	ab	
Cyrtopodion agamuroides					+		ab								
Cyrtopodion kachhense kachhense							b	+				+	+	c	

Taxa	Habitat Types														
	1	2	3	4	5	6	7	8	9	10	11	12	13	14	15
Cyrtopodion kachhense ingoldbyi							a								
Cyrtopodion kohsulaimanai							a								
Cyrtopodion montiumsalsorum							a								
Cyrtopodion potoharensis							a								
Cyrtopodion scabrum		+		+	+	+	ab	+		+	+			abc	
Cyrtopodion watsoni			+		+	+	a	+		+	+		+	a	
Hemidactylus brookii			+	+	+	+	ab	+		+	+	+	+	ab	m
Hemidactylus flaviviridis				+	+	+	ab	+		+	+	+	+	a	m
Hemidactylus frenatus							b								m
Hemidactylus leschenaultii							b								m
Hemidactylus persicus					+	+	ab							d	m
Hemidactylus triedrus							b						+		m
Hemidactylus turcicus						+	b								
Indogekko fortmunroi					+	+	a	+							
Indogekko indusoani							a	+							
Indogekko rhodocaudus					+										
Indogekko rohtasfortai		+					a			+					
Mediodactylus walli			+												
Ptyodactylus homolepis							b								m
Rhinogekko femoralis														d	
Rhinogekko misonnei														d	
Siwaligekko battalensis			+	+											
Siwaligekko dattanensis			+	+											
Siwaligekko mintoni		+		+											
Teratolepis fasciata															
Teratoscincus microlepis														d	
Teratoscincus scincus keyserlingi														d	
Tropiocolotes depressus														d	
Tropiocolotes persicus persicus														d	
Tropiocolotes persicus euphorbiacola							c								
LACERTIDAE															
Acanthodactylus blanfordii														d	m
Acanthodactylus cantoris		+					ab	+		+	+	+	+	abcd	
Acanthodactylus micropholis							b						+	d	m
Eremias acutirostris														d	
Eremias aporosceles														d	
Eremias fasciata							a	+						d	
Eremias persica					+	+								cd	
Eremias scripta														d	
Eremias velox persica							a							d	
Mesalina brevirostris							ab						+	d	m

Taxa	Habitat Types														
	1	2	3	4	5	6	7	8	9	10	11	12	13	14	15
Mesalina watsonana							ab				+	+	+	abcd	mi
Ophisops elegans								+							
Ophisops jerdonii				+	+		ab			+			+		
SCINCIDAE															
Ablepharus grayanus							ab					+			
Ablepharus pannonicus			+	+			ab					+	+		
Chalcides ocellatus							b								mi
Eurylepis taeniolatus							ab			+	+	+	+		m
Eutropis dissimilis				+	+		+	a			+	+		+	
Eutorpis macularia							b						+		m
Lygosoma punctata							a	+		+	+	+	+		m
Novoeumeces blythianus							a		+				+	d	m
Novoeumeces indothalensis												+		a	
Novoeumeces schneiderii zarudnyi														d	mi
Ophiomorus blanfordi														d	m
Ophiomorus brevipes														d	
Ophiomorus raithmai							b								
Ophiomorus tridactylus					+	+	b				+	+	+	abd	m
Scincella himalayana				+			a			+					
Scincella ladacensis		+	+	+											
UROMASTYDAE															
Uromastyx asmussi					+										
Uromastyx hardwickii							ab				+	+	+		m
VARANIDAE															
Varanus bengalensis			+	+	+		ab	+			+	+	+	ab	m
Varanus flavescens				+			b					+			
Varanus griseus caspius							ab				+	+			
Varanus griseus koniecznyi				+	+										
LEPTOTYPHLOPIDAE															
Leptotyphlops blanfordii				+	+		b						+		
Leptotyphlops macrorhynchus							a				+	+			
TYPHLOPIDAE															
Ramphotyphlops braminus				+	+		ab	+		+	+	+	+		m
Typhlops ahsanuli				+											
Typhlops diardii platyventris				+											
Typhlops ductuliformes				+			ab	+				+	+		
Typhlops madgemintonae madgemintonae				+								+			
Typhlops madgemintonae shermanai				+											
BOIDAE															
Eryx conicus							b					+			
Eryx johnii			+	+			ab	+		+	+	+	+	abc	m

Chapter 9. Herpetology of Habitat Types

Taxa	Habitat Types														
	1	2	3	4	5	6	7	8	9	10	11	12	13	14	15
Eryx tataricus					+	+								d	
Python molurus							b	+							
COLUBRIDAE															
Amphiesma platyceps								+							
Amphiesma sieboldii								+							
Amphiesma stolatum				+			ab	+		+	+	+	+		
Argyrogena fasciolatus							b								
Boiga melanocephala						+								d	m
Boiga trigonata			+	+			ab	+		+	+	+	+		m
Coluber karelini karelini					+	+								cd	
Coluber karelini mintonorum						+								d	
Eirenis persica					+	+	a								
Enhydris pakistanica												+			
Hemorrhois ravergieri							+								
Lycodon aulicus aulicus				+			ab	+			+				
Lycodon striatus striatus							ab			+	+	+			
Lycodon striatus bicolor				+								+			
Lycodon travancoricus												+			
Lytorhynchus maynardi							b							d	m
Lytorhynchus paradoxus							b				+	+	+		
Lytorhynchus ridgewayi						+								d	
Oligodon arnensis				+		+	a	+		+	+		+		
Oligodon taeniolatus							ab	+		+	+		+		
Platyceps rhodorachis rhodorachis			+	+	+	+	ab	+		+				abc	mi
Platyceps rhodorachis ladacensis			+	+	+		ab		+	+					i
Platyceps rhodorachis kashmirensis				+	+										
Platyceps ventromaculatus ventromaculatus					+	+	ab	+		+	+	+	+		m
Platyceps ventromaculatus bengalensis						+				+					
Platyceps ventromaculatus indusai					+				+		+	+			
Psammophis condanarus							b			+	+	+	ab		
Psammophis leithii							ab			+	+	+			
Psammophis lineolatus						+									
Psammophis schokari							ab			+		+	abc		m
Ptyas mucosus			+	+	+		ab	+	+	+	+	+			m
Sibynophis sagittarius									+						
Spalerosophis arenarius					+		ab			+		+	abd		m
Spalerosophis diadema		+	+	+	+	+	ab	+	+	+	+	+			m
Spalerosophis schirazianus					+	+								cd	

Taxa	Habitat Types														
	1	2	3	4	5	6	7	8	9	10	11	12	13	14	15
Telescopus rhinopoma							a		+						
Xenochrophis cerasogaster												+	+		
Xenochrophis piscator piscator															
Xenochrophis sanctijohannis		+	+	+		+	ab	+		+	+	+	+	cd	
Xenochrophis tessellata		+													
ELAPIDAE															
Bungarus caeruleus caeruleus			+	+		+	ab	+	+	+	+	+	+		m
Bungarus sindanus sindanus							b						+	ab	
Bungarus sindanus razai			+				a	+						a	
Naja naja				+	+	+	ab	+	+	+	+	+	+		
Naja oxiana		+	+	+	+	+	ab	+	+	+	+	+	+	acd	m
HYDROPHIIDAE															
Astrotia stokesii															s
Enhydrina schistosa															s
Hydrophis caerulescens															s
Hydrophis cyanocinctus															s
Hydrophis fasciatus															s
Hydrophis lapemoides															s
Hydrophis mamillaris															s
Hydrophis ornatus															s
Hydrophis spiralis															s
Lapemis curtus															s
Microcephalophis cantoris															s
Microcephalophis gracilis															s
Pelamis platurus															s
Praescutata viperina															s
VIPERIDAE															
Daboia russelii russelii							ab			+	+	+	+		
Echis carinatus astolae															i
Echis carinatus multisquamatus				+		+	a	+		+		+			
Echis carinatus sochureki					+		b				+	+			
Eristicophis macmahonii														d	
Pseudocerastes bicornis															
Pseudocerastes persicus				+	+										
Vipera lebetina				+	+		a								
CROTALIDAE															
Gloydius himalayanus		+	+	+											

Chapter 10
Altitudinal Distribution of the Herpetofauna of Pakistan

In Chapter 8, habitat distribution and associated herpetological assemblages were discussed. In this section the distribution pattern of amphibians and reptiles according to the peculiar altitudinal gradients occurring in Pakistan is being taken up (see Table 10.1). Altitudinal gradient has played an important role in establishing terrestrial vertebrate assemblages (M.S. Khan and A.Q. Khan, 2000), particularly the herpetofaunal assemblages discussed in Chapter 8 (see also Table 9.1). From this altitudinal study important information is gleaned about the environmental preferences as to temperature, altitude, rainfall, seasonal changes, habitat, distribution, and range of individual species. Similar studies have been made for Nepal (Swan and Leviton, 1962), for Iran (Anderson, 1968), and India (Waltner, 1974a, b, 1975a, b).

The following altitudinal gradient complex runs across Pakistan in different directions due to the complex local topography (see Figure 33):

1. North-South Altitudinal Gradient
The North-South Altitudinal Gradient starts from several parallel glacier-capped east-west Greater Himalayan ranges, ranging 6000–8610 m altitude, with narrow intervening river valleys, spread over Gilgit Agency, Kohistan, and Baltistan; and the Lesser Himalayas, 1800–4600 m, which is a maze of folded, faulted, and overthrust mountain complexes.

There is a marked temperature, humidity, and precipitation gradient from north to south. It encompasses Rawalpindi, Abbotabad, and Manshera Districts (Chapter 9, habitat types 1, 2, 3, 4, and 7b).

2. North-East-West Altitudinal Gradient
The sub-Himalayas or Siwaliks consist of low ranges, 600–1200 m, of folded and badly faulted rocks, most deeply dissected, and covering Rawalpindi District. The gradient drops farther in the Potwar Plateau and Salt Range which consist of the east-west Kala Chitta Ranges, Margala Hills (450–900 m). The area is cut into deep valleys. The main Potwar Plateau is an undulating area, 300–600 m, consisting of high and low small hills of bare rock. The Soan River dominates the topography; its tributaries have developed deep gullies and ravines to form typical badland topography, called "khaderas," interspersed with large tracts of mostly cultivated alluvial flats. The Salt Range comprises several parallel ranges rising to an average height of 750–900 m. Several small rivers have cut deep ravines and gorges, creating badlands, and several solution lakes with lush vegetation occur in the area (Chapter 9, habitat types 7a, 9, and 10).

3. North-West-East Altitudinal Gradient
This latitudinal gradient is formed by the piedmont of the Hindu Kush Ranges and the intervening valleys. Several ranges branch out southward to enclose the Chitral, Swat, and Dir Valleys. It continues along the western border as the Safed Koh Range (3600 m), Kohat Hills (1600 m), and Waziristan Hills (1500–3000 m). Local rivers have carved three large valleys: Vale of Peshawar (over 300 m), Kohat Valley (460 m), and Bannu Valley (150 m), all opening toward the south (Chapter 9, habitat type 8).

4. West-East Gradient
The Sulaiman and Kirthar Ranges are located between the Baluchistan Plateau and the Indus Plains, 6000 m high, extending southward near the Arabian Sea, where they decrease to 300 m elevation; comprising the extensive Piedmont Plains between the Indus River and the mountains, formed as alluvial fans by the rivers flowing eastward down the mountains (Chapter 9, habitat type 14d).

5. The Baluchistan Plateau Gradient Complex
The Baluchistan Plateau lies to the west of the Sulaiman-Kirthar Mountains. Here the altitudinal gradient rises and falls several times due to a complex of hills, mountains, and basins: Makran Coast Range (6000 m); Siahan, Ras Koh, and Chagai (3010 m); and Toba Kakar Range

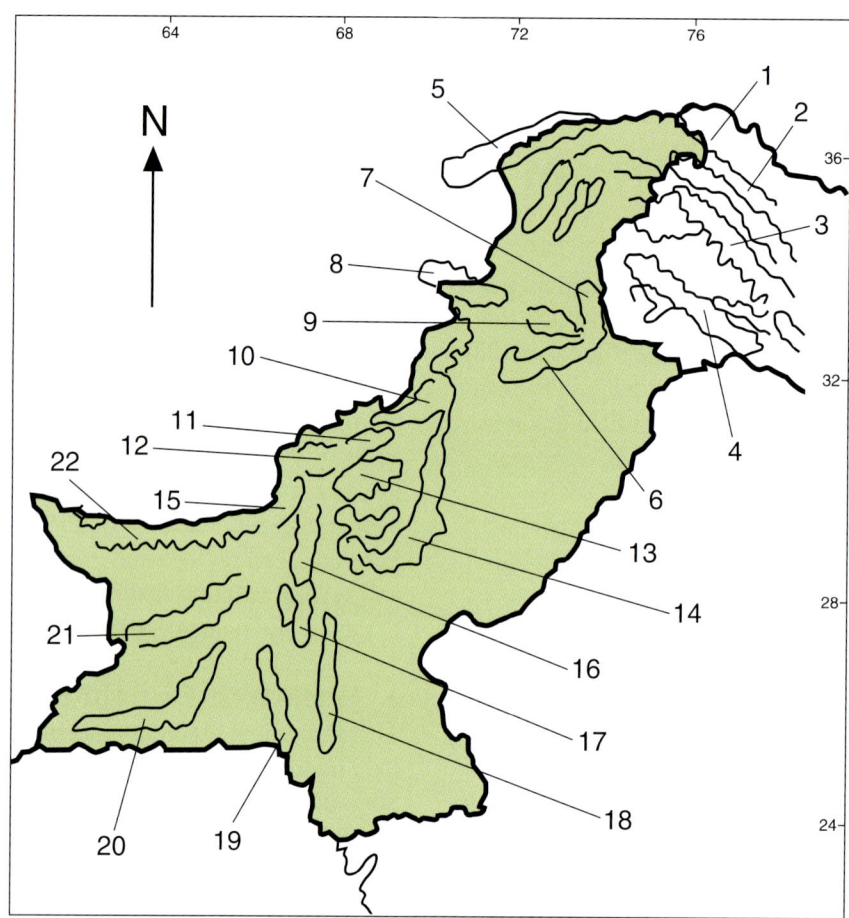

Figure 33. Pakistan, main mountain ranges:
1. Karakorum Range; 2. Kailas Range; 3. Main Himalayan Range; 4. Pir Panjal Range; 5. Hindu Kush Range; 6. Salt Range; 7. Murree Hills; 8. Safed Koh; 9. Kala Chitta Hills; 10. Torghar Hills; 11. Kaliphat Mountain; 12. Takhatu Mountain; 13. Zarghun Mountain; 14. Sulaiman Range; 15. Chiltan Mountain; 16. Koh-i-Maran; 17. Gishk Hills; 18. Kirthar Range; 19. Pab Range; 20. Makran Coast Range; 21. Siahan Range; 22. Chagai Hills.

(1500 m). The basins are located between the hills and mountains. The rivers flowing through the basins are without water most of the year, and the narrow channels are mostly shallow, but in some areas are deep. The foothills sloping into the valley floor are skirted with alluvial fans and talus cones are strewn with boulders. The gravel near the hills slopes gently down into the valley bottom which is covered with fine sands, silts, and clay, and is cultivated. Important basins are Zhob Valley, Baji Valley, Quetta Valley, and Mastung Valley. Between the Chagai Hills and Siahan Range there is a large basin occupied by an extensive desert the "Dest-i-Lut," with extensive sand plains and dunes (Khan, 1980b).

The eastern coast of Baluchistan is occupied by the triangular alluvial Las Bela Plains. The Makran coast is a narrow strip; at several places craggy scarps of hills project into the Arabian Sea, and extensive sand deposits dominate the landscape (Chapter 9, habitat types 5, 6, 7b, 14e, and 15).

6. The Indus Plain Gradient

The catchment area of the Indus River is 963,500 sq km (372,000 sq miles), in the form of a leveled plain. It slopes down, north to south, from 300 m to 75 m near Punjnad. There are a few outcrops of Precambrian rock near Sargodha, Chiniot, and Sangla—the Karana Hills. A few hills of Tertiary limestone are also present at Sukkhur as the Khairpur Hills and Ganjo Takar Hills at Hyderabad, Sind (Chapter 9, habitat types 11, 12, 13, and 14a, b, c).

Chapter 10. Altitudinal Distribution of the Herpetofauna of Pakistan

Table 10.1. Altitudinal distribution of amphibians and reptiles of Pakistan.

0 = Sea level; 1 = 100; 2 = 400; 3 = 800; 4 = 1000; 5 = 1400; 6 = 1800; 7 = 2000; 8 = 2400; 9 = 2800; 10 = 3200; 11 = 3600; 12 = 4000; 13 = 4400 meters; ▬ = Range of taxa.

TAXA	0	1	2	3	4	5	6	7	8	9	10	11	12	13
Bufonidae														
Bufo himalayanus								X	X	X	X	X		
Bufo latastii									X	X				
Bufo melanostictus hazarensis					X	X	X							
Bufo olivaceus			X	X										
Bufo pseudoraddei pseudoraddei										X	X			
Bufo pseudoraddei baturae											X	X		
Bufo siacheninsis									X	X				
Bufo stomaticus		X	X	X	X	X	X	X						
Bufo surdus			X	X										
Bufo viridis zugmayeri			X	X	X	X	X							
Megophryidae														
Scutiger nyingchiensis								X	X	X				
Microhylidae														
Microhyla ornata				X	X	X	X							
Uperodon systoma				X	X									
Ranidae														
Euphlyctis cyanophlyctis cyanophlyctis		X	X	X	X	X	X							
Euphlyctis cyanophlyctis microspinulata				X	X	X								
Euphlyctis cyanophlyctis seistanica				X	X									
Hoplobatrachus tigerinus			X	X	X	X								
Fejervarya limnocharis			X	X	X									
Fejervarya syhadrensis			X	X										
Paa barmoachensis					X	X	X							
Paa hazarensis					X									
Paa sternosignata			X	X	X	X								
Paa vicina						X								
Sphaeroteca breviceps			X	X	X	X	X							
CHELONIIDAE														
Caretta caretta	X													
Chelonia mydas	X													
Eretmochelys imbricata	X													
Lepidochelys olivacea	X													

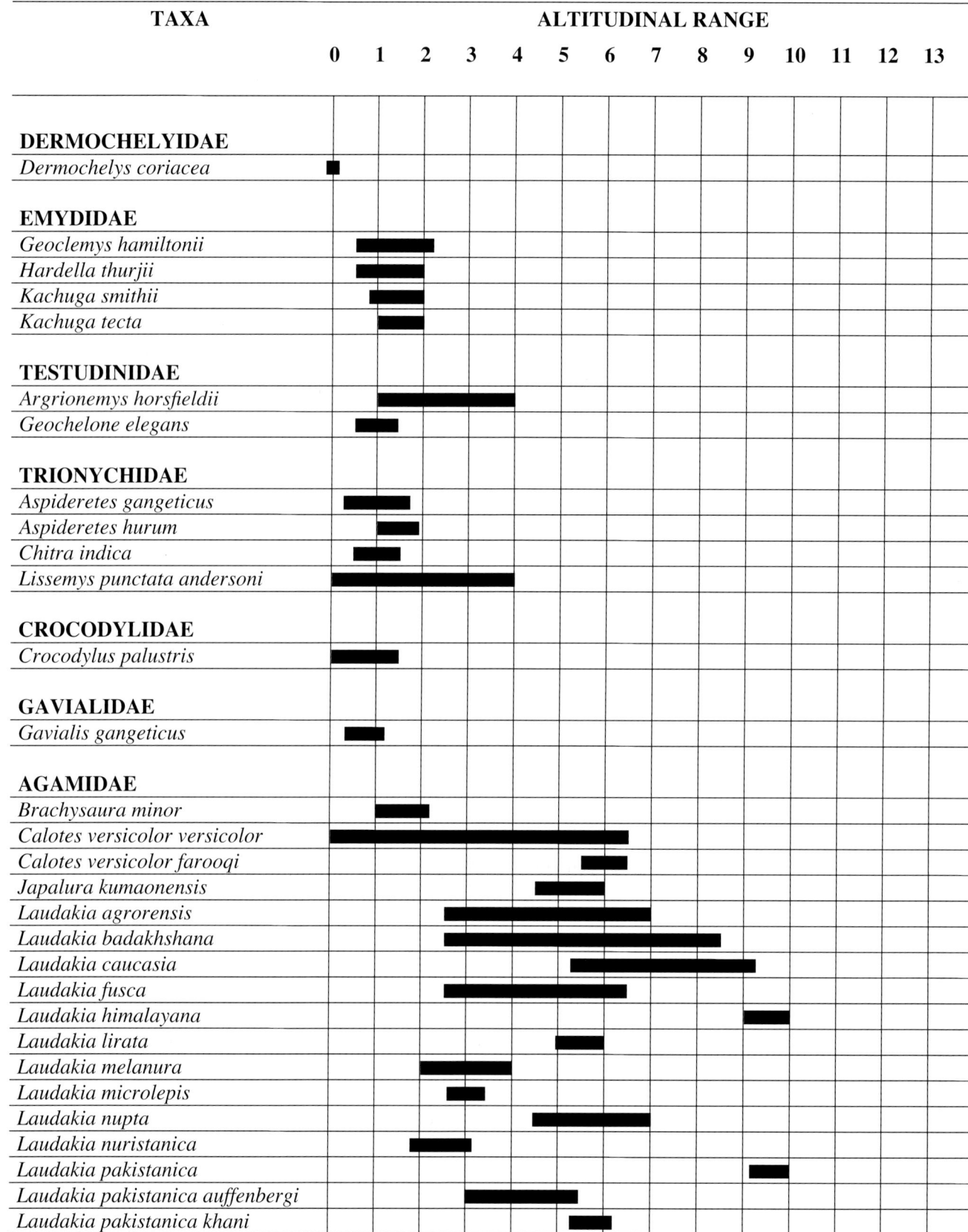

Chapter 10. Altitudinal Distribution of the Herpetofauna of Pakistan

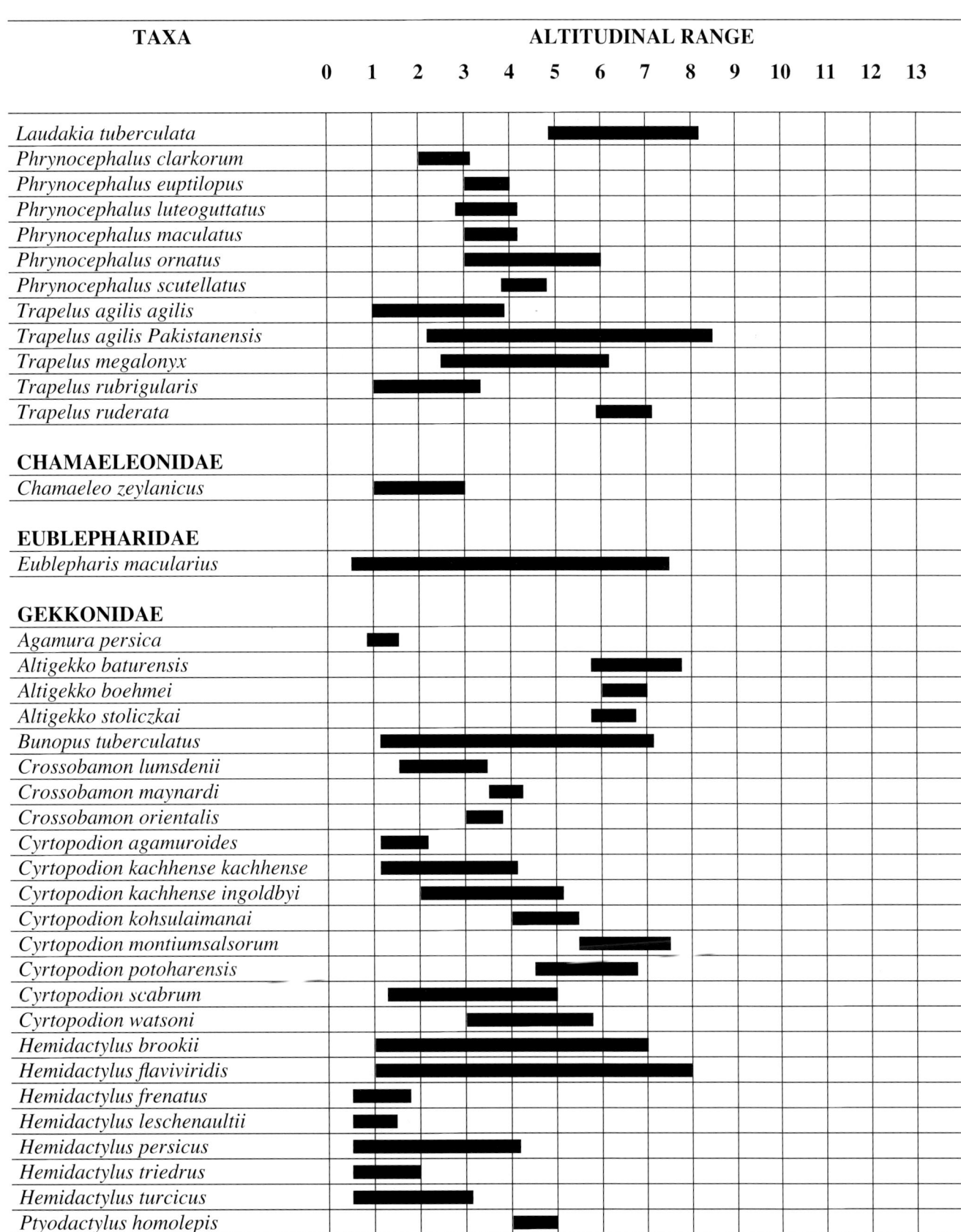

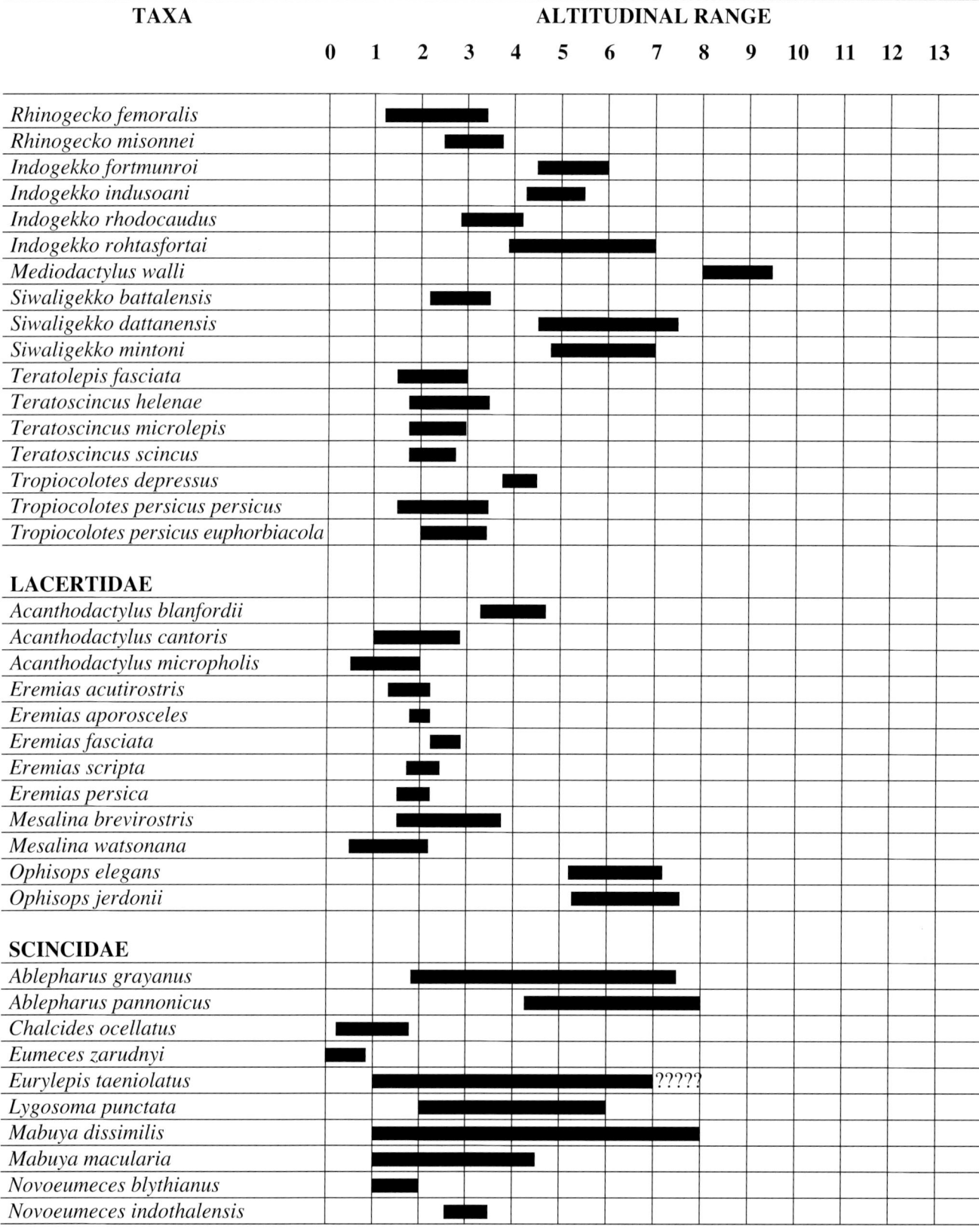

Chapter 10. Altitudinal Distribution of the Herpetofauna of Pakistan

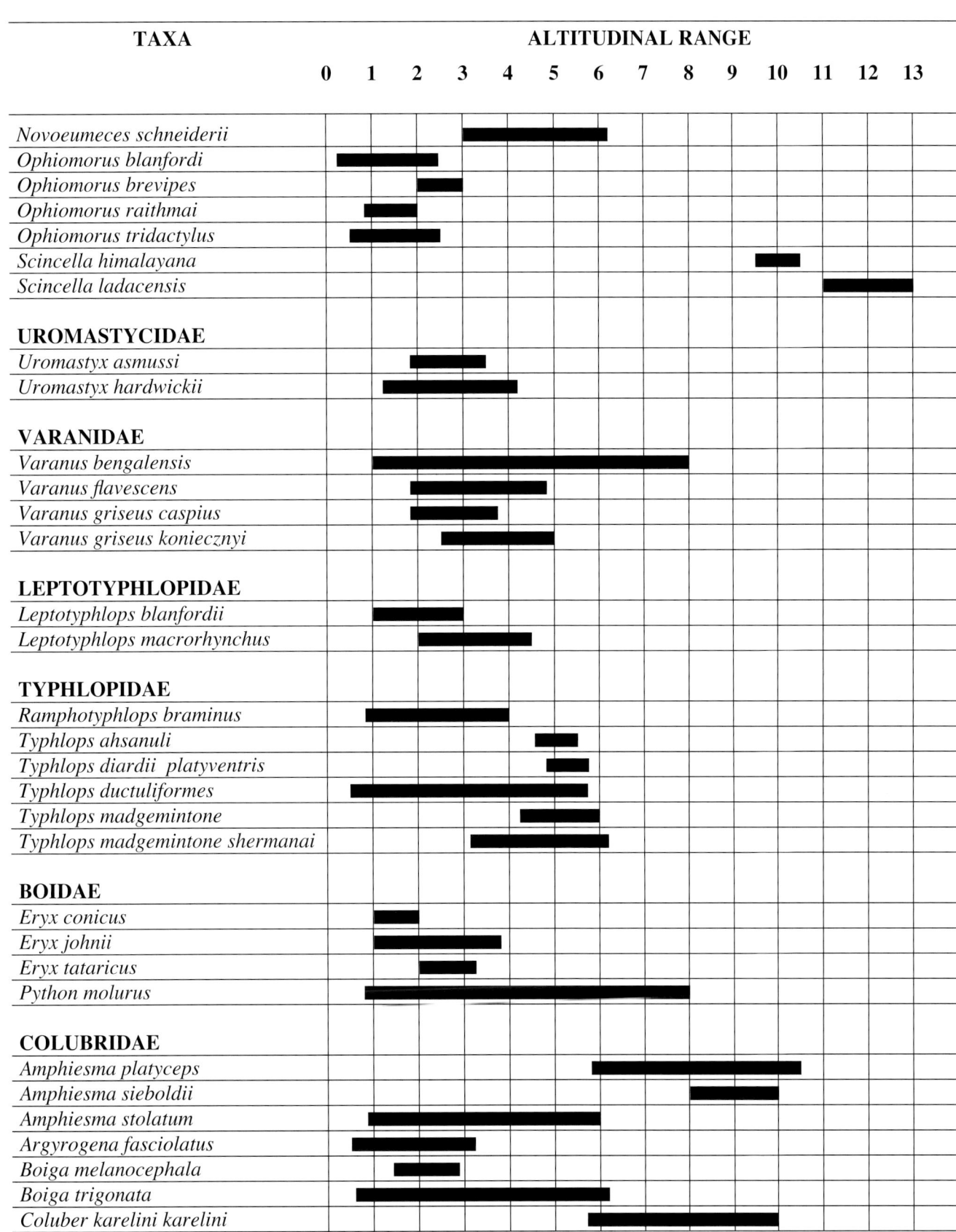

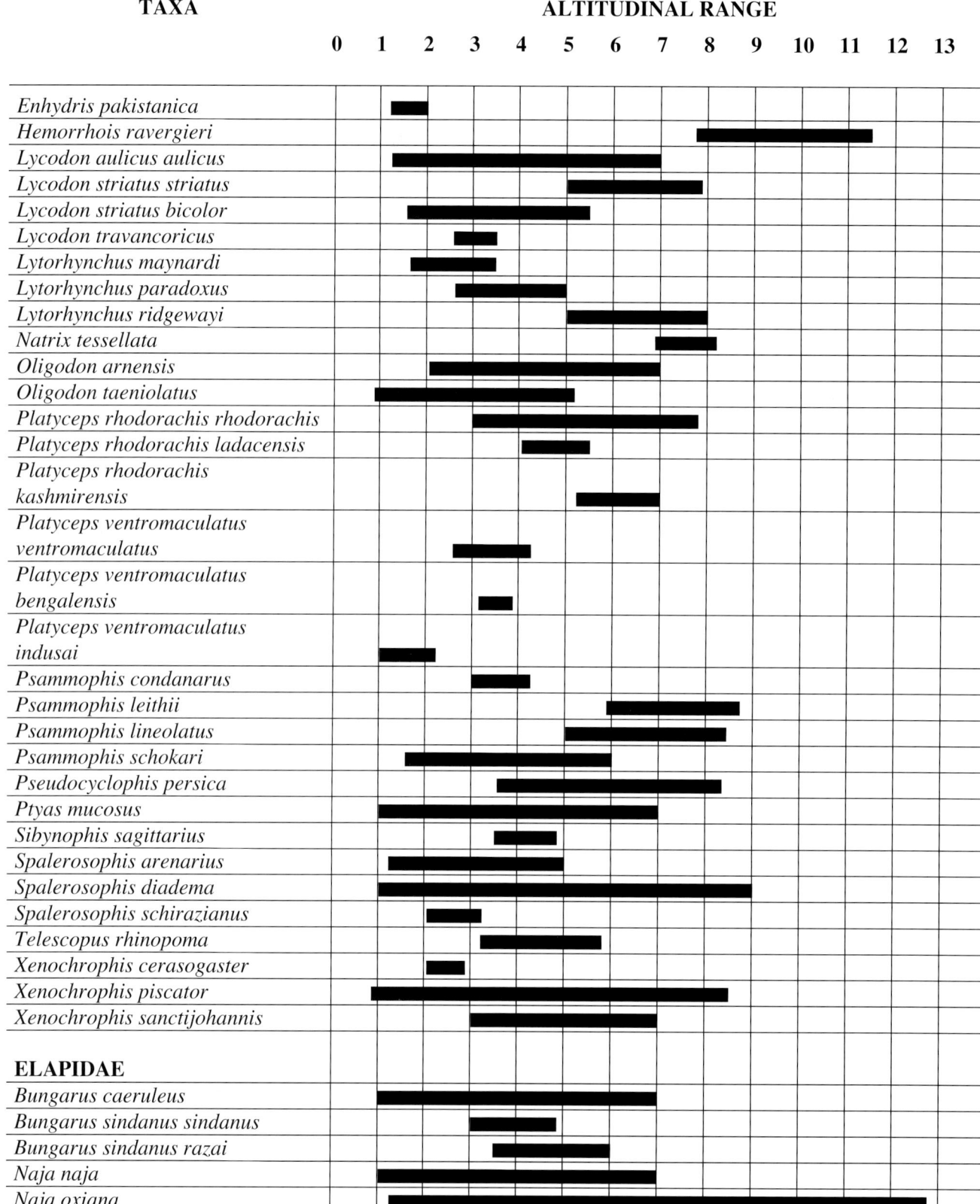

Chapter 10. Altitudinal Distribution of the Herpetofauna of Pakistan

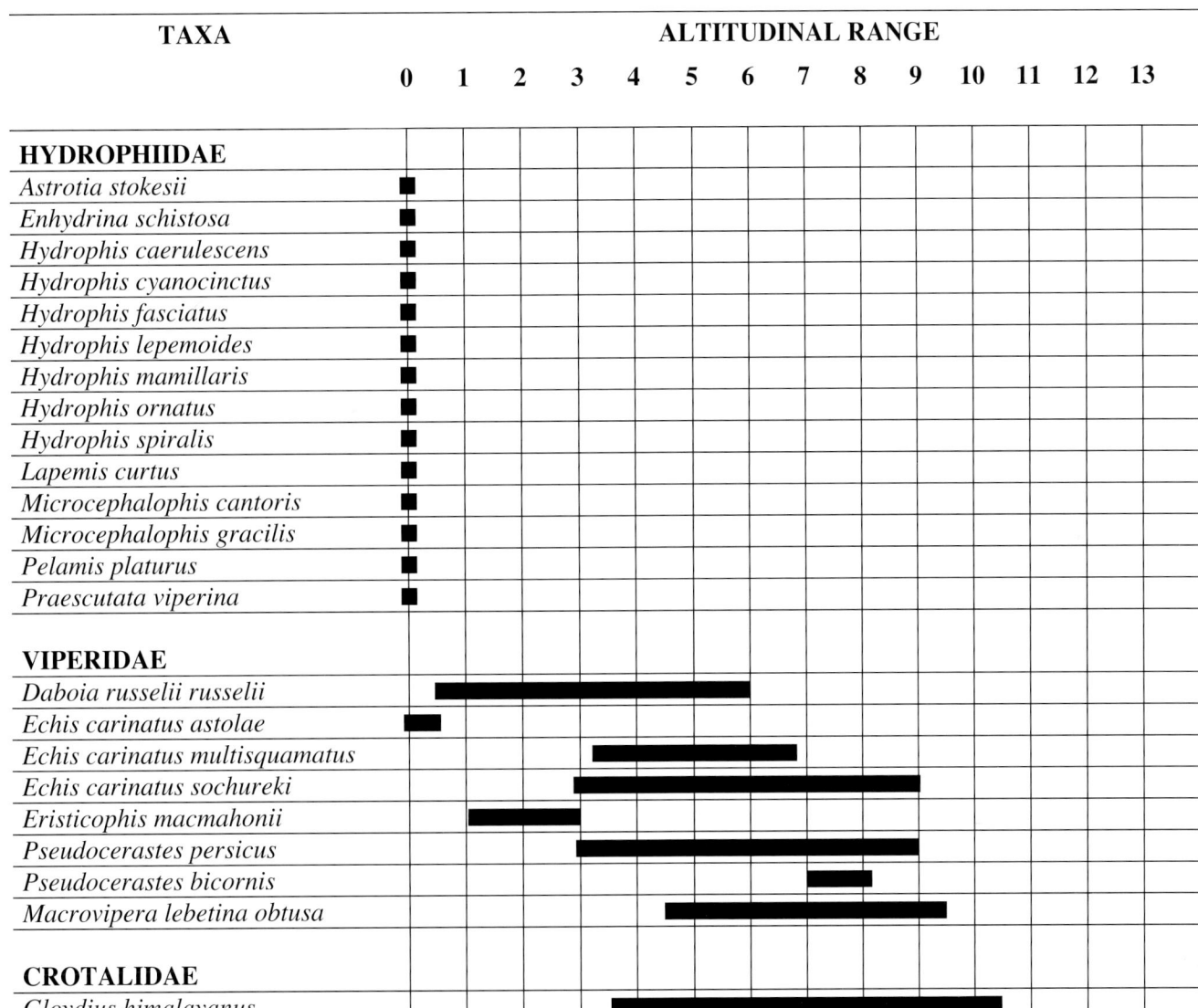

Chapter 11
Snakebite Problem in Pakistan

The worldwide mortality from venomous snakes is estimated to be between 30,000 and 40,000 per year (Swaroop and Grab, 1954). The greatest number of snakebite cases reported is from Southeast Asia, about 25,000 per year. Heaviest mortality is reported from Sri Lanka, India, Bangladesh, and Pakistan (Swaroop and Grab, 1954; Christy, 1967). No reliable snakebite data are available for Pakistan; however, Khan's (1990c) estimates are within 1000 deaths per year. For neighboring India, Whitaker (1978) estimates 6000–9000 deaths and Murthy (1990) raises the figures to 10,000–15,000 per year. In Pakistan most of the snakebite reports are from the most populated parts, Punjab and the Indus Delta, where about 95% of the country's agricultural activity takes place. Khan's (1990c) per year estimates are based partly on data gathered from annual press reports: Punjab 150, Sind 500, NWFP, and Baluchistan are less than 50 combined.

When the British invaded India during the late 18th century, they were confronted with an acute snakebite problem. Within a few years, extensive work was carried out on the taxonomy of snakes of the subcontinent and the nature of their venoms. Several publications, mostly in book form appeared within a few years since; at that time scientific journals were not so popular and were generally unavailable. The best known works of that time on Indian snakes were written mostly by people in the medical profession: Russell (1796, 1801–1809), Günther (1864), Fayrer (1872), Nicholson (1874), Theobald (1876); Ewart (1878), A.J. Wall (1883); F. Wall (1908b). Later, several publications have appeared in India on the taxonomy and identification of venomous snakes of India (Gharpurey, 1935; Deoras, 1965; Whitaker, 1978). The manual by Minton et al. (1968) contains useful information about venomous snakes of the world. Khan (1983, 1990c) deals exclusively with the venomous snakes of Pakistan and their distribution.

Pakistan is primarily an agricultural country. About 80% of its population lives in rural areas, where nearly everyone is engaged daily in some agriculture related activity. In the areas along the eastern half of Punjab, the main causes of casualties are *Bungarus caeruleus, Naja naja, and* Daboia russelii—bites by Echis *carinatus* are rare—whereas in arid northwestern Punjab, the main offenders are *Echis carinatus* and *Naja oxiana,* with bites by *Bungarus caeruleus* being rare. In the reclaimed inter-river tracts in Punjab, *Echis carinatus, Naja naja,* and *Bungarus caeruleus* are the main cause and *Daboia russelii* becomes rare. The picture changes in the Salt Range, where mostly *Echis carinatus* is involved, while *Bungarus caeruleus* and *Naja naja* fall into the second category; here *Daboia russelii* becomes rare. In the lower Sind mesic areas most snakebites are by *Bungarus caeruleus, Naja naja,* and *Daboia russelii,* whereas *Echis carinatus* confines itself to more desertic areas. Bites by *Eristicophis macmahonii, Pseudocerastes persicus, Gloydius himalayanus,* and *Macrovipara lebetina* are rare, since their habitat is rarely trespassed (Table 11.1).

The year-round prevailing aridity necessitates constant irrigation activity, which renders these areas humid and cooler during summer. Humidity attracts frogs, mice, rats, moles, and several species of resident and migratory birds, which thrive on insects and stored food grains. Thatched mud houses with untended growth of bushes and hedges around villages provide ideal nesting sites for birds, and provide ideal refuge to several fossorial secretive snakes which accompany human settlement since they are dependent on anthropogenic ecology which attracts their prey. Rodents invade houses and become resident in holes close to the stored food grains. Resident populations of mice and rats attract several species of snakes and varanid lizards. Snakes usually become resident after consuming the occupant of the hole, and thus are brought into constant encounters with inhabitants of the house, resulting in fatal snakebites. Snakebite results in agony, pain, and death to the victim, anguish and suffering to the family, often followed by long-lasting financial difficulties.

Chapter 11. Snakebite Problem in Pakistan

Table 11.1. Zoogeographical Distribution of Terrestrial Venomous Snakes in Pakistan (+ = present; − = absent)

Species	NWFP	Punjab	Sind	Baluchistan
Bungarus caeruleus	+	+	+	Makran coast
Daboia russelii	southeastern	central	southern along Indus	−
Echis carinatus	+	+	+	+
Eristicophis macmahonii	−	−	−	Iran-Afghan border
Gloydius himalayanus	northern	northeastern	−	−
Macrovipera lebetina	+	−	−	Afghan-Pakistan border
Naja naja	+	+	+	+
Pseudocerastes persicus	southwestern	−	−	Iran-Afghan border

Proper identification of the snake is necessary for the effective treatment of snake envenomation. In this chapter aids are provided for specific identification of the venomous snakes of Pakistan (Tables 11.2 and 11.3), and common symptoms of their bites (Table 11.4).

Table 11.2 refers to the plates and line drawings showing the salient morphological features of the terrestrial venomous snakes of Pakistan helpful in their specific identification.

Key to the Identification of Terrestrial Venomous Snakes of Pakistan

There is no clear-cut morphological distinction between venomous and nonvenomous snakes of Pakistan (Table 11.3). When handling an apparently dead snake for checking its identification, precaution should be taken to keep away from its mouth; sometimes a presumed

Table 11.2. Plates and Figures Pertaining to Terrestrial Venomous Snakes of Pakistan

Family	Species	Plate(s)	Figure(s)
Family Crotalidae	*Gloydius himalayanus*	169 [page 242]	32A, B
Family Elapidae	*Bungarus caeruleus caeruleus*	150A, B[page 223]	30A, Iii
Family Elapidae	*Bungarus sindanus sindanus*	151[page 224]	
Family Elapidae	*Naja naja naja*	152[page 225]	30Bi, ii
Family Elapidae	*Naja oxiana*	153[page 226]	
Family Viperidae	*Daboia russelii russelii*	163A, B[page 234, 235]	31D
Family Viperidae	*Echis carinatus multisquamatus*	164[page 237]	31Bi, ii
Family Viperidae	*Echis carinatus sochureki*	165[page 238]	
Family Viperidae	*Eristicophis macmahonii*	166A, B[page 242]	31A
Family Viperidae	*Macrovipara lebetina*	167[page 240]	
Family Viperidae	*Pseudocerastes persicus*	168[page 241]	31C

dead snake may bite due to reflex action which persists for some time in an apparently dead snake.

The following key is based on the most apparent morphological characters on the body of a Pakistani venomous snake, needing no closer inspection of the specimen.

1. Head covered with large, symmetrically arranged, smooth scales 2
Head covered with small, irregularly arranged, keeled scales 6
2. Pit between eye and nostril; body scales keeled ....*Gloydius himalayanus* (Pit-viper)
Pit between eye and nostril absent..3
3. Loreal scale present ....Family Colubridae (Coluber snakes, nonvenomous)
Loreal scale absent 4
4. Nostrils on dorsal side of the snout; tail flat ....Hydrophiidae (Sea snakes)
Nostrils anterolateral; tail round ending at a point 5
5. Scales of median dorsal row distinctly broad ....Genus *Bungarus* (Krait)
Scales on body dorsum similar ....Genus *Naja* (Cobra)
6. Rostral scale flanked by enlarged, butterfly-shaped scales ....*Eristicophis macmahonii*
No such enlarged scales along rostral scale 7
7. Subcaudals undivided, in a row ....*Echis carinatus* (Saw-scale viper)
Subcaudals divided, in a double row..8
8. A supraocular aggregation of scales forming a horn-like projection ....*Pseudocerastes persicus*
No supraocular scale aggregation forming horn 9
9. Head thick, elongated; nostrils distinctly enlarged, crescent-like ....*Daboia russelii*
Head squarish, nostril small, normal ....*Macrovipara lebetina*

Nonvenomous Snakes Often Confused With Venomous Snakes

Several nonvenomous snakes closely resemble and mimic venomous sympatric species. Table 11.3 sorts out these confusing species.

Table 11.3 Nonvenomous Snakes Often Confused With Venomous Snakes

Venomous Species	Nonvenomous Species
Echis carinatus	*Eryx johnii, E. tataricus, E. conicus, Boiga trigonata, B. melanocephala*
Naja naja	*Ptyas mucosus*
Naja oxiana	*Spalerosophis diadema*
Bungarus caeruleus	*Lycodon aulicus, Lycodon striatus, Argyrogena fasciolata, Platyceps rhodorachis, Pseudocyclophis persicus*
Daboia russelii	*Spalerosophis diadema*

Table 11.4 Common Local Symptoms of Snake Poisoning

Symptoms	Krait	Cobra	Viper
Severe pain and numbness	+	+	–
Swelling of the whole limb	–	1–2 hours	1–2 hours
Discoloration	–	+	+
Blisters	–	+	extending onto body
Local necrosis	–	+	+
Drowsiness	+	+	–
Difficulty in swallowing and in opening mouth and eyes	+	+	–
Severe abdominal pain	+	–	–
Limb weakness	+	+	+
Salivation and difficulty in breathing	+	+	–
Bleeding	–	–	+
Death	quick	quick	delayed

Chapter 12
Threats to the Herpetofauna of Pakistan

The Plains of Punjab and Sind have a very old history of human interference in their natural resources which dates back almost to 4000–1500 BC. At that time the plains were the seat of a highly developed urban society—the Indus civilization—which flourished during 3000 BC; the archeological sites at Mohenjo Daro and Harappa, Sind, speak of its glory. By about 1700 BC, the civilization was in a rapid decline, so that by 1500 BC it had practically been wiped out. The apparent causes of decline are repeated flooding of towns, which were located mostly on the riverbanks; and adverse ecological changes due to felling and destruction of natural vegetation to make room for agriculture which resulted in aridity, boosting the onslaught of desert habitat, which was later followed by tropical thorn forests.

Apart from destruction of original natural vegetation and establishment of agri-orientated habitat, perhaps the other best-known artifacts of the vanished civilization are a number of small steatite seals discovered during excavations. They are distinctive in kind and unique in quality, with depiction of a wide variety of animals which once roamed the land and inland waters of the Indus Valley: otters, alligators, elephants, tigers, rhinoceros, wild ox, and antelopes (Kureshy, 1986; Roberts, 1991; F.K. Khan, 1996).

The present-day Indus Valley is an arid grassland with scattered pockets of subtropical thorn forest cover (Chapter 9). Due to human activity over the last 1000 years, most of the tropical thorn forest cover has disappeared. Now it survives in small pockets where human interference is restricted: around airfields, graveyards, and uncultivated lands such as saline flats or "pats." Now under growing pressure of rising human population, these residual areas are also under the constant mechanical action of reclamation and industrialization (Khan, 1980b). Vast fields of plain land are being constantly and consistently reclaimed to suite the growth of cash crops and the setting up industrial complexes (Khan, 1990b, 1991d).

In Punjab, the subsoil bedrock does not allow natural subsoil drainage. Continuous irrigation activity and profuse canalization since 1859 has raised the subsoil water table, increasing soil moisture and soil salt concentration, which has resulted in waterlogging and salinity problem. Effects of salinity and waterlogging are evident by the presence of scattered saline water lakes dotting the landscape in the Punjab Plains. Several resident secretive amphibian and reptilian species have been exterminated or are on the verge of extermination. These species linger on here and there in small scattered tracts of fast-diminishing original thorn forest habitat. The thick jungles along the water courses have fallen rapidly to the onslaught of agriculture and industrialization. However, in the lower Indus Valley, where the subsoil drainage is normal, the semidesertic, dry, arid habitat is changed to humid, lush, green grain-producing fields, and is the most fertile part of the country, yielding about 80% of country's produce. Here the xerophylic reptiles, mainly Palearctic species, have shrunken westward, making room for the mesophyllic, predominantly Oriental ones that have recently expanded their range westward. Thus the whole scenario of fauna and flora in the Indus Valley has changed (Khan, 1985a).

The arthropod fauna, upon which resident amphibians and reptiles feed, is gone, since the vegetation which supported it has been killed by the saline soil. Extensive land tilling has destroyed their burrows and excavated them out, while the irrigation water has flooded their burrows and hideouts. Helpless creatures, once driven out, are at the mercy of their enemies that have extensively killed them; so much so that, within past few years, several species have been exterminated from the part of the valley where they once abounded. Moreover, due to extensive interference, natural populations of amphibians and reptiles have been broken into small groups, which continue to survive against high odds in small tracts where the original habitat still lingers on, but they are much exposed and are under constant threat of being killed. The inter-riverine forests, which 100 hundred years ago were natural habitat of *Crocodylus palustris*, *Gavialis gangeticus*, *Python molurus*, otters, large cats, elephants, and rhinos, have been cleared, resulting in complete extermination of most of these species.

Pakistani coastal beaches are used by several species of sea turtles, like *Lepidochelys olivacea*, *Chelonia mydas*, and *Dermochelys coriacea*, for egg laying. Beach-visiting female turtles are a bonanza for poachers; they are hunted and their nests plundered for eggs by men and beasts equally, so that most of the sea turtle populations are declining worldwide, and some species are on the verge of extinction (Das, 1991). The turtles of inland waters are under similar hazardous pressure due to habitat destruction and frequent human intrusion in their ecosystem. The heavily silted shallow rivers in Punjab almost regularly overflow during the rainy season, disturbing river habitat and disrupting turtle populations. Eggs of freshwater turtles are hunted, turtle soup is a popular dish and recommended by local physicians—"hakims"—as a cure for several ailments (Vohora and Khan, 1979; Das, 1991). The flood-disturbed turtles are carried away by floodwater, and when the water recedes they roam about in search of suitable habitat. In the effort, several are crushed under the flow of unheeding, rather aggressive traffic. Several suffer mortal injuries from playful adults and children (see Plates 179, 180). Most of the hatchlings are devoured by jackals, foxes, mongooses, kites, falcons, etc., and few manage to survive in surroundings that grow ever more hazardous. There is no study available to demonstrate the effects of the runoff of hazardous pesticides and chemicals—which are being used freely in fields—on the freshwater turtle fauna in Pakistan.

Several nomadic snake charmer tribes are actively participating in destruction and depletion of reptilian populations (Plate 170). They deal in the trade of reptiles and other wild animals. Locally known as "sanyasies," "gagras," and "Tapri-was," they have menaced the natural reptilian population throughout Pakistan (Minton and Minton, 1964; Khan, 1993d). They mercilessly and endlessly hunt several reptilian species like *Varanus bengalensis*, *Varanus griseus*, *Uromastyx hardwickii*, *Uromastyx asmussi*, *Trapelus agilis*, *Python molurus*, *Ptyas mucosus*, *Spalerosophis diadema*, etc., lured by the high prices which skins of these species fetch. Moreover, the body parts of these reptiles are in great demand in local markets, as native physicians use them in preparation of recipes for treatment of several common ailments (Konieczny, 1969a; Vohora and Khan, 1979; Khan, 1993d, 2000b).

Though legislation exists against trade in animals and animal products, the unlawful export of live reptiles and their products is playing havoc with the already decimated natural population of local herpetofauna. Several rare species are being exported illegally to the western countries where they are in great demand as pets. A friend in Belgium recently accosted a dealer in Holland who had 200 *Echis carinatus* and more than 100 *Eristicophis macmahonii* (a very rare snake found only in the Chagai Desert, adjoining Iran, and Afghanistan) and about 30 *Naja naja* from Pakistan, available for sale. The poor animals were in very bad health and were being kept under unhygienic conditions.

To boost the yield of cash crops, improved long-acting pesticides are being used to control crop pests. The effects of pesticides on field animals, toads, frogs, and skinks, *Mabuya dissimilis* and *Eurylepis taeniolatus*, are well illustrated by the number of dead animals lying around the sprayed fields (Khan, 1990b). Most of the victims are time-tested friends of local farmers. Moreover, tadpoles and fishes have been found killed in nearby ponds and puddles receiving most of the runoff water from sprayed fields; birds die by eating sprayed insects and caterpillars (Khan, 1990b).

Scincid species belonging to genera *Mabuya*, *Eurylepis*, *Novoeumeces*, and *Ophiomorus* are in particularly great demand in local markets, as they are sought after at a high price by "hakims." To prepare a catch for market, the animals are mercilessly eviscerated alive, dried in the sun, and sold at high prices as "reg-mahi," an important ingredient of recipes which are said to be strong sex promoters (M.S. Khan and M.R.Z. Khan, 1997).

The body fat of several reptiles is said to have curative properties for several diseases and is widely used in preparation of balms, while *Uromastyx hardwickii* "oil" is considered to have special aphrodisiac properties (Vohora and Khan, 1979; Khan, 1991d). It is extracted from living lizards, the spines of which are

Plate 170. A band of snake charmers with their catch and snake-catching paraphernalia, Sind, Pakistan (Photo courtesy of S.A. Minton; [Minton 1983; Minton and Minton 1964]).

broken to prevent their escape. The belly of the poor animal is slit open and pressed on a hot plate. The animal struggles helplessly as its body fat simmers out, and it slowly succumbs to the heinous treatment (Khan, 2000b).

Uromastyx hardwickii and *Hoplobatrachus tigerinus* are widely used in the educational and research institutions throughout India and Pakistan. Mostly these animals are dissected to demonstrate vertebrate anatomy, and used in physiological experiments. The animal catchers collect *Uromastyx hardwickii* by digging it out of its burrow, while *Hoplobatrachus tigerinus* is caught from ponds and puddles. In Baluchistan *Euphlyctis cyanophlyctis microspinulata* is used for this purpose, since *Hoplobatrachus tigerinus* is not found there. Local populations of these animals are being increasingly depleted, and from certain areas around Lahore and in different parts of Punjab they are almost exterminated (Khan, 1990b, 1991d).

Local venomous snakes *Bungarus caeruleus, Naja naja, Naja oxiana, Echis carinatus,* and *Daboia russelii* are supplied in the hundreds to health institutions for extraction of venom and production of antivenin, not taking into account the damage which is done to the natural populations and ecosystem as a whole. The snakes are not fed and are kept in congested pens under awful unhygienic conditions. Those that succumb to adverse conditions are thrown away and burned. On the other hand, the general public kills snakes and other reptiles on sight, since all species are regarded as venomous and all are victims of the philosophy "kill it before it harms you." Lots of amphibians and reptiles are killed on roads. Khan (1990b) recorded data pertaining to the road-killed common toad, *Bufo stomaticus*. However, several species, i.e., *Euphlyctis cyanophlyctis, Fejervarya limnocharis, Calotes versicolor, Varanus bengalensis, Varanus griseus,* boids, colubrids, and freshwater turtles are killed by playful drivers just for fun and for the amusement of the passengers. The vehicle is purposely maneuvered to crush the animal when it chances to cross the road at the same time (Khan, 1993d). In the countryside, the strength of a turtle shell is usually tested by pelting it with heavy stones or passing a vehicle over it, until the poor animal is crushed dead or is mortally injured (Plates 178, 179,

Chapter 12. Threats to the Herpetofauna of Pakistan

180). Due to pressures from all sides, the populations of the resident reptiles are fast depleting, as demonstrated by the record of killed/alive reptiles received by the author in 1964 and 1998 (Table 12.1). Note the decrease in receipts from 243 in 1964 to 44 in 1998.

Overall, the herpetofauna of Pakistan is under great stress, demanding immediate efforts on the part of conservationists to educate the general public, and

Plate 179. An injured common freshwater turtle, *Lissemys punctata,* was brought to me by a group of children who found the poor animal close to a pond and pelted it with stones. Seeing it injured, they panicked, fearing God's reprisal, and brought it to the Herpetological Laboratory for treatment. Antibiotic ointments were applied to the wound and the turtle was released on the lawn of my house.

Plate 177. A male garden lizard, *Calotes versicolor,* killed while crossing a road, a common road-side mortality incident during the breeding season of the lizard, April to September.

Plate 178. A fresh, road-killed, rare, large Bengal monitor, *Varanus bengalensis,* picked from the roadside in the hills at the Punjab-Baluchistan border. Note the blood-covered head. The poor animal was purposely crushed under a truck during the night.

Plate 180. Despite our efforts, the turtle (Plate 179) was not found. During clearing, about 4 months later, its bones were recovered from a shallow pit under the grass. Note the hole and cracks made by merciless pelting.

make them aware of the importance of these animals in the ecosystem. Moreover, there is dire need for the organization of a force of honest law-enforcement workers for strict implementation of the conservation laws. In addition to 24 species of Pakistani herpetofauna listed in IUCN Red Data Book (Table 12.2), several other Pakistani species come under the definition of endangered species, i.e., any species of plant or animal that is threatened with extinction (see note under Table 12.2).

Table 12.1. Record of Dead/Alive Snakes Received From Rabwah City and Suburbs at Herpetological Laboratory, Rabwah, Pakistan, During 1964 and 1998

Taxa	1964	1998
Family LEPTOTYPHLOPIDAE		
Leptotyphlops macrorhynchus	12	2
Family TYPHLOPIDAE		
Ramphotyphlops braminus	36	4
Family BOIDAE		
Eryx johnii	10	2
Python molurus	0	1 (stranded in flood water under bridge on Chenab River)
Family COLUBRIDAE		
Amphiesma stolatum	58	6
Boiga trigonata	7	2
Platyceps rhodorachis rhodorachis	5	0
Platyceps ventromaculatus ventromaculatus	10	3
Platyceps ventromaculatus indusai	2	1
Lycodon aulicus aulicus	12	3
Lycodon striatus bicolor	4	1
Lytorhynchus paradoxus	2	0
Oligodon taeniolatus	5	2
Ptyas mucosus	18	5
Spalerosophis diadema	17	3
Xenochrophis piscator	9	2
Family ELAPIDAE		
Bungarus caeruleus caeruleus	15	3
Naja naja naja	1	2
Naja oxiana	4	1
Family VIPERIDAE		
Echis carinatus multisquamatus	11	2
Echis carinatus sochureki	5	0

Chapter 12. Threats to the Herpetofauna of Pakistan

Table 12.2 List of Threatened Pakistani Amphibian and Reptilian Species Included in IUCN Red Data Book

Amphibians	**Snakes**
Hoplobatrachus tigerinus	*Python molurus*
Reptiles	*Ptyas mucosus*
Dermochelys coriacea	*Spalerosophis diadema*
Geoclemys hamiltonii	*Xenochrophis piscator*
Kachuga tecta	*Naja naja*
Lissemys punctata	*Naja oxiana*
Argrionemys horsfieldii	*Daboia russelii*
Geochelone elegans	
Aspideretes gangeticus	
Aspideretes hurum	
Crocodylus palustris	
Gavialis gangeticus	
Uromastyx asmussi	
Uromastyx hardwickii	
Varanus bengalensis	
Varanus flavescens	
Varanus griseus caspius	
Varanus griseus koniecznyi	

Additional Species Recommended for Inclusion in IUCN Listing

The way amphibians and reptiles are persecuted and mercilessly killed in Pakistan, the IUCN needs to recommend that all 223 species be placed in the Red Data Book; however, the following species need immediate attention to conserve their rapidly depleting populations:

All species of genera *Eurylepis*, *Novoeumeces*, *Mabuya*, *Scincella*, and *Ophiomorus*	Threatened by extensive use in health recipes as "reg mahi"
Tropicolotes persicus euphorbiacola, *Enhydris pakistanica*, and *Xenochrophis cerasogaster*	Threatened by disturbance in their restricted specialized habitat
Eristicophis macmahonii	Threatened by poaching
Spalerosophis arenarius	Threatened by extensive killing for skin

Bibliography

Abdulali, H. 1935. A dhaman (*Ptyas mucosus*) "rattling" its tail. J. Bombay Nat. Hist. Soc. 37:958.

Abdulali, H. 1960. Notes on the spiny tailed lizard, *Uromastix hardwickii* Gray. J. Bombay Nat. Hist. Soc. 57:421–423.

Abdulali, H. 1962. An account of a trip to the Barapede Cave, Talewandi, Belgaum District, Mysore State, with some notes on reptiles and amphibians. J. Bombay. Nat. Hist. Soc. 59:228–237.

Acton, H.W. 1921. In: *The Practice of Medicine in the Tropics*. (W. Byam and R.G. Archibald, Eds.) 1:757. Henry Frowde and Hodder and Stoughton, London.

Adamson, H. and I. Shaw, 1981. *A Traveller's Guide to Pakistan*. The Asian Study Group, Islamabad.

Adler, K., and E.M. Zhao, 1995.The proper generic name of for the Asian wolf snakes: *Lycodon* (Serpentes: Colubridae). Sichuan J. Zool. 14(2):75.

Ahmad, K.S. 1951. Climate regions of West Pakistan. Pak. Geo. Rev. 6:1–35.

Ahmed, A. 1985a. A preliminary report on the crocodiles of Pakistan. Newsletter, Crocodile Specialist Group 4:5–9, Dec. 1985.

Ahmed, A. 1985b. The distribution and population of crocodiles in the province of Sind and Baluchistan (Pakistan). J. Bombay Nat. Hist. Soc. 220–222.

Ahmed, A. 1989. *Pakistan Tourism Directory 1989*. Holiday Weekly, Karachi-5.

Aiyar, T.V.R. 1907. Notes on some sea snakes caught at Madras. Proc. Asiatic Soc. Bengal, ii:69–72.

Akhtar, S.A. 1972. Zoogeography of the termites of Pakistan. Pak. J. Zool. 6 (1 & 2):85–104.

Akhtar, S.A., and M. Ahmed, 1997. Some features of zoogeographical interest in the biodiversity of termites of Pakistan, pp.213–227, in: *Biodiversity of Pakistan* (S.A. Mufti, C.A. Woods, and S.A. Hasan, Eds.). Pakistan Nat. Hist. Mus. Islamabad and Gainesville, Florida.

Akram, M.S. 1982. The snakes of District Faisalabad. M.Sc. thesis, Zoology Department, Agriculture University, Faisalabad.

Alcock, A.W., and F. Finn, 1896. An account of the amphibians and reptiles collected by Dr. F.P. Maynard, Captain A.H. McMahon, C.I.E., and the members of the Afghan-Baluch Boundary Commission of 1896. J. Asiatic. Soc. Bengal 65, part 2, (4):550–566.

Ananjeva, N.B., and T.M. Sokolova, 1990. [The position of the genus *Phrynocephalus* Kaup, 1825 in agamid systems]. Trudy. Zool. Inst. Acad. Nauk. USSR Leningrad 207:12–21 (in Russian; English abstract).

Anderson, J. 1871. A list of the reptilian accessions to the Indian Museum, Calcutta, from 1865 to 1870, with a description of some new species. J. Asiatic Society Bengal 40:12–39.

Anderson, J. 1872. On some Persian, Himalayan and other reptiles. Proc. Zool. Soc. London 1872:371–404.

Anderson, J.1895. On a collection of reptiles and batrachians made by Colonel Yerbury at Aden and its neighbourhood. Proc. Zool. Soc. London 1872:635–663.

Anderson, J.A. 1964. A report on the gecko *Teratolepis fasciata* (Blyth, 1853). J. Bombay Nat. Hist. Soc. 61:161–171.

Anderson, J.A., and S.A. Minton, 1963. Two noteworthy herpetological records from the Thar Parker Desert, West Pakistan. Herpetologica 19(2):152.

Anderson, S.C. 1961. A note on the synonymy of *Microgecko* Nikolsky with *Tropiocolotes* Peters. Wassman J. Biol. 19:287–289.

Anderson, S.C. 1963. Amphibians and reptiles from Iran. Proc. Calif. Acad. Sci., ser. 4, 31(16):417–498.

Anderson, S.C. 1968. Zoogeographical analysis of the lizard fauna of Iran. Chapter 10:305–371, In: *The Cambridge History of Iran*. Vol.1, W.B. Fisher (Ed.), Cambridge University Press, London and New York.

Anderson, S.C. 1999. *The Lizards of Iran*. Society for the study of Amphibians and Reptiles, Ithaca, New York.

Anderson, S.C., and A.E. Leviton, 1969. Amphibians and reptiles collected by the Street Expedition to Afghanistan, 1965. Proc. California Acad. Sci. (4th Ser.) 37(2):25–56.

Andrushko, A.M., and N.E. Mikkau, 1964. Distribution and life history of *Lytorhynchus ridgewayi* Boulenger, 1887 with ecological and geographical review of the genus *Lytorhynchus* Peters, 1862. Westnik Lenningrad. Univers.19: Nr.9, Biol. Ser.:1–19.

Annandale, N. 1907. The distribution of *Bufo andersoni*. Rec. Indian Mus.1:171–172.

Annandale, N. 1913. The Indian geckoes of the genus *Gymnodactylus*. Rec. Indian Mus. 9:309–362.

Annandale, N., and C.R.N. Rao, 1918. The tadpoles of the families Ranidae and Bufonidae found in the plains of India. Rec. Indian Mus. 15:25–40.

Arnold, E.N. 1983. Osteology, genitalia and the relationships of *Acanthodactylus* (Reptilia: Lacertidae). Bull. British Mus. (Nat. Hist.). Zool. Ser. 44(5):291–339.

Arnold, E.N. 1986. The hemipenis of lacertid lizards (Reptilia: Lacertidae): structure, variation and systematic implications. J. Nat. Hist. 20:1221–1257.

Asana, J.J., and R.G. Khardi, 1937. The chromosomes of *Rana tigerina*. Curr. Sci. 5:649.

Ataev, Ch.A. 1985. Reptiles of mountainous Turkmenistan. Ashkabad Ilym, pp. 1–410.

Auffenberg, K. 1997. The biogeography of land snails of Pakistan. pp. 253–275, in: *Biodiversity of Pakistan*. (S.A. Mufti, C.A., Woods, and S.A. Hasan). Pakistan. Mus. Nat. Hist. Islamabad and Gainesville, Florida.

Auffenberg, W. 1980a. Behaviour of *Lissemys punctata* (Reptilia, Testudinata, Trionychidae) in a drying lake in Rajasthan, India. J. Bombay Nat. Hist. Soc. 78:487–493.

Auffenberg, W. 1980b. Autecological notes on *Xenochrophis piscator* (Reptilia: Serpentes) from Keoladeo Ghana sanctuary. InT.J. Eco. Environ. Sci. 6:77–82.

Auffenberg, W. 1981. Combat behaviour in *Varanus bengalensis* (Sauria, Varanidae). J. Bombay. Nat. Hist. Soc. 78(1):54–72.

Auffenberg, W. 1983a. The burrow of *Varanus bengalensis*: Characteristics and use. Rec. Zool. Survey. India, 80(3–4):375–385.

Auffenberg, W. 1983b. Courtship behavior in *Varanus bengalensis* (Sauria: Varanidae). in: *Advances in Herpetology and Evolutionary Biology*. Mus. Comp. Zoology, Cambridge:537–551. (A.C.J. Rhodin and K. Miyata, Eds.)

Auffenberg, W., 1984. Notes on feeding behaviour of *Varanus bengalensis* (Sauria, Varanidae). J. Bombay, Nat. Hist. Soc. 80:286–302.

Auffenberg, W. and N. Ahmad, 1991. Studies of Pakistan reptiles: Notes on *Kachuga smithi*. Hamadryad 16(1–2):25–29.

Auffenberg, W., and H. Rehman, 1991. Studies on Pakistan Reptiles. Pt. 1. The genus *Echis* (Viperidae). Bull. Florida Mus. Nat. Hist. 35(5):263–314.

Auffenberg, W., and H. Rehman, 1993. Studies on Pakistan Reptiles. Pt. 3. *Calotes versicolor*. Asiat. Herpetol Res., 5:14–30.

Auffenberg, W., and H. Rehman, 1997. Geographic variations in *Bufo stomaticus*, with remarks on *Bufo olivaceus*: biogeographical and systematic implications, pp. 351–372. in: *Biodiversity of Pakistan*. (S.A. Mufti, C.A. Woods, and S.A. Hasan, Eds.) Pakistan. Mus. Nat. Hist., Islamabad and Gainesville, FL.

Axelrod, D.I. 1960. The evolution of flowering plants. in: *Evolution after Darwin*. (Sol Tax, Ed.) Vol.1:27–305. Univ. Chicago Press, Chicago.

Baig, K.J. 1988. New record of *Agama nuristanica* (Sauria: Agamidae) from Pakistan. Biologia (Lahore), 34(1):199–200.

Baig, K.J. 1989. A new species of *Agama* (Sauria, Agamidae) from northern Pakistan. Bull. Kitakyushu Mus. Nat. Hist. 9:117–122.

Baig, K.J. 1990. *Japalura kumaonensis*, first record of new genus and species from Pakistan. Herpetol. Rev., 21(1):22.

Baig, K.J. 1997. Distribution of *Laudakia* (Sauria: Agamidae) and its origin. pp. 373–381. in: *Biodiversity of Pakistan*. (S.A. Mufti, C.A., Woods, and S.A. Hasan, Eds.) Pakistan Mus. Nat. Hist. Islamabad and Gainesville, FL.

Baig, K.J. 1998. A new species of *Tenuidactylus* (Sauria: geckonidae) from Baluchistan, Pakistan. Hamadryad 23(2):127–132.

Baig, K.J. 1999. Description and ecology of a new subspecies of black rock agama, *Laudakia melanura* (Sauria: Agamidae) from Baluchistan, Pakistan. Russian J. Herpetology 6(2): 81–86.

Baig, K.J., and W. Böhme. 1996. Description of two new subspecies of *Laudakia pakistanica* (Sauria: Agamidae). Russian J. Herpetol., 3(1):1–10.

Baig, K.J., and L. Gvozdik. 1998. *Uperodon systoma* (Schneider): Record of a new microhylid frog from Pakistan. Pakistan J. Zool. 30:155–156.

Balletto, E.M., M.A. Cherchi, and J. Gasperetti. 1985. Amphibians of the Arabian peninsula. Faun. Saudi Arab. 7:318–392.

Basu, D. 1998. Female reproductive cycle in *Hardella thurjii* Gray from northern India. Hamadryad 22(2):95–106.

Bauer, A.M., A.P. Russell, and R.E. Shadwick. 1993. Skin mechanics and morphology of the gecko *Teratoscincus scincus*. Amphibia-Reptilia 14(4): 321–331.

Beg, A.R. 1975. Wildlife habitats of Pakistan. Pak. Forest Inst. Bull. No. 5 (Botanical Branch), pp.1–57.

Beveridge, A.S. 1979. *Babur-Nama*. Sangemeel Publication, Lahore.

Bhaduri, J.L. 1944. On two salientian tadpoles, *Rana blanfordii* Boulenger and *Bufo himalayanus* Günther, from the Ha Valley, Bhutan, Eastern Himalayas. J. Royal Asiatic Soc. Bengal, 10: 53–57.

Biswas, S. and D.P. Sanyal. 1977. Fauna of Rajasthan, India. Part: Reptilia. Rec. Zool. Survey India 73:247–269.

Blair, W.F. 1961. Calling and spawning seasons in a mixed population of anurans. Ecology 42:99–110.

Blair, W.F. 1964. Evolution at populational and intrapopulational levels: isolating mechanisms and interspecies interactions in anuran amphibians. Quart. Rev. Biol. 39:333–344.

Blanford, W.T. 1874a. Descriptions of new lizards from Persia and Baluchistan. Ann. Mag. Nat. Hist. London, ser. 4, 13(78):453–455.

Blanford, W.T. 1874b. Descriptions of new reptilia and amphibia from Persia and Baluchistan. Ann. Mag. Nat. Hist. London, ser. 4,14:31–35.

Blanford, W.T. 1876a. *Eastern Persia, an Account of the Journeys of the Persian Boundary Commission, 1870–1872.* Vol. 2, The Zoology and Geology. Macmillan and Company, London.

Blanford, W.T. 1876b. On some new lizards from Sind with descriptions of new species of *Ptyodactylus, Stenodactylus* and *Trapelus*. J. Asiatic Soc. Bengal 45: pt.2:18–26.

Blanford, W.T. 1879. Notes on a collection of reptiles made by Major O.B. St. John at Ajmere in Rajputana. J. Asiatic. Soc. Bengal 48, pt. 2:119–127.

Blyth, E. 1854. Report of the curator, Zoological DepartmenT.J. Asiat. Soc. Bengal 23(7):737–740.

Bobrov, V.V. 1997. On the boundary between the Palearctic and Indomalayan faunistic kingdoms in the mainland part of Asia. With special reference to the distribution of lizards (Reptilia, Sauria). Biology Bull. 24(5):476–487.

Bobrov, V.V. 1998. On the boundary between the Mediterranean and Sahara-Gobi Faunistic areas of the Palearctic with special reference to the distribution of lizards (Reptilia, Sauria). Biology Bull. 25(5):474–484.

Bole Gowda, B.N. 1948. The spermatogenesis of *Uperodon systoma*. Unpublished M.Sc. dissertation, Mysore University, Mysore.

Börner, A-R. 1974. Ein neuer Lidgecko der Gattung *Eublepharis* Gray, 1827. Mis. Art. Saurologica (4):5–14.

Börner, A-R. 1976. Second Contribution to the systematics of the southwest Asian lizards of the geckonid genus *Eublepharis* Gray, 1827: Materials from the Indian subcontinent. Saurologica (2):1–15.

Börner, A-R. 1981. Third contribution to the systematics of the southwest Asia lizards of the geckonid genus *Eublepharis* Gray, 1827: Further materials from the Indian Subcontinent. Saurologica (3):1–7.

Boulenger, G.A. 1890. *Fauna of British India, including Ceylon and Burma.* Reptilia and Batrachia. Taylor and Francis, London.

Boulenger, G.A. 1920. A monograph of the South Asian, Papuan, Melanesian and Australian frogs of the genus *Rana*. Rec. Indian Mus., Calcutta 20:1–226.

Brandstaetter, F. 1992. Observations on the territory-marking behaviour of the gecko *Eublepharis macularius*. Hamadryad 17:17–20.

Buffetaut, E., and R. Ingavat, 1985. The Mesozoic vertebrates of Thailand. Sci. American, August 1985:64–70.

Carr, A., L. Ogren, and C. McVea. 1980. Apparent hibernation by the Atlantic loggerhead turtle *Caretta caretta* off Cape Canaveral, Florida. Biol. Conserv. 19:7–14.

Champion, H.G., S.K. Seth, and G.M. Khattak, 1966. *Forest types of Pakistan*. Pak. Forest Inst. Peshawar.

Chatterjee, K., and S.K. Barik.1970. A study of chromosomes of both sexes of the Himalayan toad, *Bufo himalayanus*. Genen phaenen. 14:1–3.

Cherlin, V.A. 1983. New facts on the taxonomy of snakes of the genus *Echis*. Vest. Zool. 1983(2):42–46. (Smithsonian Herp. Inform. Ser. No. 61, 1984 [English translation]).

Cherlin, V.A. 1984. New facts on the taxonomy of snakes of the genus *Echis*. Translation of Russian text by F.S.H. Owusu, Smithsonian Herpetol. Info. Service No.:1–7.

Cherlin, V.A. 1990. [Taxonomic revision of of the snake genus *Echis* (Viperidae). II. An analysis of the history of study and synonymy and description of new forms.] Proc. Zool. Inst. U.S.S.R. Acad Sci. Leningrad 204: 193–223. (In Russian.)

Cherlin, V.A. and Borkin, L.J. 1990. Taxonomic revision of the snake genus *Echis* (Viperidae). I. Analysis of the history of study and synononymy.] Proc. Zool. Inst. USSR Acad Sci. Leningrad, Proc. Zool. Inst. USSR Acad. Sci. Leningrad, 207:175–192. ([.)

Christy, N.P., Ed. 1967. Poisoning by venomous animals. American J. Medicine 42:107–128.

Church, G. 1959. Size variation in *Bufo melanostictus* from Java and Bali (Amphibia). Treubia 25:113–126.

Church, G. 1960. Annual and lunar periodicity in the sexual cycle of the Javanese toad, *Bufo melanostictus* Schneider. Zoologica 44:181–188.

Clark, R.J. 1990. A report on herpetological observations in Afghanistan. Brit. Herp Soc. Bull. 33:20–42.

Clark, R.J., E. Clark, S.C. Anderson, and A.E. Leviton. 1969. Report on a collection of amphibians and reptiles from Afghanistan. Proc. Calif. Acad. Sci. (4th Ser.) 36:279–316.

Constable, J.D. 1949. Reptiles from the Indian peninsula in the Museum of Comparative Zoology. Bull. Mus. Comp. Zool. Harvard 103:59–160.

Courtillot, V., and G.E. Vink, 1983. How continents break up. Sci. American 49(1):41–48.

Daniel, J.C. 1963a. Field guide to the amphibians of western India. Part I. J. Bombay Nat. Hist. Soc. 60:415–438.

Daniel, J.C. 1963b. Field guide to the amphibians of western India. Part 2. J. Bombay Nat. Hist. Soc. 60:690–702.

Daniel, J.C. 1975. Field guide to the amphibians of western India. Part 3. J. Bombay Nat. Hist. Soc. 72:506–522.

Daniel, J.C. 1983. *The Book of Indian Reptiles*. Bombay Nat. Hist. Soc.

Darlington, P. J., Jr. 1957. *Zoogeography: The Geographical Distribution of Animals*. John Wiley, New York.

Das, I. 1991. *Colour Guide to the Turtles and Tortoises of the Indian Subcontinent*. R&A Publishing Limited, Portshead, England.

Das, I. 1998. *The Serpent's Tongue: a Contribution to the Ethnoherpetology of India and Adjacent Ccountries*. Edition Chimaira, Frankfurt am Main.

Das, I., and S.K. Dutta, 1998. Checklist of the amphibians of India, with English common names. Hamadryad 23(1):63–68.

Dattatri, S. 1985. In search of Himalayan pit viper. Hamadryad 10(1):10–11.

Deoras, P.J. 1965. *Snakes of India*. National Book Trust, New Delhi, India.

Dotsenko, I.B. 1985. [Revision of the genus *Eirenis* (Reptilia, Colubridae). Communication 1. Resurrection of the genus *Pseudocyclophis* Boettger, 1888.] Vestn. Zool. 1985(4):41–44. (In Russian.)

Dubois, A. 1975. Un nouveau sous-genre (*Paa*) et trois nouvelles espèces du genere *Rana*. Remarques sur la phylogenie des Ranidés (Amphibiens, Anoures). Bull. Mus. Nat. D'Histor. Natur. 3e ser. 231: 1093–1115.

Dubois, A. 1976. Les grenouilles de sousgenre *Paa* du Nepal famille Ranidae, genre *Rana* (famille Ranidae, genere *Rana*). Cah. Nep. Doc. 6:1–vi + 1–275.

Dubois, A. 1981. Liste des genres et sous-generes nominaux de ranoidea (Amphibien, Anoures) du monde, avec identification de leure especes-type-consequences nomenclaturales. Monnitore Zool. Italino 13:225–284.

Dubois, A. 1983a. Classification et nomenclature supra-generique des amphibiens anoures. Extrait du Bull. Mensuel de la Soc. Linn. de Lyon 52:270–276.

Dubois, A. 1983b. Note preliminaire sur le groupe de *Rana (Tomopterna) breviceps* Schneider, 1799 (Amphibiens, Anoures), avec diagnose d'une sous-espèce nouvelle de Ceylon. Alytes 2:163–170.

Dubois, A. 1984. Miscellanea nomenclatorica batrachologica, (IV). Alytes 160–162.

Dubois, A. 1986a. Diagnose preliminaire sur le groupe de *Rana limnocharis* (Amphibiens, Anoures) du snd de l'Inde. Alytes, 4:113–118.

Dubois, A. 1986b. Miscellanea Taxinomica batrachologia (I). Alytes 5:7–95.

Dubois, A. 1992. Notes sur la classification des Ranidae (Amphibiens Anoures). Bull. Mens. Soc. Linn. Lyon 61(10): 305–352.

Dubois, A. and M.S. Khan. 1979. A new species of frog (genus *Rana*, subgenus *Paa*) from northern Pakistan (Amphibia, Anura). J. Herpetol. 13(4): 403–410.

Dubois, A., and J. Martens. 1977. Sur les crapauds du groupe de *Bufo viridis* (Amphibiens, Anoures) de l'Himalaya Occidental (Cachemire et Ladakh). Bulletin de la Societe Zoologique de France 102(4):459–465.

Duda, P.L., and V.K. Gupta. 1982. Trans abdominal migration of ova in some fresh water turtles. Pro. Indian Acad. Sci. (Anim. Sci.) 91:189–197.

Duméril, A., and G. Bibron. 1834–1854. *Erpetologie generale*. Librairie Encyclopédique de Roret, Paris, Vol.1–9.

Eckert, S.N., and C. Luginbuhl. 1988. Death of a giant. Mar. Turtle Newsl. (43):2–3.

Eckert, S.N., D.W. Nellis, K.L Eckert, and G.L. Kooyman. 1986. Diving pattern of two leatherback sea turtles (*Dermochelys coriacea*) during interesting interval at Sandy Point, St. Croix, U.S. Virgin Islands. Herpetologica 42:381–388.

Eiselt, J., and J.F. Schmidtler. 1973. Froschlurche aus dem Iran unter Berucksichtigung ausseriranischer Populationsgruppen. Ann. Naturhistor. Mus. Wien. 77:181–243.

Eremchenko, V.K., and N.N. Szczerbak. 1986. [Ablepharine lizards in the Fauna of the USSR and Neighboring Countries.] Acad. Nauk Kirg. SSR, Ylym, Frunze, 171 pp. (In Russian.)

Estes, R.D., K. De Queiroz, and J.A Gautier, 1988. Phylogenetic relationships within squamata, pp.119–211. *In*: Phylogenetic relationships of the lizard families. R.D. Estes and G.K. Pregill, Eds. Stanford University Press, Stanford.

Ewart, J. 1878. *The Poisonous Snakes of India: For the Use of Officials and Others Residing in the Indian Empire*. J.&A. Churchill, London.

Fayrer, J. 1872. *The Thanatophidia of India, With an Account of Their Poison and a Series of Experiments*. Churchill, London.

Fenton, L.L. 1910. The snakes of Kashmir. J. Bombay Nat. Hist. Soc. 19:1002–1004.

Ferguson, H.S. 1904. A list of Travancore batrachians. J. Bombay Mus. Nat. Hist. Soc. 1904:499–509.

Firdous, F. 1989. Male leatherback strands in Karachi. Mar. Turtle Newsl. (47):14–15.

Fitzinger, L.J. 1823. *In*: Lichtenstein, M.H.C., Universitat zu Berlin nebst Beschreibung vieler biserunbekannter Arten von Saugethieren, Vogeln, Amphibien und

Fischen. Verzeichniss der Doubletten des zoologischen Museums. Berlin.

Fitzinger, L.J. 1843. *Systema reptilium*. Fasciculus Primus. Amblyglossae (Conspectus Geographicus). Braunmüller and Seidel, Vienna.

Frost, D.R., and R.E. Etheridge. 1989. A phylogenetic analysis and taxonomy of iguanian lizards (Reptilia: Squamata). University of Kansas Museum of Natural History Misc. Publ. 81:ii+65.

Fühn, I.E. 1969. Revision and redefinition of the genus *Ablepharus* Lichtenstein, 1823 (Reptilia, Scincidae). Revue Rouma. De Biol. ser. Zool. 14(1):23–41.

Gans, C., and M. Latifi. 1973. Another case of presumptive mimicry in snakes. Copeia 1973(4): 801–802.

Gasperetti, J. 1988. Snakes of Arabia. Fauna of Saudi Arabia National Commission for Wildlife Conservation and Development, Riyadh, Saudi Arabia 9:169–450

Ghalib, S.A., and S.S.H. Zaidi. 1976. Observations on the survey and breeding of marine turtles of the Karachi coast. Agri. Pakistan 27:87–96.

Gharpurey, K.G. 1960. *The Snakes of India and Pakistan*. Popular Book Depot, Grant Road, Bombay.

Gloyd, H.K., and R. Conant, 1982. The classification of the *Agkistrodon halys* Complex. Jap. J. Herp. 9(3):75–78.

Gloyd, H.K., and R. Conant, 1990. *Snakes of the Agkistrodon Complex: A Monographic Review*. Soc. Study of Amphib. and Reptiles, Athens, Ohio.

Golubev, M., M.S. Khan, and S.C. Anderson. 1995. On the systematics of some Palearctic geckos. Abstracts of the Second Asian Herpetological Meeting. Ashgabat, 6–10 September, 1995:23–24.

Golubev, M., and Szczerbak, N. 1981. [A new species of genus *Gymnodactylus* Spix 1823 (Reptilia, Sauria, Gekkonidae)—from Pakistan.] Vestnik Zool. 1981 (3):40–45. (In Russian.)

Greer, A.E. 1970. A subfamilial classification of scincid lizards. Bull. Mus. Comp. Zool., Harvard University 139:151–183.

Griffith, H., A. Ngo, and R.W. Murphy. 2000. A cladistic evaluation of the cosmopolitan genus *Eumeces* Wiegmann (Reptilia, Squamata, Scincidae). Russian Journal of Herpetology 7(1):1–16.

Grismer, L.L. 1988. The phylogeny, taxonomy classification, and biography of eublepharid geckos (Reptilia: Squamata). *In*: R.D. Estes and G.K. Pregill, Eds. *Phylogenetic Relationships of the Lizard Families*. Stanford University Press, Stanford, pp. 367–469.

Groombridge, B. 1987. *A Preliminary Marine Turtle Survey on the Makran Coast, Baluchistan, Pakistan (With Notes on Birds and Mammals)*. IUCN, Cambridge.

Gruber, U. 1981. Notes on the herpetofauna of Kashmir and Ladakh. British J. Herpetol. 6(5):145–150.

Günther, A. 1864. *The Reptiles of British India*. Ray Society, London.

Gupta, V.K. 1987. Retention of egg by the emydine turtle *Kachuga tectum tectum* and *Kachuga smithii*. J. Bombay Nat. Hist. Soc. 64:445–446.

Gvozdik, L., and H. Radek. 1997. A small collection of amphibians from Baluchistan and Punjab, Pakistan, in the Silesian Museum, Opava. Cas. Slez. Muz., Opava (A)46:203–208.

Hahn, D.E., and W. Wallach. 1998. Comments of the systematics of Old World *Leptotyphlops* (Serpentes: Leptotyphlopidae), with description of a new species. Hamadryad 23(1):50–62.

Hardwicke, T. and J.E. Gray. 1827. A synopsis of the species of saurian reptiles, collected in India by Major General Hardwicke. Zool. Jour. 3(10):213–229.

Heaney, L.R. 1985. Zoogeographic evidence for middle and late Pleistocene land bridges to the Philippine Islands. *In:* Modern Quaternary Research in South-East Asia, 9 (G. Bartstra and W.A. Casparie, Eds.) A.A. Balkema, Rotterdam, pp. 127–143.

Heaney, L.R.1991. A synopsis of climatic and vegetational change in Southeast Asia. Climatic Change 19:53–61.

Hemmer, H., J.F. Schmidtler, and W. Böhme. 1978. Zur Systematik zentralasiatischer Grunkroten (*Bufo viridis*-Komplex) (Amphibia, Salientia, Bufonidae). Zool. Abh. Staat. Mus. Tierkd. Dresden 34(24):349–384.

Holmes, D.L. 1965. *Principles of Physical Geology*. 2nd. ed. Ronald Press, New York.

Hora, S.L., and B. Chopra. 1923. Reptilia and Batrachia of the Salt Range. Rec. Ind. Mus. 25:369–376.

Inger, R.F. 1957. A new gecko of the genus *Cyrtodactylus* with a key to the species from Borneo and the Phillipine Islands. Sarawak Mus. J. 8:261–264.

Inger, R.F., and P.K. Chin. 1962. The fresh-water fishes of North Borneo. Fieldiana Zool. 45:1–268.

Ingoldby, C.M. 1922. A new stone gecko from Himalayas. J. Bombay Nat. Hist. Soc. 28:1051.

Ingoldby, C.M., and J.B. Proctor, 1923. Notes on a collection of reptilia from Waziristan and the adjoining portion of the N.W. Frontier Province. J. Bombay Nat. Hist. Soc. 24:127–130.

Jaeger, J.J., V. Courtillot, and P. Tapponnier. 1989. Paleontological view of the ages of the Deccan Traps, the Cretaceous/Tertiary boundary, and the India-Asia collision. Geology 17:316–319.

Jan, G. 1865. Prime linee dúna fauna della Persia occidentale. *In*: Note di un viaggio in Persia nel 1862, di Filippio de Filippi. F. De Filippi. G. Daelli and Co. Editori, Milan, pp. 352–357.

Jerdon, T.C. 1853. Catalogue of the reptiles inhabiting the peninsula of India. J. Asiatic Soc. Bengal 1853:462–469, 532–534.

Kabraji, A.M., and F. Firdous, 1984. Conservation of turtles: Hawkesbay and Sandspit, Pakistan. World Wildlife Fund and Sind Wildlife Management Board, Karachi 1–52.

Khan, A.Q., and M.S. Khan, 1996. Snakes of State of Azad Jammu and Kashmir. Proc. Pakistan Zool. Congr. 16:173–182.

Khan, F.K. 1996. *A Geography of Pakistan: Environment, People and Economy*. Oxford Univ. Press, Lahore.

Khan, M.S. 1965. A normal table of *Bufo stomaticus*. Biologia 11:1–39.

Khan, M.S. 1968a. Amphibian fauna of District Jhang with notes on habits. Pakistan J. Sci.20(5–6):227–233.

Khan, M.S. 1968b. Morphogenesis of digestive tract of *Bufo stomaticus*. Pakistan J. Scient. Res. 20:93–106.

Khan, M.S. 1969. A normal table of *Rana tigerina* Daudin. 1. Early development (Stages 1–27). Pakistan J. Sci. 21:36–50.

Khan, M.S. 1972a. Checklist and key to the lizards of Jhang District, West Pakistan. Herpetologica 28(2):94–98.

Khan, M.S. 1972b. The "commonest toad" of West Pakistan and a note on *Bufo melanostictus* Schneider. Biologia 18(2):131–133.

Khan, M.S. 1973. Food of tiger frog *Rana tigerina* Daudin. Biologia 19(1–2):93–107.

Khan, M.S. 1974. Discovery of *Microhyla ornata* (Duméril and Bibron) from the Punjab, Pakistan. Biologia 20(2):179–180.

Khan, M.S. 1976. An annotated checklist and key to the amphibians of Pakistan. Biologia 22(2):201–210.

Khan, M.S. 1977. Checklist and key to the snakes of District Jhang, Punjab, Pakistan. Biologia 23(2):145–157.

Khan, M.S. 1979. On a collection of amphibians from northern Punjab and Azad Kashmir, with ecological notes. Biologia 25(1–2):37–50.

Khan, M.S. 1980a. A new species of gecko from northern Pakistan. Pakistan J. Zool. 12(1):11–16.

Khan, M.S. 1980b. Affinities and Zoogeography of herpetiles of Pakistan. Biologia 26(1–2):113–171.

Khan, M.S. 1982a. An annotated checklist and key to the reptiles of Pakistan. Part III: Serpentes (Ophidia). Biologia 28(2):215–254.

Khan, M.S. 1982b. Collection, preservation and identification of amphibian eggs from the plains of Pakistan. Pakistan. J. Zool.14(2):241–243.

Khan, M.S. 1982c. Key for the identification of amphibian tadpoles from the plains of Pakistan. Pakistan J. Zool. 14(2):133–145.

Khan, M.S. 1983. Venomous terrestrial snakes of Pakistan. The Snake 15(2):101–105.

Khan, M.S. 1984a. A cobra with an unusual hood pattern. The Snake 16(2):131–134.

Khan, M.S. 1984b. Rediscovery and validity of *Bungarus sindanus* Boulenger. The Snake 16(1):43–48.

Khan, M.S. 1984c. Validity of the natricine taxon *Natrix sanctijohannis* Boulenger. J. Herpetol. 18(2):198–200.

Khan, M.S. 1985a. An interesting collection of amphibians and reptiles from Cholistan Desert, Punjab, Pakistan. J. Bombay. Nat. Hist. Soc. 82(1):144–148.

Khan, M.S. 1985b. Taxonomic notes on *Bungarus caeruleus* (Schneider) and *Bungarus sindanus* Boulenger. The Snake 17(1):71–78.

Khan, M.S. 1986. A noteworthy collection of amphibians and reptiles from northwestern Punjab, Pakistan. The Snake 18(2):118–125.

Khan, M.S. 1987. Checklist, distribution and zoogeographical affinities of amphibians and reptiles of Baluchistan. Proceedings of the 7th Pakistan Congr. Zool.:105–112.

Khan, M.S. 1988. A new cyrtodactylid gecko from northwestern Punjab, Pakistan. J. Herpetol. 22(2):241–243.

Khan, M.S. 1989. Rediscovery and redescription of the highland ground gecko, *Tenuidactylus montiumsalsorum* (Annandale 1913). Herpetologica 45(1):46–54.

Khan, M.S. 1990a. Discovery of a new gecko! Natura (WWF-Pakistan Newsletter), Lahore 9:2.

Khan, M.S. 1990b. The impact of human activities on the status and distribution of amphibians in Pakistan. Hamadryad 15(1):21–24.

Khan, M.S. 1990c. Venomous terrestrial snakes of Pakistan and snake bite problem, in: *Snakes of Medical Importance (Asia-Pacific Region)*: pp. 419–446. (P. Gopalakrishnakone and L.M. Chou, Eds.) National University of Singapore and International Society of Toxicology (Asia-Pacific Section), Singapore.

Khan, M.S. 1991a. A new *Tenuidactylus* gecko from the Sulaiman Range, Punjab, Pakistan. J. Herpetol. 25(2):199–204.

Khan, M.S. 1991b. Additions of new species to the herpetofauna of Pakistan. Species No. 17:56.

Khan, M.S. 1991c. Amphibians, lizards, turtles and snakes. Chapter 3, in: *Pakistan ki Jangli Hayat (Wildlife of Pakistan)*: pp. 61–124. Publication No.241. Urdu Science Board, 299 Upper Mall, Lahore (in Urdu).

Khan, M.S. 1991d. Endangered species of reptiles of Pakistan and suggested conservation measures, in: Handbook published to mark second seminar on "Nature Conservation and Environmental Protection," 12 March, 1991, Islamabad. Pakistan Wildlife Conservation Foundation.

Khan, M.S. 1991e. Morphoanatomical specializations of the buccopharyngeal region of the anuran larvae and its bearing on the mode of larval feeding. Ph.D. Diss., University of the Punjab, Lahore, Pakistan.

Khan, M.S. 1991f. New additions to the herpetofauna of Pakistan. Hamadryad 16 (1–2):48–49.

Khan, M.S. 1992. Validity of the mountain gecko *Gymnodactylus walli* Ingoldby, 1922. Herptol. J. 2:106–109.

Khan, M.S. 1993a. A new angular-toed gecko from Pakistan, with remarks on the taxonomy and a key to the species belonging to genus *Cyrtodactylus* (Reptilia: Sauria: Geckkonidae). Pakistan J. Zool.25 (1):67–73.

Khan, M.S. 1993b. A new sandstone gecko from Fort Munro, Dera Ghazi Khan District, Punjab, Pakistan. Pakistan J. Zool. 25(3): 217–221.

Khan, M.S. 1993c. Hemipenis morphology of *Varanus flavescens* (Hardwicke and Gray, 1927), and its phylogenetic implications. Pakistan J. Zool. 25(2):135–138.

Khan, M.S. 1993d. *Sar Zameen-a-Pakistan kay Saamp* [Snakes of Pakistan]. Urdu Science Board, 299 Upper Mall, Lahore (in Urdu).

Khan, M.S. 1994a. Key for identification of amphibians and reptiles of Pakistan. Pakistan J. Zool. 26:249–255.

Khan, M.S. 1994b. A revised checklist and key to the amphibians of Pakistan. Hamadryad 19:11–14.

Khan, M.S. 1994c. Validity and redescription of *Tenuidactylus yarkandensis* (J. Anderson, 1872). Pakistan J. Zool., 26(2): 139–143.

Khan, M.S. 1995. A report on an unborn litter of chain viper *Daboia russellii* (Shaw and Noder, 1977). Pakistan J. Zool. 27(2): 119–122.

Khan, M.S. 1996a. The oropharyngeal morphology and feeding habits of tadpole of tiger frog *Rana tigerina* Daudin. Russian J. Herpetol. 3(2):163–171.

Khan, M.S. 1996b. Oropharyngeal morphology of tadpole of southern cricket frog *Rana syhadrensis* Annandale, 1919, and its ecological correlates. Pakistan J. Zool. 28:133–138.

Khan, M.S. 1997a. Biodiversity of geckonid fauna of Pakistan. pp. 383–389. in: *Biodiversity of Pakistan*.(S.A. Mufti, C.A. Woods, S.A. Hasan, Eds.) Pakistan Mus. Nat. Hist. Islamabad and Gainesville, Florida.

Khan, M.S. 1997b. A report on an aberrant specimen of Punjab krait *Bungarus sindanus razai* Khan 1985 (Ophidia, Elapidae), from Azad Kashmir. Pakistan J. Zool. 29(3): 203–205.

Khan, M.S. 1997c. A new subspecies of common skittering frog *Euphlyctis cyanophlyctis* (Schneider 1799) from Baluchistan, Pakistan. Pakistan J. Zool. 29(2):107–112.

Khan, M.S. 1997d. A new toad from the foot of Siachin Glacier, Baltistan, northeastern Pakistan. Pakistan J. Zool. 29(1):43–48.

Khan, M.S. 1997e. Taxonomic notes on Pakistani snakes of the *Coluber-karelini-rhodorachis-ventromaculatus* species complex: a new approach to the problem. Asiatic. Herpetol. Res.1997, 7:51–60.

Khan, M.S. 1997f. Validity, generic designation, and taxonomy of western rock gecko *Gymnodactylus ingoldbyi* Proctor 1923. Russian J. Herpetol. 4:83–88.

Khan, M.S. 1998a. Country report for Pakistan. Herpetofauna of Pakistan: present status, distribution and conservation. *In*: Biology and Conservation of the Amphibians, Reptiles and their Habitats in South Asia. Proceedings of the International Conference on the Biology and Conservation of the South Asian Amphibians and Reptiles, 1–5 August 1996. A. De Silva, Ed. Amphibia and Reptile Research Organization of Sri Lanka, Peradeniya, pp. 47–50.

Khan, M.S. 1998b. Exploitation of herpetofauna of Pakistan, *In*: Biology and Conservation of the Amphibians, Reptiles and their Habitats in South Asia. A. De Silva, Ed. Proceedings of the International Conference on the Biology and Conservation of the South Asian Amphibians and Reptiles, 1–5 August 1996. Amphibia and Reptile Research Organization of Sri Lanka, Peradeniya, pp. 302.

Khan, M.S. 1998c. Notes on *Typhlops diardii* Schlegel, 1839, with description of a new subspecies (Squamata, Serpentes, Scolecophidia). Pakistan J. Zool. 30(3):213–221.

Khan, M.S. 1998d. Oropharyngeal morphology and feeding specializations of amphibian tadpoles, in: Biology and Conservation of the Amphibians, Reptiles and their Habitats in South Asia. Proceedings of the International Conference on the Biology and Conservation of the South Asian Amphibians and Reptiles, 1–5 August 1996. A. De Silva, Ed. Amphibia and Reptile Research Organization of Sri Lanka, Peradeniya, pp. 47–50.

Khan, M.S. 1998e. Status of amphibian fauna of Pakistan, *In*: Biology and Conservation of the Amphibians, Reptiles and Their Habitats in South Asia. Proceedings of the International Conference on the Biology and Conservation of the South Asian Amphibians and Reptiles, 1–5 August 1996. A. De Silva, Ed. Amphibia and Reptile Research Organization of Sri Lanka, Peradeniya, pp.137–139.

Khan, M.S. 1999a. A checklist and key to the phrynocephalid lizards of Pakistan, with ethnological notes (Squamata, Agamidae). Pakistan J. Zool. 31:17–24.

Khan, M.S. 1999b. Food particle retrieval in amphibian tadpoles. Zoo's Print J., 14: 17–20.

Khan, M.S. 1999c. Herpetology of habitat types of Pakistan. Pakistan J. Zool., 31: 275–289.

Khan, M.S. 1999d. Two new species and a subspecies of blind snakes of genus *Typhlops* from Azad Kashmir and Punjab, Pakistan (Serpentes, Typhlopidae). Russian J. Herpetol. 6(3): 231–240.

Khan, M.S. 1999e. *Typhlops ductuliformes* a new species of blind snakes from Pakistan and a note on *T. porrectus* Stoliczka 1871, (Squamata, Serpentes, Scolecophidia). Pakistan J. Zool. 31(4):385–390.

Khan, M.S. 2000a. Redescription and generic redesignation of *Gymnodactylus stoliczkai* Steindachner, 1869. Pakistan J. Zool., 32(2):157–163.

Khan, M.S. 2000b. *Sar Zameen-a-Pakistan kay maindak aur Khazinday* (Frogs and lizards of Pakistan). Urdu Science Board, 299 Upper Mall, Lahore, Pakistan.

Khan, M.S. 2001a. Morphology and feeding ecology of *Microhyla ornata* tadpole. Asiatic Herp. Res. 9: 130–138.

Khan, M.S. 2001b. Notes on cranial-ridged toads of Pakistan and description of a new subspecies (Amphibia: Bufonidae). Pakistan J. Zool., 33(4): 293–298.

Khan, M.S. 2001c. Recent advances in the taxonomic status of ranid frogs of Pakistan. Pakistan J. Zool., 33(3):169–171.

Khan, M.S. 2001d. Taxonomic notes on angular-toed gekkota of Pakistan, with description of a new species of genus *Cyrtopodion*. Pakistan J. Zool. 33(1):13–24.

Khan, M.S. 2002a. *A guide to the snakes of Pakistan*. Edition Chimaira, Frankfurt am Main.

Khan, M.S. 2002b. Key and checklist to the lizards of Pakistan (Reptilia: Squamata: Sauria). Herpetozoa 15, (3–4): 99–119.

Khan, M.S. 2002c. Notes on saw-scale viper *Echis carinatus* and its status in Pakistan. 34(3):181–188.

Khan, M.S. 2003a. Anmerkungen zur Morphologie, Verbreitung und den Habitatpraferenzen einiger pakistanischer geckos [Notes on the morphology, distribution and habitat preferences of some Pakistani geckos]. Sauria, Berlin, 25(3): 35–47. Erratum: Sauria, Berlin, 25(4): 27.

Khan, M.S. 2003b. Morphology of the *Limnonectes* tadpole, with notes on its feeding ecology and on the breeding habits of *Limnonectes* frogs in riparian Punjab. Bulletin Chicago Herpetological Society, 38(9):177–179.

Khan, M.S. 2003c. Morphology of riparian tadpoles: *Euphlyctis cyanophlyctis* (Schneider, 1799). Bulletin of the Chicago Herpetological Society. 38(5):95–98.

Khan, M.S. 2003d. Morphology of the tadpole of *Microhyla ornata*, with notes on its feeding ecology and breeding habits. Bulletin Chicago Herpetological Society, 38(3):40–51.

Khan, M.S. 2003e. Notes on circum Indus geckos of genus *Cyrtopodion* (Squamata: Gekkonidae). Gekkota 4.

Khan, M.S. 2003f. Questions of generic designation of angular-toed geckos of Pakistan with descriptions of three new genera (Reptilia: Gekkonidae). J. Nat. Hist. Wildlife (Karachi), Vol. 2(2).

Khan, M.S. 2003g. Up-to-date checklist of amphibians and reptiles of Pakistan. J. Nat. Hist Wildlife (Karachi), 2(1):11–17.

Khan, M.S. 2005a. Addition of a frog of the family Megophryidae to the amphibian fauna of Pakistan. Bulletin of the Chicago Herpetological Society, 40(4): 70–71.

Khan, M.S. 2005b. An overview of the angular-toed geckos of Pakistan (Squamata: Sekkonidae). Gekko, 4(2):20–30.

Khan, M.S. 2005c. Notes on new texa of typhlopid snakes from Pakistan (Serpentes: Typhlopidae) Bulletin Chicago Herpetological Society, 40(8):145–147.

Khan, M.S., and N. Ahmed. 1987. On a collection of amphibians and reptiles from Baluchistan, Pakistan. Pakistan J. Zool.19(4):361–370.

Khan, M.S., and K.J. Baig, 1988. Checklist of the amphibians and reptiles of District Jhelum, Punjab, Pakistan. The Snake 20:156–161.

Khan, M.S., and K.J. Baig. 1992. A new *Tenuidactylus* gecko from northeastern Gilgit Agency, North Pakistan. Pakistan J. Zool. 24(4):273–277.

Khan, M.S., and A.Q. Khan. 2000. Three new subspecies of snakes of genus *Coluber* from Pakistan. Pakistan J. Zool., 32(1):49–52.

Khan, M.S., and M.R.Z. Khan. 1997. A new skink from the Thal Desert of Pakistan. Asiat. Herpetol. Res. 7: 61–67.

Khan, M.S., and S.A. Malik. 1987a. Buccopharyngeal morphology of tadpole larva of *Rana hazarensis* Dubois and Khan 1979, and its Torrenticole adaptations. Biologia 33(2):45–60.

Khan, M.S., and S.A. Malik. 1987b. Reproductive strategies in a subtropical anuran population in arid Punjab, Pakistan. Biologia 33(2): 279–303.

Khan, M.S., and M.R. Mirza. 1976. An annotated checklist and key to the reptiles of Pakistan. Part I: Chelonia and Crocodilia. Biologia 22(2):211–219.

Khan, M.S., and M.R. Mirza. 1977. An annotated checklist and key to the reptiles of Pakistan. Part II: Sauria (Lacertilia). Biologia 23(1):41–64.

Khan, M.S. and S.A. Mufti. 1994a. Buccopharyngeal specializations of tadpole of *Bufo stomaticus* and its ecological correlates. Pakistan J. Zool. 26(4):285–292.

Khan, M.S., and S.A. Mufti. 1994b. Oral disc morphology of amphibian tadpole and its functional correlates. Pakistan J. Zool. 26(4):25–30.

Khan, M.S., and S.A. Mufti. 1995. Oropharyngeal morphology of detritivorous tadpole of *Rana cyanophlyctis* Schneider, and its ecological correlates. Pakistan J. Zool. 27(1):43–49.

Khan, M.S., and H. Rosler. 1999. Redescription and generic reallocation of Ladakhian gecko *Gymnodactylus stoliczkai* Steindachner, 1969 [sic]. Asiatic Herpetol. Res. 8:60–68.

Khan, M.S., and R. Tasnim. 1986a. Balling and caudal luring in young *Bungarus caeruleus*. The Snake 18(1):42–46.

Khan, M.S., and R. Tasnim. 1986b. Notes on the Himalayan pit viper, *Agkistrodon himalayanus* (Günther). Litteratura Serpentium (Eng. Ed.) 6:46–55.

Khan, M.S., and R. Tasnim. 1987a. A field guide to the identification of herps of Pakistan. Part I, Amphibia. Biological Society of Pakistan, Monograph No. 14:1–28.

Khan, M.S., and R. Tasnim. 1987b. Observations on distress behaviour of an injured *Eryx johnii*. The Snake 19(2):144–145.

Khan, M.S., and R. Tasnim. 1989. A new frog from southwestern Azad Kashmir genus (*Rana,* subgenus *Paa*). J. Herpetol. 23:419–423.

Khan, M.S., and R. Tasnim. 1990a. A field guide to the identification of herps of Pakistan. Part: II Chelonia. Biological Society of Pakistan, Monograph No.15:1–15.

Khan, M.S. and R. Tasnim. 1990b. A new gecko of the genus *Tenuidactylus* from northeastern Punjab, Pakistan, and southwestern Azad Kashmir. Herpetologica 46(2):142–148.

Khan, W.A. 1997. Lizards of south Waziristan Agency, Northwestern Frontier Province, Pakistan, some aspects of systematics and biology. M.Sc. thesis, Department of Zoology, Government College, Lahore, submitted Nov. 1997.

Khozatsky, L.I., and M. Mlynarski. 1966. *Agrionemys*-nouveau genre de tortues terrestres (Testudinidae). Bull. Acad. Pol. Sci. Ser. Sci. Biol. (2):123–125.

King, L.C. 1967. *The Morphology of the Earth*. 2nd ed. Oliver and Boyd, London.

Kirtisinghe, P. 1957. *The Amphibia of Ceylon*. Published by the author, 2 Charles Circus, Colombo 3, Ceylon.

Kluge, A.G. 1967. Higher taxonomic categories of gekkonid lizards and their evolution. Bulletin of the American Museum of Natural History 135(1):1–59.

Kluge, A.G. 1983. Cladistic relationships among gekkonid lizards. Copeia 1983(2):465–475.

Kluge, A.G. 1987. Cladistic relationships in the Gekkonoidea (Squamata: Sauria). Miscellaneous Publications, Mus. Zool., Univ. Michigan (173):iv+1–54.

Kluge, A.G. 2001. Gekkotan lizard taxonomy. Hamadryad 26(1):1–209.

Konieczny, M.G. 1969a. *Bedrohte* Reptilien-Arten. *In*: Mertens, R. Die Amphibien und Reptilien West-Pakistans. Stutt. Beitt. Naturk. 197:1–96.

Konieczny, M.G. 1969b. Liste der Fundorte mit geographischen Bemerkungen. *In*: Mertens, R. Die Amphibien und Reptilien West-Pakistans. Stutt. Beitt. Naturk. 197:1–96.

Kral, B. 1969. Notes on the herpetofauna of certain provinces of Afghanistan. Zoology Listy 18(1): 55–66.

Kramer, E., and H. Schnurrenberger. 1963. Systematik, Verbreitung und Okologie der Libyschen Schlangen. Rev. Suisse Zool. 70 Fasc. 3(27):453–568.

Kullmann, E. 1974. Die Tierwelt Ostafghanistans in ihren geographischen Beziehungen. Freunde des Kolner Zoo., 13: 3–25.

Kumar, S. 1992. Indian monitor (*Varanus bengalensis*) feeding on blackbuck (*Antilope cervicapra*) kill. Hamadryad 17: 48.

Kureshy, K.U. 1986. *Geography of Pakistan*. National Book Service, Lahore.

Leviton, A., and S.C. Anderson. 1961. Further remarks on the amphibians and reptiles of Afghanistan. Wassman J. Biol. 19(2):269–276.

Leviton, A., and S.C. Anderson. 1972. Description of a new species of *Tropiocolotes* (Reptiliua: geckonidae) with a revised key to the genus. Occas. Pap. Calif. Acad. Sci. (96):1–7.

Loveridge, A. 1959. Reptiles and amphibians collected by the expedition in Saudi Arabia and in Baluchistan and Bahawalpur, West Pakistan *In*: Henry Fitch. An Anthropological Reconnaisance in West Pakistan. Pap. Peabody Mus. Arch. Ethnol. Harvard University 52:226–227.

Macey, J.R., Y. Wang, N.B. Ananjeva, A. Larson, and T.J. Papenfuss. 1999. Vicariant patterns of fragmentation among gekkonid lizards of the genus *Teratoscincus* produced by the Indian Collision: A molecular phylogenetic perspective and an area cladogram for Central Asia. Molec. Phylog. Evolu. 12(3):320–332.

Macey, J.R., J.A. Schulte, H.G. Kami, N.B. Ananjeva, A. Larson, and T.J. Papenfuss. 2000. Testing hypothesis of vicariance in the agamid lizards *Laudakia caucasia* from mountain ranges on the northern Iranian Plateau. Molec. Phylog. Evolu. 14(3):479–483.

Mahendra, B.C. 1936. Contribution to the bionomics, anatomy, reproduction and development of the Indian house-gecko, *Hemidactylus flaviviridis* Rüppell. Part 1. Proc. Ind. Acad. Sci. 4:250–281.

Mahendra, B.C. 1984. *Handbook of the Snakes of India, Ceylon, Burma, Bangladesh, and Pakistan*. The Ann. Zool. 22 (1984B):1–412.

Malnate, E.V. 1966. *Amphiesma platyceps* (Blyth) and *Amphiesma sieboldii* (Günther): Sibling species (Reptilia: Serpentes). J. Bombay Nat. Hist. Soc. 63(1):1–17.

Malnate, E.V., and S.A. Minton. 1965. A redescription of the natricine snake *Xenochrophis cerasogaster*, with comments on its taxonomic status. Proc. Acad. nat. Sci. Phila. 117(2):19–43.

Marx, H. 1959. Review of the colubrid snake *Spalerosophis*. Fieldiana: Zool. Chicago 39(30):347–361.

Marx, H. 1988. The colubrid snake, *Psammophis schokari*, from Arabian Peninsula. Fieldiana: Zool. 40:1–16.

Matz, G. 1978. Des geckos sans ventouses les Eublepharines. Aquarama 12(43):44–45 and 88.

Mausfeld, P., and A. Schmitz. 2003. Molecular phylogeography, interspecific variations and speciation of the Asian scincid lizard genus *Eutropis* Fitzinger, 1843 (Squamata: Reptilia: Scincidae), taxonomic and biographic implications. Organisms Diversity and Evolution, 3:161–171.

Maxson, L.R. 1984. Molecular probes of phylogeny and biogeography in the toads of the widespread genus *Bufo*. Mol. Biol. Evol. 1:345–356.

Mayer, W., and G. Benyr. 1994. Albumin-Evolution und Phylogenese in der Familie Lacertidae (Reptilia: Sauria). Annalen des Naturhist. Mus. Wien, 96B:621–648.

McCann, C. 1938. The reptiles and amphibians of Cutch State. J. Bombay Nat. Hist. Soc. 40:425–427.

McDiarmid, R.W., J.A. Campbell, and T.A. Touré. 1999. *Snake Species of the World. A Taxonomic and Geographic Reference*. Vol. 1. The Herpetologists' League, Washington, DC.

McDowell, S.B. 1972. The genera of sea-snakes of the *Hydrophis* group (Serpentes: Elapidae). Trans. Zool. Soc. Lond. 32(2):189–247.

McMahon, A.H. 1901a. Notes on the fauna of Chitral. J. Asiatic Soc. Bengal 70(2):1–6.

McMahon, A.H. 1901b. Notes on the fauna of Dir and Swat. J. Asiatic Soc. Bengal 70(2):7–12.

Mertens, R. 1942. Die Familie der Warane (Varanidae). Abb. Senckn. Natur. Gesells. 162, 165–166:1–391.

Mertens, R. 1959a. Eine neue Wassnernatter aus West Pakistan. Senckenb. Biol. 40(3–4):117–120.

Mertens, R. 1959b. Uber einige seltene Eidechsen aus West Pakistan. Aquar. Terr. Zeit. 12(10):307–310.

Mertens, R. 1965. Wenig bekannte "Seitenwinder" unter den Wustenottern Asiens. Natur und Museum, 59(8): 346–352.

Mertens, R. 1969a. Die Amphibien und Reptilien West-Pakistans. Stuttg. Beitr. Naturk. (197):1–96.

Mertens, R. 1969b. Eine neue Rasse der Dachschildkrote, *Kachuga tecta*. Senckenb. Biol. 50:23–30.

Mertens, R. 1970. Die Amphibien und Reptilien West-Pakistans. 1. Nachtrag. Stuttg. Beitr. Naturk. (216):1–5.

Mertens, R. 1971. Die Amphibien und Reptilien West-Pakistans. 2. Nachtrag. Senckenb. Biol.52 (1–2):7–15.

Mertens, R. 1972. Nachtrage zum Krokodil-Katalog der senckenbergischen Sammlungen. 3. Nachtrag. Senckenb. Biol. 53:21–35.

Mertens, R. 1974. Die Amphibien und Reptilien West-Pakistans. Senckenb. Biol. 55(1–3):35–38.

Minton, M.R. 1983. Snake people of Sind. Animal Kingdom, 86(1):40–46.

Minton, S.A. 1962. An annotated key to the amphibians and reptiles of Sind and Las Bela, West Pakistan. Am. Mus. Novit. (2081):1–21.

Minton, S.A. 1966. A contribution to the herpetology of West Pakistan. Bull. Amer. Mus. Nat. Hist. 134(2): 31–184.

Minton, S.A. 1990. Venomous bites by nonvenomous snakes: An annotated bibliography of colubrid envenomation. J. Wild. Med. 1:119–127.

Minton, S.A., and J.A. Anderson, 1962. A record of the turtle, *Hardella thurjii*, from salt water. Herpetologica 18; 126.

Minton, S.A., and J.A. Anderson, 1963. Feeding habits of the kukri snake, *Oligodon taeniolatus*. Herpetologica 19(2):147.

Minton, S.A., and J.A. Anderson, 1965. A new dwarf gecko (*Tropiocolotes*) from Baluchistan. Herpetologica 21(1):59–61.

Minton, S.A., and M.R. Minton. 1964. The snake charmers of Sind. Bull. Philadelphia Herp. Soc. 1964: 33–38.

Minton, S.A., and M.R. Minton, 1969. *Venomous reptiles*. Charles Scribner's, New York.

Minton, S.A., S.C. Anderson, and J.A. Anderson, 1970. Remarks on some geckos from Southwest Asia, with description of three new forms and a key to the genus *Tropiocolotes*. Proc. Calif. Acad. Sci. (4th Ser.) 37:333–362.

Minton, S.A., A.G. Dowling, and F.E. Russell, 1968. *Poisonous snakes of the world*. U.S. Department of the Navy Bureau of Medicine and Surgery, Washington, DC.

Mirza, M.R. 1975. Fresh water fishes and zoogeography of Pakistan. Bijd. Tot. de Dierk., 45(2):144–180.

Moll, E.O. 1986. Survey of the freshwater turtles of India. Part I: The genus *Kachuga*. Bombay J. Nat. Hist. Soc. 83:538–552.

Moll, E.O. 1987. Survey of the freshwater turtles of India. Part II: The genus *Kachuga*. Bombay. Nat. Hist. Soc. 84:7–25.

Moody, S.M. 1980. Phylogenetic and Historical Biogeographical Relationships of the genera in the family Agamidae (Reptilia: Lacertilia) Doctoral Dissertation, University of Michigan, Ann Arbor.

Moody, S.M. 1987. A preliminary cladistic study of the lizard genus *Uromastyx* (Agamidae, sensu lato) with checklist and diagnostic key to the species. Proc. 4th.

Ordinary meeting of the Societas Europea Herpetologica, Nijmegen, pp. 285–288.

Moses, S.T. 1948. Crocodiles in India. Bull. 15, Department of Fisheries, Baroda.

Mufti, S.A., C.A. Woods, and S.A. Hasan (Eds.). 1997. *Biodiversity of Pakistan*. Pakistan Nat. Hist. Mus. Islamabad and Gainesville, Florida.

Mukerji, D.D. 1931. Some observations on the burrowing toad *Cacopus globulosum* Günther. J. Proc. Asiatic Soc. Bengal, (N.S.)27:97–100.

Murray, J.A. 1874. Additions to the reptilian fauna of Sind. Ann. Mag. Nat. Hist., London (ser.5) 14:106–108.

Murray, J.A.. 1884. *The Vertebrate Zoology of Sind: A Systematic Account, With Descriptions of all the Known Mammals, Birds and Reptiles Inhabiting the Pprovince; Observations on the Habits, and c; Tables of the Geographical Distribution in Persia, Beloochistan, and Afghanistan; Punjab, North-west Provinces, and the Peninsula of India Generally*. Education Society's Press, Byculla, Bombay, and Richardson and Co., London.

Murray, J.A. 1886. *The Reptiles of Sind: A Systematic Account*. Education Society's Press, Bombay, and Richardson and Co., London.

Murray, J.A. 1892. *The Zoology of Baloochistan and Southern Afghanistan*. Education Society's Press, Byculla, Bombay and Trubner and Co., London.

Murthy, T.S.N. 1990. Venomous snakes of medical importance in India Part A. in: *Snakes of Medical Importance* (Asia-Pacific Region), pp:281–297. (P. Gopalakrishnakone and L.M. Chou, Eds.). Venom and Toxin Research Group, National University of Singapore, Singapore.

Murthy, T.S.N., and R.A.A. Arockiasamy. 1977. Observations on the spiny-tailed lizard, *Uromastix hardwickii* Gray in captivity. Geobios 4:167–168.

Murthy, T.S.N., and B.B. Sharma, 1976. A contribution to the herpetology of Jammu and Kashmir. British J. Herpetol. 5(6):533–538.

Murthy, T.S.N., B.B. Sharma, and T. Sharma. 1979. Second Report on the herpetofauna of Jammu and Kashmir. The Snake 11(2):234–241.

Myers, G.S. 1947. Murray's Reptiles of Sind, with a note on three forgotten descriptions of Indian sea-snakes published therein. Herpetologica, Lawrence 3(5):167–168.

Natarajan, R. 1953. A note on the chromosomes of *Cacopus systoma*. Proc. 40th Indian Sci. Cong. Part 3:180–181.

Natarajan, R. 1958. Contribution to the cytology of Indian Anura: II. On the number and morphology of the chromosomes of three species of *Rana* (Ranidae). J. Zool. Soc. India 9:114–119.

Nicholson, E. 1874. *Indian Snakes: An Elementary Treatise on Ophilology*. Higginbotham and Co., Madras.

Norman, B.R. 2003. A new geographical record of the introduced house gecko, *Hemidactylus frenatus*, at Cabo San Lucas, Baja California Sur, Mexico, with notes on other species observed. Bull. Chicago Herp. Soc. 38(5):98–100.

Obst, F.J. 1983. Zur Kentnis de Schlangengattung *Vipera* (Reptilia, Serpentes, Viperidae). Zool. Abh. Staatl. Mus. Tierkde. Dresden 38(13):229–235.

Parshad, B. 1914a. Notes on the spiny tailed lizard (*Uromastix hardwickii*). J. Bombay Nat. Hist. Soc. 23:370.

Parshad, B. 1914b. Notes on aquatic chelonia of the Indus system. Rec. Indian Mus. 10:267–271.

Parshad, B. 1916. Some observations on a common house-lizard (*Hemidactylus flaviviridis*) J. Bombay Nat. Hist. Soc. 24:834–838.

Pasteur, G. 1981. A survey of the species groups of the old world scincid genus *Chalcides*. J. Herpetol. 15:1–16.

Perveen, Z. and Auffenberg, W. 1988. New reptile records for Pakistan. Herp. Review, 19(3): 61.

Powell, C.M. 1979. A speculative tectonic history of Pakistan and surroundings: some constraints from the Indian ocean. in. *Geodynamics of Pakistan*. Geological Survey of Pakistan, Sp. Memo. :5–25.

Polunin, O., and A. Stainton, 1985. *Flowers of the Himalayas*. Oxford University Press, Oxford.

Prakash, I. 1982. The Thar: A desert alive. Sanctuary Asia 2:130–137.

Prakash, S. 1988. Genetic studies on *Rana limnocharis*. Unpublished Ph.D. dissertation, North Eastern Hill University, Shillong.

Prater, S.H. 1965. *The Book of Indian Animals*. 2nd ed. Bombay Natural History Society, Bombay.

Proctor, J.B. 1923. *In*: C.M. Ingoldby, and J.B. Proctor. Notes on a collection of reptilia from Waziristan and the adjoining portion of the Northwest Frontier Provinces. J. Bombay Nat. Hist. Soc. 29:117–135.

Queiroz, K. de, and J. Gauthier. 1994. Toward a phylogenetic system of biological nomenclature. Trends in Ecology and Evolution 9(1):27–31.

Rao, C.R.N. 1917. On the occurrence of iridocytes in the larva of *Microhyla ornata* Boulenger. Rec. Indian Mus. 13:281–292.

Rao, C.R.N. 1918. Notes on the tadpoles of Indian Engystomatidae. Rec. Indian Mus. 15:41–45.

Rao, M.V., and R. Suba. 1974. Bahaviour of the agamid garden lizard *Calotes versicolor*. J. Bombay Nat. Hist. Soc. 71:148–150.

Rastegar-Pouyani, N. 1999. Two new subspecies of *Trapelus agilis* complex (Sauria, Agamidae) from lowland southwestern Iran and southeastern Pakistan. Asiatic Herpetol. Research 8:90–101.

Rastegar-Pouyani, N. 2000. Taxonomic status of *Trapelus ruderatus* (Olivier) and *T. persicus* (Blanford), and validity of *T. lessonae* (DeFilippi). Amphibia-Reptilia 21:91–102.

Roberts, T.J. 1975. A note on *Testudo horsfieldi* Gray, the Afghan tortoise or Horsefield's four-toed tortoise. J. Bombay Nat. Hist Soc. 72:206–209.

Roberts, T.J. 1977. *Mammals of Pakistan*. Ernest Benn, London.

Roberts, T.J. 1991. *Birds of Pakistan*. Vol. 1. Oxford University Press, London.

Roberts, T.J. 1992. *Birds of Pakistan*. Vol. 2. Oxford University Press, London.

Ruddiman, W.F., and J.E. Kutzbach. 1991. Plateau uplift and climatic change. Scient. American 1991(3):42–50.

Russell, A.P., and A. Bauer. 2002. Underwood's classification of the geckos: A 21st century appreciation. Bull. Nat. Hist. Museum, London (Zoology) 68(2): 113–121.

Russell, P. 1796. *An Account of Indian Serpents Collected on the Coast of Coromandel; Containing Descriptions and Drawings of Each Species, Together With Experiments and Remarks on Their Several Poisons.* George Nicol, London.

Russell, P. 1801–1809. *A Continuation of an Account of Indian Serpents; Containing Descriptions and Figures From Specimens and Drawings, Transmitted From Various Parts of India.* G. and W. Nicol, London.

Salvador, A. 1982. A revision of the lizards of the genus *Acanthodactylus* (Sauria: Lacertidae). Bonner Zool. Monog.: Zool. Forsch. Mus. Alexander Koenig, Bonn 16:1–167.

Savage, J.M. 1973. The geographic distribution of frogs: patterns and predictions. *In: Evolutionary Biology of the anurans: contemporary research and major problems,* chapter 13:351–445. J.L. Vial, Ed. University Missouri Press, Columbia.

Schmidtler, J.J., and J.F. Schmidtler. 1969. Uber *Bufo surdus*; mit einem Schlüssel und Anmerkungen zu den übrigen Kröten Irans und West-Pakistans. Salamandra 5(3–4):113–123.

Schweinfurth, U. 1957. Die horizontale und vertikale Verbreitung der Vegetation im Himalaya. Bonner Geographic. Abh. 20:1–373.

Sclater, P. 1858. On the general geographical distribution of the members of the class Aves. J. Proc. Linn. Soc. London (Zoology) Vol. 2:130–145.

Sengör, A.M.C. 1985. The story of Tethys: How many wives did Okeanos have? Episodes 8(1):3–12.

Sengör, A.M.C., D. Altiner, A. Cin, T. Ustaomer, and K.J. Hsu. 1988. Origin and assembly of the Tethyside orogenic collage at the expanse of Gondwana Land. In Gondwana and the Tethys. M.G. Audley-Charles and A. Hallam, Eds., pp. 119–181. Geol. Soc. Spec. Publ. No. 37. Oxford Univ. Press, Oxford.

Sharma, G.P., R.C. Sobto, and E. Rahman. 1977. Karyological analysis of three anurans from north India. Proc. 64th Indian Sci. Cong. Part 3:171.

Shockley, C.H. 1949. Herpetological notes for Ras Jiunri, Baluchistan. Herpetologica 5:121–123.

Simcox, A.H.A. 1906. The crocodile, its food and muscular vitality. J. Bombay Nat. Hist. Soc. 16:375–376.

Smith, D.S., and S.A. Hasan, 1997. A preliminary survey of diversity and distribution of butterflies of northern Pakistan: Gilgit to Khunjerab, pp. 205–211, in: *Biodiversity of Pakistan*. S.A. Mufti, C.A. Woods, and S.A. Hasan, Eds., Pakistan. Mus. Nat. Hist. Islamabad and Gainesville, Florida.

Smith, M.A. 1931. *The Fauna of British India, Including Ceylon and Burma*. Reptilia and Amphibia. Vol. I: Loricata, Testudines. Taylor and Francis Ltd., London.

Smith, M.A. 1935. *The Fauna of British India, Including Ceylon and Burma*. Reptilia and Amphibia. Vol. II: Sauria. Taylor and Francis Ltd. London.

Smith, M.A. 1943. *The fauna of British India including Ceylon and Burma*. Reptilia and Amphibia. Vol. III: Serpentes. Taylor and Francis, London.

Steindachner, F. 1867. Amphibien. *In*: Reise der österreichischen Fregatte Novara-Expedition um der Erde in den Jahren 1857, 1858, 1859 unter den Befehlen des Commodore B. von Wüllerstorf-Urbair, Zoologischer Theil. Kaiserl.-Königl. Hof –Staatsdruckerei, Vienna I(4):1–70.

Steindachner, F. 1869. Reptilia. *In*: Reise der österreichischen Fregatte Novara-Expedition um der Erde in den Jahren 1857, 1858, 1859 unter den Befehlen des Commodore B. von Wüllerstorf-Urbair. Zoologischer Theil. Vienna 1(3):1–98.

Sternberg, J. 1981. The worldwide distribution of sea turtle nesting beaches. Center for Environmental Education, Washington, DC.

Stöck, M., M. Schmid, C. Steinlein, and W-R. Grosse. 1999. Mosaicism in somatic triploid specimens of the *Bufo viridis* complex in the Karakoram with examination of calls, morphology and taxonomic conclusions. Ital. J. Zool. 66:215–232.

Stoliczka, F. 1870. Observations on some Indian and Malayan Amphibia and Reptilia. J. Asiatic Soc. Bengal 39(2):159–228.

Stoliczka, F. 1871. Notes on some Indian and Burmese ophidians. J. Asiatic Soc. Bengal 40(2):421–445.

Stoliczka, F. 1872a. Observation on some Indian and Malayan Amphibia and Reptilia. J. Asiatic Soc. Bengal 39:134–228.

Stoliczka, F. 1872b. Notes on some new species of Reptilia and Amphibia, collected by Dr. W. Waagen in North-Western Panjab. Proc. Asiatic Soc. Bengal 1872:124–131.

Swan, L.W., and A.E. Leviton. 1962. The herpetology of Nepal: a history, checklist and zoogeographical analysis of the herpetofauna. Proc. Calif. Acad. Sci. ser. 4, 32:103–147.

Swaroop. S., and B. Grab. 1954. Snakebite mortality in the world. Bull. World Health. Org. 10:35–76.

Szczerbak, N.N. 1974. [*The Palearctic Desert Lizards*]. Akad. Nauk Ukrain. SSR Inst. Zool. Nauk. Dumka, Kiev. (In Russian.)

Szczerbak, N.N. 1986. Review of the geckonidae of the fauna of the USSR and neighbouring countries. *In: Studies in Herpetology,* Rocek Z., Ed. Charles University, Prague, pp. 705–710.

Szczerbak, N.N. 1988. [Palaearctic rock geckos (*Tenuidactylus*)]. Vest. Zool. 4: 84 (in Russian).

Szczerbak, N.N. 1989. Catalogue of the African sand lizards (Reptilia: Sauria: Eremiainae: *Lampreremias, Pseuderemias, Taenieremias, Mesalina, Meroles*). Herpetozoa 1(3–4):119–132.

Szczerbak, N.N. 1990. [Systematics and geographic variabilityu of *Eumeces taeniolatus* (Sauria, Scincidae)]. Vestnik Zoologii 1990(3):33–40. (In Russian.)

Szczerbak, N.N. 1991. Eine neue Gecko-Art aus Pakistan: *Alsophylax* (*Altiphylax*) boehmei sp. nov. Salamandra 27:53–57.

Szczerbak, N.N., and M.L. Golubev. 1977. [Materials to the systematics of Palearctic geckos (genera *Gymnodactylus, Bunopus, Alsophylax*)], 120–123. In: Herpetological collected papers. Proc. Zool. Ins. Acad. Sci. U.S.S.R. Vol. 74. (In Russian.)

Szczerbak, N.N., and M.L. Golubev. 1984. [On generic assignment and intergeneric structure of the Palearctic *Cyrtodactylus* lizard species]. Vest. Zool. 2:50–56 (In Russian.)

Szczerbak, N.N., and M.L. Golubev, 1986. [*The Gecko Fauna of the USSR and Adjacent Countries*]. Sci. Acad. Uckr. SSR Zool. Inst. 1986:1–232. (In Russian.)

Szczerbak, N.N., and M.L. Golubev, 1996. *Gecko Fauna of the USSR and Contiguous Regions.* English edition. Soc. Study Amphib. Rept., Ithaca, New York.

Taylor, E.H. "1935" (1936). A taxonomic study of the cosmopolitan scincoid genus *Eumeces*. University Kansas Science Bull. (23):1–643.

Telford, S.R. 1980. Notes on *Agkistrodon himalayanum* from Pakistan's Kaghan Valley. Copeia, 1980(1):154–155.

Theobald, W. 1876. Descriptive catalogue of the reptiles of British India. Thacker, Spink and Co. Calcutta.

Tuck, Robert G. 1971. Rediscovery and redescription of the Khuzistan dwarf gecko *Microgecko helenae* Nikolsky (Sauria: Geckkonidae). Proc. Biol. Soc. Wash. 83(42):477–482.

U.S. Fish and Wildlife Service, 2000. Endangered Species. CITES Appendices: Appendices I, II, III. Office of Scientific Authority, 1999.

Vohora, S.B., and S.Y. Khan. 1979. Animal origin drugs used in Unani medicine. Vikas Publishing House Pvt. Ltd., New Delhi.

Voris, H.K. 2000. Maps of Pleistocene sea levels in Southeast Asia: Shorelines, river systems and time durations. J. Biogeography 27:1153–1167.

Vyas, R. 1990. Notes on the capture of the spiny-tailed lizard (*Uromastyx hardwickii*) in Gujarat. Hamadryad 15(15): 28.

Vyas, R. 1998. Captive breeding of the saw-scaled viper (*Echis carinatus*). Hamadryad 22: 115–117.

Vyas, R., and B.H. Patel. 1992. Studies on the reproduction of Indian softshell turtle, *Asperetes gangeticus*. Hamadryad 17:32–34).

Wadia, D.N. 1966. *Geology of India*. Third edition. Macmillan and Co, London.

Wall, A.J. 1883. Indian snake poisons, their nature and effects. London.

Wall, F. 1899. Notes on 26 specimens of the Pohur or Himalayan pit viper (*Agkistrodon himalayanus*). J. Bombay Nat. Hist. Soc.12:411–414.

Wall, F. 1902. Aids to the differentiation of snakes. J. Bombay Nat. Hist. Soc.14:337–343.

Wall, F. 1906. A popular treatise on the common Indian snakes, *Zamenis mucosus*. J. Bombay Nat. Hist. Soc. 17:258–273.

Wall, F. 1907a. A new krait from Oudh (*Bungarus walli*) J. Bombay Nat. Hist. Soc. 17:608–611.

Wall, F. 1907b. Notes on snakes collected in Fyzabad. J. Bombay Nat. Hist. Soc. 18:101–120.

Wall, F. 1907c. A popular treatise on the common Indian snakes *Dipsas trigonata*. J. Bombay Nat. Hist. Soc. 18:543–55.

Wall, F. 1907d. A popular treatise on the common Indian snakes *Tropidonotus piscator*. J. Bombay Nat. Hist. Soc. 17:857–870.

Wall, F. 1908a. The poisonous terrestrial snakes of our British Indian Dominions (including Ceylon) and how to recognize them, with symptoms of snake poisoning and treatment. Bombay Nat. Hist. Soc.

Wall, F. 1908b. A popular treatise on the common Indian snakes *Bungarus*. J. Bombay Nat. Hist. Soc., 18: 711–735.

Bibliography

Wall, F. 1909a. A popular treatise on the common Indian snakes *Lycodon aulicus*. J. Bombay Nat. Hist. Soc. 19: 87–101.

Wall, F. 1909b. A popular treatise on the common Indian snakes *Lycodon striatus*. J. Bombay Nat. Hist. Soc. 19:102–106.

Wall, F. 1910. A popular treatise on the common Indian snakes. *Ancistrodon himalayanus*. J. Bombay Nat. Hist. Soc. 20:65–72.

Wall, F. 1911a. Remarks on the snake collection in the Quetta museum. J. Bombay Nat. Hist. Soc. 20:1033–1042.

Wall, F. 1911b. Reptiles collected in Chitral. J. Bombay Nat. Hist. Soc. 21:132–145.

Wall, F. 1914. A new snake from Baluchistan *Dipsadomorphus jollyi*. J. Bombay Nat. Hist. Soc. 23:167–168.

Wall, F. 1923. A handlist of the snakes of Indian Empire. Part-I. J. Bombay Nat. Hist Soc. 29(2):345–361.

Waltner, R.C. 1974a. Geographical and altitudinal distribution of amphibians and reptiles in the Himalayas. Part I. Cheetal 16(1):17–25.

Waltner, R.C. 1974b. Geographical and altitudinal distribution of amphibians and reptiles in the Himalayas. Part II. Cheetal 16:28–36.

Waltner, R.C. 1975a. Geographical and altitudinal distribution of amphibians and reptiles in the Himalayas. Part III. Cheetal 16(3):14–19.

Waltner, R.C. 1975b. Geographical and altitudinal distribution of amphibians and reptiles in the Himalayas. Part IV. Cheetal 17:12–17.

Wallace, A.R. 1876. *The Geographical Distribution of Animals*. 2 vols. Macmillan, London.

Wallace, A.R. 1881. *Island Life or the Phenomena and Causes of Insular Faunas and Floras Including Revision and Attempted Solution of the Problem of Geological Climates*. Macmillan, London.

Welch, K.R.G., A.S. Cooke, and A.S. Wright, 1990. *Lizards of the Orient: A Checklist*. Krieger Publishing Co. Malabar, Florida.

Whitaker, R. 1978. *Common Indian Snakes. A Field Guide*. Macmillan, New Delhi.

Wilson, L.D. 1967. Generic reallocation and review of *Coluber fasciolatus* Shaw (Serpentes, Colubridae). Herpetologica 23(4):260–275.

Wüster, W. 1998a. The cobras of the genus *Naja* in India. Hamadryad 23(1):15–32.

Wüster, W. 1998b. The genus *Daboia* (Serpentes, Viperidae): Russell's viper. Hamadryad 23(1):33–40.

Yadav, J.S., and R.K. Pillai. 1975. Somatic karyotypes of two Indian species of frogs (Anura, Amphibia). Cytobios 13:109–115.

Zarudny, N. 1904. [Reptiles and fishes of east Persia.] Mem. Imper. Russian Geogr. Sci. J. 36(3):1–42 (In Russian.)

Suggested Reading

Abercromby, A.F. 1910. *The Snakes of Ceylon*. London.

Acharji, M.N., and M.B. Kripalani. 1951. On a collection of Reptiles and Batrachia from Kangra and Kulu Valleys, western Himalayas. Rec. Indian Mus. Calcutta 44:175–184.

Acharji, M.N. and H.C. Ray. 1936. A new species of *Oligodon* from the United Provinces (India). Rec. Indian Mus. Calcutta 38:519–520.

Agrawal, H.P. 1979. A check-list of reptiles of Himachal Pradesh, India. Indian J. Zootomy 20:115–124.

Anderson, S.C. 1974. Preliminary key to the turtles, lizards and amphisbaenians of Iran. Fieldiana: Zool. 65(4):27–44.

Auffenberg, W. 1979. Intersexual differences in behaviour of captive *Varanus bengalensis* (Reptilia, Lacertilia, Varanidae). J. Herpetol. 13(3):313–315.

Bogert, C.M. 1943. Dentitional phenomenon in cobras and other elapids with notes on adaptive modifications of fangs. Bull. Amer. Mus. Nat. Hist.81:285–360.

Chin-shiang, W., and W. Yu-hsi M. Snakes of Taiwan. Quar. J. Taiwan Mus. 9:1–86.

Cochran, D.M. 1961. *Living Amphibians of the world*. Doubleday, Garden City, New York.

Daniel, J.C., and A.G. Sekar. 1989. Field guide to the amphibians of western India. Part. 4. J. Bombay Nat. Hist. Soc. 86:194–204.

Daudin, F.M. 1803. *Histoire naturelle generale et particuliere des reptiles*. Paris Vol. 1–8.

De Silva, A. 1990. *Colour Guide to the Snakes of Sri Lanka*. R&A Publishing Limited, Portishead, England.

Disi, A.M., D. Modrý, P. Nečas, and L. Rifai, 2001. *Amphibians and Reptiles of the Hashemite Kingdom of Jordan: an Atlas and Field Guide*. Edition Chimaira, Frankfurt am Main.

Dunson, W.A., Ed. 1975. *The Biology of Sea Snakes*. University Park Press, Baltimore.

Dutta, S.K. 1992. Amphibians of India: Updated species list with distribution record. Hamadryad 17: 1–13.

Dutta, S.K. 1997. *Amphibians of India and Sri Lanka* (Checklist and Bibliography). Odyssey Publ. House, Orrisa.

Flower, S.S. 1899. Notes on a second collection of batrachians made in the Malay Peninsula and Siam, from November 1896 to September 1898, with the list of the species recorded from these countries. Proc. Zool. Soc., London 1899:885–918.

Günther, A. 1869.Report on two collections of Indian reptiles. Proc. Zool. Soc. London1869:500–506.

Hora, S.L. 1926. Notes on lizards in the Indian Museum. Part 1. Geckonidae. Rec. Ind. Mus. 28:187–193.

Hora, S.L. 1927. Notes on lizards in the Indian Museum. Part 2. Agamidae. Rec. Ind. Mus.:215–220.

Hora, S.L. 1927. Notes on lizards in the Indian Museum. Part 3. On the unnamed collection of the family Scincidae. Rec. Ind. Mus. 29:1–6.

Iverson, J.B. 1986. *A Checklist with Distribution Maps of the Turtles of the World*. Paust Printing (privately printed by author), Richmond, Indiana.

Joger, U. 1984. The venomous snakes of near and Middle East. Tübinger Atlas vorder Orients, Wiesb. Reichert, Germany, Reihe A(12):1–115.

Joger, U. 1991. A molecular phylogeny of agamid lizards. Copeia, 1991(3):616–622.

Khalaf, K.T. 1959. *Reptiles of Iraq, with Some Notes on the Amphibians*. Ar-Rabitta Press, Baghdad.

Khan, M.S. 1971. An interesting abnormality in the arterial system of *Uromastyx hardwickii* Grey [sic] and its possible evolutionary significance. Pak. J. Sci., 23(1–2):78–80.

Khan, M.S. 1978. A double-headed monster of *Hemidactylus flaviviridis* Rüppell. Biologia 24:31–36.

Krishnan, H. 1992. The common monitor. Cobra 8:3–6.

Lanworn, R.A. 1972. *The Book of Reptiles*. Hamlyn, London.

Latifi, M. 1991. *Snakes of Iran*. English edition. Society for the study of amphibians and reptiles, Oxford, Ohio.

Leviton, A., and S.C. Anderson. 1972. Description of a new species of *Tropiocolotes* (Reptilia: geckonidae) with a revised key to the genus. Occas. Pap. Calif. Acad. Sci. (96):1–7.

Leviton, A.E., S.C. Anderson, K.K. Adler, and S.A. Minton. 1992. *Handbook to Middle East Amphibians and Reptiles*. Cont. Herpt. No. 8. Soc. Study of Amphibians and Reptiles, Oxford, Ohio.

Manna, G.K., and S.P. Bhunyan. 1966. A study of somatic chromosomes of both sexes of the common Indian toad, *Bufo melanostictus*. Caryologia 19:403–411.

Suggested Reading

Matz, G. 1974. Les Boides ou serpents constricteurs 4. *Eryx* Daudin 1803. Aquarama 8(26):53–55.

McCann, C. 1932. Notes on Indian batrachians. J. Bombay Nat. Hist. Soc. 36:152–180.

McFarland, F.H., W.N. Pough, T.J. Cade, and J.B. Heiser. 1979. *Vertebrate life*. Collier MacMillan International.

Mertens, R. 1954a. Als Herpetologie in Pakistan. Aquar. Terr. Zeit. 7(1–4):18–21, 42–46, 168–171, 103–107.

Mertens, R. 1954b. Uber die Rassen des Wustenwarans (*Varanus griseus*). Sencken. Biol. 35(5–6):353–357.

Mirza, Z.B. 1998. *Illustrated Handbook of Animal Biodiversity of Pakistan*. Centre for Environmental Research and Conservation, Islamabad, Pakistan, pp. 99.

Murray, J.A. 1884. Additions to the present knowledge of the vertebrate zoology of Persia. Ann. Mag. Nat. Hist. London (ser. 4) 14:101–106.

Schätti, B., and L.D. Wilson, 1986. *Coluber* Linnaeus. Cat. American Amph. Rept. 399:1–4.

Stoliczka, F. 1914. Lizards of the Simla Hill States. Rec. Indian Mus. Calcutta 10:367–369.

Tremper, R.L. 1997. Designer leopard geckos. Reptiles 5(3):16–18.

Underwood, G., and A.F. Stimson, 1990. A classification of pythons (Serpentes, Pythoninae). J. Zool. London. 221:265–603.

Werner, Y.L., and N. Sivan, 1992. Systematics and zoogeography of *Cerastes* (Ophidia: Viperidae) in the Levant: 2. Taxonomy, ecology, and zoogeography. The Snake, 24(1): 34–49.

INDEX

Figures represent page number for each category

Taxonomic category	Account	Image	Dist. Map	Figure
Family BUFONIDAE	42			19,20,21
Bufo	42	42	43	
Bufo himalayanus	42	42	43	
Bufo latastii	43		43	
Bufo melanostictus hazarensis	43	44	45	19
Bufo olivaceus	45	46		
Bufo pseudoraddei baturae	47		47	
Bufo pseudoraddei pseudoraddei	46	46	47	
Bufo siacheninsis	47	48	49	
Bufo stomaticus	49	49	50	19
Bufo surdus	50		51	
Bufo viridis zugmayeri	52	52	52	
Family MEGOPHRYIDAE				
Stutiger				
Scutiger nyingchiensis				
Family MICROHYLIDAE	54			
Microhyla	54			
Microhyla ornata	54	54	55	20
Uperodon	55			
Uperodon systoma	55	55	56	
Family RANIDAE	56			
Euphlyctis	56			
Euphlyctis cyanophlyctis cyanophlyctis	56	57	58	20,21
Euphlyctis cyanophlyctis microspinulata	58	58	58	
Euphlyctis cyanophlyctis seistanica	58		58	
Hoplobatrachus	60			
Hoplobatrachus tigerinus	60	61	62	21
Fejervarya	58			
Fejervarya limnocharis	58	59	59	
Fejervarya syhadrensis	60	60	60	21
Paa	62			
Paa barmoachensis	62	62	63	
Paa hazarensis	63	64	64	20
Paa sternosignata	64	65	65	
Paa vicina	65		66	
Sphaeroteca	66			
Sphaeroteca breviceps	66	66	67	20
CHELONIIDAE	68			
Caretta	68			

Index

Caretta caretta	68		68	22,24
Chelonia	69			
Chelonia mydas	69	69	70	22
Eretmochelys	70			
Eretmochelys imbricata	70	71	71	22,23,40
Lepidochelys	72			
Lepidochelys olivacea	72		72	24
DERMOCHELYIDAE	72			
Dermochelys	72			
Dermochelys coriacea	72		72	22,23,40
Family EMYDIDAE	73			
Geoclemys	73			
Geoclemys hamiltonii	73	73	73	
Hardella	74			
Hardella thurjii	74	75	76	
Kachuga	76			
Kachuga smithii	76	76	76	
Kachuga tecta	77	77	77	
Family TESTUDINIDAE	77			
Agrionemys	77			
Agrionemys horsfieldii	78	78	78	23
Geochelone	79			
Geochelone elegans	79	79	80	
Family TRIONYCHIDAE	80			
Aspideretes	80			
Aspideretes gangeticus	80	80	80	23
Aspideretes hurum	81	81	81	
Chitra	82			
Chitra indica	82	82	82	
Lissemys	83			
Lissemys punctata andersoni	83	83	83	23,25
Family CROCODYLIDAE	85			
Crocodylus	85			
Crocodylus palustris	85	85	86	
Family GAVIALIDAE	87			
Gavialis	87			
Gavialis gangeticus	87	87	88	
Family AGAMIDAE	89			
Brachysaura	89			
Brachysaura minor	89	89	90	
Calotes	90			
Calotes versicolor versicolor	90	90	91	
Calotes versicolor farooqi	92	92	91	
Japalura	92			
Japalura kumaonensis	92	92	92	
Laudakia	93			
Laudakia agrorensis	93	93	94	
Laudakia badakhshana	94		95	

Laudakia caucasia	95	95	95	
Laudakia fusca	96	96	97	
Laudakia himalayana	97	97	98	
Laudakia lirata	98		98	
Laudakia melanura nasiri	100		100	
Laudakia melanura melanura	100	99	100	
Laudakia microlepis	101		101	
Laudakia nupta	101	101	102	
Laudakia nuristanica	102	102	103	
Laudakia pakistanica pakistanica	104	104	104	
Laudakia pakistanica auffenbergi	104	104	105	
Laudakia pakistanica khani	105	105	105	
Laudakia tuberculata	106	106	106	
Phrynocephalus	107		31	
Phrynocephalus clarkorum	107		107	
Phrynocephalus euptilopus	107		108	
Phrynocephalus luteoguttatus	108	108	109	
Phrynocephalus maculatus	109	109	110	
Phrynocephalus ornatus	110	110	111	
Phrynocephalus scutellatus	111	111	112	
Trapelus	112			
Trapelus agilis agilis	112		113	
Trapelus agilis pakistanensis	114	112	114	
Trapelus megalonyx	114	114	115	
Trapelus rubrigularis	115	115	116	
Trapelus ruderatus baluchianus	116	116	116	
Family CHAMAELEONIDAE	117			
Chamaeleo	117			
Chamaeleo zeylanicus	117		117	
Family EUBLEPHARIDAE	118			
Eublepharis	118			
Eublepharis macularius	118	118	118	
Family GEKKONIDAE	119			
Agamura	121			
Agamura persica	121	121	121	
Altigekko	122			
Altigekko baturensis	122	122	122	
Altigekko boehmei	123		123	
Altigekko stoliczkai	123		123	
Bunopus	124			
Bunopus tuberculatus	124	124	124	
Crossobamon	124			
Crossobamon lumsdenii	125		125	
Crossobamon maynardi	125		125	
Crossobamon orientalis	126	126	126	
Cyrtopodion	127			
Cyrtopodion agamuroides	127	127	127	
Cyrtopodion kachhense kachhense	128	128	128	27

Index

Cyrtopodion kachhense ingoldbyi	129	129	128	
Cyrtopodion kohsulaimanai	129	129	130	
Cyrtopodion montiumsalsorum	130	130	131	26
Cyrtopodion potoharensis	131	131	131	
Cyrtopodion scabrum	132	132	132	27
Cyrtopodion watsoni	132		132	
Hemidactylus	133			
Hemidactylus brookii	133	133	134	
Hemidactylus flaviviridis	134	133	135	
Hemidactylus frenatus	135	136	136	
Hemidactylus leschenaultii	136	136	137	
Hemidactylus persicus	137	137	137	
Hemidactylus triedrus	138	138	139	
Hemidactylus turcicus				
Indogekko	139			
Indogekko fortmunroi	140	140	140	
Indogekko indusoani	140	140	140	
Indogekko rhodocaudus	141		141	
Indogekko rohtasfortai	142	142	142	
Mediodactylus	142			
Mediodactylus walli	143	143	143	
Ptyodactylus	144			
Ptyodactylus homolepis	144		144	26
Rhinogecko	144			
Rhinogecko femoralis	144		145	
Rhinogecko misonnei	145		145	
Siwaligekko	145			
Siwaligekko battalensis	145	146	146	
Siwaligekko dattanensis	146	146	147	
Siwaligekko mintoni	147	147	147	
Teratolepis	148			
Teratolepis fasciata	148		148	26
Teratoscincus	149			
Teratoscincus microlepis	149	149	149	
Teratoscincus scincus keyserlingi	150	150	150	26,27
Tropiocolotes	151			
Tropiocolotes depressus	151		151	
Tropiocolotes persicus persicus	151	151	152	
Tropiocolotes persicus euphorbiacola	152		152	
Family LACERTIDAE	153			
Acanthodactylus	153			
Acanthodactylus blanfordii	153		153	
Acanthodactylus cantoris	153	154	154	
Acanthodactylus micropholis	154	155	155	
Eremias	155			
Eremias acutirostris	155	155	156	
Eremias aporosceles	156		156	
Eremias fasciata	156		157	

Eremias persica	157	157	158	
Eremias scripta	158	158	158	
Mesalina	158			
Mesalina brevirostris	158	159	159	
Mesalina watsonana	159	159	160	
Ophisops	160			
Ophisops elegans	160	160	160	
Ophisops jerdonii	161	161	161	
Family SCINCIDAE	162			
Ablepharus	162			
Ablepharus grayanus	162	162	162	
Ablepharus pannonicus	163		163	
Chalcides	163			
Chalcides ocellatus	163	163	164	
Eurylepis	166			
Eurylepis taeniolatus taeniolatus	166	166	167	
Eutrophis	164			
Eutrophis carinata	164	164	165	
Eutrophis dissimilis	164	164	165	
Eutrophis macularia	165	165	165	
Lygosoma	167			
Lygosoma punctata	167	167	167	
Novoeumeces	168			
Novoeumeces blythianus	168	168	169	
Novoeumeces indothalensis	169	169	169	169
Novoeumeces schneiderii zarudnyi	170		170	
Ophiomorus	170			
Ophiomorus blanfordi	170		171	
Ophiomorus brevipes	171		171	
Ophiomorus raithmai	171		172	
Ophiomorus tridactylus	172	172	172	
Scincella	172			
Scincella himalayana	172	172	172	
Scincella ladacensis	173		174	
Family UROMASTYCIDAE	174			
Uromastyx	174			
Uromastyx asmussi	174	174	174	
Uromastyx hardwickii	175	175	176	
Family VARANIDAE	176			
Varanus	176			
Varanus bengalensis	176	176	177	
Varanus flavescens	177	177	178	
Varanus griseus	178			
Varanus griseus caspius	178	178	179	
Varanus griseus koniecznyi	179		179	
Family LEPTOTYPHLOPIDAE	180			
Leptotyphlops	180			
Leptotyphlops blanfordii	180		180	

Index

Leptotyphlops macrorhynchus	181	181	181	32
Family TYPHLOPIDAE	181			32
Ramphotyphlops	181			
Ramphotyphlops braminus	181	182	182	
Typhlops	182			
Typhlops ahsanuli	183		183	
Typhlops diardii platyventris	183	184	184	
Typhlops ductuliformes	184	184	184	
Typhlops madgemintonae madgemintonae	185	185	185	
Typhlops madgemintonae shermanai	186		186	
Family BOIDAE	186			
Eryx	186			33
Eryx conicus	187	187	187	
Eryx johnii	187	187	188	33
Eryx tataricus speciosus	188	188	189	
Python	189			33
Python molurus	189	189	190	33
Family COLUBRIDAE	190			41
Amphiesma	190			
Amphiesma platyceps	190		191	
Amphiesma sieboldii	191			
Amphiesma stolatum	191	191	192	
Argyrogena	192			
Argyrogena fasciolata	192	192	193	35
Boiga	193			
Boiga melanocephala	193		193	
Boiga trigonata	194	194	194	36
Coluber	195			
Coluber karelini karelini	195	195	195	38
Coluber karelini mintonorum	196			
Enhydris	196			
Enhydris pakistanica	196	196	196	35
Hemorrhois	197			
Hemorrhois ravergieri	197	197	197	
Lycodon	198			36
Lycodon aulicus aulicus	198	198	199	36
Lycodon striatus bicolor	200		200	
Lycodon striatus striatus	199	199	200	
Lycodon travancoricus	200		200	
Lytorhynchus	201			36
Lytorhynchus maynardi	201	201	201	36
Lytorhynchus paradoxus	202	202	202	36
Lytorhynchus ridgewayi	203	203	203	36
Oligodon	204			36
Oligodon arnensis	204	204	204	36
Oligodon taeniolatus	205	204	205	36
Platyceps	205			
Platyceps rhodorachis kashmirensis	206	207	206	

Platyceps rhodorachis ladacensis	207	206		
Platyceps rhodorachis rhodorachis	206	206	206	37
Platyceps ventromaculatus bengalensis	208	208		
Platyceps ventromaculatus indusai	208	208	208	
Platyceps ventromaculatus ventromaculatus	207		208	38
Psammophis	209			
Psammophis condanarus condanarus	209	209	209	
Psammophis leithii leithii	210	210	210	37
Psammophis lineolatus lineolatus	211	211	211	
Psammophis schokari schokari	211	212	212	
Pseudocyclophis	212			
Pseudocyclophis persicus	213	213	213	
Ptyas	213			
Ptyas mucosus	213	214	215	37
Sibynophis	215			
Sibynophis sagittarius	215		215	
Spalerosophis	216			
Spalerosophis arenarius	216	216	216	
Spalerosophis diadema diadema	216	217	217	35
Spalerosophis diadema atriceps	217	218		
Spalerosophis schirazianus	218	218	218	
Telescopus	219			
Telescopus rhinopoma	219		219	
Xenochrophis	219			
Xenochrophis cerasogaster cerasogaster	219	219	220	37
Xenochrophis piscator piscator	220	220	221	38
Xenochrophis sanctijohannis	221	221	222	
Xenochrophis tessellata	222		222	
Family ELAPIDAE	222			39
Bungarus	223			
Bungarus caeruleus caeruleus	223	223	224	39
Bungarus sindanus razai	224		225	
Bungarus sindanus sindanus	224		224	
Naja	225			
Naja naja naja	225	225	226	39
Naja oxiana	226	226	227	
Family HYDROPHIIDAE	227			38
Astrotia	227			
Astrotia stokesii	227	227		
Enhydrina	228			
Enhydrina schistosa	228	228		
Hydrophis	228			
Hydrophis caerulescens	228			
Hydrophis cyanocinctus	229	229		
Hydrophis fasciatus	229	230		
Hydrophis lapemoides	230			
Hydrophis mamillaris	230	230		
Hydrophis ornatus	230	231		

Index

Hydrophis spiralis	231	231		
Lapemis	231			
Lapemis curtus	232			
Microcephalophis	232			
Microcephalophis cantoris	232			
Microcephalophis gracilis	232			
Pelamis	233			
Pelamis platurus	233	233		
Praescutata	233			
Praescutata viperina	233			
Family VIPERIDAE	234			
Daboia	234			
Daboia russelii russelii	234	234	235	40
Echis	235			
Echis carinatus	236			
Echis carinatus astolae	236			
Echis carinatus multisquamatus	237	237	237	
Echis carinatus sochureki	237	238	238	
Eristicophis	238			40
Eristicophis macmahonii	238	239	239	40
Macrovipera	239			
Macrovipera lebetina obtusa	239	240	240	
Pseudocerastes	240			40
Pseudocerastes bicornis	240	241	241	40
Pseudocerastes persicus	240	241	241	40
Family CROTALIDAE	241			
Gloydius	241			
Gloydius himalayanus	241	241	241	